Aussortiert:

Math. Inst. d. Univ. Bonn
Bibliothek

Variational and Quasivariational Inequalities

Variational and Quasivariational Inequalities

Applications to Free Boundary Problems

CLAUDIO BAIOCCHI
and
ANTÓNIO CAPELO

Istituto di Matematica dell' Università di Pavia
and
Istituto di Analisi Numerica del Consiglio Nazionale delle Ricerche

Translated by

LAKSHMI JAYAKAR

A Wiley–Interscience Publication

JOHN WILEY AND SONS
Chichester • New York • Brisbane • Toronto • Singapore

Copyright © 1984 by John Wiley & Sons Ltd.

All rights reserved.

No part of this book may be reproduced by any means, nor transmitted, nor translated into a machine language without the written permission of the publisher.

Library of Congress Cataloging in Publication Data:

Baiocchi, C.
   Variational and quasivariational inequalities.
   Translation of: Disequazioni variazionali e quasivariazionali.
   'Wiley–Interscience.'
   Bibliography: p.
   Includes index.
   1. Variational inequalities (Mathematics) 2. Boundary value problems. I. Capelo, A. II. Title.
QA316.B2513 1983 515'.26 83–6731

ISBN 0 471 90201 2

British Library Cataloguing in Publication Data:

Baiocchi, Claudio
   Variational and quasivariational inequalities.
   1. Calculus of variations 2. Inequalities (Mathematics)
   I. Title II. Capelo, António
   515'.64 QA316

ISBN 0 471 90201 2

Printed in Northern Ireland at The Universities Press (Belfast) Ltd.
Bound at the Pitman Press Ltd., Bath, Avon.

Contents

Preface ix

PART I VARIATIONAL PROBLEMS

Chapter 1 Introduction 3
  1.1 Example of a variational problem. The notion of well-posed problem 3
  1.2 Some existence results 6

Chapter 2 Minimization of Convex Functionals 9
  2.1 Fundamental theorem 9
  2.2 Variational formulation of a minimization problem 15
  2.3 Projections on convex sets 18

Chapter 3 Variational Inequalities 21
  3.1 Fundamental theorem 21
  3.2 Minimization of convex functionals 30
    3.2.1 Gâteaux derivative 30
    3.2.2 General results 34
    3.2.3 The concept of subdifferential 37

Chapter 4 Transposition of Operators. Applications 41
  4.1 The concept of transposition. Fundamental properties . . . 41
  4.2 Applications of the concept of transposition 45
    4.2.1 Some examples of function spaces important in the applications. Examples of continuous linear operators 47
    4.2.2 Dual spaces and transposed operators 55

Chapter 5 Sobolev Spaces 67
  5.1 On the necessity for new function spaces 67
  5.2 Spaces $W^{s,p}(\mathbb{R}^n)$ 69
  5.3 Spaces $W^{s,p}(\Omega)$ 84
  5.4 Spaces $W^{s,p}(\Gamma)$ 93
  5.5 Normal contractions and Dirichlet spaces 99

Chapter 6 Examples of Variational Problems in One Dimension . . 104
6.1 The obstacle problem. Generalities about regularity results . 104
6.2 Some considerations regarding second order linear problems . 112

Chapter 7 Examples of Variational Problems in Several Dimensions 119
7.1 General comments on differential operators 119
7.2 Linear problems 129
    7.2.1 Introduction. The homogeneous Dirichlet problem . . . 129
    7.2.2 General formulation in variational terms 137
    7.2.3 Examples of boundary value problems 143
7.3 Nonlinear problems 156
7.4 Regularity results 169
    7.4.1 Regularity of the solutions of differential equations . . . 169
    7.4.2 Regularity of solutions of variational inequalities 174

Chapter 8 Variational Formulation of a Free-Boundary Problem . 183
8.1 Generalities. The physical problem 183
8.2 Transformation of the problem 187
8.3 Quasivariational problems 190

PART II QUASIVARIATIONAL PROBLEMS

Chapter 9 Fixed Point Theorems 199
9.1 Introduction 199
9.2 Fixed point theorems for Lipschitz continuous applications . 201
    9.2.1 Single-valued applications: Banach theorem 201
    9.2.2 Multi-valued applications 204
9.3 Fixed point theorems for continuous applications 205
    9.3.1 The Knaster–Kuratowski–Mazurkiewicz lemma and the Fan lemma 205
    9.3.2 Single-valued applications: the theorems of Brouwer, Schauder and Tychonov 210
    9.3.3 Multi-valued applications 216
9.4 Fixed point theorems for monotone applications 222

Chapter 10 Some Results on the Existence of Solutions of Variational Inequalities 226
10.1 General results of existence 226
10.2 Particular cases. I 232
10.3 Particular cases. II 234

Chapter 11 Quasivariational Inequalities 237
11.1 Introduction 237

| 11.2 | Techniques of monotonicity | 244 |
| 11.3 | Techniques of compactness | 253 |

Chapter 12 Free-Boundary Problems 262
- 12.1 Introduction 262
- 12.2 The physical problem 264
- 12.3 The mathematical problem 276
  - 12.3.1 Rigorous formulation of the free-boundary problem . . 277
  - 12.3.2 Transformation of the free-boundary problem 280
  - 12.3.3 Study of a linear mixed problem 286
  - 12.3.4 Study of a nonlinear mixed problem 291
  - 12.3.5 Study of a quasivariational problem. Existence of the solution of a free-boundary problem 294
  - 12.3.6 Uniqueness and regularity of the solution of a free-boundary problem 302
  - 12.3.7 Dam with vertical walls 304

Chapter 13 Free-Boundary Problems and Variational Inequalities . 317

PART III TECHNICAL TOOLS

Chapter 14 Seminorms 325

Chapter 15 Regularization and Partition of the Unit 335
- 15.1 Regularization 335
- 15.2 Partition of the unit 342

Chapter 16 On the Regularity of the Open Sets 346
- 16.1 Manifolds. Open sets of class C^k and $C^{k,\mu}$. Cone and segment properties 346
- 16.2 Distributions on a manifold. Spaces $L^p(\Gamma)$ 353

Chapter 17 The Maximum Principle and its Applications 357
- 17.1 Introduction 357
- 17.2 Maximum principle in $\mathbb{R}$ 357
- 17.3 Maximum principle in $\mathbb{C}$ 361
- 17.4 Maximum principle in $\mathbb{R}^n$ 365
- 17.5 Complements 368

Chapter 18 On Green's Formulae 370

Chapter 19 Ordered Structures 381
- 19.1 Basic definitions 381
- 19.2 Lattices 384
- 19.3 Ordered vector spaces. Vector lattices 388

19.4	Topological vector lattices. Banach–Riesz spaces	393
19.5	Hilbert pseudo-lattices	399

Chapter 20 Multi-Valued Mappings 401
   20.0 Notations 401
   20.1 Fundamental definitions 401
   20.2 Topologies in $2_\blacksquare^Y$ 405
      20.2.1 The Hausdorff metric topology 405
      20.2.2 Vietoris topologies 408
   20.3 Orders in $2_{\blacksquare i}^Y$ 414

Bibliography 417

Index 449

Special Function Spaces Index 452

Preface

In the last fifteen years, variational inequalities have gained importance in analysis, both from the theoretical and the practical points of view.

This book, some chapters of which made up the courses of functional analysis taught by C. Baiocchi at the University of Pavia in the academic years 1974–75 and 1975–76, is devoted to variational and quasivariational inequalities of elliptic type, and to the applications of this theory to the study of free-boundary problems. The book is far from being a complete treatment: subjects such as inequalities of evolution, numerical treatment of inequalities, and other types of applications of inequalities, are either completely missing or have been only barely mentioned. We have, however, tried, in the limited field which we have considered, to present an organic and self-sufficient treatment with a considerable bibliography.

The book is divided into three parts. The first two parts deal respectively with problems of variational and of quasivariational type; an understanding of these two parts requires a knowledge of calculus, the basic elements of measure theory and Lebesgue integration, and the elementary properties of Hilbert and Banach spaces. The third part is devoted to some collateral subjects which are outside the scope of the topics mentioned above but which are nevertheless necessary for the development of the theory (e.g. Green's formulae, seminorms, the maximum principle, . . .); sometimes, in parts I and II, we refer to this part III for specific results and notations.

For some of the topics, for example Sobolev spaces, we have not even tried to be self-sufficient, but in these cases we have provided the reader with precise references to enable him to enlarge and complete his knowledge of the subject.

We are grateful to the Istituto di Analisi Numerica del Consiglio Nazionale delle Ricerche (Pavia), and the Unione Matematica Italiana who made the publication of the Italian version of this book possible, and to our colleagues of the Istituto di Analisi Numerica del Consiglio Nazionale delle Ricerche and of the Istituto di Matematica dell'Università di Pavia for constructive criticism and correction of printing errors in the Italian version.

Pavia Claudio Baiocchi
January 1983 António C. Capelo

PART I

Variational Problems

1

Introduction

1.1 EXAMPLE OF A VARIATIONAL PROBLEM. THE NOTION OF WELL-POSED PROBLEM

The obstacle problem

The following example, based on a problem of physics, allows us to present a first problem of variational kind—the reason for this particular name will be seen later on. For simplicity we present a one-dimensional example, which will therefore lead to ordinary inequalities: substitution of the elastic string by an elastic membrane and the two-dimensional obstacle by a three-dimensional obstacle will give us an example in which partial derivatives are necessary.

Let us consider a body $A \subset \mathbb{R}^2$, which we shall call the *obstacle*, and two points P_1 and P_2 not belonging to A (see fig. 1.1); let us connect P_1 to P_2 by a weightless *elastic string* whose points cannot penetrate A: we are interested in studying the shape assumed by the string. With this aim we introduce a system of Cartesian axes $0xy$ with respect to which P_1 and P_2, respectively, have coordinates $(0, 0)$ and $(l, 0)$. Let us suppose that, with respect to this system of axes, the 'lower part' of the boundary of obstacle A (in the zone in which we are interested, i.e. in $[0, l]$) is a Cartesian curve of equation $y = \psi(x)$. Experience tells us that if $y = u(x)$ is the shape assumed by the string then

$$u(0) = u(l) = 0 \tag{1.1}$$

since the string connects P_1 and P_2,

$$u(x) \leq \psi(x) \tag{1.2}$$

because the string does not penetrate the obstacle,

$$u''(x) \geq 0 \tag{1.3}$$

because the string being elastic and weightless must assume a convex shape, and

$$u(x) < \psi(x) \Rightarrow u''(x) = 0 \tag{1.4}$$

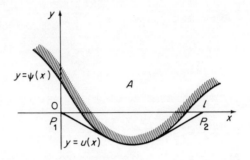

Figure 1.1

(i.e. where it does not touch the obstacle the string takes a linear shape) since the string tends to assume the shape with the minimum possible length (in particular were there no obstacle this would be l); (1.1), (1.2), (1.3), and (1.4) are equivalent to (1.1), (1.2), (1.3), and (1.5), with

$$[u(x)-\psi(x)]u''(x)=0, \tag{1.5}$$

because if (1.4) is true then either $u(x)-\psi(x)=0$ or $u''(x)=0$, and hence (1.5) is true, and if this is true then when $u(x)\neq\psi(x)$ we must have $u''(x)=0$, and (1.2) implies that $u(x)\neq\psi(x)$ only if $u(x)<\psi(x)$.

Expressions (1.1), (1.2), (1.3), and (1.5) constitute a mathematical formulation of the physical problem that we are dealing with, and the search for the function u that satisfies them constitutes a *variational problem*, with which, as we will see later, a *variational inequality* is associated. In the future we will refer to this problem as **problem 1.0**.

Spaces of data and spaces of unknowns

Problem 1.0, however, is not *complete* since we have not indicated the regularity which the unknown u must satisfy (i.e., we have not defined the function space to which u must belong) nor the regularity which we attribute to the data function ψ (i.e., the function space from which ψ is taken). It is clear that given ψ we can determine its regularity by simple inspection, but if we want to solve once and for all the 'problem of the string and the obstacle' we cannot fix ψ but only indicate some of its properties—i.e. indicate the space to which it belongs.

The matter of completeness of a problem—a concept not to be confused with that of well-posedness which we will speak of later—is anything but academic, as can be seen from the following considerations. Physics tells us that u must be continuous (otherwise there would be a break in the string) but not, for example, that it must be differentiable: we will have a non-differentiable shape of the string in the case of fig. 1.2, and if in this case we look for u in $C^0([0, l])$, as would be natural, (1.3) for example makes no sense—the shape of the string continues to be convex but the mathematical

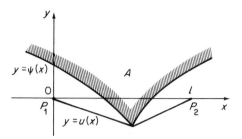

Figure 1.2

statement of this fact has to be changed. It is then most important to indicate the spaces in which the problem is posed, so that one knows how the written expressions are to be interpreted. Thus, for example, if ψ is given in $C^2([0, l])$ and we look for u in $C^2([0, l])$ then the expressions (1.1), (1.2), (1.3), and (1.5) can be interpreted in the usual sense; if, on the other hand, ψ is given in $C^1([0, l])$ and we look for u among the functions of $C^1([0, l])$ which have absolutely continuous derivative, then (1.1) and (1.2) can be still interpreted in the usual sense but (1.3) and (1.5) must be taken to be valid almost everywhere (a.e.) (we notice that if $u \in C^1([0, l])$ and $u' \in AC([0, l])$ then in fact u'' exists a.e.). The problems 'given $\psi \in C^2([0, l])$ (or: $\psi \in C^1([0, l])$) find $u \in C^2([0, l])$ (or: $u \in \{u \in C^1([0, l]): u' \in AC([0, l])\}$) which satisfies (1.1), (1.2), (1.3), and (1.5) in the usual sense (resp.: (1.1) and (1.2) in the usual sense and (1.3) and (1.5) almost everywhere)' are complete problems, even if in both cases the spaces chosen are not the most suitable ones for the treatment of the problem (which so formulated may not have a solution—on the other hand, the assumptions on ψ are too restrictive).

Well-posed problems

We observe, however, that in order to have a well formulated problem it is not enough to define the spaces where the expressions make sense. Thus we introduce the:

Notion of well-posed problem. The problem 'given the set $\{d_n \in \mathcal{D}_n\}_{n=1,\dots,N}$ of data d_n in topological spaces $\mathcal{D}_n$, find a set $\{u_m \in \mathcal{U}_m\}_{m=1,\dots,M}$ of unknowns u_m in topological spaces $\mathcal{U}_m$, which satisfy a set $\{C_k\}_{k=1,\dots,K}$ of conditions which connect, in the sense of the spaces $\mathcal{D}_n$ and $\mathcal{U}_m$, the unknowns to the data' is *well-posed* if for every set of data $\{d_n\}$ just one set $\{u_m\}$ of unknowns exists (which will be called solutions) that satisfy the conditions $\{C_k\}$, and further $\{u_m\}$ varies continuously with $\{d_n\}$ relative to the topologies of the spaces $\prod_{n=1}^{N} \mathcal{D}_n$ and $\prod_{m=1}^{M} \mathcal{U}_m$.

Note that in many problems (generally in nonlinear problems) the data $\{d_n\}$ can be 'hidden' in the conditions $\{C_k\}$; on the other hand, linear problems can also involve problems of continuous dependence of solutions on 'coefficients' that appear in $\{C_k\}$.

One of the fundamental objectives of functional analysis is to define spaces in which problems can be well-posed, or for each problem find spaces where the problem is well posed, i.e. those in which we can prove existence, uniqueness and continuous dependence of solutions on the data. It must be noted that a problem can be well posed relative to different spaces: we must in principle select the formulation that is least demanding on the data, but we must take into account the possibility of proving 'regularity results' (a concept which will be explained later).

1.2 SOME EXISTENCE RESULTS

We now consider some examples of problems for which we can ensure the existence of solutions, and at the same time mention some fundamental results which will be useful later; for the results which are given without proof see any textbook of functional analysis (e.g. Yosida, 1971 or Kolmogorov–Fomin, 1957).

Hahn–Banach theorem

The first problem that we will consider is the following:

PROBLEM 1.1. Let B be a Banach space and let $y \in B \setminus \{0\}$; find $L \in B'$ such that $L(y) = 1$ (as usual we denote by B' the dual of B).

That this problem has at least one solution is an elementary corollary of

THEOREM 1.1 (Hahn–Banach theorem). If B is a normed space and V one of its linear varieties then each continuous linear functional on V can be extended as a continuous linear functional on B with the same norm.

The Hahn–Banach theorem also solves (as is mentioned in its enunciation) the problem of the extension of continuous linear functionals: we will exploit it frequently.

Riesz theorem

Let us proceed to the problem of the representation of continuous linear functionals.

PROBLEM 1.2. Let H be a Hilbert space and $L \in H'$; find $u_L \in H$ such that $\forall v \in H \; L(v) = (u_L, v)_H$.

The question of the existence of a solution of this problem is solved by

THEOREM 1.2 (Riesz theorem). *If H is a Hilbert space, for each $L \in H'$ there exists one and only one $u_L \in H$ such that $\forall v \in H\ L(v) = (u_L, v)_H$; further u_L satisfies $\|u_L\|_H = \|L\|_{H'}$.*

It is interesting to note that this theorem not only gives us the existence and the uniqueness of the solution of problem 1.2 but also the continuous dependence on the data (since the application that to every L associates u_L is an isometry): the problem of the representation of continuous linear functionals is well-posed in each Hilbert space with the topology generated by the norm.

Banach's fixed-point theorem

The very important theorems of Hahn–Banach and of Riesz are results of 'vectorial-topological type': let us now consider a result which does not depend on a vectorial structure.

Let (S, d) be a complete non-empty metric space and $T: S \to S$ a *contraction operator*, i.e. an operator such that

$$\exists k < 1 \quad \forall x, y \in S \quad d(T(x), T(y)) \leq k d(x, y). \tag{1.6}$$

With these data let us consider the

PROBLEM 1.3. Find $\bar{x} \in S$ such that $\bar{x} = T(\bar{x})$ (i.e. determine a fixed point for T).

This problem has a solution, as can be seen from

THEOREM 1.3 (Banach theorem for contractions). *If (S, d) is a complete non-empty metric space and $T: S \to S$ is a contraction, there is one and only one fixed point for T.*

Proof. Let us begin by showing the uniqueness, which we do by *reductio ad absurdum*. Let us suppose that T has two distinct fixed points $\bar{x}$ and $\bar{\bar{x}}$: we can write $\bar{x} = T(\bar{x})$ and $\bar{\bar{x}} = T(\bar{\bar{x}})$ and, since T is a contraction, $d(\bar{x}, \bar{\bar{x}}) = d(T(\bar{x}), T(\bar{\bar{x}})) \leq k d(\bar{x}, \bar{\bar{x}}) < d(\bar{x}, \bar{\bar{x}})$, which is absurd. To show existence let us consider the sequence defined by the recurrence relation

$$x_n \triangleq T(x_{n-1}) (n = 1, 2, \ldots), \tag{1.7}$$

and show that no matter how we fix $x^* = x_0$ this sequence converges to a fixed point for T. Now, noting that

$$x_n = T(x_{n-1}) = T^2(x_{n-2}) = \ldots = T^n(x_0) \tag{1.8}$$

and that from (1.6), for each n,

$$\forall x, y \in S \quad d(T^n(x), T^n(y)) \leq k^n d(x, y), \tag{1.9}$$

we can write successively, with $m \geq n$,

$$d(x_n, x_m) = d(T^n(x_0), T^m(x_0)) \leq k^n d(x_0, x_{m-n})$$
$$\leq k^n[d(x_0, x_1) + d(x_1, x_2) + \ldots + d(x_{m-n-1}, x_{m-n})]$$
$$\leq k^n d(x_0, x_1)[1 + k + \ldots + k^{m-n-1}] = d(x_0, x_1)\frac{k^n - k^m}{1-k},$$
(1.10)

which shows that $\{x_n\}$ is a Cauchy sequence. Since (S, d) is complete, x_n converges: let $\bar{x}$ be its limit. To see that $\bar{x}$ is a fixed point for T is is enough to note that this operator is continuous (from (1.6)) and therefore $T(\bar{x}) = T(\lim_n x_n) = \lim_n T(x_n) = \lim_n x_{n+1} = \bar{x}$. ∎

Note that the Banach theorem tells us that problem 1.3 is well-posed in non-empty complete metric spaces, because it demonstrates the existence and the uniqueness of the fixed point and, since there are no data, the problem of continuous dependence of the solution on the data does not arise (see, however, what we say immediately after the notion of well-posed problem).

It is interesting to note that the proof of the existence of a fixed point has a very important characteristic: it is of constructive type. In fact, we have shown the existence of the fixed point indicating a process by which it can be found. It is true that the solution, $\bar{x}$, is obtained through countable many operations, but we can approximate $\bar{x}$ as closely as we want with a finite number of operations and we can estimate the error which we make: so, if we take as an approximate solution $\bar{\bar{x}} = x_n$, the error $d(\bar{x}, \bar{\bar{x}})$ is not greater than $d(x_1, x_0) k^n / 1 - k$.

As a first application of the Banach theorem we note that from it follows the proof of the following result (Picard's theorem): *'given Ω, open set in $\mathbb{R}^2$, $f : \Omega \to \mathbb{R}$ continuous in the pair of variables and Lipschitz continuous with respect to y, and given $(x_0, y_0) \in \Omega$, the Cauchy problem "$y' = f(x, y)$, $y(x_0) = y_0$" is well posed locally'* (i.e., there exists $\varepsilon > 0$ such that in $]x_0 - \varepsilon, x_0 + \varepsilon[$ there exists a unique solution of the problem and this solution depends continuously on the data). In Chapter 3 we will see other applications of the Banach theorem.

In Chapter 9 we will present other fixed point theorems.

2
Minimization of Convex Functionals

2.1 FUNDAMENTAL THEOREM

The general problem of minimization

In this section we will look for the sufficient conditions such that the following problem of minimization has one (or one and only one) solution.

PROBLEM 2.1. Given a real vector space E, a function $f: E \to \mathbb{R}$ and a set $X \subset E$, find the minimum of f in X, i.e. find $x_0 \in X$ such that $f(x_0) = \inf_{x \in X} f(x)$.

The conditions we are looking for will naturally involve hypotheses on E, f, and X.

The search of 'good' hypothesis

In the particular case when $E = \mathbb{R}^n$, if f is continuous and X is compact, the theorem of Weierstrass ensures the existence of a minimum of f in X. In fact this theorem also ensures the existence of a maximum, which leads us to suspect that its assumptions are too strong; let us analyse them in order to obtain some useful information for our problem.

The function f is assumed to be continuous, and hence $\mathbb{R}^n$ must be taken to be a topological space. This presents the first problem: 'independently of whether the assumption *f is continuous* is necessary or not, will it be strictly necessary to exploit the topological structure of $\mathbb{R}^n$?'. The function $f_1 : \mathbb{R} \to \mathbb{R}$ defined by $f_1(x) = |x|$ for $x \neq 0$ and $f(0) = 1$ (see fig. 2.1) enables us to give an answer in the affirmative: f_1 does not have a minimum in $\mathbb{R}$ (the greatest lower bound of $f_1(\mathbb{R})$ is 0, which is not achieved) and the characterization of functions such as f_1 requires a topological structure. We conclude, therefore, that we must provide E with a topology; it will be useful to have a structure of a vector space on E too, so that we require:

(H1) E is a topological vector space

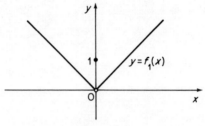

Figure 2.1

On the other hand the function $f_2:\mathbb{R}\to\mathbb{R}$ defined by $f_2(x)=|x|$ if $x\neq 0$ and $f_2(0)=-1$ (see fig. 2.2) which has a minimum (-1, for $x=0$) and is not continuous shows us that continuity is not strictly necessary.

At this point let us open a parenthesis to recall the concept of lower (and upper) semicontinuity. If T is a topological space, a function $f:T\to\mathbb{R}$ is *lower semicontinuous* (l.s.c.) in $x_0\in T$ if, indicating by $\mathcal{N}(x_0)$ the family of the open neighbourhoods of x_0, is true that $\forall\varepsilon>0\ \exists U_\varepsilon\in\mathcal{N}(x_0)\ \forall y\in U_\varepsilon f(y)\geq f(x_0)-\varepsilon$. When the topology of T can be interpreted in terms of the convergence of sequences (like the metric topologies, of which the so-called natural topology of $\mathbb{R}^n$ is an example) we can say that '$f:T\to\mathbb{R}$ is l.s.c. in $x_0\in T$ if for every $x_n\xrightarrow[T]{}x_0$ is true that $\lim'_n f(x_n)\geq f(x_0)$' (we will use the symbols $\lim'$ and $\lim''$ to indicate respectively the minimum limit and the maximum limit; other frequent notations are $\underline{\lim}$ and lim inf for $\lim'$ and $\overline{\lim}$ and lim sup for $\lim''$). Functions are defined *upper semicontinuous* (u.s.c.) in an analogous manner.

Let us now return to our problem. Actually the function f_2 is l.s.c. (which is not true for function f_1) and we will then require the following second hypothesis:

(H2) f is l.s.c. with respect to the topology of E.

(Or better, relative to a natural topology of E: we will be fundamentally interested in vector spaces with infinite dimensions, in which, contrary to those of finite dimensions, one can introduce several 'natural' topologies.)

Let us analyse now the assumption of the compactness of X. The function

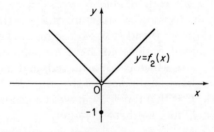

Figure 2.2

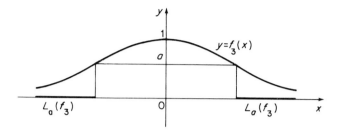

Figure 2.3

$f_3 : \mathbb{R} \to \mathbb{R}$ defined by $f_3(x) = e^{-x^2}$ (see fig. 2.3) shows that it is necessary, unless we require something more on f as suggested by the function $f_4 : \mathbb{R} \to \mathbb{R}$ defined by $f_4(x) = 1 - e^{-x^2}$ (see fig. 2.4) (or, but this is clearly unnatural, one can make $X = \mathbb{R}$ compact—choosing for example the Alexandroff's compactification—and put $f_3(\infty) = 0 : f_3$ would then have one minimum 0, taken at $x = \infty$).

Let us study these two functions f_3 and f_4 and try to discover what is responsible for the difference in their behaviour. Let us then consider their level sets: remember that given a function $f : T \to \mathbb{R}$ and $a \in \mathbb{R}$ the *level set* of f relative to a is the set

$$L_a(f) = f^{-1}(\,]-\infty, a]) = \{x \in T : f(x) \leq a\}. \tag{2.1}$$

Now, the level sets of f_4 are compact (if $a < 1$) while those of f_3 are not: and this is the difference (for our purposes) between the two functions. To see that f_4 has a minimum in $\mathbb{R}$, it is enough to note that if $a_1 < a_2$ then $L_{a_1}(f_4) \subset L_{a_2}(f_4)$ and hence the family of compact sets (we limit ourselves to the non-empty members of this family, as is natural) $\{L_a(f_4)\}, a \in \,]-\infty, 1[$, has the finite intersection property, and hence a non-empty intersection: the points of this intersection, equal to $\{0\}$, are those where the minimum is taken.

We wished to minimize f_4 on $\mathbb{R}$: had we tried to do this, for example, on $X = \,]-\infty, 0[$ we would have concluded that there is no minimum (the greatest lower bound would be 0, a value which is never assumed in points of X)—in fact, the sets which must be compact are $L_a(f) \cap X$. From these

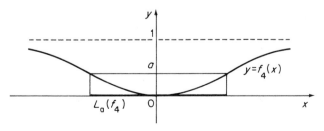

Figure 2.4

considerations, it is natural to require

(H3) X is closed and non-empty

(let us remark that if $X = \emptyset$ the problem has no sense: $f(X) = \emptyset \subset \mathbb{R}$ and inf $\emptyset$ does not exist in $\mathbb{R}$) and

(H4) $\exists \bar{a} \in \mathbb{R}\ L_{\bar{a}}(f) \cap X$ is a non-empty compact set.

In both hypotheses (H3) and (H4) it is understood that X is closed or compact with respect to the same topology for which f is lower semicontinuous.

We see that it is not necessary that for $a < \bar{a}$ the sets $L_a(f) \cap X$ be compact since (H2) implies that $L_a(f)$ are closed sets (notice that $E \setminus L_a(f)$ are open sets). It is interesting to note that the functions which are continuous for the topology of $\mathbb{R}$ which has as a sub-base the family $\{x \in \mathbb{R}: x > a, a \in \mathbb{R}\}$ are l.s.c. in the sense previously defined. For further information on l.s.c. functions, see Dieudonné, 1968c, p. 23; Ekeland–Temam, 1974, p. 9).

The hypothesis (H4), however, is difficult to test and it is convenient therefore to find conditions which imply it: if the topology of E is such that (as for finite-dimensional spaces) compact sets are bounded and closed sets, it is sufficient to require that the sets $L_a(f) \cap X$ be bounded, and for this latter condition to be true it is sufficient that

(H5) X is bounded

or, alternatively, a *coercivity* of f (for simplicity we suppose that E is a normed space: see the hypothesis (H9) which follows)

(H6) $\lim_{\|x\|_E \to +\infty,\ x \in X} f(x) = +\infty.$

The coercivity of f implies that the sets $L_a(f)$ are bounded, since if there is an a such that $L_a(f)$ is not bounded we could extract from this set a sequence x_n such that $\|x_n\| \xrightarrow[E]{} +\infty$, which, together with (H6), contradicts $f(x_n) \leq a$. This condition excludes functions of the type of function f_3 (fig. 2.3) and of function f_4 (fig. 2.4) from our study.

We will see now, with the help of a particular case, which spaces we can take into consideration so that the boundedness of the closed sets $L_a(f) \cap X$ implies their compactness: we will thus obtain not only a characterization of the suitable spaces E but also other conditions to impose on f and on X. Let $E = H$ be a separable Hilbert space of infinite dimension. Here we have two natural topologies, the weak one and the strong one: with respect to the latter, closed and bounded sets are not necessarily compact (for example the unit sphere $\{x \in H: \|x\|_H \leq 1\}$ is not compact), but *bounded weakly closed sets are weakly compact* (what is true for any reflexive normed space, see e.g. Yosida, 1971, p. 140). One would then work with the weak topology; but the condition that f be weakly lower semicontinuous implies that (H2) is difficult to test.

The problem is simplified if we add the assumption of convexity, since *convex sets are strongly closed if and only if they are weakly closed* (as is true for any topology of a pair of dual spaces, see e.g. Robertson–Robertson, 1966, p. 34). Now for the sets $L_a(f) \cap X$ to be convex it is sufficient that X and $L_a(f)$ are so, and hence we can write

(H7) X is a convex set

and

(H8) $f: E \to X$ is a convex function,

by virtue of

THEOREM 2.1. *If $f: E \to \mathbb{R}$ is a convex function then $\forall a \in \mathbb{R}$ $L_a(f)$ is a convex set.*

Proof. If $x_1, x_2 \in L_a(f)$ then $f(x_1) \leq a$ and $f(x_2) \leq a$, hence $\forall \lambda \in [0, 1] f((1-\lambda)x_1 + \lambda x_2) \leq (1-\lambda)f(x_1) + \lambda f(x_2) \leq (1-\lambda)a + \lambda a = a$. ∎

As stated above, the characteristics of the topologies of the Hilbert space H (which we have taken as a basis of our reasoning since it is the most immediate generalization of finite-dimensional spaces) which we have effectively exploited are true in any reflexive normed space; so, it is natural to impose

(H9) E is a reflexive Banach space,

where we impose also the condition of completeness of E since it is inconvenient to work in non-complete spaces.

The fundamental theorem

Summarizing the hypotheses (H1), ..., (H9) we can state the following fundamental result:

THEOREM 2.2 (Theorem of the minimization of convex functionals). *If E is a reflexive Banach space, $f: E \to \mathbb{R}$ is convex and l.s.c., $X \neq \emptyset$ is a closed convex subset of E and X is bounded or f is coercive, then the problem 2.1 has a solution. This solution is unique if f is strictly convex.*

Proof. Let us remark that it is not necessary to define the topology respect to which f is l.s.c. since *convex functions are weakly l.s.c. iff they are strongly l.s.c.* (see, e.g., Ekeland–Temam, 1974, p. 10). Now, the sets $L_a(f)$ are strongly closed since f is strongly l.s.c. and they are convex since f is convex: they are therefore weakly closed and hence so are the sets $X \cap L_a(f)$, which, because of the boundedness of X or because of the coercivity of f, are bounded, and hence, because of the reflexivity of the normed space E, are weakly compact. The existence of the minimum is assured by the fact that this family of weakly compact sets has the finite intersection property (as is obvious, there is at least one non-empty set $X \cap L_a(f)$). If f is strictly convex, the minimum, m, is achieved at only one point of X, for were it achieved at x_1 and x_2 we would write (note that $\frac{1}{2}x_1 + \frac{1}{2}x_2 \in X$ since X is convex) $f((x_1+x_2)/2) < \frac{1}{2}f(x_1) + \frac{1}{2}f(x_2) = m$, which is absurd. ∎

Minimization of quadratic functionals

The theorem 2.2 assures us that the following problem has one and only one solution:

PROBLEM 2.2. Let H be a Hilbert space, $K \subset H$ a non-empty closed convex set, $L \in H'$ and $f: H \to \mathbb{R}$ defined by

$$f(v) = \tfrac{1}{2}\|v\|_H^2 - L(v); \tag{2.2}$$

find $u_0 \in K$ such that $f(u_0) = \inf_{v \in K} f(v)$.

In fact the assumptions of the theorem are satisfied since H is reflexive, $K \neq \emptyset$ is a closed convex set and f is strictly convex and (since $f(v) \geq \tfrac{1}{2}\|v\|_H^2 - \|L\|_{H'}\|v\|_H$) coercive. On the other hand the theorem does not give us a procedure for finding the solution: in fact, it is a non-constructive result. This non-constructivity is a consequence of the fact that we exploit the finite intersection property of the family of weakly compact sets $X \cap L_a(f)$, which guarantees that the intersection is non-empty but does not show us how to obtain such an intersection.

Given the theoretical importance of problem 2.2, we will give a proof (also non-constructive) of the existence of a solution independently of theorem 2.2. For this purpose, let

$$i = \inf_{v \in K} f(v), \tag{2.3}$$

and, keeping in mind the definition of greatest lower bound, let us write (remark that $i > -\infty$)

$$\forall n \in \mathbb{N} \quad \exists v_n \in K \quad f(v_n) \leq i + \frac{1}{n}. \tag{2.4}$$

Note that were we to know how to 'construct' this sequence v_n the proof which follows would also be 'constructive'. From (2.2) and (2.4) follows

$$\tfrac{1}{2}\|v_n\|_H^2 - L(v_n) \leq i + \frac{1}{n}, \tag{2.5}$$

and hence

$$\tfrac{1}{2}\|v_n\|_H^2 \leq i + \frac{1}{n} + \|L\|_{H'}\|v_n\|_H \tag{2.6}$$

from which it follows that $\forall n \, \|v_n\|_H \leq M \in \mathbb{R}$. Now, *from a sequence bounded in the norm we can extract a weakly convergent subsequence*: let v_{n_k} be the subsequence extracted and v_∞ its (weak) limit. Since K is weakly closed, $v_\infty \in K$. We see now that v_∞ is a point where the minimum is taken: we know that $v_{n_k} \rightharpoonup v_\infty$ implies that $\lim' \|v_{n_k}\|_H \geq \|v_\infty\|_H$ (see, e.g., Yosida, 1971, p. 120) and that $L(v_{n_k}) \to L(v_\infty)$ (from the definition of weak convergence), and hence applying the operator $\lim'$ to

$$i + \frac{1}{n_k} \geq f(v_{n_k}) = \tfrac{1}{2} \|v_{n_k}\|_H^2 - L(v_{n_k}) \tag{2.7}$$

we get

$$i \geq \tfrac{1}{2} \|v_\infty\|_H^2 - L(v_\infty) = f(v_\infty), \tag{2.8}$$

and thus, from (2.3), $f(v_\infty) = i$. We have thus shown the existence of a solution to problem 2.2.

The above proof can be improved by showing that $v_n \rightharpoonup v_\infty$. With this aim let us show that for the whole of v_n (and not only for certain extracted subsequences) $v_n \rightharpoonup v_\infty$ is true, which we will do by *reductio ad absurdum*. Let us suppose then that v_n does not converge weakly to v_∞; this would mean that

$$\exists f \in H' \quad \exists \varepsilon > 0 \quad \exists \{n_h\}_{h=1,2,\ldots} \quad \forall h \, |f(v_{n_h} - v_\infty)| > \varepsilon, \tag{2.9}$$

but, since $\|v_{n_h}\|_H \leq M$, we can extract from v_{n_h} a weakly convergent subsequence: let $v_{n_{h_m}}$ be this subsequence and v'_∞ its weak limit. We can repeat for $v_{n_{h_m}}$ the reasoning applied to v_{n_k} and conclude thus that $f(v'_\infty) = i$, which is absurd since the minimum of f is unique and from (2.9) it follows that $v'_\infty \neq v_\infty$. To see that $v_n \to v_\infty$ it is sufficient to remark that $\lim \|v_n\|_H = \|v_\infty\|_H$ (since $\lim'' \|v_n\|_H \geq \lim' \|v_n\|_H \geq \|v_\infty\|_H$ and $\lim'' \|v_n\|_H > \|v_\infty\|_H$ cannot be true since from (2.7) one would have $i > f(v_\infty)$) and to recall that in a Hilbert space *the norm convergence associated with weak convergence implies strong convergence* (besides, this result is true under more general conditions on the space H: in fact it is enough that H be a locally uniformly convex normed space (see Day, 1958, p. 113).

2.2 VARIATIONAL FORMULATION OF A MINIMIZATION PROBLEM

Variational inequalities

Let us now consider the

PROBLEM 2.3. Given a Hilbert space H, $L \in H'$ and $K \neq \emptyset$ a closed convex set of H, find $u_0 \in K$ such that

$$(u_0, u_0 - v)_H \leq L(u_0 - v) \quad \forall v \in K. \tag{2.10}$$

Our objective is to prove the following

THEOREM 2.3. *The problems 2.2 and 2.3 are equivalent, i.e.* u_0 *is a solution of the problem 2.2 if and only if it is a solution of the problem 2.3.*

Proof. We first show that if u_0 is a solution of problem 2.2 it solves also problem 2.3. If u_0 solves the problem 2.2 then $u_0 \in K$ and, if v is any element of K, we can consider the application

$$[0, 1] \ni \lambda \mapsto F(\lambda) = f(\lambda u_0 + (1-\lambda)v) \in \mathbb{R} \tag{2.11}$$

(note that $\lambda u_0 + (1-\lambda)v \in K$) which has a minimum at $\lambda = 1$, a fact which is characterized by

$$\left[\frac{d}{d\lambda} F(\lambda)\right]_{\lambda=1} \leq 0 \tag{2.12}$$

(note that 1 is an extreme of the interval in which $F(\lambda)$ is defined and that $F(\lambda)$ is differentiable) or, keeping in mind (2.2), by

$$\left[\frac{d}{d\lambda} (\tfrac{1}{2}\|\lambda u_0 + (1-\lambda)v\|_H^2 - L(\lambda u_0 + (1-\lambda)v))\right]_{\lambda=1} \leq 0 \tag{2.13}$$

or again, after some calculations, by

$$(u_0, u_0 - v)_H \leq L(u_0 - v). \tag{2.14}$$

To show that problem 2.3 $\Rightarrow$ problem 2.2 it is sufficient to show that the solution of problem 2.3 is unique (since that of problem 2.2 is so, and as we have seen, problem 2.2 $\Rightarrow$ problem 2.3) which we will do by *reductio ad absurdum*. Let u_1 and u_2 be two solutions of the problem 2.3. We can write

$$(u_1, u_1 - v)_H \leq L(u_1 - v) \quad \forall v \in K \tag{2.15}$$

$$(u_2, u_2 - v)_H \leq L(u_2 - v) \quad \forall v \in K, \tag{2.16}$$

and hence, putting $v = u_2$ in (2.15) and $v = u_1$ in (2.16) (a trick which is necessary because we cannot subtract—as one is tempted to do—corresponding members of (2.15) and (2.16) since they are inequalities)

$$(u_1, u_1 - u_2)_H \leq L(u_1 - u_2) \tag{2.17}$$

$$(u_2, u_2 - u_1)_H \leq L(u_2 - u_1), \tag{2.18}$$

and summing these two inequalities we get, given that L is linear,

$$(u_1 - u_2, u_1 - u_2)_H \leq 0, \tag{2.19}$$

which shows that $u_1 = u_2$. ∎

We call problem 2.3 a *variational formulation of the minimization problem 2.2* since in (2.11) one 'varies' the argument of the functional f, and we call (2.10) the *variational inequality* or the *Euler inequality* for the minimization problem (note that (2.10) constitutes in effect a system of card K inequalities).

Variational equations

In the particular case in which $K = H$ the problem 2.3 is written as

PROBLEM 2.4. Given a Hilbert space H and $L \in H'$ find $u_0 \in H$ such that

$$(u_0, w)_H = L(w) \quad \forall w \in H. \tag{2.20}$$

In fact, since $K = H$, we have $\forall w \in H \,\exists v \in K\; u_0 - v = w$ and $\forall w \in H \,\exists v \in K\; u_0 - v = -w$.

Now problem 2.4 is nothing but problem 1.2 and, since from theorem 2.2 (or from the proof relative to problem 2.2) and from theorem 2.3 we know that problem 2.4 has one and only one solution, we conclude that we have proved the Riesz theorem. Let us remark that if we wish to prove theorem 2.3 again in this particular case, (2.11) becomes

$$\mathbb{R} \ni \lambda \mapsto f(\lambda u_0 + (1-\lambda)v) \in \mathbb{R} \tag{2.11 bis}$$

and (2.12) becomes

$$\left[\frac{\mathrm{d}}{\mathrm{d}\lambda} F(\lambda)\right]_{\lambda=1} = 0. \tag{2.12 bis}$$

In agreement with the nomenclature introduced above, we call (2.20) the *variational equation* or the *Euler equation* relative to the minimization problem 'Problem 2.2 with $K = H$'.

Geometrical interpretation

The Riesz theorem enables us to give an interesting geometrical interpretation of problems 2.2 and 2.3. In fact, this theorem allows us to write (2.2) in the form

$$f(v) = \tfrac{1}{2}\|v\|_H^2 - (u^*, v)_H \tag{2.21}$$

(with $u^* = u_L$), and since minimizing $f(v)$ in K is obviously equivalent to minimizing, again on K,

$$F(v) = f(v) + \tfrac{1}{2}\|u^*\|_H^2 = \tfrac{1}{2}\|v\|_H^2 - (u^*, v)_H + \tfrac{1}{2}\|u^*\|_H^2 = \tfrac{1}{2}\|v - u^*\|_H^2 \tag{2.22}$$

(i.e. to minimize the distance from u^* to K: see fig. 2.5), we can say that problems 2.2 and 2.3 are equivalent to finding the *projection* u_0 of u^* on K, which is therefore characterized by

$$u_0 \in K; \quad (u_0 - u^*, u_0 - v)_H \leq 0 \quad \forall v \in K, \tag{2.23}$$

which shows that $u_0 - u^*$ and $u_0 - v$ subtend an obtuse angle between them.

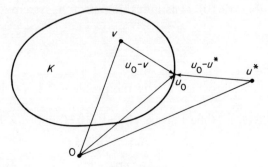

Figure 2.5

2.3 PROJECTIONS ON CONVEX SETS

Since we will use later the concept of projection on a convex set of a Hilbert space, we should establish here the notation and present some properties of these projections.

DEFINITION 2.1. Let H be a Hilbert space and $K \subset H$ a non-empty closed convex set. If $u \in H$, by *projection* of u on K we mean the element $P_K(u) \in K$ such that

$$\|u - P_K(u)\|_H \leq \|u - v\|_H \quad \forall v \in K. \tag{2.24}$$

In other words we can say that $P_K(u)$ is the element of K closest to u; let us remark also that (2.24) is merely (2.23) written in a different way.

This definition is coherent since, given u in H, its projection on K exists and is unique, as can be seen from the existence and uniqueness of the solution of problem 2.3. On the other hand, it is curious that, among the weakly closed sets of a Hilbert space, the convex sets are the only ones for which a coherent definition of projection can be given.

Projection operators

The projection $P_K(u)$ of u on K can then be interpreted as the result of applying to u the operator $P_K: H \to K$. Some properties of P_K are given in the following theorem; before stating it, however, let us recall that an operator T defined on a metric space (S, d) is called *non-expansive* if (cf. (1.6))

$$\forall x, y \in S \quad d(T(x), T(y)) \leq d(x, y), \tag{2.25}$$

and let us introduce the concept of monotonicity for operators (which is an extension of the concept of an increasing monotone function):

DEFINITION 2.2. Let H be a Hilbert space and $A: H \to H$ an operator (not

necessarily linear). A is called *monotone* if
$$(A(v_1) - A(v_2), v_1 - v_2)_H \geq 0 \quad \forall v_1, v_2 \in H, \tag{2.26}$$
and *strictly monotone* if it is monotone and
$$(A(v_1) - A(v_2), v_1 - v_2)_H = 0 \Rightarrow v_1 = v_2. \tag{2.27}$$

THEOREM 2.1. *Let H be a Hilbert space, $K \subset H$ a non-empty closed convex set and P_K the projection operator on K. P_K is non-expansive, monotone but not strictly monotone, and strongly continuous.*

Proof. Let us first note that the characterization of the projection given by (2.23) can be written, with the notation introduced in the definition 2.1, in the form
$$(P_K(u) - u, P_K(u) - v)_H \leq 0 \quad \forall v \in K. \tag{2.28}$$
To show that P_K is monotone let us fix u_1 and u_2 in H and write
$$(P_K(u_1) - u_1, P_K(u_1) - v)_H \leq 0 \quad \forall v \in K \tag{2.29}$$
$$(P_K(u_2) - u_2, P_K(u_2) - v)_H \leq 0 \quad \forall v \in K. \tag{2.30}$$
Putting $v = P_K(u_2)$ in (2.29) and $v = P_K(u_1)$ in (2.30) (we can take these particular values of v because $P_K(u_i) \in K$) we have
$$(P_K(u_1) - u_1 - P_K(u_2) + u_2, P_K(u_1) - P_K(u_2))_H \leq 0 \tag{2.31}$$
and therefore
$$\|P_K(u_1) - P_K(u_2)\|_H^2 \leq (u_1 - u_2, P_K(u_1) - P_K(u_2))_H \quad \forall u_1, u_2 \in H, \tag{2.32}$$
which in particular implies the monotonicity of P_K. Further, if $K = H$ then $P_K = I$ (identity) and one has strict monotonicity; but in general, for $K \neq H$, P_K is not injective (and hence not strictly monotone): if $u_0 \notin K$ then $P_K(u_0) \neq u_0$ but $P_K(P_K(u_0)) = P_K(u_0)$. To show that P_K is not expansive it is enough to apply the Schwarz inequality to (2.32) and obtain thus
$$\|P_K(u_1) - P_K(u_2)\|_H^2 \leq \|u_1 - u_2\|_H \|P_K(u_1) - P_K(u_2)\|_H \tag{2.33}$$
or, dividing by $\|P_K(u_1) - P_K(u_2)\|_H$ (notice that if $\|P_K(u_1) - P_K(u_2)\|_H = 0$ then (2.25) is true),
$$\|P_K(u_1) - P_K(u_2)\|_H \leq \|u_1 - u_2\|_H. \tag{2.34}$$
The (strong) continuity follows immediately from (2.34). ∎

The geometrical interpretation of the non-expansivity of P_K is shown in fig. 2.6: for the pair u_1, u_2 the sign $<$ holds in (2.34), i.e. the operator reduces the distance—and for the pair u_3, u_4 the sign $=$ holds, i.e. the operator conserves the distance.

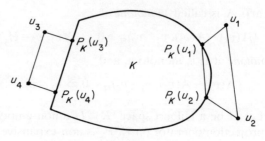

Figure 2.6

Finally, note that the set of fixed points of P_K is K and further, as shown in Zarantonello, 1971, p. 246, the set of fixed points for a non-expansive operator is always a closed convex set.

REMARK 2.1 (Projections on convex cones). In the particular case in which K is a convex cone with its vertex at the origin (i.e. such that $K+K \subset K$ and $\lambda K \subset K \; \forall \lambda \geq 0$—as usual, if A and B are two subsets of a real (or complex) vector space, and λ, $\mu \in \mathbb{R}$ (or $\mathbb{C}$) we put $\lambda A + \mu B = \{\lambda a + \mu b : a \in A, b \in B\}$) the inequality (2.28) is equivalent to the pair of conditions

$$(P_K(u) - u, v)_H \geq 0 \quad \forall v \in K \tag{2.35}$$

$$(P_K(u) - u, P_K(u))_H = 0. \tag{2.36}$$

In fact, if (2.36) and (2.35) are satisfied then their difference (2.28) is also satisfied. Conversely, if (2.28) is satisfied and K is a cone then (2.35) and (2.36) are satisfied since, putting $v = P_K(u) + w$ (note that $K + K \subset K$) in (2.28), we have

$$(P_K(u) - u, P_K(u) - P_K(u) - w)_H \leq 0 \quad \forall w \in K, \tag{2.37}$$

which is (2.35), and putting then $v = 0$ in (2.28) we get $(P_K(u) - u, P_K(u))_H \leq 0$, and this together with (2.37) (with $w = P_K(u)$) gives (2.36).

For a more detailed study of projections on cones, see Zarantonello, 1974.

3
Variational Inequalities

3.1 FUNDAMENTAL THEOREM

Bilinear forms

The functional (2.2) considered in the problem 2.2 is a particular case of *quadratic functionals* of the type

$$f(v) = \tfrac{1}{2} a(v, v) - L(v), \qquad (3.1)$$

where $L \in H'$ and $a : H \times H \to \mathbb{R}$ is a *bilinear form*, i.e.

$$a(\mu_1 u_1 + \mu_2 u_2, \lambda_1 v_1 + \lambda_2 v_2) = \sum_{i,j=1}^{2} \mu_i \lambda_j a(u_i, v_j) \qquad (3.2)$$

and is *continuous*, i.e.,

$$\exists M \in \mathbb{R} \ |a(u, v)| \leq M \|u\|_H \|v\|_H \qquad \forall u, v \in H. \qquad (3.3)$$

It is in fact sufficient to put $a(u, v) = (u, v)_H$, and a is a continuous bilinear form (with $M = 1$); this form is also *symmetric*, i.e.

$$a(u, v) = a(v, u) \qquad \forall u, v \in H, \qquad (3.4)$$

and has the property of *coercivity* on H (with $\alpha = 1$)

$$\exists \alpha \in \mathbb{R}, \alpha > 0 \quad a(u, u) \geq \alpha \|u\|_H^2 \qquad \forall u \in H. \qquad (3.5)$$

On the other hand, if $a(u, v)$ is a coercive symmetric and continuous bilinear form, the functional (3.1) is no more general than (2.2) since, in these conditions, we can write it in the form

$$f(v) = \tfrac{1}{2} \|\|v\|\|_H^2 - L(v), \qquad (3.6)$$

where $\|\|\cdot\|\|_H$ is the norm associated with the inner product

$$((u, v))_H = a(u, v) \qquad \forall u, v \in H, \qquad (3.7)$$

and, since the norms $\|\cdot\|_H$ and $\|\|\cdot\|\|_H$ are equivalent, the continuous linear functionals for the two norms are the same (the dual does not change with respect to equivalent norms and therefore equivalent strong topologies). Let us note that the equivalence between the original norm $\|\cdot\|_H$ and the new one $\|\|\cdot\|\|_H$ (note that $((\,,\,))_H$ is obviously an inner product and hence $\|\|\cdot\|\|_H$ is in fact a norm) can be seen from the fact that $\alpha \|v\|_H^2 \leq a(v,v) = ((v,v))_H = \|\|v\|\|_H^2 = ((v,v))_H = a(v,v) \leq M \|v\|_H^2$.

Minimization problems and variational problems

We have seen that problems 2.2 and 2.3 are equivalent, and it is therefore natural to ask the question regarding the equivalence of the following problems (not necessarily well posed):

PROBLEM 3.1. Let H be a Hilbert space, $K \subset H$ a non-empty closed convex set and $f: H \to \mathbb{R}$ defined in (3.1), where $L \in H'$ and a is a continuous bilinear form; find $u_0 \in K$ such that

$$f(u_0) \leq f(v) \qquad \forall v \in K. \tag{3.8}$$

PROBLEM 3.2. Let H be a Hilbert space, $K \subset H$ a non-empty closed convex set, $L \in H'$ and a a continuous bilinear form; find $u_0 \in K$ such that

$$a(u_0, u_0 - v) \leq L(u_0 - v) \qquad \forall v \in K. \tag{3.9}$$

Following the reasoning in the proof of theorem 2.3, let us consider the application

$$[0, 1] \ni \lambda \mapsto F(\lambda) = f((1-\lambda)u_0 + \lambda v) \in \mathbb{R}, \tag{3.10}$$

which has a minimum at $\lambda = 0$ (if f has one at $u = u_0$), i.e.

$$F(\lambda) \geq F(0) \qquad \forall \lambda \in [0, 1] \quad \forall v \in K, \tag{3.11}$$

which is characterized by the fact that

$$\left[\frac{d}{d\lambda} F(\lambda)\right]_{\lambda=0} \geq 0. \tag{3.12}$$

Given the linearity of L and the bilinearity of a, we can write,

$$\frac{d}{d\lambda} [\tfrac{1}{2} a((1-\lambda)u_0 + \lambda v, (1-\lambda)u_0 + \lambda v) - L((1-\lambda)u_0 + \lambda v)]_{\lambda=0}$$

$$= \tfrac{1}{2} a(u_0, v - u_0) + \tfrac{1}{2} a(v - u_0, u_0) - L(v - u_0), \tag{3.13}$$

and hence

$$\tfrac{1}{2} a(u_0, u_0 - v) + \tfrac{1}{2} a(u_0 - v, u_0) \leq L(u_0 - v). \tag{3.14}$$

If a is symmetric (3.14) is identical with (3.9) and all we have done is to prove once again the first part of theorem 2.3. If a is not symmetric, let us decompose a into its *symmetric part* a_S and its *antisymmetric part* a_A:

$$a(u, v) = \frac{a(u, v) + a(v, u)}{2} + \frac{a(u, v) - a(v, u)}{2}$$

$$= a_S(u, v) + a_A(u, v). \tag{3.15}$$

Since
$$a(u, u) = a_S(u, u) \quad \forall u \in H, \tag{3.16}$$
we can write the functional (3.1) in the form
$$f(v) = \tfrac{1}{2} a_S(v, v) - L(v) \tag{3.17}$$
and, redoing all the calculations in this new (but equivalent) way of writing the functional, we obtain, instead of (3.14),
$$\tfrac{1}{2} a_S(u_0, u_0 - v) + \tfrac{1}{2} a_S(u_0 - v, u_0) \leq L(u_0 - v), \tag{3.18}$$
or
$$a_S(u_0, u_0 - v) \leq L(u_0 - v) \quad \forall v \in K. \tag{3.19}$$

We can then say that if the problem 3.1 has a solution, this is characterized by (3.19); we can also conclude that if a is not symmetric the problems 3.1 and 3.2 are not equivalent.

Using further the analogy between the question of this equivalence and theorem 2.3 we see now under what conditions problem 3.2 has a unique solution; if u_1 and u_2 are two solutions of problem 3.2, we can write

$$a(u_1, u_1 - v) \leq L(u_1 - v) \quad \forall v \in K \tag{3.20}$$
$$a(u_2, u_2 - v) \leq L(u_2 - v) \quad \forall v \in K \tag{3.21}$$

and, putting $v = u_2$ in (3.20), $v = u_1$ in (3.21) and summing the two expressions thus obtained after the usual change of sign in (3.21),

$$a(u_1 - u_2, u_1 - u_2) \leq 0. \tag{3.22}$$

Imposing on a an hypothesis of the type

$$a(u, u) \leq 0 \Rightarrow u = 0 \quad \forall u \in K - K \tag{3.23}$$

or, equivalently,

$$a(u, u) \geq 0 \quad \forall u \in K - K \quad \text{and} \quad a(u, u) = 0 \quad \text{iff } u = 0, \tag{3.24}$$

we can state, from (3.22), that if a solution of problem 3.2 exists, it is unique.

However (3.24) does not imply that such a solution exists. In order to find what conditions have to be imposed on the form a so that we have existence of a solution, let us come back for a moment to the functional (3.1); whether a is symmetric or not, this functional can be written in the form

$$f(v) = \tfrac{1}{2} |||v|||_{H,S}^2 - L(v), \tag{3.25}$$

where the meaning of $|||\cdot|||_{H,S}$ is obvious. Now for (3.25) to satisfy the hypotheses of theorem 2.2, and hence to ensure the existence of the solution of problem 3.1, it is sufficient that a (or a_S) satisfy an hypothesis of the type

$$a(u, u) \to +\infty \quad \text{when} \quad \|u\|_H \to +\infty, \ u \in K. \tag{3.26}$$

In fact, we impose something stronger—the coercivity of a on $K-K$:

$$\exists \alpha > 0 \quad a(u-v, u-v) \geq \alpha \|u-v\|_H^2 \quad \forall u, v \in K. \tag{3.27}$$

Note that (3.27) implies (3.23); to see that (3.27) implies (3.26) it is enough to see that

$$\begin{aligned}
a(u, u) &= a(u-w, u-w) + a(u, w) + a(w, u) - a(w, w) \\
&\geq a(u-w, u-w) - M \|w\|_H (\|w\|_H + 2 \|u\|_H) \\
&\geq a(u-w, u-w) - M \|w\|_H (3 \|w\|_H + 2 \|u-w\|_H) \\
&\geq \alpha \|u-w\|_H^2 - 2M \|w\|_H \|u-w\|_H - 3M \|w\|_H^2,
\end{aligned}$$

where w is fixed in K, and that when $\|u\|_H \to +\infty$ also $\|u-w\|_H \to \infty$.

Lions–Stampacchia theorem

In the next (most important) theorem we will show that under the assumption (3.27), problem 3.2 has only one solution, u_0; under the same assumption problem 3.1 also has only one solution, u_0^*, but u_0 and u_0^* are in no way related, unless a is symmetric, in which case $u_0 \equiv u_0^*$.

THEOREM 3.1 (Lions–Stampacchia theorem). *Let H be a Hilbert space, $K \subset H$ a non-empty closed convex set, $L \in H'$ and $a: H \times H \to \mathbb{R}$ a continuous bilinear form on H and coercive on $K-K$. There is then one and only one $u_0 \in K$ such that*

$$a(u_0, u_0 - v) \leq L(u_0 - v) \quad \forall v \in K, \tag{3.28}$$

and further the application which associates u_0 to every L is continuous (i.e. the problem 3.2 is well posed).

Proof. The continuous dependence of u_0 on the datum L can be easily seen by rewriting (3.20) and (3.21) with L_1 and L_2 in place of L, respectively, and putting $v = u_2$ in (3.20) and $v = u_1$ in (3.21): we have then, instead of (3.22),

$$a(u_1 - u_2, u_1 - u_2) \leq L_1(u_1 - u_2) + L_2(u_2 - u_1) = (L_1 - L_2)(u_1 - u_2), \tag{3.29}$$

and hence

$$\alpha \|u_1 - u_2\|_H^2 \leq \|L_1 - L_2\|_{H'} \|u_1 - u_2\|_H, \tag{3.30}$$

which gives us the continuous (further: Lipschitz continuous) dependence and also the uniqueness of the solution. ∎

For the existence, given the importance of the theorem, we give two proofs:

First proof. Let us decompose the form a in its symmetric and antisymmetric parts,

$$a(u, v) = a_S(u, v) + a_A(u, v), \tag{3.31}$$

and introduce the new bilinear form

$$a_t(u, v) = a_S(u, v) + ta_A(u, v), \tag{3.32}$$

which is dependent on the parameter $t \in \mathbb{R}$ and for every t is continuous (in fact $|a_t(u, v)| \leq M(\|u\|_H \|v\|_H + |t| \|u\|_H \|v\|_H)$. Let us now consider the

PROBLEM 3.3. Find $u_0 \in K$ such that

$$a_t(u_0, u_0 - v) \leq L(u_0 - v) \quad \forall v \in K. \tag{3.33}$$

We will later show that there exists $\delta > 0$, independent of t_0, such that if problem 3.3 has a solution for $t = t_0$ then the same problem has a solution for every $t \in [t_0 - \delta, t_0 + \delta]$. This proves the theorem since we know that problem 3.3 has a solution for $t = t_0 = 0$ (we are then concerned with the problem 2.3, since a_0 is a symmetric form) and, since δ is independent of t_0, we can extend for the whole of $\mathbb{R}$ the set of t for which problem 3.3 has a solution; in particular, with a finite number of extensions we can get to $t = 1$, a value of t which transforms problem 3.3 to problem 3.2, since $a_1 = a$ (this technique of proof is for obvious reasons called *the method of extension with respect to the parameter*).

Let us consider the problem:

PROBLEM 3.4. Find $u_0 \in K$ such that

$$a_{t_0}(u_0, u_0 - v) \leq L(u_0 - v) \quad \forall v \in K. \tag{3.34}$$

Let us suppose then, according to what we have already stated, that this problem has a solution, and let us write (3.33) in the manifestly equivalent form

$$a_{t_0}(u_0, u_0 - v) \leq L(u_0 - v) + (t_0 - t)a_A(u_0, u_0 - v) \quad \forall v \in K. \tag{3.35}$$

Now, for every $w \in H$ fixed, let us consider the problem

PROBLEM 3.5. Find $u \in K$ such that

$$a_{t_0}(u, u - v) \leq L(u - v) + (t_0 - t)a_A(w, u - v) \quad \forall v \in K. \tag{3.36}$$

This problem has one and only one solution for every w since, the latter being fixed, the second member of (3.36) is a continuous linear form in $u - v$ and hence problem 3.5 is nothing else but problem 3.4: we have thus constructed an application—let us call it τ—which associates with every $w \in H$ the solution $u = \tau(w)$ of problem 3.5. If u_0 satisfies (3.33) then it also satisfies (3.35), and it is then obvious that u_0 is a solution of the problem 3.3 if and only if $u_0 = \tau(u_0)$, i.e., if and only if u_0 is a fixed point for τ.

We will now show that if $|t - t_0| \leq \delta$ (with $\delta > 0$ to the determined) then τ is a contraction and hence has a fixed point (Banach's theorem). Let then w_1

and w_2 be two elements of H and $u_1 = \tau(w_1)$ and $u_2 = \tau(w_2)$; hence u_1 and u_2 satisfy the conditions

$$a_{t_0}(u_1, u_1 - v) \leq L(u_1 - v) + (t_0 - t)a_A(w_1, u_1 - v) \quad \forall v \in K \quad (3.37)$$

$$a_{t_0}(u_2, u_2 - v) \leq L(u_2 - v) + (t_0 - t)a_A(w_2, u_2 - v) \quad \forall v \in K. \quad (3.38)$$

Putting $v = u_2$ in (3.37) and $v = u_1$ in (3.38) (note that these substitutions are possible because $u_1, u_2 \in K$) we have

$$a_{t_0}(u_1, u_1 - u_2) \leq L(u_1 - u_2) + (t_0 - t)a_A(w_1, u_1 - u_2) \quad (3.39)$$

$$a_{t_0}(u_2, u_2 - u_1) \leq L(u_2 - u_1) + (t_0 - t)a_A(w_2, u_2 - u_1) \quad (3.40)$$

and summing them we have

$$a_{t_0}(u_1 - u_2, u_1 - u_2) \leq (t_0 - t)a_A(w_1 - w_2, u_1 - u_2) \quad (3.41)$$

or, since

$$a_{t_0}(u, u) = a(u, u) \quad \forall u \in K \quad (3.42)$$

as we can easily see,

$$a(u_1 - u_2, u_1 - u_2) \leq (t_0 - t)a_A(w_1 - w_2, u_1 - u_2). \quad (3.43)$$

Keeping in mind now that a_A is continuous (with constant M) and that a is coercive we can write

$$\alpha \|u_1 - u_2\|_H^2 \leq |t_0 - t| M \|w_1 - w_2\|_H \|u_1 - u_2\|_H \quad (3.44)$$

and hence, dividing by $\|u_1 - u_2\|_H$ (if $\|u_1 - u_2\|_H = 0$ we have evidently (3.45)),

$$\|\tau(w_1) - \tau(w_2)\|_H \leq |t_0 - t| \frac{M}{\alpha} \|w_1 - w_2\|_H, \quad (3.45)$$

which tells us that τ is a contraction as long as t is such that $|t_0 - t| M/\alpha < 1$: so that it is enough to put $\delta = \alpha/2M$ and choose $|t - t_0| \leq \delta$ (and δ is independent of t_0 as was required). ∎

Second proof. The previous proof is constructive (if we know how to solve a problem of the type of problem 2.3) but gives us a computational algorithm which is slow, particularly if α/M is small. We will now provide another proof which gives us a much faster algorithm. We need to write the functional L and the form a as scalar products, which we will do by using the Riesz theorem. In the following we will use the notations $L(v)$ and ${}_{H'}\langle L, v\rangle_H$ to indicate the value taken by $L \in H'$ in $v \in H$, choosing one or the other as is more convenient.

The theorem of Riesz states that there exists $u_L \in H$ such that

$${}_{H'}\langle L, v\rangle_H = (u_L, v)_H \quad \forall v \in H, \quad (3.46)$$

and hence directly resolves the question of the representation of L.

Let us now consider, for u fixed in H, the application $H \ni v \mapsto a(u, v) \in \mathbb{R}$:

denoting this application by $A(u)$ we can state that, for every $u \in H$,

$$[A(u)](v) \triangleq a(u, v) \quad \forall v \in H. \tag{3.47}$$

For every $u \in H$, the application $A(u)$ is linear:

$$[A(u)](\lambda_1 v_1 + \lambda_2 v_2) = a(u, \lambda_1 v_1 + \lambda_2 v_2) = \lambda_1 [A(u)](v_1) + \lambda_2 [A(u)](v_2)$$
$$\forall \lambda_1, \lambda_2 \in \mathbb{R} \quad \forall v_1, v_2 \in H, \tag{3.48}$$

and continuous:

$$|[A(u)](v)| = |a(u, v)| \leq M \|u\|_H \|v\|_H, \tag{3.49}$$

and hence we can write (3.47) in the form

$$_{H'}\langle A(u), v \rangle_H = a(u, v) \quad \forall v \in H. \tag{3.50}$$

The application $A(u)$ can be interpreted as the value taken on u by an operator $A : H \to H'$, which is linear:

$$_{H'}\langle A(\lambda_1 u_1 + \lambda_2 u_2), v \rangle_H = a(\lambda_1 u_1 + \lambda_2 u_2, v)$$
$$= {}_{H'}\langle \lambda_1 A(u_1) + \lambda_2 A(u_2), v \rangle_H \quad \forall v \in H \tag{3.51}$$

and continuous, since from (3.49) we get

$$\|A(u)\|_{H'} = \sup_{\|v\|_H = 1} |[A(u)](v)| \leq M \|u\|_H \tag{3.52}$$

and hence

$$\|A\|_{\mathcal{L}(H, H')} \leq M. \tag{3.53}$$

Let us still remark that, given a linear operator $A : H \to H'$ which satisfies (3.53), the equation (3.50) defines a continuous bilinear form (with constant M).

Now, since $A(u) \in H'$, the Riesz theorem ensures the existence of $\mathcal{A}(u) \in H$ such that

$$_{H'}\langle A(u), v \rangle_H = (\mathcal{A}(u), v)_H \quad \forall v \in H, \tag{3.54}$$

and hence also solves the problem of the representation of the form a:

$$a(u, v) = (\mathcal{A}(u), v)_H \quad \forall v \in H. \tag{3.55}$$

The vector $\mathcal{A}(u)$ can be interpreted as the result of applying on u a linear and continuous operator $\mathcal{A} : H \to H$, with

$$\|\mathcal{A}\|_{\mathcal{L}(H, H)} \leq M. \tag{3.56}$$

On the other hand, it can be easily seen that given a linear operator $\mathcal{A} : H \to H$ which satisfies (3.56), the equation (3.55) defines a continuous bilinear form (with constant M).

The operators $\mathcal{A}$ and A are related by $A = J\mathcal{A}$, where $J : H \to H'$, defined by $_{H'}\langle J(u), v \rangle_H = (u, v)_H$ (note the analogy with (3.50)...), is the *Riesz operator* or *canonical injection* of H in H'. The operator J is the inverse operator of the one naturally defined by the theorem of Riesz: see Stampacchia, 1969, p. 108; Lions–Stampacchia, 1967, p. 495; Mosco, 1973, p. 526. If H and H' are identified, J coincides with the identity operator I, and A and $\mathscr{A}$ are automatically identified also. Though it is always possible to make this identification between H and H', it is not always convenient since H and H' are isomorphic as Hilbert spaces but they have on them different structures which can be lost. The phenomenon whereby these structures are lost in making the identification between H and H' will be clarified later with the example of Sobolev spaces.

Let us now return to the proof of the theorem. The expressions (3.46) and (3.55) allow us to write (3.28) in the form

$$(\mathscr{A}(u_0), u_0 - v)_H \leq (u_L, u_0 - v)_H \quad \forall v \in K, \tag{3.57}$$

or

$$(\mathscr{A}(u_0) - u_L, u_0 - v)_H \leq 0 \quad \forall v \in K. \tag{3.58}$$

Adding and subtracting u_0, we can write

$$(u_0 - u_0 + \mathscr{A}(u_0) - u_L, u_0 - v)_H \leq 0 \quad \forall v \in K, \tag{3.59}$$

which recalls (2.23) with $u^* = u_0 - \mathscr{A}(u_0) + u_L$ and induces us to find the projection of $u_0 - \mathscr{A}(u_0) + u_L$ on K. We are dealing, however, with an *implicit projection* since $u_0 - \mathscr{A}(u_0) + u_L$ already depends on the solution u_0. We have then to interpret this projection in a convenient manner.

Let $\rho > 0$ be a real parameter and let us consider the operator

$$T_\rho(\square) = I(\square) - \rho(\mathscr{A}(\square) - u_L)), \tag{3.60}$$

which allows us to write (3.59) in the equivalent form

$$(u_0 - T_\rho(u_0), u_0 - v)_H \leq 0 \quad \forall v \in K, \tag{3.61}$$

which suggests that u_0 is a fixed point of $P_K \circ T_\rho$. In fact, if $u_0 = (P_K \circ T_\rho)(u_0) = P_K(T_\rho(u_0))$ then (3.61) is satisfied since (2.28) allows us to write

$$(P_K(T_\rho(u_0)) - T_\rho(u_0), P_K(T_\rho(u_0)) - v)_H = (u_0 - T_\rho(u_0), u_0 - v)_H \leq 0 \quad \forall v \in K. \tag{3.62}$$

We will now show that $P_K \circ T_\rho$ is a contraction for convenient ρ, and thus that there is such a fixed point. If $v_1, v_2 \in K$, keeping in mind that P_K is non-expansive (theorem 2.4), we can write

$$\|P_K(T_\rho(v_1)) - P_K(T_\rho(v_2))\|_H^2 \leq \|T_\rho(v_1) - T_\rho(v_2)\|_H^2$$
$$= (v_1 - \rho[\mathscr{A}(v_1) - u_L] - v_2 + \rho[\mathscr{A}(v_2) - u_L],$$
$$v_1 - \rho[\mathscr{A}(v_1) - u_L] - v_2 + \rho[\mathscr{A}(v_2) - u_L])_H$$
$$= (v_1 - v_2 - \rho[\mathscr{A}(v_1) - \mathscr{A}(v_2)],$$
$$v_1 - v_2 - \rho[\mathscr{A}(v_1) - \mathscr{A}(v_2)])_H \tag{3.63}$$

and hence
$$\|P_K(T_\rho(v_1)) - P_K(T_\rho(v_2))\|_H^2 \leq \|v_1 - v_2\|_H^2 + \rho^2 \|\mathcal{A}(v_1) - \mathcal{A}(v_2)\|_H^2$$
$$- 2\rho(v_1 - v_2, \mathcal{A}(v_1) - \mathcal{A}(v_2))_H. \quad (3.64)$$

We have now to find a suitable upper bound for the right-hand member of (3.64). We will leave the first part as it is:
$$\|v_1 - v_2\|_H^2. \quad (3.65)$$

We will upper estimate the second part keeping in mind (3.56):
$$\rho^2 \|\mathcal{A}(v_1) - \mathcal{A}(v_2)\|_H^2 \leq \rho^2 M^2 \|v_1 - v_2\|_H^2 \quad (3.66)$$

(note that it is not necessary here that $\mathcal{A}$ be linear but only that it be Lipschitz continuous in K, i.e. that $\exists M \in \mathbb{R} \ \|\mathcal{A}(v_1) - \mathcal{A}(v_2)\|_H \leq M \|v_1 - v_2\|_H$ $\forall v_1, v_2 \in K$). For the third part, we need a lower bound since it has a minus sign. A crude way of doing this is to put it equal to zero, since we know that
$$(\mathcal{A}(v_1) - \mathcal{A}(v_2), v_1 - v_2)_H = a(v_1 - v_2, v_1 - v_2) \geq 0, \quad (3.67)$$

but the 'tout court' elimination of this part does not lead us to the conclusion that $P_K \circ T_\rho$ is a contraction: we must look for a finer estimate (lower estimate in this case). Now we know that there is a stronger form of (3.67), namely
$$(\mathcal{A}(v_1) - \mathcal{A}(v_2), v_1 - v_2)_H = a(v_1 - v_2, v_1 - v_2) \geq \alpha \|v_1 - v_2\|_H^2 \quad (3.68)$$

since a is coercive on $K - K$ (note that $v_1 - v_2 \in K$), and we can thus write
$$-2\rho(\mathcal{A}(v_1) - \mathcal{A}(v_2), v_1 - v_2)_H \leq -2\rho\alpha \|v_1 - v_2\|_H^2 \quad (3.69)$$

(here too we do not require linearity of $\mathcal{A}$ but merely its 'strong' monotonicity). Putting together (3.65), (3.66), and (3.69) we can write
$$\|P_K(T_\rho(v_1)) - P_K(T_\rho(v_2))\|_H^2 \leq (1 + M^2\rho^2 - 2\rho\alpha) \|v_1 - v_2\|_H^2 \quad (3.70)$$

and hence $P_K \circ T_\rho$ is a contraction as long as $\rho < 2\alpha/M^2$. Lastly, according to the Banach theorem for contractions, the sequence
$$u_{n+1} \triangleq (P_K \circ T_\rho)(u_n) = P_K(u_n - \rho[\mathcal{A}(u_n) - u_L]) \quad (3.71)$$

converges to the solution of problem 3.2 if ρ satisfies the same condition. The optimal value of ρ, i.e. that value which gives us the fastest convergence in the sense that a smaller number of steps gives us a solution approximate to the same accuracy, is $\rho = \alpha/M^2$, as can be easily seen. ∎

This second proof of the theorem allows us to state the following more general result:

THEOREM 3.2. *Let H be a Hilbert space, $K \subset H$ a non-empty closed convex set and $\mathcal{A} : H \to H$ an operator (not necessarily linear: the linear case corresponds to theorem 3.1), Lipschitz continuous in K and strongly* *monotone* in K in the sense that there exists an $\alpha > 0$ such that

$$(\mathscr{A}(u) - \mathscr{A}(v), u - v)_H \geq \alpha \|u - v\|_H^2 \quad \forall u, v \in K; \tag{3.72}$$

then for every $u_L \in H$ there is one and only one $u_0 \in K$ such that

$$(\mathscr{A}(u_0) - u_L, u_0 - v)_H \leq 0 \quad \forall v \in K \tag{3.73}$$

and further u_0 depends continuously on u_L.

Lax–Milgram lemma

If $K = H$, problem 3.2 becomes, as is easy to see:

PROBLEM 3.6. Find $u_0 \in H$ such that

$$a(u_0, w) = L(w) \quad \forall w \in H. \tag{3.74}$$

The result which states that problem 3.6 is well-posed, immediate corollary of the Lions–Stampacchia theorem, is known in the literature as the Lax–Milgram lemma. Direct proofs of this result, fundamental in the (variational) study of partial differential equations, can be found in Visik, 1951, Lax–Milgram, 1954, Lions, 1955. If in (3.74) the form a is symmetric, the Lax–Milgram lemma is nothing but the Riesz theorem.

3.2 MINIMIZATION OF CONVEX FUNCTIONALS

3.2.1 GÂTEAUX DERIVATIVE

Functionals defined on Hilbert spaces

The problem 3.2 is a generalization of problem 2.3 since it imposes weaker conditions on the form a. In this section we will make another type of generalization, working on the functional L.

Due to the linearity of L, the expression (3.28) can be written in the form

$$a(u_0, u_0 - v) - L(u_0) \leq -L(v) \quad \forall v \in K, \tag{3.75}$$

which suggests that we can consider expressions of the type

$$a(u, u - v) + j(u) \leq j(v), \tag{3.76}$$

with $j: H \to \mathbb{R}$ not necessarily linear. Besides, theorems 2.2 and 2.3 suggest that, by imposing the assumptions of theorem 3.1 together with that of symmetry on a, and assumptions of the type 'convexity' and 'l.s.c.' on j, it is natural that a problem associated with (3.76) in the usual way can be interpreted as a problem of minimization of the functional

$$f(u) = \tfrac{1}{2} a(u, u) + j(u). \tag{3.77}$$

More generally we consider functionals of the form

$$f(u) = h(u) + j(u), \qquad (3.78)$$

where $h(u)$ (which in (3.77) corresponds to $\tfrac{1}{2}a(u, u)$ and $j(u)$ are functionals in u on which we will make assumptions of a different kind.

Before looking for appropriate assumptions on h and j such that, for example, the minimization problem associated with f has one and only one solution, we must specify the framework in which we are working. Theorem 2.2 has been proved under rather general conditions, especially as regards the topological vector space in which the functional is defined (which we have assumed to be a reflexive Banach space, but the result is true under even more general conditions); but subsequently we have always worked in the context of Hilbert spaces: why? Because the Hilbertian structure being richer permits us to work more easily (among other things the assumptions of theorem 2.2 have a simple interpretation in such a space) and to obtain a greater number of interesting results (in Stampacchia, 1969, p. 120) problems of the type we have considered in theorem 3.2 are considered in the context of Banach spaces.

In this and in the following sections we will still work in the Hilbertian framework, but we will extend the concept of functional, by letting them take the value $+\infty$, i.e., by *functional* we refer to an application $f: C \to]-\infty, +\infty] = \mathbb{R} \cup \{+\infty\}$, where C is a subset of a Hilbert space H. In $]-\infty, +\infty]$ we will consider the usual structure of ordered set, with the usual algebraic rules. We will not consider functionals which can take the value $-\infty$ since, as we will see, we are interested essentially in convex functionals, and convex functionals which take the value $-\infty$ are very particular ones: *an l.s.c. convex functional which takes the value $-\infty$ in a point cannot take real values* (see e.g., Ekeland–Temam, 1974, pp. 8–10).

It is convenient to introduce here some nomenclature:

DEFINITION 3.1. A functional $f: C \to]-\infty, +\infty]$ is said to be *proper* if there exists $u \in C$ such that $f(u) < +\infty$, and we will call the *effective domain* of f the set $\mathrm{Dom}\, f = \{u \in C : f(u) \neq +\infty\}$ (thus a functional is proper iff $\mathrm{Dom}\, f \neq \varnothing$).

A minimization problem

Let us now consider the minimization problem.

PROBLEM 3.7. Given a Hilbert space H, $K \subset H$ a non-empty closed convex set and $f: K \to]-\infty, +\infty]$ a functional of the form $f(u) = h(u) + j(u)$, find $u_0 \in K$ such that

$$f(u_0) \leq f(v) \qquad \forall v \in K. \qquad (3.79)$$

Our objective now is to look for the conditions that have to be imposed on h and j so that this problem has one and only one solution. The case of the functions defined on $\mathbb{R}^n$ suggests that we can try to use a concept of the same type as that of derivative, in the context perhaps of variational inequalities as suggested by the technique of variation of the functional which we have applied above. On the other hand, the second proof of theorem 3.1 shows us that the role played by L is secondary, and that there is perhaps no need of imposing much on j which has now taken its place: we will impose then only on h the condition that it be sufficiently regular, say differentiable in a sense which we will soon specify.

Gâteaux derivative and Gâteaux differential

At this point the question arises naturally: does a concept of derivative for functionals defined on a Hilbert space exist? The answer is yes. There are various concepts that generalise in one way or another the concept of derivative of functions of a real variable; here we will use just one, the Gâteaux derivative, which is a further generalisation of the concept of directional derivative of functions from $\mathbb{R}^n$ in $\mathbb{R}$.

DEFINITION 3.2. Let H be a Hilbert space and $h : H \to \mathbb{R}$ a (real) functional. The functional h is said to have a *derivative in the Gâteaux sense* (*G-derivative*) in $u \in H$ if

$$\exists h'(u) \in H' \quad \forall v \in H \frac{h(u+\lambda v)-h(u)}{\lambda} \to {}_{H'}\langle h'(u), v\rangle_H \quad \text{when} \quad \lambda \to 0$$

(3.80)

and $h'(u)$, which we also denote by $\nabla h(u)$, will be called the *Gâteaux derivative* (*G-derivative*) or the *gradient* of h in u. If for every $u \in H$ holds (3.80), the functional h is said to be *differentiable in the Gâteaux sense* (*G-differentiable*) in H and the operator $D_G : H \to H'$ which with every u associates $D_G(u) = h'(u) = \nabla h(u)$ is said to be the *Gâteaux differential* (*G-differential*) of h in H.

This definition deserves some comment. In the first place it is coherent, since the G-derivative of a functional h in a point, if it exists, is unique. Secondly, it should not surprise us that the derivative is an element of the dual space since this is true also for functions from $\mathbb{R}^n$ in $\mathbb{R}$ (in fact, the gradient of $f : \mathbb{R}^n \to \mathbb{R}$ is a vector of $(\mathbb{R}^n)'$, but we identify systematically $\mathbb{R}^n$ and $(\mathbb{R}^n)'$ . . .).

A concept which is strictly connected to the derivative in the Gâteaux sense is that of the *derivative in the Fréchet sense*: h will be said to have a *F-derivative* in $u \in H$ if there exists $\Phi \in H'$ such that $h(u+v) = h(u) + \Phi(v) + o(\|v\|) \|v\|$. *If h has a F-derivative it has also a G-derivative and the two derivatives are the same* (see e.g. J. Schwartz, 1969, ch. 1; see also Robert, 1974 for an abstract formulation of the connection between these concepts, and also for other notions of derivative).

G-derivative of a quadratic functional

We will now give an example, which will be of use to us in what follows. Let a be a bilinear form and let us consider the functional of H in $\mathbb{R}$ defined by

$$h(u) = \tfrac{1}{2} a(u, u). \tag{3.81}$$

We are interested in finding out, when it exists, the G-derivative of h in $u_0 \in H$. Making use of the bilinearity of the form a, we can write

$$\frac{1}{2\lambda}[a(u_0+\lambda v, u_0+\lambda v) - a(u_0, u_0)] = \frac{1}{2\lambda}[\lambda a(u_0, v) + \lambda a(v, u_0) + \lambda^2 a(v, v)]$$

$$= a_S(u_0, v) + \frac{\lambda}{2} a(v, v) \tag{3.82}$$

and, since this expression converges to $a_S(u_0, v)$ as $\lambda \to 0$, we might be tempted to say that the G-derivative of h in u_0 is given by

$$_{H'}\langle h'(u_0), v\rangle_H = {}_{H'}\langle \nabla h(u_0), v\rangle_H = a_S(u_0, v) \quad \forall v \in H. \tag{3.83}$$

But this is false, under the single assumption of bilinearity of a, since '$v \mapsto a_S(u_0, v)$' need not be an element of H'. It is instead true if a is a bilinear form whose symmetric part is continuous in the second variable:

$$\forall u \quad \exists M_2 \quad |a_S(u, v)| \leq M_2 \|v\|_H; \tag{3.84}$$

but, then, given its symmetry, it is also continuous in the first:

$$\forall v \quad \exists M_1 \quad |a_S(u, v)| \leq M_1 \|u\|_H, \tag{3.85}$$

and since from (3.84) and (3.85) (applying the Banach–Steinhaus theorem) it follows that

$$\exists M \quad \forall u, v \in H \quad |a_S(u, v)| \leq M \|u\|_H \|v\|_H, \tag{3.86}$$

we can, without loss of generality, impose the condition that a_S be continuous (in the complex of the variables). On the other hand, since in (3.81) we make use only of the symmetric part of a, we can assume that a is symmetric.

Summarizing, we can state that *if a is a continuous and symmetric bilinear form then the G-derivative in $u_0 \in H$ of the functional $h(u)$ is given by*

$$_{H'}\langle h'(u_0), v\rangle_H = a(u_0, v) \quad \forall v \in H. \tag{3.87}$$

Further, keeping in mind (3.50), we can put

$$_{H'}\langle h'(u_0), v\rangle_H = a(u_0, v) = {}_{H'}\langle A(u_0), v\rangle_H \quad \forall v \in H \tag{3.88}$$

and say that the G-differential of h is A (and that, since $A \in \mathcal{L}(H, H')$, the G-derivative of h in u_0 varies continuously with $u_0 \ldots$).

3.2.2 General results

On the equivalence of two problems

Equation (3.88), together with (3.73) and (3.75), suggests that we consider expressions of the type

$$_{H'}\langle \nabla h(u_0), u_0 - v \rangle_H + j(u_0) \leq j(v) \quad \forall v \in K, \quad (3.89)$$

and theorem 2.3 suggests that we look for the conditions under which a solution of a problem such as the following problem 3.8 can be interpreted as a solution of the minimization problem 3.7:

PROBLEM 3.8. Given a Hilbert space H, $K \subset H$ a non-empty closed convex set and $f: H \to \mathbb{R} \cup \{+\infty\}$ a functional of the form $f(u) = h(u) + j(u)$, find $u_0 \in K$ such that (3.89) holds.

Clearly, for problem 3.7 to make sense h and j must be proper on K, and for the problem 3.8 to make sense j must be proper on K and h must be G-differentiable on K (which, among other things, implies that h be finite on K). On the other hand, the problems so far treated suggests that we impose that f be convex on K, which we fulfil by requiring the convexity of h and j. Thus we can state the following theorem:

THEOREM 3.3. Let H be a Hilbert space and $K \subset H$ a non-empty closed convex set. If $f: K \to \mathbb{R} \cup \{+\infty\}$ is a functional of the form $f(u) = h(u) + j(u)$, with h finite, convex and G-differentiable on K and j convex and proper on K, then u_0 is a solution of problem 3.7 iff it is a solution of problem 3.8. That is,

$$u_0 \in K; \quad \forall v \in K \quad h(u_0) + j(u_0) \leq h(v) + j(v) \quad (3.90)$$

and

$$u_0 \in K; \quad \forall v \in K \quad _{H'}\langle \nabla h(u_0), u_0 - v \rangle_H + j(u_0) \leq j(v) \quad (3.91)$$

are equivalent.

Proof. We will first show that (3.90) implies (3.91). Let us suppose that $v \in \text{Dom } f$ and let us write (3.91) in the form

$$_{H'}\langle \nabla h(u_0), u_0 - v \rangle_H + j(u_0) - j(v) \leq 0 \quad \forall v \in K \quad (3.92)$$

(note that if $v \notin \text{Dom } f$ the expressions are clearly both true; on the other hand, note that $u_0 \in \text{Dom } f$). Since (3.90) is true we can write

$$h(u_0) + j(u_0) \leq h((1-\lambda)u_0 + \lambda v) + j((1-\lambda)u_0 + \lambda v) \quad \forall v \in K \quad \forall \lambda \in [0,1], \quad (3.93)$$

and making use of the convexity of j we can write

$$h(u_0) + j(u_0) - h((1-\lambda)u_0 + \lambda v) - (1-\lambda)j(u_0) - \lambda j(v) \leq 0$$
$$\forall v \in K \quad \forall \lambda \in [0,1] \quad (3.94)$$

and then, dividing by $\lambda \neq 0$,

$$\frac{h((1-\lambda)u_0+\lambda v)-h(u_0)}{-\lambda}+j(u_0)-j(v)\leq 0 \qquad \forall v \in K \quad \forall \lambda \in \,]0,1] \quad (3.95)$$

and taking the limit as $\lambda \to 0^+$ we have (3.92).
Conversely, since if h is convex and G-differentiable then

$$_{H'}\langle \nabla h(u_0), u_0-v\rangle_H \geq h(u_0)-h(v) \qquad (3.96)$$

(see, e.g., Ekeland–Temam, 1974, p. 24), and keeping in mind (3.92) we can write

$$h(u_0)-h(v)+j(u_0)-j(v)\leq 0 \qquad \forall v \in K, \qquad (3.97)$$

which is (3.90). ∎

Note that in this proof we have not made use of the hypothesis that K is closed nor that it is non-empty (in this case (3.90) and (3.91) are both false and thus equivalent); the convexity of K has instead been used in (3.93) (and on the other hand the consideration of convex functions defined on non-convex sets is of no interest). Besides, we have shown neither the existence nor the uniqueness of either of the problems—we have limited ourselves to proving their equivalence.

Indicatrix functions

Before looking for further conditions on h and j such that the solution of problems 3.7 and 3.8 exists and is unique, we will show that the problem of minimization of a convex functional $f: H \to \mathbb{R} \cup \{+\infty\}$ on a convex set $K \subset H$ can always be interpreted as a problem of the minimization of f on H, with the help of a simple trick based on the indicatrix functions which we will now define.

DEFINITION 3.3. Let $K \subset H$ be a convex set. The *indicatrix* of K is the function defined by

$$j_K(u) = \begin{cases} 0 & \text{if } u \in K \\ +\infty & \text{if } u \notin K. \end{cases} \qquad (3.98)$$

The function j_K is convex in H since the inequality

$$j_K(\lambda u+(1-\lambda)v)\leq \lambda j_K(u)+(1-\lambda)j_K(v) \qquad (3.99)$$

is trivially true if $u \notin K$ or if $v \notin K$, and if $u \in K$ and $v \in K$ then (since K is convex) $\lambda u+(1-\lambda)v \in K$ and therefore both members of (3.99) are zero. On the other hand j_K is l.s.c. iff K is closed, as can be seen e.g. in Ekeland–Temam, 1974, p. 9.

Thus let us consider the following pair of problems and prove the theorem which states their equivalence:

PROBLEM 3.9. Find $u_0 \in K$ such that
$$f(u_0) \leq f(v) \qquad \forall v \in K. \tag{3.100}$$

PROBLEM 3.10. Find $u_0 \in H$ such that
$$f(u_0) + j_K(u_0) \leq f(v) + j_K(v) \qquad \forall v \in H. \tag{3.101}$$

THEOREM 3.4. *Let H be a Hilbert space, $K \subset H$ a non-empty convex set and $f: H \to \mathbb{R} \cup \{+\infty\}$ a convex functional. The problems 3.9 and 3.10 are equivalent, i.e., minimizing f on K and $f + j_K$ on H are equivalent.*

Proof. Let us note that if f is not defined on all H we can extend it with the value $+\infty$ outside its domain, which preserves its convexity.

If u_0 is a solution of problem 3.9 then $j_K(u_0) = 0$; consequently (3.101) is satisfied since if $v \in K$ then $j_K(v) = 0$ and (3.101) reduces to (3.100), and if $v \notin K$ then $j_K(v) = +\infty$ and (3.101) is trivially satisfied.

Conversely, if $u_0 \in H$ is a solution of problem 3.10 let us suppose, to avoid triviality, that f is proper: then $u_0 \in K$ and further (3.101) implies that

$$f(u_0) + j_K(u_0) \leq f(v) + j_K(v) \qquad \forall v \in K, \tag{3.102}$$

which reduces to (3.100). ∎

A fundamental result of existence and uniqueness

Let us now look for sufficient conditions such that the solution of problem 3.8 exists and is unique. Clearly it would be enough to assume h and j l.s.c. and $f = h + j$ coercive (or K bounded) since in this way we would satisfy the assumptions of theorem 2.2, but this is obviously too generic. It is more interesting to look for such assumptions in a particular case.

Let us consider then the functional defined on H by

$$f(u) = \tfrac{1}{2} a(u, u) + j_K(u) + L(u) + j(u), \tag{3.103}$$

where a is a continuous and symmetric bilinear form, $L \in H'$, j_K is the indicatrix function of a closed convex set K and j is an l.s.c. proper convex functional; with f we associate the problem:

PROBLEM 3.11. Find $u_0 \in H$ such that $f(u_0) \leq f(v) \; \forall v \in H$.

To ensure the convexity of f it is sufficient to assume that

$$a(u, u) \geq 0 \qquad \forall u \in H, \tag{3.104}$$

and so, satisfying the assumptions of theorem 3.3, ensuring that problem 3.11 is equivalent to the problem:

PROBLEM 3.12. Find $u_0 \in H$ such that

$$a(u_0, u_0 - v) + j_K(u_0) + L(u_0) + j(u_0) \leq j_K(v) + L(v) + j(v) \quad \forall v \in H. \quad (3.105)$$

The expression (3.105) is a generalization of (3.75) in the sense that if j is linear we have once again the latter, but where the form a is supposed to be symmetric.

Under what conditions does problem 3.11 have a solution? To satisfy the assumptions of theorem 2.2 it is enough either that: (i) *a be coercive*, since in that case

$$|f(u)| \to +\infty \quad \text{when} \quad \|u\|_H \to +\infty \quad (3.106)$$

(besides, if a is coercive then f is strictly convex and hence the solution of the problem is unique); or that: (ii) *K be bounded*; or again that: (iii) $j(u) \to +\infty$ when $\|u\|_H \to +\infty$ *and a be non-negative*. Case (i) is the most important and we will restate it:

THEOREM 3.5. Let H be a Hilbert space, $K \subset H$ a non-empty closed convex set, a a symmetric bilinear form, continuous on H and coercive on $K - K$, j an l.s.c. proper convex functional. There exists one and only one $u_0 \in K$ such that

$$a(u_0, u_0 - v) + j(u_0) \leq j(v) \quad \forall v \in K. \quad (3.107)$$

3.2.3 THE CONCEPT OF SUBDIFFERENTIAL

It is important to note that variational inequalities do not appear only in the presence of convex constraints (e.g. minimization of functionals on convex sets), but also when the functionals are non-differentiable. In fact in formulating a problem of minimization of a convex functional, such as problem 3.7, an inequality arises, inequality (3.79), which, if f is G-differentiable, can be written equivalently in the form $_H\langle f'(u_0), u_0 - v\rangle_H \leq 0$; this inequality becomes an equation if the problem is formulated on $K = H$, and we can therefore say that the minimization problem without constraints for differentiable functionals leads us to the problem of the search for zeros of the derivative. However, if f is not G-differentiable then the inequality equivalent to (3.79) is of the type of inequality (3.91), where h is the sum of the G-differentiable parts of f and j that of the other parts, and thus even if $K = H$ we have no longer an equation in the usual sense of this term. In this section we introduce some concepts which enable us to write those inequalities as 'equations' of a certain kind.

Right and left derivatives and subdifferential

As we know, if $f : \mathbb{R} \to \mathbb{R}$ is a convex function then in every point of $\mathbb{R}$ there exists the right derivative, f'_R, and the left derivative, f'_L, and we have the property of monotonicity

$$\forall x, y \in \mathbb{R} \qquad y > x \Rightarrow f'_L(x) \leq f'_R(x) \leq f'_L(y). \qquad (3.108)$$

On the other hand, the graph of f always lies 'above' its right and left tangents, and the set of points where there is no derivative, which is the set

$$C = \{x \in \mathbb{R} : f'_L(x) \neq f'_R(x)\}, \qquad (3.109)$$

has a cardinality which is not greater than $\aleph_0$. Thus, e.g., the function $f : \mathbb{R} \to \mathbb{R}$ defined by $f(x) = |x|$ is differentiable in $\mathbb{R} \setminus \{0\}$ (and therefore $C = \{0\}$) and $f'_L(0) = -1$ and $f'_R(0) = 1$; these values (-1 and 1) are the angular coefficients of the lateral tangents to the graph of f at the point $(0, 0)$. Let us now consider the set of angular coefficients of the lines which 'touch' the graph of f at $(0, 0)$ but which do not intersect it (see fig. 3.1); one can easily see that these angular coefficients, x', satisfy the relationship

$$f(0) - f(y) \leq x'(0 - y) \qquad \forall y \in \mathbb{R}, \qquad (3.110)$$

which we write as

$$f(0) - f(y) \leq_{\mathbb{R}} \langle x', 0 - y \rangle_{\mathbb{R}} \qquad \forall y \in \mathbb{R} \qquad (3.111)$$

to underline the fact that although the x' are real numbers they are considered as angular coefficients of lines, or briefly: they are elements of $\mathbb{R}'$. The graph of f', which is nothing but the set of points of the form (x, angular coefficient of the tangent to the graph of f at $(x, f(x))$), extended with the angular coefficients of the lateral tangents and 'subtangents' at $(0, 0)$ is shown in fig. 3.2.

The inequality (3.111) recalls (3.96) and we are thus led to introducing the definition:

DEFINITION 3.4. Let $f : H \to \mathbb{R} \cup \{+\infty\}$ be a convex functional and $u \in H$. The functional f is said to be subdifferentiable in u if the set

$$\partial f(u) = \{h' \in H' : f(u) - f(v) \leq_H \langle h', u - v \rangle_H \qquad \forall v \in H\} \qquad (3.112)$$

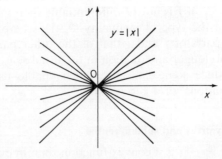

Figure 3.1

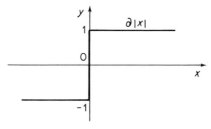

Figure 3.2

is non-empty. The set $\partial f(u)$ is said to be the *subderivative* of f in u and its elements are called *subgradients* of f at u. If $\forall u \in H$ $\partial f(u) \neq \varnothing$ we say that f is subdifferentiable in H, and the application $\partial f : H \to 2^{H'}$ which with every $u \in H$ associates $\partial f(u) \in 2^{H'}$ is called the *subdiffierential* of f.

Note that ∂f can be interpreted both as a single-valued operator from H in $2^{H'}$ or as a multi-valued operator of H in H'.... As is obvious, if f is G-differentiable in u, then $\partial f(u) = \{f'(u)\}$. With the notation just introduced, if f is the function $f(x) = |x|$ which we have previously considered, we can write $\partial f(x) = \{\text{sgn } x\}$ if $x \neq 0$, $\partial f(0) = [-1, 1]$.

Expression (3.108) and the graph of fig. 3.2 suggest that ∂f is a monotone operator in the sense of the following definition (which is a generalization of definition 2.2):

DEFINITION 3.5. Let H be a Hilbert space and f an operator from H in $2^{H'}$. The operator f is said to be *monotone* if

$$\forall u, v \in H \quad \forall \xi \in f(u), \eta \in f(v) \quad {}_{H'}\langle \xi - \eta, u - v \rangle_H \geq 0, \quad (3.113)$$

and f is said to be *maximal monotone* if it is monotone and there does not exist $\bar{f} : H \to 2^{H'}$ such that $\bar{f}$ is monotone and $G_r(f) = \{(u, f(u)) : u \in H\} \subsetneq G_r(\bar{f}) = \{(u, \bar{f}(u)) : u \in H\}$ (i.e. the graph of f does not have any proper extension which is the graph of a monotone operator).

In fact the following theorem holds:

THEOREM 3.6. If $f : H \to \mathbb{R} \cup \{+\infty\}$ is a convex functional then its subdifferential ∂f is a monotone operator, i.e.

$$\forall u, v \in H \quad \forall \xi \in \partial f(u), \eta \in \partial f(v) \quad {}_{H'}\langle \xi - \eta, u - v \rangle_H \geq 0, \quad (3.114)$$

which can be written more concisely as

$$\forall u, v \in H \quad {}_{H'}\langle \partial f(u) - \partial f(v), u - v \rangle_H \geq 0. \quad (3.115)$$

Proof. If $\partial f(u)$ or $\partial f(v)$ is empty, (3.114) is obvious; otherwise, taking ξ in

$\partial f(u)$ and η in $\partial f(v)$, we can write

$$_{H'}\langle \xi, u-v \rangle_H \geq f(u)-f(v) \qquad \forall v \in H \tag{3.116}$$

$$_{H'}\langle \eta, v-u \rangle_H \geq f(v)-f(u) \qquad \forall u \in H \tag{3.117}$$

and summing, after having made the usual change of sign, we have

$$_{H'}\langle \xi-\eta, u-v \rangle_H \geq 0 \tag{3.118}$$

which, since ξ, η, u, and v are arbitrary, is (3.114). ∎

It can also be shown that f is maximal monotone (see Moreau, 1967, p. 85); for other proprieties of subdifferential operators, see Ekeland–Temam, 1974, Ioffe–Levin, 1972, Zowe, 1974.

Multi-valued equations

Let us now consider the inequality

$$a(u, u-v)+h(u) \leq h(v) \qquad \forall v \in H, \tag{3.119}$$

which can be written in the form

$$_{H'}\langle A(u), u-v \rangle_H \leq h(v)-h(u) \qquad \forall v \in H, \tag{3.120}$$

where A is the operator (cf. (3.50)) associated with the symmetric continuous bilinear form a, or again in the form

$$h(u)-h(v) \leq {}_{H'}\langle -A(u), u-v \rangle_H \qquad \forall v \in H. \tag{3.121}$$

From (3.121) we see that

$$-A(u) \in \partial h(u), \tag{3.122}$$

or that

$$0 \in (\partial h + A)(u), \tag{3.123}$$

which is an 'equation' associated with the multi-valued operator $\partial h + A$: it was to this type of 'equation' that we referred at the beginning of this section. Clearly, if h is G-differentiable then (3.123) can be written as

$$0 = (\nabla h + A)(u), \tag{3.124}$$

which is an equation in the usual sense (we do not have an inequality here since (3.119) is formulated in all of $H \ldots$; for constrained problems a formulation of the type (3.123) is possible, but this is complicated by the fact that, in general, the relation $\partial(f+g) = \partial f + \partial g$ is false).

4
Transposition of Operators. Applications

4.1 THE CONCEPT OF TRANSPOSITION FUNDAMENTAL PROPERTIES

The transposition of continuous linear operators defined between topological vector spaces (t.v.s.) provides one of the most fruitful processes of generalization in analysis. In this section we will define this operation and give some abstract results relative to it; in the following sections we will describe some of its applications to very important concrete cases.

All the vector spaces we consider are, except when the contrary is explicitly stated, vector spaces over the field $\mathbb{C}$ of complex numbers.

Transpose of an operator. Linearity and continuity

DEFINITION 4.1. Let E and F be two t.v.s. and $T \in \mathscr{L}(E, F)$. The *transpose* of the operator T is the operator ${}^t T : F' \to E'$ defined by

$$\forall f' \in F' \quad \forall e \in E \quad {}_{E'}\langle {}^t T(f'), e \rangle_E \triangleq {}_{F'}\langle f', T(e) \rangle_F. \tag{4.1}$$

In the literature this operator has also been called the *adjoint* of T; we will not use this nomenclature because in the particular case in which E and F are finite-dimensional vector spaces, if T is represented (assuming that there are two bases fixed in E and F) by means of a matrix $[T_{ij}]$, the operator ${}^t T$ is represented (relative to their dual bases) by means of the transposed matrix of $[T_{ij}]$ and not by means of the adjoint matrix, which means something entirely different—for the proof of this, see e.g., Halmos, 1958, p. 81 (who, however, calls ${}^t T$ the adjoint of the operator $T \ldots$).

Note that definition 4.1 makes sense since, as we will now prove (as an example...), $\forall f' \in F' \ {}^t T(f') \in E'$ (and thus in particular we can use the notation $\langle , \rangle$ in the first term of (4.1), as we have done). In effect, having fixed f' in F', ${}^t T(f')$ is a *functional* on E because of

$$\forall e \in E \quad {}_{E'}\langle {}^t T(f'), e \rangle_E \triangleq {}_{F'}\langle f', T(e) \rangle_F \in \mathbb{C}, \tag{4.2}$$

is *linear* since, f' and T being linear,

$$\forall \lambda_1, \lambda_2 \in \mathbb{C} \quad \forall e_1, e_2 \in E$$

$$_{E'}\langle {}^tT(f'), \lambda_1 e_1 + \lambda_2 e_2 \rangle_E \triangleq {}_{F'}\langle f', T(\lambda_1 e_1 + \lambda_2 e_2) \rangle_F$$
$$= {}_{F'}\langle f', \lambda_1 T(e_1) + \lambda_2 T(e_2) \rangle_F \quad (4.3)$$
$$= \lambda_{1F'}\langle f', T(e_1) \rangle_F + \lambda_{2F'}\langle f', T(e_2) \rangle_F$$
$$\triangleq \lambda_{1E'}\langle {}^tT(f'), e_1 \rangle_E + \lambda_{2E'}\langle {}^tT(f'), e_2 \rangle_E,$$

and is *continuous* because, for every $e \in E$, $_{E'}\langle {}^tT(f'), e \rangle_E$ it is the value on e of $f' \circ T$, which is the composition of a continuous functional with a continuous operator.

THEOREM 4.1. If $T \in \mathscr{L}(E, F)$ and E' and F' have the same type of duality topology (e.g. E' and F' have both the strong topology) ${}^tT \in \mathscr{L}(F', E')$.

Proof. The linearity of ${}^tT$ can be seen immediately:

$$\forall \lambda_1, \lambda_2 \in \mathbb{C} \quad \forall f'_1, f'_2 \in F'$$

$$_{E'}\langle {}^tT(\lambda_1 f'_1 + \lambda_2 f'_2), e \rangle_E \triangleq {}_{F'}\langle \lambda_1 f'_1 + \lambda_2 f'_2, T(e) \rangle_F$$
$$= {}_{F'}\langle \lambda_1 f'_1, T(e) \rangle_F + {}_{F'}\langle \lambda_2 f'_2, T(e) \rangle_F \quad (4.4)$$
$$\triangleq {}_{E'}\langle \lambda_1 {}^tT(f'_1), e \rangle_E + {}_{E'}\langle \lambda_2 {}^tT(f'_2), e \rangle_E$$
$$= {}_{E'}\langle \lambda_1 {}^tT(f'_1) + \lambda_2 {}^tT(f'_2), e \rangle_E \quad \forall e \in E.$$

In order to prove that ${}^tT$ is continuous in the general case we refer the reader to Treves, 1967a, p. 199; we deal here with the particular case in which E and F are normed spaces and E' and F' have both their strong topologies. Given that f' and T are continuous we can write successively

$$|_{E'}\langle {}^tT(f'), e \rangle_E| = |_{F'}\langle f', T(e) \rangle_F| \leq \|f'\|_{F'}\|T(e)\|_F \leq \|f'\|_{F'}\|T\|_{\mathscr{L}(E,F)}\|e\|_E \quad (4.5)$$

hence

$$\|{}^tT(f')\|_{E'} \leq \|T\|_{\mathscr{L}(E,F)}\|f'\|_{F'}, \quad (4.6)$$

and finally

$$\|{}^tT\|_{\mathscr{L}(F',E')} \leq \|T\|_{\mathscr{L}(E,F)} \quad (4.7)$$

which shows that ${}^tT$ is continuous (it can be shown moreover that $\|T\|_{\mathscr{L}(E,F)} = \|{}^tT\|_{\mathscr{L}(F',E')}$: see, for example, Treves, 1967a, p. 242; Horváth, 1966, p. 257). ∎

Properties of the transpose

The previous theorem shows that there are properties of T which are preserved by ${}^tT$. It is natural then to look for other such properties, or more generally see what happens to certain properties of T in terms of ${}^tT$: this line of research leads us to interesting results.

Let us suppose, e.g. that T is *injective* (i.e. that $T(e) = 0 \Rightarrow e = 0$) and let us see what properties we can deduce for ${}^tT$ from this.

In general the injectivity of ${}^tT$, i.e. ${}^tT(f') = 0 \Rightarrow f' = 0$, does not follow from that of T; in fact, from the definition of ${}^tT$, ${}^tT(f') = 0$ means that

$$_{E'}\langle {}^tT(f'), e\rangle_E \triangleq {}_{F'}\langle f', T(e)\rangle_F = 0 \qquad \forall e \in E, \tag{4.8}$$

which in turn implies that $f' = 0$ on $T(E)$, but not necessarily on F. We have been unable to obtain any important information regarding the injectivity of ${}^tT$, but (4.8) helps us to reason in an alternative manner: if T is not injective then there exists $\bar{e} \neq 0$ such that $T(\bar{e}) = 0$, and under these conditions from (4.8) it follows that

$$_{E'}\langle {}^tT(f'), \bar{e}\rangle_E = 0 \qquad \forall f' \in F', \tag{4.9}$$

i.e. that $\bar{e}$ is *orthogonal* to the image of F' by means of ${}^tT$ and hence the latter is not dense in E' (let us remark that the concept of orthogonality does not presuppose a Hilbertian structure: if E is a t.v.s. and A is a linear subspace of E, the *orthogonal* of A is the linear subspace A° of E' made up of the continuous linear functionals on E which are identically zero on A; in the particular case in which E is a Hilbert space, the Hilbertian orthogonal $A^\perp$ of A coincides with A°, if an identification between E and its dual is done). The negative converse of this conclusion is as follows: '*if ${}^tT$ has a dense image then T is injective*', and the great importance of this result leads us to look for others like it. In the following theorem we collect some similar results; before this, however, we want to present some considerations regarding the concept of injection, among other things in order to establish the notation which we will use in what follows.

Injectivity, immersions and identifications

If E and F are two topological vector spaces, we call $j_{EF} : E \to F$ an *injection* of E in F if j_{EF} is an injective linear application of E in F. If j_{EF} is a continuous application the injection is said to be a *continuous injection*, and if $j_{EF}(E)$ is dense in F the injection is said to be a *dense injection*. Whenever between two t.v.s. E and F there is a continuous injection j_{EF}, it is possible to *identify* E and $j_{EF}(E) \subset F$, since j_{EF} establishes a vectorial–topological isomorphism between E and $j_{EF}(E) \subset F$ if the latter space is provided with the topology determined by E (and not the one induced by F): under these conditions we write $E \subset F$.

This type of identification is very frequent in analysis; among others let us quote the identification which one frequently establishes between a Hilbert space H and its dual H' and that between a t.v.s. E and a part of its strong bidual E''. In the latter case $j_{EE''}$ is then the so-called *canonical injection* of E in E''; in the former case the identification is in agreement with the general theorem of the representation of functionals in Hilbert spaces, and if H is real then $j_{HH'}$ is the Riesz *operator* (however if H is complex $j_{HH'}$ does not identify H with its dual but to its antidual). In the following sections we will see further examples.

Let us now suppose that $E \subset F$, right so or unless an identification is to be established. In this case there exists a chosen canonical of j_{EF}: the *set immersion* $i_{EF}: E \to F$ defined by

$$\forall x \in E \qquad i_{EF}(x) = x \in F. \tag{4.10}$$

If i_{EF} is continuous (which means that the topology of E is finer than that induced in E by the topology of F), we write $E \hookrightarrow F$, and if i_{EF} is dense (which means that the closure of E in the topology of F is equal to F) we write $E \subset^{ds} F$. Let us remark at this point that the distinction between '$E \subset F$ right so' and '$E \subset F$ after identification' is often hairsplitting as can be seen from the following example: if the set $\mathbb{R}$ of real numbers is constructed from the set $\mathbb{Q}$ of rational numbers (through, e.g., the completation process of Cantor), we identify $\mathbb{Q}$ with a part of $\mathbb{R}$ (that which is 'obtained' from the convergent sequences in $\mathbb{Q}$), but if $\mathbb{R}$ is defined axiomatically $\mathbb{Q}$ is directly defined as a subset of $\mathbb{R}$. In order to make this distinction it is enough, when necessary, to substitute i_{EF} by $i_{[j_{EF}(E)]F} \circ j_{EF}$—note that the essence of the identification lies exactly in the confounding of these two applications.

Properties of the transpose again

THEOREM 4.2. If E and F are two t.v.s. and $T \in \mathscr{L}(E, F)$ then

(i) $T(E) \overset{ds}{\subset} F \Leftrightarrow {}^tT$ is injective,

(ii) ${}^tT(F') \overset{ds}{\subset} E' \Rightarrow T$ is injective, and the converse is true if E is reflexive.

Proof. (i) In the first place, let us prove that $T(E) \subset^{ds} F$ implies that ${}^tT$ is injective. To say that ${}^tT$ is injective means that

$$(\forall e \in E \quad {}_E\langle {}^tT(f'), e\rangle_E \triangleq {}_F\langle f', T(e)\rangle_F = 0) \Rightarrow f' = 0, \tag{4.11}$$

and this follows immediately since from ${}_F\langle f', T(e)\rangle_F = 0 \; \forall e \in E$ it follows that $f' = 0$ on $T(E)$ and hence on F (since from the density of $T(E)$ in F it follows that the only extension to F of an identically null functional on $T(E)$ is the functional which is identically null on F). Conversely, let us assume that ${}^tT$ is injective and let us show that $T(E) \subset^{ds} F$, i.e., that if $f' \in F'$ is identically null on $T(E)$ then f' is identically null on F. Now, to say that $\forall f \in T(E) {}_F\langle f', f\rangle_F = 0$ is equivalent to saying that ${}_F\langle f', T(e)\rangle_F = 0 \; \forall e \in E$ and this, from the definition of ${}^tT$, is equivalent to saying that $\forall e \in E \; {}_E\langle {}^tT(f'), e\rangle_E = 0$, or that ${}^tT(f') = 0$, and hence, from the injectivity of ${}^tT$, to saying that f' is the zero of F'.

(ii) We have already seen that ${}^tT(f') \subset^{ds} E'$ implies that T is injective. To show the converse, in the case where E is reflexive, we introduce the transpose operator of the operator ^tT, which theorem 4.1 enables us to do (let us remark that if the assumption of reflexivity is not satisfied, the result does not hold: we will see a counter-example in the following chapter). From definition 4.1, it follows that if E'' and F'' are the biduals (which we assume to be the strong ones) of E and F, respectively, then the transpose of ^tT is the operator $^t(^tT): E'' \to F''$ characterized by

$$\forall e'' \in E'' \quad \forall f' \in F' \quad {}_{F''}\langle {}^t({}^tT)(e''), f'\rangle_{F'} \triangleq {}_{E''}\langle e'', {}^tT(f')\rangle_{E'}. \quad (4.12)$$

If $j_{EE''}$ and $j_{FF''}$ are the canonical injections of E in E'' and of F in F'', respectively, we can write (4.12) in the form

$$\forall e \in E \quad \forall f' \in F' \quad {}_{F''}\langle {}^t({}^tT)(j_{EE''}e), f'\rangle_{F'} = {}_{E''}\langle j_{EE''}e, {}^tT(f')\rangle_{E'}, \quad (4.13)$$

and since

$$_{E''}\langle j_{EE''}e, {}^tT(f')\rangle_{E'} = {}_{E'}\langle {}^tT(f'), e\rangle_E = {}_{F'}\langle f', T(e)\rangle_F = {}_{F''}\langle j_{FF''}T(e), f'\rangle_{F'} \quad (4.14)$$

it follows that

$$^t(^tT) \circ j_{EE''} = j_{FF''} \circ T. \quad (4.15)$$

If E is reflexive $j_{EE''}$ has an inverse $j_{EE''}^{-1} = j_{E''E}$ and (4.15) becomes

$$^t(^tT) = j_{FF''} \circ T \circ j_{E''E}, \quad (4.16)$$

which shows that if T is injective $^t(^tT)$ is also injective: we are then under the conditions of part $(i, \Leftarrow)$ of the theorem and hence $^tT(F') \subset^{ds} E'$. ∎

For a more detailed study of the transposition of operators, with applications to the theory of differential operators, see, among others, Treves, 1967b, p. 47 *et seq.* and Browder, 1959b; 1962.

4.2 APPLICATIONS OF THE CONCEPT OF TRANSPOSITION

Injections between t.v.s. again

As a first example of application of the concept of transposition of operators, let us once again take up the discussion on injections between t.v.s. which we considered in the previous section.

Let then E and F be two t.v.s. such that $E \subset^{ds} F$, i.e., that the immersion i_{EF} is continuous and dense. The operator $i_{EF}: E \to F$ is an element of $\mathcal{L}(E, F)$ and it is meaningful therefore to consider $^t(i_{EF}): F' \to E'$ which, in turn, is an operator which is linear and continuous between F' and E' (see theorem 4.1) and injective (see theorem 4.2, part $(i, \Rightarrow)$). Thus F' can be *identified* with $^t(i_{EF})(F') \subset E'$ and we can write $F' \subset E'$.

The case of normed spaces

Let us analyse the meaning of this inclusion in the particularly simple case in which E and F are normed spaces. Let us start by noting that in this case the fact that the immersion of E in F is continuous (in other words that the topology of E is finer than that induced by F on E) can be expressed very elegantly by

$$\exists c \in \mathbb{R} \quad \forall e \in E \; \|e\|_F \leq c \|e\|_E. \tag{4.17}$$

An element f' of F' is a continuous linear functional '$F \ni f \mapsto {}_{F'}\langle f', f \rangle_F \in \mathbb{C}$' and restricting this functional to E (note that $E \subset F$) we obtain a functional '$E \ni e \mapsto {}_{F'}\langle f', e \rangle_F \in \mathbb{C}$' also linear and continuous (in the topology of E) since, from the continuity of f', we have

$$\|f'\|_{E'} = \sup_{\substack{e \neq 0 \\ e \in E}} \frac{|{}_{F'}\langle f', e \rangle_F|}{\|e\|_E} \leq \sup_{\substack{e \neq 0 \\ e \in E}} c \frac{|{}_{F'}\langle f', e \rangle_F|}{\|e\|_F} \leq c \|f'\|_{F'}, \tag{4.18}$$

which tells us precisely that the restriction of f' to E is continuous in the topology of E—or rather we can write

$$\exists c \in \mathbb{R} \quad \forall f' \in F' \quad \|f'\|_{E'} \leq c \|f'\|_{F'}, \tag{4.19}$$

which shows us that the strong topology of F' is finer than that induced on F' by the strong topology of E'. *The identification of F' with a part of E' consists of the identification of the continuous linear functions on F with their restrictions to E.* If, however, the assumption regarding the density of the immersion of E in F is not true the above identification is no longer possible; in fact, if i_{EF} is not dense then $\exists f' \in F'$ such that $f' \neq 0$ and $\langle f', e \rangle = 0 \; \forall e \in E$ and when we consider the restriction of f' to E we obtain the zero of E' which, on the other hand, can be obtained starting from the zero of F': ${}^t(^i EF)$ is not injective and hence this operator is not an injection between F' and E'.

In order to give concrete examples of applications of the transposition we need two ingredients: topological vector spaces and continuous linear operators between these spaces. In section 4.2.1 we will introduce some t.v.s. which are useful for the applications and we will define some continuous linear operators on them. Even though we give (without proofs) some properties of the spaces considered, this section should be considered simply as a (very incomplete) glossary of definitions; for a more detailed study of the spaces considered here see, among others, Horváth, 1966; Treves, 1967a; Garnir–De Wilde–Schmets, 1972. In section 4.2.2 we will very briefly characterize (again without proofs) the duals of these spaces and we will study the transpose of the previously defined operators. Incidentally, we will deal with *distributions*; we will not, however, develop the theory of distributions and we will merely introduce the fundamental nomenclature; for a study of this theory, see, among others, L. Schwartz, 1966; Horváth, 1966; Treves, 1967a; Choquet-Bruhat, 1973, Hörmander, 1964; Guelfand-

Chilov, 1962, 1964, 1965; Guelfand–Vilenkin, 1967; Guelfand–Graev–Vilenkin, 1970.

4.2.1 SOME EXAMPLES OF FUNCTION SPACES IMPORTANT IN THE APPLICATIONS. EXAMPLES OF CONTINUOUS LINEAR OPERATORS

A few words about notations

Let A be an arbitrary set. On the set $\mathscr{F}(A) = \mathbb{C}^A$ of functions defined on A and valued in $\mathbb{C}$, we introduce a *structure of vector space over the field* $\mathbb{C}$ by means of the usual formulae:

$$\forall f, g \in \mathscr{F}(A) \quad \forall x \in A \quad (f+g)(x) = f(x) + g(x) \tag{4.20}$$

$$\forall f \in \mathscr{F}(A) \quad \forall \lambda \in \mathbb{C} \quad \forall x \in A \quad (\lambda f)(x) = \lambda g(x), \tag{4.21}$$

which define, respectively, the sum of two functions and the product of a function by a scalar. All vector spaces we will consider in this section are vector subspaces of $\mathscr{F}(A)$, with an appropriate A, or are obtained from subspaces of $\mathscr{F}(A)$ by means of the quotient by an appropriate equivalence relation. In any case, a linear structure is naturally defined on them, and we will not define it anew each time.

If $A \subset \mathbb{R}^n$ and $f \in \mathscr{F}(A)$ we denote by $\operatorname{supp} f$ the *support* of f, i.e. the closure in A (provided with the topology induced by $\mathbb{R}^n$) of the set $\{x \in A : f(x) \neq 0\}$.

We will use the usual compact notation for derivatives. If $\alpha \equiv (\alpha_1, \ldots, \alpha_n) \in \mathbb{N}^n$ we will denote by $|\alpha| = \alpha_1 + \ldots + \alpha_n$ the order of the multi-index α and by D^α the operator $D^\alpha \equiv \partial^{|\alpha|}/\partial x_1^{\alpha_1} \ldots \partial x_n^{\alpha_n}$ (with the convention that $\partial^0/\partial x_i^0 \equiv$ identity; and in particular if $|\alpha| = 0$ then $D^\alpha \equiv$ identity). Further, regarding multi-indices, if $\alpha \equiv (\alpha_1, \ldots, \alpha_n)$ and $\beta \equiv (\beta_1, \ldots, \beta_n)$ we will write

$$\binom{\alpha}{\beta} = \binom{\alpha_1}{\beta_1} \cdots \binom{\alpha_n}{\beta_n}.$$

Let Ω be an open set of $\mathbb{R}^n$: with the notation $K \subset\subset \Omega$ we mean not only that $K \subset \Omega$ but also that K is a compact set of $\mathbb{R}^n$.

Spaces $C^m(\Omega)$ and $\mathscr{E}^m(\Omega)$

DEFINITION 4.2. Let Ω be an open set of $\mathbb{R}^n$ and let $m \in \mathbb{N}$ or $m = +\infty$. We denote by $C^m(\Omega)$ the space of the functions defined and continuous in Ω together with their derivatives of order $\leq m$, and by $\mathscr{E}^m(\Omega)$ the space $C^m(\Omega)$ provided with the family of seminorms

$$\forall f \in C^m(\Omega) \quad |f|_{K,m} = \sum_{|\alpha| \leq m} \sup_{x \in K} |D^\alpha f(x)| \quad (m \in \mathbb{N}) \tag{4.22}$$

where $K \subset\subset \Omega$ is arbitrary. In what follows we will write $\mathscr{E}^\infty(\Omega) = \mathscr{E}(\Omega)$.

Thus in other words $C^m(\Omega)$ is the space of the functions $f \in \mathscr{F}(\Omega)$ such that $D^\alpha f$ exists and is continuous in Ω for all $\alpha \in \mathbb{N}^n$ such that $|\alpha| \leq m$. The space of the functions which are indefinitely differentiable can be defined by the algebraic relationship $C^\infty(\Omega) = \bigcap_{m \in \mathbb{N}} C^m(\Omega)$ (and let us remark that (4.22), letting m vary in $\mathbb{N}$, also provides the topology of $\mathscr{E}^\infty(\Omega)$). If $f \in C^m(\Omega)$ we will also say that f is of *class* C^m in Ω.

The spaces $\mathscr{E}^m(\Omega)$ are Fréchet spaces (see, e.g., Yosida, 1971, p. 27 where a metric is explicitly defined); their topology is known as the 'topology of uniform convergence (of the functions and of their derivatives of order $\leq m$) on the compact sets', since *a sequence $f_k (k \in \mathbb{N})$ converges to f in $\mathscr{E}^m(\Omega)$ iff the sequences $D^\alpha f_k$ converges uniformly on every $K \subset\subset \Omega$ to $D^\alpha f$, for all $|\alpha| \leq m$*.

Spaces $C^m(\bar{\Omega})$

The space of the functions of class C^m in Ω which are continuous with their derivatives 'up to the boundary $\partial\Omega$ of Ω' is also often taken into consideration. We will specify this concept by considering the space $C^m(\bar{\Omega})$ of the extensions by continuity to $\bar{\Omega} = \Omega \cup \partial\Omega$ of the functions which, together with their derivatives of order $\leq m$, have such an extension, and we will provide it with the family of seminorms (4.22) where now $K \subset\subset \bar{\Omega}$. If Ω is bounded, these functions are essentially those bounded and uniformly continuous in Ω together with their derivatives of order $\leq m$. If Ω is bounded and m is finite, $C^m(\bar{\Omega})$ is a Banach space with the norm

$$\forall f \in C^m(\bar{\Omega}) \qquad \|f\|_{C^m(\bar{\Omega})} = \sum_{|\alpha| \leq m} \sup_{x \in \bar{\Omega}} |D^\alpha f(x)|, \qquad (4.23)$$

otherwise it is a Fréchet space. Let us remark that if f is the restriction to $\bar{\Omega}$ of $\bar{f} \in C^m(\mathbb{R}^n)$ then $f \in C^m(\bar{\Omega})$; the converse holds only if Ω is 'regular'.

Spaces $C^{m,\mu}(\Omega)$ **and** $C^{m,\mu}(\bar{\Omega})$

Another space which is closely related to $C^m(\Omega)$ is that of the functions with mth order Hölder continuous derivative. Let us recall that if A is a subset of $\mathbb{R}^n$, $f \in \mathscr{F}(A)$ is said to be *Hölder continuous of index* $\mu \in \,]0,1]$ in A if

$$\exists M \in \mathbb{R} \quad \forall x', x'' \in A \qquad |f(x') - f(x'')| \leq M \|x' - x''\|^\mu, \qquad (4.24)$$

and the quantity

$$[f]_{\mu,A} = \sup_{\substack{x' \neq x'' \\ x', x'' \in A}} \frac{|f(x') - f(x'')|}{\|x' - x''\|^\mu} \qquad (4.25)$$

is called the *Hölder modulus* of f. The Hölder continuous functions are obviously continuous functions in A, and the Hölder continuous functions of index 1 are in particular called *Lipschitz continuous* functions.

By $C^{m,\mu}(\Omega)$, $m \in \mathbb{N}$ and $\mu \in \,]0,1]$, we denote the space of the functions of class C^m in Ω which have mth order Hölder continuous derivatives of index

μ in every $K \subset\subset \Omega$, provided with the family of seminorms

$$\forall f \in C^{m,\mu}(\Omega) \; |f|_{K,\mu,m} = \sum_{|\alpha| \leq m} \sup_{x \in K} |D^\alpha f(x)| + \sum_{|\alpha|=m} [D^\alpha f]_{\mu,K}. \quad (4.26)$$

Further, by $C^{m,\mu}(\bar\Omega)$ we denote the subspace of $C^m(\bar\Omega)$ made up of functions which have mth order Hölder continuous derivatives of index μ in $\bar\Omega$: if Ω is bounded, $\bar\Omega$ is compact and (4.26), with $\bar\Omega = K$, defines a norm with respect to which $C^{m,\mu}(\bar\Omega)$ is a Banach space. If $f \in C^{m,\mu}(\Omega)$ and $[D^\alpha f]_{\mu,K}$ remains bounded when K varies then f is said to be uniformly of class $C^{m,\mu}$—it follows immediately that if $f \in C^{m,\mu}(\Omega)$ then f is the extension to $\bar\Omega$ of a function of this type. Let us remark that if $m \in \mathbb{N}$ and $0 < \nu \leq \lambda \leq 1$ then $C^{m+1}(\bar\Omega) \subsetneq C^m(\bar\Omega)$ and $C^{m,\lambda}(\bar\Omega) \subsetneq C^{m,\nu}(\bar\Omega)$, and if Ω is a convex set (for example) then $C^{m+1}(\bar\Omega) \subsetneq C^{m,\nu}(\bar\Omega)$ (see, e.g., Adams, 1975, p. 11).

Spaces $C_0^m(\Omega)$ and $\mathcal{D}^m(\Omega)$

DEFINITION 4.3. Let Ω be an open set of $\mathbb{R}^n$ and $m \in \mathbb{N}$ or $m = +\infty$. We denote by $C_0^m(\Omega)$ the space of functions $f \in C^m(\Omega)$ such that $\operatorname{supp} f \subset\subset \Omega$ and by $\mathcal{D}^m(\Omega)$ the space $C_0^m(\Omega)$ provided with the family of seminorms

$$\forall f \in C_0^m(\Omega) \; |f|_{\theta,m} = \sum_{|\alpha| \leq m} \sup_{x \in \Omega} |\theta_\alpha(x) D^\alpha f(x)| \quad (m \in \mathbb{N}), \quad (4.27)$$

where $\theta = \{\theta_\alpha\}_{\alpha \in \mathbb{N}^n}$ is a family of continuous functions in Ω such that the family $\{\operatorname{supp} \theta_\alpha\}_{\alpha \in \mathbb{N}^n}$ is *locally finite* (i.e., for all $x \in \Omega$ there exists a neighbourhood $U(x)$ of x such that $\operatorname{card}\{\alpha \in \mathbb{N}^n : \operatorname{supp} \theta_\alpha \cap U(x) \neq \varnothing\}$ is finite) and $\bigcup_\alpha (\operatorname{supp} \theta_\alpha)^\circ = \Omega$. In what follows we will write $\mathcal{D}^\infty(\Omega) = \mathcal{D}(\Omega)$ (with the topology defined by (4.27) with m varying in $\mathbb{N}$).

As a vector space, $C_0^m(\Omega)$ is a subspace of $C^m(\Omega)$ and hence it is naturally a t.v.s. with the topology induced by $\mathcal{E}^m(\Omega)$; the t.v.s. so obtained is not, however, complete (for a counterexample, see Choquet–Bruhat, 1973, p. 13) and for this reason we have provided $C_0^m(\Omega)$ with a finer topology. This topology can be characterized by the fact that $A \subset \mathcal{D}^m(\Omega)$ is open iff $A \cap C_K^m(\Omega)$ is open in $C_K^m(\Omega)$ for every $K \subset\subset \Omega$, where $C_K^m(\Omega)$ is the space of the functions $f \in C^m(\Omega)$ such that $\operatorname{supp} f \subset K$, provided with the topology determined by the family (4.22) with K fixed (see e.g. Horváth, 1966, p. 171. In other words, the topology of $\mathcal{D}^m(\Omega)$ is the *inductive limit* of the topologies of the spaces $C_K^m(\Omega)$; for further information on these topologies see, e.g., Dieudonné–Schwartz, 1950; Silva, 1955. From the above characterization it follows in particular that *a sequence f_k ($k \in \mathbb{N}$) converges to f in $\mathcal{D}^m(\Omega)$ iff $\exists K \subset\subset \Omega$ $\forall k \in \mathbb{N}$ $\operatorname{supp} f_k \subset K$ and the sequences $D^\alpha f_k$ converges uniformly on K to $D^\alpha f$, for every $|\alpha| \leq m$* (see, e.g., Yosida, 1971, p. 28).

The spaces $\mathcal{D}^m(\Omega)$ are complete locally convex t.v.s. (see L. Schwartz, 1966, p. 66; Dieudonné–Schwartz, 1950, p. 87) and they are not metrizable (see Dieudonné–Schwartz, 1950, p. 70). The space $\mathcal{D}(\Omega)$ is of special interest in analysis: is the only one of the $\mathcal{D}^m(\Omega)$ which is reflexive (see, e.g., L.

Schwartz, 1966, p. 75) and its elements are called *test functions*. $\mathscr{D}(\Omega)$ is the 'smallest' function space we will consider: it is not however too small, in the sense that it does not reduce to the identically zero function (see theorem 15.1)

Spaces $\mathscr{D}^m(\bar{\Omega})$

In what follows we will denote by $\mathscr{D}^m(\bar{\Omega})$ ($\Omega \neq \mathbb{R}^n$) the space of restrictions to $\bar{\Omega}$ of the functions of $C_0^m(\mathbb{R}^n)$. We will provide it with the topology defined by (4.22), so that if Ω is bounded and regular (e.g. of class C^∞, see Chapter 16) we have, set-wise and topologically, $\mathscr{D}^m(\bar{\Omega}) = C^m(\bar{\Omega})$.

Spaces $\mathscr{S}^m(\mathbb{R}^n)$

DEFINITION 4.4. Let $m \in \mathbb{N}$ or $m = +\infty$. We denote by $\mathscr{S}^m(\mathbb{R}^n)$ the space of the functions $f \in C^m(\mathbb{R}^n)$ such that

$$\forall \varepsilon > 0 \quad \forall k \in \mathbb{Z} \quad \forall |\alpha| \leq m \quad \exists M \in \mathbb{R} \; \|x\| > M \Rightarrow |(1+\|x\|^2)^k D^\alpha f(x)| < \varepsilon \tag{4.28}$$

provided with the family of seminorms

$$\forall f \in \mathscr{S}^m(\mathbb{R}^n) \; |f|_{k,\alpha} = \sup_{x \in \mathbb{R}^n} |(1+\|x\|^2)^k D^\alpha f(x)| \tag{4.29}$$

where $k \in \mathbb{Z}$ and $|\alpha| \leq m$. In what follows we will put $\mathscr{S}^\infty(\mathbb{R}^n) = \mathscr{S}(\mathbb{R}^n)$.

The spaces $\mathscr{S}^m(\mathbb{R}^n)$, whose elements are called *rapidly decreasing functions at infinity together with their derivatives of order $\leq m$*, are Fréchet spaces (see, e.g., Horváth, 1966, pp. 116 and 154). Frequently associated with these spaces are the spaces of *multiplicators* $\mathscr{O}_M$ and $\mathscr{O}_C$ which we will not deal with (see the references indicated for the theory of distributions).

Spaces $L^p(A)$

DEFINITION 4.5. Let A be a measurable subset of $\mathbb{R}^n$. We denote by $L^p(A)$, $p \in [1, +\infty[$, the space of (class of) functions $f \in \mathscr{F}(A)$ measurable in A and such that $|f|^p$ is summable in A, provided with the norm

$$\forall f \in L^p(A) \quad \|f\|_{L^p(A)} = \left(\int_A |f(x)|^p \, dx \right)^{1/p}. \tag{4.30}$$

We denote by $L^\infty(A)$ the space of (class of) functions $f \in \mathscr{F}(A)$ measurable and essentially bounded in A, provided with the norm

$$\forall f \in L^\infty(A) \; \|f\|_{L^\infty(A)} = \operatorname*{ess\,sup}_{A} |f(x)|. \tag{4.31}$$

Let us remember that two functions are in the same class iff they are equal almost everywhere (a.e.) with respect to the ordinary Lebesgue measure (μ), the only one we will consider. As is usual in the theory of integration we will not distinguish between a class of functions equal a.e. and a member of this class (i.e. we will pass to the quotient...); thus, a function $f \in L^p(A)$ is defined except for a set of measure zero and to say, e.g., that $f \in L^p(A)$ is continuous in A is to be interpreted in the sense that f is equal a.e. to a continuous function in A.

The spaces $L^p(A)$, $p \in [1, +\infty]$, are Banach spaces (see, e.g., Treves, 1967a, pp. 103 and 106), and reflexive if $p \neq 1, +\infty$; the space $L^2(A)$ is a Hilbert space with the inner product

$$\forall f, g \in L^2(A) \qquad (f, g)_{L^2(A)} = \int_A f(x)\overline{g(x)}\, dx. \tag{4.32}$$

One can also consider the spaces $L^p(A)$ with $0 < p < 1$, but (4.30) is no longer a norm; with the metric $\rho(f, g) = \int_A |f(x) - g(x)|^p\, dx$, $0 < p < 1$, one obtains complete metric spaces (see, e.g., Hewitt–Stromberg, 1965, p. 200; see also Dunford–Schwartz, 1958, p. 171).

In general we will take A to be an open set or a compact set; let us remark that if $A \subset \mathbb{R}^n$ and ∂A has zero measure ($\mu(\partial A) = 0$) then there is effectively no difference between $L^p(\bar{A})$ and $L^p(A)$, since if $f \in L^p(A)$ then any (!) extension of f to $\bar{A}$ belongs to $L^p(\bar{A})$.

Spaces $L^p_{\text{comp}}(\Omega)$ and $L^p_{\text{loc}}(\Omega)$

DEFINITION 4.6. Let Ω be an open set of $\mathbb{R}^n$ and $p \in [1, +\infty]$. We denote by $L^p_{\text{comp}}(\Omega)$ the space of (class of) functions $f \in L^p(\Omega)$ such that supp $f \subset\subset \Omega$, provided with the inductive limit topology of the Banach topologies of the spaces $L^p_K(\Omega)$ of the functions $f \in L^p(\Omega)$ such that supp $f \subset K \subset\subset \Omega$. We denote by $L^p_{\text{loc}}(\Omega)$ the space of the functions $f \in \mathscr{F}(\Omega)$ such that $f|_K \in L^p(K)$ for every $K \subset\subset \Omega$, provided with the family of seminorms

$$\forall f \in L^p_{\text{loc}}(\Omega) \; |f|_K = \|f|_K\|_{L^p(K)}. \tag{4.33}$$

The spaces $L^p_{\text{comp}}(\Omega)$ ('comp' for 'compact') are complete locally convex t.v.s. (but not Banach spaces!). The spaces $L^p_{\text{loc}}(\Omega)$ ('loc' for 'locally') are Fréchet spaces. The space $L^1_{\text{loc}}(\Omega)$ of the *functions locally summable in* Ω is of special interest, because it is, of all those we will consider, the 'biggest' t.v.s. whose elements are functions defined at the points of a set (a.e.!).

Spaces diagram

For ease of reference, we have included in a single diagram the immersions that exist between the previously defined spaces. The diagram is 'transitive': if a space can be connected to another by a chain of continuous (dense) immersions then the immersion of the first in the second is continuous

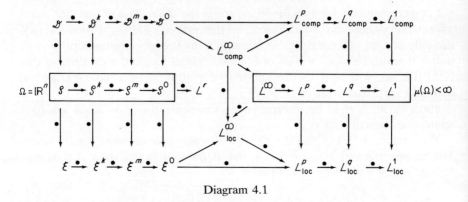

Diagram 4.1

(dense). The part of the diagram relative to the spaces $\mathscr{S}^m(\mathbb{R}^n)$ should be ignored if $\Omega \neq \mathbb{R}^n$ (but not the immersions which through the spaces $\mathscr{S}^m(\mathbb{R}^n)$ connects other spaces), as should be the inclusions between the spaces $L^p(\Omega)$ in the case in which Ω has an infinite measure (but not those between $L^p_{\text{comp}}(\Omega)$, $L^p(\Omega)$, and $L^p_{\text{loc}}(\Omega)$). All the spaces in the diagram refer to the same open set Ω, the indication of which has been excluded for clarity; for the same reason we have written $\to$ instead of $\hookrightarrow$ and $\to$ instead of $\hookrightarrow^{ds}$. In the diagram $k > m > 0$ are natural numbers and $p > q > 1$ and $r \geq 1$ are real numbers.

We will not prove the results shown in the diagram. We will merely note, to show a concrete case of (4.17), that if $\mu(\Omega) < +\infty$ and $q < p$ then $\|f\|_{L^q(\Omega)} \leq (\mu(\Omega))^{(1/p)-(1/q)} \|f\|_{L^p(\Omega)}$, as follows immediately from the Hölder inequality: this proves that the inclusion of $L^p(\Omega)$ in $L^q(\Omega)$ is continuous. On the other hand, the most important results regarding density are proved in Chapter 15 and the results regarding continuity follow easily from the following two theorems (see, e.g., Treves, 1967a, pp. 64 and 128; Horváth, 1966, p. 97).

THEOREM 4.3. Let E and F be two locally convex t.v.s. whose topologies are determined, respectively, by the families of continuous seminorms $\{e_\gamma\}_{\gamma \in \Gamma}$ and $\{f_\lambda\}_{\lambda \in \Lambda}$. A linear application $L: E \to F$ is continuous iff

$$\forall \lambda \in \Lambda \quad \exists c \in \mathbb{R} \quad \exists \{\gamma_i \in \Gamma\}_{i=1,\ldots,m} \quad \forall x \in E \quad f_\lambda(L(x)) \leq c \sum_{i=1}^m e_{\gamma_i}(x).$$
(4.34)

THEOREM 4.4. Let E and F be two locally convex t.v.s. and let us suppose that E is the inductive limit of a family $\{E_\alpha\}_{\alpha \in A}$ of Fréchet spaces. A linear application $L: E \to F$ is continuous iff for all $\alpha \in A$ the application $L|_{E_\alpha}: E_\alpha \to F$ is continuous.

We present now some examples of continuous linear operators defined on the spaces previously considered.

Product

EXAMPLE 4.1. Let $\psi \in C^\infty(\Omega)$ be a fixed function. The operator

$$^\cdot\psi : \mathcal{D}(\Omega) \to \mathcal{D}(\Omega) \qquad (4.35)$$

which to every $\phi \in \mathcal{D}(\Omega)$ associates its *product* with ψ, i.e. $^\cdot\psi(\phi) \triangleq \psi\phi$, is linear and continuous.

Linearity follows immediately:

$$\forall \lambda_1, \lambda_2 \in \mathbb{C} \quad \forall \phi_1, \phi_2 \in \mathcal{D}(\Omega)$$
$$^\cdot\psi(\lambda_1\phi_1 + \lambda_2\phi_2) = \psi(\lambda_1\phi_1 + \lambda_2\phi_2)$$
$$= \lambda_1\psi\phi_1 + \lambda_2\psi\phi_2 = \lambda_1{}^\cdot\psi(\phi_1) + \lambda_2{}^\cdot\psi(\phi_2). \qquad (4.36)$$

To prove the continuity of $^\cdot\psi$ it is sufficient to prove (theorem 4.4) that for every $K \subset\subset \Omega$ the operator $^\cdot\psi|_{C_K^\infty(\Omega)} : C_K^\infty(\Omega) \to \mathcal{D}(\Omega)$ is continuous. On the other hand (theorem 4.3) to prove this it is enough to construct, for any $m \in \mathbb{N}$, a family of functions $\theta = \{\theta_\alpha\}$ under the conditions of definition 4.3 and such that for every $\phi \in C_K^\infty(\Omega)$

$$\sum_{|\alpha| \leq m} \sup_{x \in K} |D^\alpha(\psi\phi)(x)| \leq \sum_{|\alpha| \leq m} \sup_{x \in \Omega} |\theta_\alpha(x) D^\alpha \phi(x)|. \qquad (4.37)$$

Expanding the first term of (4.37) with the help of the *Leibniz formula* (see, e.g., Horváth, 1966, p. 101)

$$D^\alpha(\psi\phi) = \sum_{|\beta| \leq |\alpha|} \binom{\alpha}{\beta} D^\beta \phi D^{\alpha-\beta} \psi \qquad (4.38)$$

one can easily see that such a family can be obtained from the derivatives of ψ by appropriate modifications—the details are left to the reader.

Translation

EXAMPLE 4.2. If $\phi \in \mathcal{D}(\mathbb{R}^n)$ and $h \in \mathbb{R}^n$, the function $\tau_h \phi \in \mathcal{D}(\mathbb{R}^n)$ defined by

$$\forall x \in \mathbb{R}^n (\tau_h \phi)(x) \triangleq \phi(x - h) \qquad (4.39)$$

is called the *h-translation* of ϕ. For fixed h, (4.39) defines a continuous linear operator

$$\tau_h : \mathcal{D}(\mathbb{R}^n) \to \mathcal{D}(\mathbb{R}^n). \qquad (4.40)$$

Derivative

EXAMPLE 4.3. For every $\alpha \in \mathbb{N}^n$, the operator

$$D^\alpha : \mathcal{D}(\Omega) \to \mathcal{D}(\Omega) \qquad (4.41)$$

which to every $\phi \in \mathcal{D}(\Omega)$ associates its *partial derivative* of order $|\alpha|$

$$D^\alpha \phi(x) \triangleq \frac{\partial^{|\alpha|}\phi(x)}{\partial x_1^{\alpha_1} \ldots \partial x_n^{\alpha_n}} \qquad (4.42)$$

is linear and continuous. Let us remark that D^α operates effectively between $\mathcal{D}(\Omega)$ and $\mathcal{D}(\Omega)$ since the derivative does not enlarge the support of the functions.

Change of variable

EXAMPLE 4.4. Let Ω_1 and Ω_2 be two open sets of $\mathbb{R}^n$ and $\psi: \Omega_1 \to \Omega_2$ a *diffeomorphism* of class C^∞ from Ω_1 into Ω_2 (i.e., ψ is a bijective application of class C^∞ together with its inverse). The operator

$$°\psi : \mathcal{D}(\Omega_2) \to \mathcal{D}(\Omega_1) \qquad (4.43)$$

which to each $\phi \in \mathcal{D}(\Omega_2)$ associates $°\psi(\phi) \in \mathcal{D}(\Omega_1)$ defined by

$$°\psi(\phi)(x) \triangleq (\phi \circ \psi)(x) = \phi[\psi(x)] \qquad \forall x \in \Omega_1 \qquad (4.44)$$

is linear and continuous.

Let us remark that, denoting by $|J\psi|$ the modulus of the Jacobian of the transformation ψ, the following formula holds:

$$\int_{\Omega_2} \phi(y)\, dy = \int_{\Omega_1} \phi[\psi(x)] |J\psi|\, dx \qquad (4.45)$$

(see, e.g., Narasimhan, 1968, p. 100).

Convolution

EXAMPLE 4.5. Let $\psi \in C_0^\infty(\mathbb{R}^n)$ be a fixed function. The operator

$$*\psi : \mathcal{D}(\mathbb{R}^n) \to \mathcal{D}(\mathbb{R}^n) \qquad (4.46)$$

which to every $\phi \in \mathcal{D}(\mathbb{R}^n)$ associates its *convolution product* with ψ, $*\psi(\phi) = \phi * \psi$ defined by

$$(\phi * \psi)(x) \triangleq \int_{\mathbb{R}^n} \phi(y)\psi(x-y)\, dy, \qquad (4.47)$$

is linear and continuous.

Fourier transform

EXAMPLE 4.6. Let $\phi \in \mathcal{S}(\mathbb{R}^n)$. The function $\hat{\phi} \in \mathcal{S}(\mathbb{R}^n)$ defined by

$$\hat{\phi}(y) \triangleq \int_{\mathbb{R}^n} e^{-2\pi i(x,y)} \phi(x)\, dx, \qquad (4.48)$$

where $(x, y) = x_1 y_1 + \ldots + x_n y_n$, is called the *Fourier transform* of ϕ (in order to see that $\hat{\phi} \in \mathscr{S}(\mathbb{R}^n)$ it must be kept in mind that $|e^{-2\pi i(x,y)}| = 1 \ldots$).

The expression (4.48) defines an operator

$$\mathscr{F} : \mathscr{S}(\mathbb{R}^n) \to \mathscr{S}(\mathbb{R}^n) \tag{4.49}$$

which to every $\phi \in \mathscr{S}(\mathbb{R}^n)$ associates its Fourier transform $\hat{\phi} = \mathscr{F}\phi$. This operator is linear and continuous; or rather, it is an (algebraic and topological) endomorphism of $\mathscr{S}(\mathbb{R}^n)$, the *inverse operator* being defined by

$$\mathscr{F}^{-1}(\hat{\phi}(y)) = \phi(x) = \int_{\mathbb{R}^n} e^{2\pi i(x,y)} \hat{\phi}(y) \, dy. \tag{4.50}$$

Most correctly, $\mathscr{F}$ is an isomorphism of $\mathscr{S}(\mathbb{R}^n)$ (current variable: x) in $\mathscr{S}(\mathbb{R}^{n'})$ (current variable: y)—we will, however, consider $\mathbb{R}^n$ and its dual $\mathbb{R}^{n'}$ as being identified through the (Riesz) operator associated with $(x, y)_{\mathbb{R}^n} \triangleq {}_{\mathbb{R}^{n'}} \langle x, y \rangle_{\mathbb{R}^n}$.

We will state without proof (see, e.g., Treves, 1967a, p. 270; Hörmander, 1964; p. 19) the following important result:

THEOREM 4.5. *If $\phi, \psi \in \mathscr{S}(\mathbb{R}^n)$ then the following relations hold:*

$$\int_{\mathbb{R}^n} \phi(x) \overline{\psi(x)} \, dx = \int_{\mathbb{R}^n} \hat{\phi}(y) \overline{\hat{\psi}(y)} \, dy, \tag{4.51}$$

$$\int_{\mathbb{R}^n} |\phi(x)|^2 \, dx = \int_{\mathbb{R}^n} |\hat{\phi}(y)|^2 \, dy, \tag{4.52}$$

$$\int_{\mathbb{R}^n} \hat{\phi}(x) \psi(x) \, dx = \int_{\mathbb{R}^n} \phi(x) \hat{\psi}(x) \, dx. \tag{4.53}$$

This theorem and the fact that $\mathscr{S}(\mathbb{R}^n) \hookrightarrow^{ds} L^2(\mathbb{R}^n)$ show that *the Fourier transform $\mathscr{F} : \mathscr{S}(\mathbb{R}^n) \to \mathscr{S}(\mathbb{R}^n)$ can be extended in only one way in an isometry of $L^2(\mathbb{R}^n)$ in $L^2(\mathbb{R}^n)$*. Using the same notation for the operator and for its extension (which is permissible since the latter is unique) and keeping in mind (4.32), we can rewrite (4.51) and (4.52) in the form

$$(\phi, \psi)_{L^2(\mathbb{R}^n)} = (\hat{\phi}, \hat{\psi})_{L^2(\mathbb{R}^n)} \tag{4.54}$$

(where ϕ and ψ are now two arbitrary elements of $L^2(\mathbb{R}^n)$) and (having eliminated the useless exponent)

$$\|\phi\|_{L^2(\mathbb{R}^n)} = \|\hat{\phi}\|_{L^2(\mathbb{R}^n)}, \tag{4.55}$$

which are known, respectively, as the *Parseval formula* and the *Plancherel formula*.

4.2.2 DUAL SPACES AND TRANSPOSED OPERATORS

Dual spaces diagram

Keeping in mind what we have said regarding the identification between dual spaces, one can construct, starting from diagram 4.1, another diagram

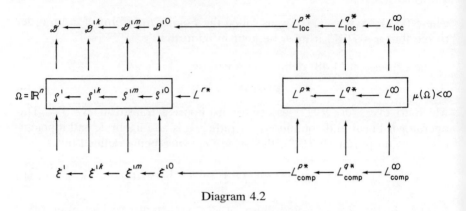

Diagram 4.2

for the dual spaces. In order to do this all we have to do is to substitute the generic space E by its strong (for example) dual E' (ignoring the spaces $L^\infty_{\text{comp}}(\Omega)$, $L^\infty(\Omega)$ and $L^\infty_{\text{loc}}(\Omega)$ since the immersions are not dense) and invert the direction of the immersions, which are no longer necessarily dense. For the spaces $L^p(\Omega)$, $L^p_{\text{comp}}(\Omega)$ and $L^p_{\text{loc}}(\Omega)$, $p \in [1, +\infty[$, we can do more since we have a simple characterization of their duals: if $p \in [1, +\infty[$ the dual of $L^p(\Omega)$ (respectively: $L^p_{\text{comp}}(\Omega)$, $L^p_{\text{loc}}(\Omega)$) is the space $L^{p^*}(\Omega)$ (respectively: $L^{p^*}_{\text{loc}}(\Omega)$, $L^{p^*}_{\text{comp}}(\Omega)$), where $p^* = p/(p-1)$ if $p \neq 1$ and $p^* = +\infty$ if $p = 1$. We have then the diagram 4.2.

The space $\mathscr{D}'(\Omega)$: Convergence

When one is dealing with a topological space, one of the most natural problems which arises is how to characterize, in a manageable fashion, the concept of convergence of sequences in the topology concerned.

We will give such a characterization for the space $\mathscr{D}'(\Omega)$, which deserves privileged treatment since it is the 'largest' of the spaces shown in diagram 4.2. A sequence u_l $(l \in \mathbb{N})$ of elements of $\mathscr{D}'(\Omega)$ converges (strongly) to $u \in \mathscr{D}'(\Omega)$ iff for every $\phi \in \mathscr{D}(\Omega)$ the numerical sequence ${}_{\mathscr{D}'(\Omega)}\langle u_l, \phi \rangle_{\mathscr{D}(\Omega)}$ converges to ${}_{\mathscr{D}'(\Omega)}\langle u, \phi \rangle_{\mathscr{D}(\Omega)}$ (see, e.g., Horváth, 1966, p. 315). This concept of convergence is the one usually found in connection with the weak topologies of the dual spaces and not with the strong ones; however, it happens that *in $\mathscr{D}'(\Omega)$ a sequence converges strongly iff it converges weakly* (see, e.g., Yosida, 1971, p. 123). Among the other properties of the strong topology of $\mathscr{D}'(\Omega)$ let us mention the following: *$\mathscr{D}'(\Omega)$ is a reflexive and complete Hausdorff space.*

Pseudo-topology

It is interesting that, given any vector space E, one can define on E a concept of convergence without first introducing a topology: one says then that a pseudo-topology is defined in E. More precisely (see e.g. Cristescu–Marinescu, 1973, p. 11; Marinescu, 1963, chs I and II, and for a more abstract treatment, Binz, 1972):

DEFINITION 4.7. By a *pseudo-topological* vector space we mean a triplet $(E, \mathscr{C}, \lim)$ where E is a vector space, $\mathscr{C} \subset E^{\mathbb{N}}$ is a vector subspace of $E^{\mathbb{N}}$ and $\lim: \mathscr{C} \to E$ is a linear operator which to every $u \in \mathscr{C}$ associates $u = \lim(u) \in E$ accordingly to the following conditions:

(i) if $\forall n \in \mathbb{N}$ $u(n) = \tilde{u}$ then $u \in \mathscr{C}$ and $\lim(u) = \tilde{u}$,
(ii) $\forall u \in \mathscr{C}$, if $\lim(u) = \tilde{u}$ and $\phi : \mathbb{N} \to \mathbb{N}$ is strictly increasing then $u \circ \phi \in \mathscr{C}$ and $\lim(u \circ \phi) = \tilde{u}$,
(iii) if $\lambda \in \mathbb{C}^{\mathbb{N}}$ is a sequence converging to $\tilde{\lambda}$ (in the topology of $\mathbb{C}$) then $\forall \tilde{u} \in E$ $\lambda \tilde{u} \in \mathscr{C}$ and $\lim(\lambda \tilde{u}) = \tilde{\lambda}\tilde{u}$.

The elements of $\mathscr{C}$ are called *convergent sequences* and $\tilde{u} = \lim(u)$ is called the *limit* of the sequence $u(=u(n) = u_n \ldots)$. As is obvious, topologies can be introduced on E relative to which the pseudo-topologically convergent sequences become convergent: the finest of these topologies is that which introduces the smallest 'number' of topologically convergent sequences which are not so pseudo-topologically, and we can call this the *topology associated* with the pseudo-topology.

Thus, if we want to avoid studying the topologies of duality, we can provide $\mathscr{D}'(\Omega)$ with the pseudo-topology determined by the concept of convergence considered above (now taken as the definition of an operator lim and of a class $\mathscr{C} \ldots$); besides, in order to avoid studying 'limit inductive' topologies, one can pseudo-topologize $\mathscr{D}(\Omega)$ itself by the use of the concept of convergence which we have already introduced in the previous section. This is sufficient for the most applications of the space $\mathscr{D}'(\Omega)$, and here we find ourselves in an optimal situation: the topology associated with the pseudo-topology of $\mathscr{D}'(\Omega)$ is the strong topology of $\mathscr{D}'(\Omega)$ and, obviously, there are no strongly convergent sequences which are not pseudo-topologically convergent (the topology is seen to be much richer only in the filters or in the nets).

Paraphrasing what is done with the normed spaces, the topology of which, being interpretable in terms of the convergence of sequences, cannot be distinguished by the pseudo-topology determined by those which are convergent, we can construct (for example) a pseudo-topological theory of continuity.

The space $\mathscr{D}'(\Omega)$: Characterization

The following result, the proof of which follows from theorems 4.3 and 4.4 (see, e.g., Hörmander, 1964, pp. 4 *et seq.*; Choquet–Bruhat, 1973, p. 19), characterizes the elements of $\mathscr{D}'(\Omega)$, i.e., characterizes those linear functionals on $\mathcal{D}(\Omega)$ which are continuous (note the clear pseudo-topological character of (ii)):

THEOREM 4.6. A linear functional $u:\mathcal{D}(\Omega)\to\mathbb{C}$ is continuous (and hence $u\in\mathcal{D}'(\Omega)$) iff one of the following conditions is satisfied:

(i) $\forall k\subset\subset\Omega\quad\exists m\in\mathbb{N}\quad\exists c>0\quad\forall\phi\in\mathcal{D}(\Omega),\operatorname{supp}\phi\subset K$

$$|_{\mathcal{D}'(\Omega)}\langle u,\phi\rangle_{\mathcal{D}(\Omega)}|\leq c\sum_{|\alpha|\leq m}\sup_{x\in\Omega}|D^{\alpha}\phi(x)|,$$

(ii) if $\phi_l\to 0$ in $\mathcal{D}(\Omega)$ then $_{\mathcal{D}'(\Omega)}\langle u,\phi_l\rangle_{\mathcal{D}(\Omega)}\to 0$ in $\mathbb{C}$.

As the first and very important application of this theorem, let us show that diagrams 4.1 and 4.2 can be 'connected', thus giving origin to a third diagram—diagram 4.3. We shall show therefore that $L^1_{\mathrm{loc}}(\Omega)\subsetneq^{ds}\mathcal{D}'(\Omega)$, but we will in fact prove only that $L^1_{\mathrm{loc}}(\Omega)\subsetneq\mathcal{D}'(\Omega)$—the density of the immersion follows from that of $\mathcal{D}(\Omega)$ in $\mathcal{D}'(\Omega)$, for the proof of which see, e.g., Horváth, 1966; p. 316; L. Schwartz, 1966, p. 75.

The immersion $L^1_{\mathrm{loc}}(\Omega)\subsetneq\mathcal{D}'(\Omega)$ must be suitably interpreted since on the one hand we have a space, $L^1_{\mathrm{loc}}(\Omega)$, whose elements are defined point-wise on Ω, and on the other we have a space, $\mathcal{D}'(\Omega)$, whose elements are defined globally on Ω, through the points of $\mathcal{D}(\Omega)$. A glance at diagram 4.2 allows us to realize that we have already a phenomenon of this kind there: $L^{p^*}(\Omega)$, e.g., appears as a subset of $\mathcal{D}'(\Omega)$. We must not forget, however, that this space $L^{p^*}(\Omega)$ arises as the dual of $L^p(\Omega)$, and that the elements of $(L^p(\Omega))'$ are continuous linear functionals on $L^p(\Omega)$, which are later identified with the elements of $L^{p^*}(\Omega)$ which represent them according to the theorem of representation of the functionals on the spaces $L^p(\Omega)$. In the formula $L^{p^*}(\Omega)\subsetneq\mathcal{D}'(\Omega)$ we must therefore see on the left the representative of a functional on $L^p(\Omega)$ and not a function of $L^{p^*}(\Omega)$.

In the case $L^1_{\mathrm{loc}}(\Omega)\subsetneq\mathcal{D}'(\Omega)$ the philosophy is the same: we identify $f\in L^1_{\mathrm{loc}}(\Omega)$ with the continuous linear functional $A(f)$ defined by

$$\mathcal{D}(\Omega)\ni\phi\mapsto\langle A(f),\phi\rangle=\int_\Omega f(x)\phi(x)\,dx. \tag{4.56}$$

Its linearity is obvious, and its continuity follows from theorem 4.6 (i) with $m=0$ and $c=\mu(K)\,\|f\|_{L^1(K)}$. Also, the application $A\equiv j_{L^1_{\mathrm{loc}}(\Omega)\mathcal{D}'(\Omega)}$ of $L^1_{\mathrm{loc}}(\Omega)$

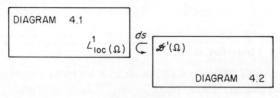

Diagram 4.3

in $\mathscr{D}'(\Omega)$ which with every f associates $A(f)$ is linear and injective (which follows from the density of the immersion of $\mathscr{D}(\Omega)$ in $L^1_{\text{loc}}(\Omega)$—for a detailed proof, see Adams, 1975, p. 59). We are therefore dealing here with a continuous injection which enables us to write $L^1_{\text{loc}}(\Omega) \hookrightarrow \mathscr{D}'(\Omega)$.

Henceforth, every time we have to deal with an element $u \in \mathscr{D}'(\Omega)$ which we know can be represented through $f \in L^1_{\text{loc}}(\Omega)$ (i.e., such that there exists an $f \in L^1_{\text{loc}}(\Omega)$ for which $u = A(f)$) we will not distinguish between $\langle u, \phi \rangle = \langle A(f), \phi \rangle = \langle f, \phi \rangle = \int_\Omega f(x)\phi(x)\,dx$ to indicate its value on $\phi \in \mathscr{D}(\Omega)$. Let us remark that to say that $\mathscr{D}(\Omega)$ is dense in $\mathscr{D}'(\Omega)$ means that every element $u \in \mathscr{D}'(\Omega)$ can be approximated, in the topology of $\mathscr{D}'(\Omega)$, by functionals of the type $\langle u_m, \phi \rangle = \int_\Omega u_m(x)\phi(x)\,dx$ with $u_m \in \mathscr{D}(\Omega)$.

The space $\mathscr{D}'(\Omega)$: Nomenclature

Before starting to study the transposes of the linear operators considered in the previous section, we must introduce some nomenclature.

An element of $\mathscr{D}'(\Omega)$ is called a *distribution* on Ω. A distribution $u \in \mathscr{D}'(\Omega)$ is said to be *null* in an open set $\tilde{\Omega} \subset \Omega$ if $_{\mathscr{D}'(\Omega)}\langle u, \phi \rangle_{\mathscr{D}(\Omega)} = 0$ for every $\phi \in \mathscr{D}(\Omega)$ with supp $\phi \subset \tilde{\Omega}$. One can easily see that if a distribution $u \in \mathscr{D}'(\Omega)$ is null on a family of open sets $\{\Omega_\lambda\}_{\lambda \in \Lambda}$ then u is null on $\bigcup_{\lambda \in \Lambda} \Omega_\lambda$—accordingly, the *support* of u, supp u, is defined to be the complement of the union of the open sets in which u is null (if u can be represented through $f \in L^1_{\text{loc}}(\Omega)$ then supp $u =$ supp f). As is natural, two distributions $u_1, u_2 \in \mathscr{D}'(\Omega)$ are said to be *equal* on $\tilde{\Omega} \subset \Omega$ if $u_1 - u_2$ is null in $\tilde{\Omega}$. It can be shown that the elements of $\mathscr{E}'(\Omega)$ are the distributions whose supports are compact (see, e.g., Horváth, 1966, p. 320; Treves, 1967a, p. 256). If in condition (i) of theorem 4.6 one can fix m independently of K, u is said to be of *finite order*, and the *order* of the distribution is defined to be smallest integer in these conditions. It can be shown (see, e.g., Horváth, 1966, p. 339; Treves, 1967a, p. 258) that if u is of order m then $u \in \mathscr{D}^{m'}(\Omega)$. The set of distributions of finite order is frequently denoted by $\mathscr{D}'^F(\Omega)$: it is the dual of the space $\mathscr{D}^F(\Omega)$ which is obtained by providing $C_0^\infty(\Omega)$ with the less fine topology which makes the immersions of $C_0^\infty(\Omega)$ in $\mathscr{D}^m(\Omega)$ continuous; it can be shown that $\mathscr{E}'(\Omega) \subset \mathscr{D}'^F(\Omega)$ (see Horváth, 1966, p. 340; Treves, 1967a, p. 259). The distributions of order zero are called *Radon measures*, and among them can be found those distributions which can be represented by locally summable functions, which are called the *densities* of the corresponding measures: the distribution associated with the function $f(x) \equiv 1$ is none other than the ordinary Lebesgue measure!; for a more detailed account of Radon measures, see L. Schwartz, 1973. A distribution u is said to be *real* if $_{\mathscr{D}'(\Omega)}\langle u, \phi \rangle_{\mathscr{D}(\Omega)} \in \mathbb{R}$ for every $\phi \in \mathscr{D}(\Omega)$ with $\phi : \Omega \to \mathbb{R}$, and a real distribution u is said to be *positive* if $_{\mathscr{D}'(\Omega)}\langle u, \phi \rangle_{\mathscr{D}(\Omega)} \geq 0$ for every $\phi \in \mathscr{D}(\Omega)$ with $\phi \geq 0$; it is interesting to note that if u is a positive (or a negative) distribution then u is a Radon measure, i.e., if u is a continuous linear functional on $\mathscr{D}(\Omega)$ to which one can attribute a sign, as we have done above, then u can be extended in only one way as a continuous linear functional on the 'much larger set' $\mathscr{D}^0(\Omega)$ (this result, the proof of which can be found in L. Schwartz, 1966, p. 29, is disappointing in that it shows that the concept of order in $\mathscr{D}'(\Omega)$ cannot be extended outside the field of the measures). Finally, the elements of $\mathscr{S}'(\mathbb{R}^n)$ are said to be *temperate distributions*.

We will next study the transposes of the operators considered in the examples 4.1–4.6 of section 4.2.1.

Product

EXAMPLE 4.1'. The transpose of the operator $\cdot \psi \in \mathscr{L}(\mathscr{D}(\Omega), \mathscr{D}(\Omega))$ is the operator ${}^t\cdot\psi : \mathscr{D}'(\Omega) \to \mathscr{D}'(\Omega)$ defined by

$$\forall u \in \mathscr{D}'(\Omega) \quad \forall \phi \in \mathscr{D}(\Omega)$$

$$_{\mathscr{D}'(\Omega)}\langle ({}^t\cdot\psi)(u), \phi \rangle_{\mathscr{D}(\Omega)} = {}_{\mathscr{D}'(\Omega)}\langle u, \cdot\psi(\phi) \rangle_{\mathscr{D}(\Omega)} = {}_{\mathscr{D}'(\Omega)}\langle u, \psi\phi \rangle_{\mathscr{D}(\Omega)}. \quad (4.57)$$

If u can be represented by means of a function f, from (4.57) it follows that

$$_{\mathscr{D}'(\Omega)}\langle f, \phi\psi \rangle_{\mathscr{D}(\Omega)} = \int_\Omega f(x)\phi(x)\psi(x)\,dx = {}_{\mathscr{D}'(\Omega)}\langle f\psi, \phi \rangle_{\mathscr{D}(\Omega)} \quad (4.58)$$

and one again has the usual concept of product of two functions. It is therefore natural to write (4.57) in the form

$$\forall u \in \mathscr{D}'(\Omega) \quad \forall \phi \in \mathscr{D}(\Omega) \quad _{\mathscr{D}'(\Omega)}\langle u\psi, \phi \rangle_{\mathscr{D}(\Omega)} \triangleq {}_{\mathscr{D}'(\Omega)}\langle u, \phi\psi \rangle_{\mathscr{D}(\Omega)} \quad (4.59)$$

and to take this expression as the definition of product of a distribution $u \in \mathscr{D}'(\Omega)$ by a function $\psi \in C^\infty(\Omega)$. $\mathscr{D}'(\Omega)$ is then a module over the algebra $C^\infty(\Omega)$.

Note that it is impossible to define an operator $\cdot : \mathscr{D}'(\Omega) \times \mathscr{D}'(\Omega) \to \mathscr{D}'(\Omega)$ such that the algebraic system $\langle \mathscr{D}'(\Omega), +, \cdot \rangle$ is a ring and $\cdot$ generalizes, in the sense we have just studied, the usual product of two functions. This negative result is not so strange considering that the product of two functions of $L^1(\Omega)$ is not necessarily a function of $L^1(\Omega)\dots$.

Translation

EXAMPLE 4.2'. Transposing the operator $\tau_h \in \mathscr{L}(\mathscr{D}(\mathbb{R}^n), \mathscr{D}(\mathbb{R}^n))$ we have ${}^t\tau_h \in \mathscr{L}(\mathscr{D}'(\mathbb{R}^n), \mathscr{D}'(\mathbb{R}^n))$ defined by

$$\forall u \in \mathscr{D}'(\mathbb{R}^n) \quad \forall \phi \in \mathscr{D}(\mathbb{R}^n) \quad _{\mathscr{D}'(\mathbb{R}^n)}\langle ({}^t\tau_h)(u), \phi \rangle_{\mathscr{D}(\mathbb{R}^n)} = {}_{\mathscr{D}'(\mathbb{R}^n)}\langle u, \tau_h\phi \rangle_{\mathscr{D}(\mathbb{R}^n)}.$$
$$(4.60)$$

Recalling the previous example, it is natural to ask whether ${}^t\tau_h$ generalizes the operator τ_h to $\mathscr{D}'(\mathbb{R}^n)$. The answer is no. Indeed, if u can be represented by a function f one must have, in order to speak of a generalization,

$$\int_{\mathbb{R}^n} f(x)\phi(x-h)\,dx = \int_{\mathbb{R}^n} f(x-h)\phi(x)\,dx \quad \forall \phi \in \mathscr{D}(\Omega), \quad (4.61)$$

which is not true: there is a change of sign. But this is exactly what suggests how the notion of translation can be generalized to $\mathscr{D}'(\mathbb{R}^n)$: instead of τ_h we must transpose the operator τ_{-h} (contragradient respect to τ_h) and thus define as the h-translation of u the distribution ${}^t\tau_{-h}(u)$, which we will henceforth denote (with a natural misuse of language) $\tau_h(u)$.

More precisely, we will define the h-translation of u by the expression

$$\forall u \in \mathscr{D}'(\mathbb{R}^n) \quad \forall \phi \in \mathscr{D}(\mathbb{R}^n) \quad {}_{\mathscr{D}'(\mathbb{R}^n)}\langle \tau_h(u), \phi \rangle_{\mathscr{D}(\mathbb{R}^n)} \triangleq {}_{\mathscr{D}'(\mathbb{R}^n)}\langle u, \tau_{-h}\phi \rangle_{\mathscr{D}(\mathbb{R}^n)}. \quad (4.62)$$

Derivative

EXAMPLE 4.3'. The transpose of $D^\alpha : \mathscr{D}(\Omega) \to \mathscr{D}(\Omega)$ is the operator ${}^tD^\alpha : \mathscr{D}'(\Omega) \to \mathscr{D}'(\Omega)$ defined by

$$\forall u \in \mathscr{D}'(\Omega) \quad \forall \phi \in \mathscr{D}(\Omega) \quad {}_{\mathscr{D}'(\Omega)}\langle ({}^tD^\alpha)(u), \phi \rangle_{\mathscr{D}(\Omega)} = {}_{\mathscr{D}'(\Omega)}\langle u, D^\alpha \phi \rangle_{\mathscr{D}(\Omega)}. \quad (4.63)$$

We have here a problem analogous to the one we encountered in the previous example: in order that we can interpret ${}^tD^\alpha$ as a generalization to $\mathscr{D}'(\Omega)$ of the derivative we must have, if $u \in \mathscr{D}'(\Omega)$ can be represented by a function f (at least of class $C^{|\alpha|} \ldots$),

$$\int_\Omega D^\alpha f(x)\phi(x)\,dx = \int_\Omega f(x)D^\alpha \phi(x)\,dx \quad \forall \phi \in \mathscr{D}(\Omega), \quad (4.64)$$

while an integration by parts (iterated) shows that we have

$$\int_\Omega D^\alpha f(x)\phi(x)\,dx = (-1)^{|\alpha|}\int_\Omega f(x)D^\alpha \phi(x)\,dx \quad \forall \phi \in \mathscr{D}(\Omega). \quad (4.65)$$

However, (4.65) shows us that we can in fact generalize the derivative by transposing the operator $(-1)^{|\alpha|}D^\alpha$ instead of D^α *tout court*. Denoting still by D^α the operator ${}^t(-1)^{|\alpha|}D^\alpha$ we can write

$$\forall u \in \mathscr{D}'(\Omega) \quad \forall \phi \in \mathscr{D}(\Omega) \quad {}_{\mathscr{D}'(\Omega)}\langle D^\alpha u, \phi \rangle_{\mathscr{D}(\Omega)} \triangleq (-1)^{|\alpha|} {}_{\mathscr{D}'(\Omega)}\langle u, D^\alpha \phi \rangle_{\mathscr{D}(\Omega)} \quad (4.66)$$

which defines the derivative (or rather, the operator D^α) in the sense of $\mathscr{D}'(\Omega)$. In general, when we use this formula we say that the derivatives are 'discharged' on the regular function . . . .

From (4.66) it follows immediately that *every distribution is indefinitely differentiable and the partial derivatives are commutative operators*.

If $u \in \mathscr{D}'(\Omega)$ can be represented through $f \in L^1_{\text{loc}}(\Omega)$ we will write, without distinguishing them, $D^\alpha u$ or $D^\alpha f$; however $D^\alpha f$ is not in general a function but a distribution (of finite order $\leq |\alpha|$)—it will be a function, and will coincide with the corresponding ordinary derivative of f if $f \in C^{|\alpha|}(\Omega)$.

The Leibniz formula (4.38) still holds if D^α are the new derivative operators, $\phi \in \mathcal{D}'(\Omega)$ and $\psi \in C^\infty(\Omega)$. Among the other properties of the ordinary derivative which are preserved we will mention the following: *if I is an open interval of $\mathbb{R}$ and $f \in \mathcal{D}'(I)$ is such that $D^k f = 0$ then f can be represented by a polynomial function of degree $\leq k - 1$*—in particular, if $k = 1$ then f can be represented by a constant (see, e.g., Horváth, 1966, p. 328). Finally, note that one can directly define the derivative of a distribution $u \in \mathcal{D}'(\mathbb{R}^n)$ as the limit of a difference quotient: the partial derivative of u with respect to x_k, $1 \leq k \leq n$, is given by

$$\frac{\partial u}{\partial x_k} = \lim_{h_k \to 0} \frac{\tau_h u - u}{-h_k}, \qquad (4.67)$$

where $h = (0, \ldots, h_k, \ldots, 0) \in \mathbb{R}^n$ and the limit is, obviously, taken in the sense of $\mathcal{D}'(\mathbb{R}^n)$ (see Horváth, 1966, p. 330; L. Schwartz, 1966, p. 77).

Change of variable

EXAMPLE 4.4'. The transpose of the operator $°\psi \in \mathcal{L}(\mathcal{D}(\Omega_2), \mathcal{D}(\Omega_1))$ is the operator ${}^{t\circ}\psi : \mathcal{D}'(\Omega_1) \to \mathcal{D}'(\Omega_2)$ defined by

$$\forall u \in \mathcal{D}'(\Omega_1) \quad \forall \phi \in \mathcal{D}(\Omega_2)$$

$$_{\mathcal{D}'(\Omega_2)}\langle {}^{t\circ}\psi(u), \phi \rangle_{\mathcal{D}(\Omega_2)} = {}_{\mathcal{D}'(\Omega_1)}\langle u, °\psi(\phi) \rangle_{\mathcal{D}(\Omega_1)} = {}_{\mathcal{D}'(\Omega_1)}\langle u, \phi \circ \psi \rangle_{\mathcal{D}(\Omega_1)}. \qquad (4.68)$$

This operator cannot be taken to be a generalization of the operator $°\psi$ since, if u can be represented by a function $f : \Omega_1 \to \mathbb{C}$, from (4.68) it follows that

$$\int_{\Omega_2} f(\psi(x))\phi(y) \, dy = \int_{\Omega_1} f(x)\phi(\psi(x)) \, dx \qquad \forall \phi \in \mathcal{D}(\Omega_2) \qquad (4.69)$$

(y being the variable in Ω_2 and x the variable in Ω_1) the first term of which does not make sense.

Now the translation considered in example 4.2 is nothing but a particular change of variable and, when we generalize it to distributions, we transpose $\tau_{-h} = \tau_h^{-1}$ and not τ_h; naturally then here too we should try to transpose the operator $°\psi^{-1}$ rather than $°\psi$ and define ${}^{t\circ}\psi^{-1}(u) = u \circ \psi$. Now the transpose of the operator $°\psi^{-1} \in \mathcal{L}(\mathcal{D}(\Omega_1), \mathcal{D}(\Omega_2))$ is the operator ${}^{t\circ}\psi^{-1} \in \mathcal{L}(\mathcal{D}'(\Omega_2), \mathcal{D}'(\Omega_1))$ defined by

$$\forall u \in \mathcal{D}'(\Omega_2) \quad \forall \phi \in \mathcal{D}(\Omega_1)$$

$$_{\mathcal{D}'(\Omega_1)}\langle {}^{t\circ}\psi^{-1}(u), \phi \rangle_{\mathcal{D}(\Omega_1)} = {}_{\mathcal{D}'(\Omega_2)}\langle u, °\psi^{-1}(\phi) \rangle_{\mathcal{D}(\Omega_2)} = {}_{\mathcal{D}'(\Omega_2)}\langle u, \phi \circ \psi^{-1} \rangle_{\mathcal{D}(\Omega_2)} \qquad (4.70)$$

and, if u can be represented by $f : \Omega_2 \to \mathbb{C}$, the analogue of (4.69) becomes

$$\int_{\Omega_1} f(\psi(x))\phi(x) \, dx = \int_{\Omega_2} f(y)\phi(\psi^{-1}(y)) \, dy \qquad (4.71)$$

which now makes sense but contradicts (4.45) (to see this put $f \equiv 1$).

The proper way of generalizing the change of variable operator to distributions is based on the transposition of $|J\psi|^\circ \psi^{-1}$ (note that the modulus of the Jacobian associated with the translation is 1, and for this reason we have not become aware of having used it), and specifically:

$$\forall u \in \mathscr{D}'(\Omega_2) \quad \forall \phi \in \mathscr{D}(\Omega_1)$$

$$_{\mathscr{D}'(\Omega_1)}\langle u \circ \psi, \phi \rangle_{\mathscr{D}(\Omega_1)} \triangleq {}_{\mathscr{D}'(\Omega_2)}\langle u, |J\psi| (\phi \circ \psi^{-1}) \rangle_{\mathscr{D}(\Omega_2)}. \quad (4.72)$$

Convolution

EXAMPLE 4.5'. The transpose of the operator $*\psi : \mathscr{D}(\mathbb{R}^n) \to \mathscr{D}(\mathbb{R}^n)$ is the operator ${}^t*\psi$ defined by

$$\forall u \in \mathscr{D}'(\mathbb{R}^n) \quad \forall \phi \in \mathscr{D}(\mathbb{R}^n)$$

$$_{\mathscr{D}'(\mathbb{R}^n)}\langle ({}^t*\psi)(u), \phi \rangle_{\mathscr{D}(\mathbb{R}^n)} = {}_{\mathscr{D}'(\mathbb{R}^n)}\langle u, *\psi(\phi) \rangle_{\mathscr{D}(\mathbb{R}^n)} = {}_{\mathscr{D}'(\mathbb{R}^n)}\langle u, \phi * \psi \rangle_{\mathscr{D}(\mathbb{R}^n)}. \quad (4.73)$$

The distribution $({}^t*\psi)(u)$ cannot be called the convolution of u with ψ since if u can be represented by a function f the distribution $({}^t*\psi)(f)$ does not correspond to $f*\psi$; one should in fact have

$$\int_{\mathbb{R}^n} f(x) \left(\int_{\mathbb{R}^n} \phi(y) \psi(x-y) \, dy \right) dx = \int_{\mathbb{R}^n} \phi(x) \left(\int_{\mathbb{R}^n} f(y) \psi(x-y) \, dy \right) dx$$

$$\forall \phi \in \mathscr{D}(\mathbb{R}^n) \quad (4.74)$$

but, instead, as can be easily seen we have

$$\int_{\mathbb{R}^n} f(x) \left(\int_{\mathbb{R}^n} \phi(y) \psi(x-y) \, dy \right) dx = \int_{\mathbb{R}^n} \phi(x) \left(\int_{\mathbb{R}^n} f(y) \psi(y-x) \, dy \right) dx$$

$$\forall \phi \in \mathscr{D}(\mathbb{R}^n). \quad (4.75)$$

Let us introduce a new function $\check{\phi}$ defined by

$$\check{\phi}(x) = \phi(-x) \quad \forall x \in \mathbb{R}^n \quad (4.76)$$

and let us define the convolution of the distribution $u \in \mathscr{D}'(\mathbb{R}^n)$ with the function $\psi \in C_0^\infty(\mathbb{R}^n)$ as the distribution $({}^t*\check{\psi})(u)$, or specifically (with the usual misuse of terminology, which is not confusing since it is a true generalization)

$$\forall u \in \mathscr{D}'(\mathbb{R}^n) \quad \forall \phi \in \mathscr{D}(\mathbb{R}^n) \quad _{\mathscr{D}'(\mathbb{R}^n)}\langle u * \psi, \phi \rangle_{\mathscr{D}(\mathbb{R}^n)} \triangleq {}_{\mathscr{D}'(\mathbb{R}^n)}\langle u, \phi * \check{\psi} \rangle_{\mathscr{D}(\mathbb{R}^n)}. \quad (4.77)$$

Let us remark that $u*\psi$ is a very particular distribution: one can show, in fact, that it can be represented by a function of $C^\infty(\mathbb{R}^n)$ (see, e.g., Hörmander, 1964, p. 14)—denoting once again this function by $u*\psi$, we can write $u*\psi \in C^\infty(\mathbb{R}^n)$; hence we can write the formula

$$(u*\psi)(x) = {}_{\mathscr{D}'(\mathbb{R}^n)}\langle u_y, \psi(x-y) \rangle_{\mathscr{D}(\mathbb{R}^n)} = {}_{\mathscr{D}'(\mathbb{R}^n)}\langle u, \tau_x \check{\psi}(y) \rangle_{\mathscr{D}(\mathbb{R}^n)} \quad (4.78)$$

(where by u_y we mean that u operates on $\psi(x-y)$ as a function of y for fixed x), which makes sense since $u*\psi$ is a function.

If in (4.78) we substitute ψ by $\psi_\varepsilon(x) = \phi(x/\varepsilon)\varepsilon^{-n}$ (with ϕ as given by (15.8)) we get a sequence of indefinitely differentiable functions which converges, in $\mathscr{D}'(\mathbb{R}^n)$, to u (see, e.g. Hörmander, 1964, p. 15; Treves, 1967a, p. 301 and the proof of theorem 15.2).

For every $\alpha \in \mathbb{N}^n$ the following formula holds:

$$D^\alpha(u*\psi) = u*D^\alpha\psi = D^\alpha u*\psi, \tag{4.79}$$

i.e., *in order to differentiate the convolution product one has only to differentiate one of its factors*, which generalizes a well-known result.

Finally, let us remark that more generally one can consider the convolution product of two distributions $u \in \mathscr{D}'(\mathbb{R}^n)$ and $v \in \mathscr{E}'(\mathbb{R}^n)$—see also the books on the Theory of Distributions quoted in the Bibliography, in particular, Horváth, 1966; Treves, 1967a; L. Schwartz, 1966.

Fourier transform

EXAMPLE 4.6'. The transpose of the operator $\mathscr{F} \in \mathscr{L}(\mathscr{S}(\mathbb{R}^n), \mathscr{S}(\mathbb{R}^n))$ is the operator ${}^t\mathscr{F} \in \mathscr{L}(\mathscr{S}'(\mathbb{R}^n), \mathscr{S}'(\mathbb{R}^n))$ defined by

$$\forall u \in \mathscr{S}'(\mathbb{R}^n) \quad \forall \phi \in \mathscr{S}(\mathbb{R}^n) \quad {}_{\mathscr{S}'(\mathbb{R}^n)}\langle ({}^t\mathscr{F})(u), \phi \rangle_{\mathscr{S}(\mathbb{R}^n)} = {}_{\mathscr{S}'(\mathbb{R}^n)}\langle u, \mathscr{F}(\phi) \rangle_{\mathscr{S}(\mathbb{R}^n)}. \tag{4.80}$$

It follows immediately that if u can be represented by $f \in \mathscr{S}(\mathbb{R}^n)$ then ${}^t\mathscr{F}(u)$ can be represented by $\mathscr{F}(f)$, and hence it is natural to call ${}^t\mathscr{F}$ the Fourier transform in $\mathscr{S}'(\mathbb{R}^n)$—we will then indicate the operator ${}^t\mathscr{F}$ simply as $\mathscr{F}$ and we will define

$$\forall u \in \mathscr{S}'(\mathbb{R}^n) \quad \forall \phi \in \mathscr{S}(\mathbb{R}^n) \quad {}_{\mathscr{S}'(\mathbb{R}^n)}\langle \mathscr{F}(u), \phi \rangle_{\mathscr{S}(\mathbb{R}^n)} \stackrel{\Delta}{=} {}_{\mathscr{S}'(\mathbb{R}^n)}\langle u, \mathscr{F}(\phi) \rangle_{\mathscr{S}(\mathbb{R}^n)}. \tag{4.81}$$

It is important to mention here that in the literature one frequently finds ${}^t\mathscr{F} = \mathscr{F}^{-1}$—however the authors of such texts define ${}^t\mathscr{F}$ as the adjoint of $\mathscr{F}$ and not as its transpose.

The Fourier transform is an isomorphism of $\mathscr{S}'(\mathbb{R}^n)$ of which we will mention, for their importance, the following properties (which hold already in $\mathscr{S}(\mathbb{R}^n)$ and, when they make sense, in $L^2(\mathbb{R}^n)\ldots$; let us remark also that $\mathscr{F}: \mathscr{S}'(\mathbb{R}^n) \to \mathscr{S}'(\mathbb{R}^n)$ also extends $\mathscr{F}: L^2(\mathbb{R}^n) \to L^2(\mathbb{R}^n)\ldots$): if $u \in \mathscr{S}'(\mathbb{R}^n)$ and $\phi \in C_0^\infty(\mathbb{R}^n)$ then

$$\mathscr{F}(u*\phi) = \mathscr{F}(u)\mathscr{F}(\phi), \tag{4.82}$$

$$\mathscr{F}(\phi u) = \mathscr{F}(u)*\mathscr{F}(\phi), \tag{4.83}$$

$$\mathscr{F}(\partial u/\partial x_k) = 2\pi i y_k \mathscr{F}(u), \tag{4.84}$$

$$\mathscr{F}(x_k u) = -\frac{1}{2\pi i}\frac{\partial}{\partial y_k}\mathscr{F}(u), \tag{4.85}$$

$$\mathscr{F}(\tau_h u) = e^{-2\pi i(h,y)}\mathscr{F}(u). \tag{4.86}$$

For the proof see, e.g., Treves, 1967a, pp. 219 and 220; Stein–Weiss, 1971, ch. 1.

Normal and local spaces

To end this chapter we give an important definition:

DEFINITION 4.8. A vector subspace E of $\mathcal{D}'(\Omega)$ is said to be a *normal space* of distributions if $\mathcal{D}(\Omega) \hookrightarrow^{ds} E \hookrightarrow \mathcal{D}'(\Omega)$, and is said to be a *local space* of distributions if E is a module over the algebra $C_0^\infty(\Omega)$.

Let us remark that the dual of a normal space of distributions is also a space of distributions (though not necessarily normal). All the spaces in diagram 4.3 are local and, with the exception of $L^\infty(\Omega)$, $L^\infty_{\text{comp}}(\Omega)$ and $L^\infty_{\text{loc}}(\Omega)$, are normal. An example of a non-local space is that of functions of $L^1(\Omega)$ with zero mean.

Dirac's distribution and Heaviside function

REMARK 4.1. Among the distributions which can not be represented by locally summable functions, we will mention, due to its importance, the Dirac's δ distribution: we will call *Dirac's distribution* concentrated in $x_0 \in \mathbb{R}^n$ the distribution $\delta_{x_0} \in \mathcal{D}'(\mathbb{R}^n)$ which to every $\phi \in \mathcal{D}(\mathbb{R}^n)$ associates $\delta_{x_0}(\phi) = \phi(x_0)$.

If $n = 1$, δ_{x_0} can be represented as the distributional derivative of the *Heaviside function* $H(x - x_0) = \chi_{]x_0, +\infty[}$. More generally, in $\mathbb{R}^n$, δ_{x_0} can be represented as the 'tensor product' of the distributional derivatives of the functions $H(x_i - x_{0_i})$, $i = 1, \ldots, n$, being the *tensor product* of two distributions the obvious generalization of the tensor product of two functions which, for example, with the functions $f: \mathbb{R} \to \mathbb{R}$ and $g: \mathbb{R} \to \mathbb{R}$ associates the function $f \otimes g: \mathbb{R}^2 \to \mathbb{R}$ defined by $(f \otimes g)(x, y) = f(x)g(y)$.

Let us remark till that $\mathcal{F}(1) = \delta_0$.

Ultradistributions. Gevrey classes

REMARK 4.2. Roughly, the 'smaller' a topological space, the 'bigger' its dual. It is natural then to think of generalizing the concept of distribution imposing further limitations on the space of test functions: one obtains thus the concept of *ultradistribution*.

Particularly important theoretically are the Gevrey ultradistributions (see Guelfand–Chilov, 1962, Lions–Magenes, 1973): if $s > 1$ is a real number, a *Gevrey ultradistribution of class s in Ω* is an element of the space $\mathcal{D}'_s(\Omega)$, the (strong) dual of the space $\mathcal{D}_s(\Omega)$ of the *Gevrey functions of class s in Ω*, which is the inductive limit as $m \to +\infty$ of the family of Banach spaces

$$\mathscr{D}_{s,L_m,K_m}(\Omega) = \left\{ \phi \in \mathscr{D}_{K_m}(\Omega) : \sup_{\substack{x,h \\ |\alpha|=h}} \frac{|D^\alpha \phi(x)|}{L_m^h (h!)^s} < +\infty \right\},$$

$$\|\phi\|_{\mathscr{D}_{s,L_m,K_m}(\Omega)} = \sup_{\substack{x,h \\ |\alpha|=h}} \frac{|D^\alpha \phi(x)|}{L_m^h (h!)^s}$$

(compare the discussion which follows definition 4.3), where $L_m \to +\infty$ is an increasing sequence of positive real numbers and K_m is an increasing sequence of compact sets invading Ω.

For applications and for other types of ultradistributions, see Lions–Magenes, 1973 and the references in that book. For other types of extensions of the concept of distribution, with applications to physics, see Silva, 1967.

5
Sobolev Spaces

5.1 ON THE NECESSITY FOR NEW FUNCTION SPACES

As we have stated (section 1.1) in connection with the concept of the well-posed problem, one of the fundamental aims of functional analysis is to construct spaces where certain problems can be well-posed. Also, we have seen by means of an example (problem 1.0) that the classical spaces (by this ambiguous statement we mean the spaces of continuous functions, differentiable functions, etc.) may not be suitable for this purpose.

We intend now to construct a family of spaces which provides the suitable framework for posing a large number of problems.

Why new spaces?

It is not simple to construct a function space, and moreover some of the spaces which we use are not, as we will see, spaces of functions in the usual sense ... : we need a guide. In order to guide us, we will study the 'pathology' of the classical spaces; in other words we will ask the following question: *why are these spaces unsuitable for posing so many problems which arise, for example, in physics?*

The classical spaces are, in the last analysis, defined on the basis of the concept of the derivative, which is a very delicate one. Thus in $C^0(\mathbb{R})$ the derivative, in the sense of an operator, is not defined everywhere, whereas in $C^1(\mathbb{R})$, where it is so defined, it does not satisfy the most rudimentary continuity property: if f_m is a sequence of functions of $C^1(\mathbb{R})$ which converges uniformly to a function f it is not always true that $f \in C^1(\mathbb{R})$ and, if $f \in C^1(\mathbb{R})$, f' is not necessarily the limit of f'_m (recall, e.g., the sequence $f_m(x) = \sin mx/\sqrt(m)$ which converges uniformly to $f = 0$).

But the deficiencies of the concept of the derivative are more evident in the *non-invertibility of the order of differentiation*: we will now point out a consequence of this. Suppose that a bidimensional physical phenomenon is modelled by the equation

$$\frac{\partial^2 u}{\partial x \, \partial y} = 0 \quad \text{in } \mathbb{R}^2, \tag{5.1}$$

where $u \equiv u(x, y): \mathbb{R}^2 \to \mathbb{R}$ and $\partial^2/\partial x \, \partial y$ is the operator defined, as usual, by the identity

$$\frac{\partial^2 v}{\partial x \, \partial y} \equiv \frac{\partial}{\partial x}\left(\frac{\partial v}{\partial y}\right) \tag{5.2}$$

(as long as the right-hand term makes sense); before going further, let us note that physically we can also consider, instead of (5.1), the equation

$$\frac{\partial^2 u}{\partial y \, \partial x} = 0 \quad \text{in } \mathbb{R}^2, \tag{5.3}$$

where $\partial^2 u/\partial y \, \partial x$ has a definition analogous to that of $\partial^2 u/\partial x \, \partial y$. Going back to (5.1), if we formulate the problem in the space of the functions $u \in C^0(\mathbb{R}^2)$ such that there exist $f, g \in C^0(\mathbb{R}^2)$ such that $f = \partial u/\partial y$ and $g = \partial f/\partial x$ (we will not bother to complete the problem...) we see that in whatever way $k \in C^0(\mathbb{R})$ and $h \in C^1(\mathbb{R})$ are taken, the function

$$\bar{u}(x, y) = k(x) + h(y) \tag{5.4}$$

is a solution of (5.1). However, $\bar{u}$ is not, in general, a solution of (5.3)!

We have then found the guide we needed: since we cannot avoid the concept of differentiation in treating differential problems, we need to modify it. *We need a 'differentiation operator' with a larger domain of definition, good continuity properties, which is independent of the order in which the 'derivative' is taken, and which, in order to be effectively useful, coincides with the usual derivative on the class of 'smooth' functions.*

We already have such an operator: the differentiation in the sense of distributions. Is then the space of distributions what we are looking for? The answer is only partly in the affirmative: this space is 'too large', and although this can be an advantage from the point of view of results of existence, it is undoubtedly a disadvantage for the study of questions regarding uniqueness and regularity. We need in fact certain subspaces of the space of distributions, and the aim of this chapter is precisely that of identifying and studying an important family of subspaces of $\mathcal{D}'(\Omega)$: the *Sobolev spaces* $W^{s,p}(\Omega)$ (often written as $L_p^s(\Omega)$, $W_p^s(\Omega)$, etc.).

Weak and strong derivatives

Let us remark that the Sobolev spaces were introduced in analysis before the introduction of the theory of distributions, from the concept of weak derivative (see Sobolev, 1963b; see also Sobolev, 1963a, p. 33 and Sobolev, 1961), which is also a particular case of the distributional derivative.

A function $f \in L^p(\Omega)$ is said to be the αth *weak derivative* of $u \in L^p(\Omega)$, in the sense of $L^p(\Omega)$, if $\int_\Omega f(x)\phi(x) \, dx = (-1)^{|\alpha|} \int_\Omega u(x) D^\alpha \phi(x) \, dx \; \forall \phi \in C_0^{|\alpha|}(\Omega)$ (cf. (4.66)). The weak derivative, if it exists, is unique. Connected to this concept of the weak derivative is that of the strong derivative: a function $g \in L^p(\Omega)$ is said to be the αth *strong derivative* of $u \in L^p(\Omega)$, in the sense of

$L^p(\Omega)$, if there exists a sequence u_m in $C^{|\alpha|}(\bar{\Omega})$ such that $u_m \to u$ in $L^p(\Omega)$ and $D^\alpha u_m \to g$ in $L^p(\Omega)$. If the strong derivative exists, then the weak derivative also exists and is identical to it (hence, the strong derivative, if it exists, is unique); conversely, under certain assumptions of regularity on Ω, one can show that, if the weak derivative exists, the strong derivative also exists (and is identical to it) (see Friedrichs, 1944; see also Meyers–Serrin, 1964)—we will take up this discussion again.

Other classes of spaces

There are many other families of spaces which are important in the study of differential problems closely connected to those considered here: *Bessel potentials* or *Lebesgue spaces, Besov spaces, weighted spaces, non-isotropic spaces*, etc. For all these spaces, see Volevich–Paneyakh, 1965; Magenes, 1964; Nikol'skiĭ, 1961; 1975; Sobolev–Nikol'skiĭ, 1970; Besov, 1964; Slobodeckiĭ, 1966; Uspenskiĭ, 1970a; 1970b; Il'in, 1969; Avantaggiati, 1976; Fučik–Kufner, 1980; Kufner–John–Fučík, 1977; Triebel, 1978 (in addition there is an extensive bibliography in Burenkov, 1968). For a more detailed study of Sobolev spaces, see, e.g. Gagliardo, 1958; 1959; Lions–Magenes, 1960; 1961a; 1961b; 1961c; 1962; 1963a; 1963b; Adams, 1975; Nečas, 1967; Lions, 1965 and, for the Hilbertain case, Lions–Magenes, 1972a and Treves, 1975.

For ease of presentation, the Sobolev spaces relative to $\mathbb{R}^n$ ($W^{s,p}(\mathbb{R}^n)$) and those relative to an arbitrary open set $\Omega \subset \mathbb{R}^n$ ($W^{s,p}(\Omega)$) will be introduced separately; further, for the same reason, we will define separately the cases 's a non-negative integer' ($s \in \mathbb{N}$), 's a positive real non-integer' ($s \in \mathbb{R}_+ \setminus \mathbb{N}$) and '$s$ strictly negative and real' ($s \in \mathbb{R}_- \setminus \{0\}$).

5.2 SPACES $W^{s,p}(\mathbb{R}^n)$

$W^{k,p}(\mathbb{R}^n)$, $k \in \mathbb{N}$

We first define the Sobolev spaces with s being a non-negative integer, or rather, adopting the traditional notation which attributes the letters i, j, k, l, m, n to the integers, the Sobolev spaces with k being a non-negative integer.

DEFINITION 5.1. Let $k \in \mathbb{N}$ and $p \in [1, +\infty]$. We denote by $W^{k,p}(\mathbb{R}^n)$ the vector space $\{u \in \mathscr{D}'(\mathbb{R}^n) : D^\alpha u \in L^p(\mathbb{R}^n), |\alpha| \leq k\}$ provided with the norm

$$\|u\|_{W^{k,p}(\mathbb{R}^n)} = \left[\sum_{|\alpha| \leq k} \|D^\alpha u\|^p_{L^p(\mathbb{R}^n)} \right]^{1/p}, \tag{5.5}$$

with the usual modification for $p = +\infty$.

We first remark that, with the vectorial structure induced by $\mathscr{D}'(\mathbb{R}^n)$, $W^{k,p}(\mathbb{R}^n)$ is a vector space on which (5.5) effectively defines a norm. If $k = 0$

we have obviously $W^{0,p}(\mathbb{R}^n) = L^p(\mathbb{R}^n)$ and in general we have $W^{k,p}(\mathbb{R}^n) \subset L^p(\mathbb{R}^n)$—it is then natural to ask whether the topology of $W^{k,p}(\mathbb{R}^n)$ has the same characteristics as those of $L^p(\mathbb{R}^n)$ (note also that the topology of $W^{k,p}(\mathbb{R}^n)$ is the least fine topology for which the applications $D^\alpha : W^{k,p}(\mathbb{R}^n) \to L^p(\mathbb{R}^n)$, $|\alpha| \leq k$, are continuous). The answer is in the affirmative and is given by:

THEOREM 5.1. *The spaces $W^{k,p}(\mathbb{R}^n)$, $k \in \mathbb{N}$ and $p \in [1, +\infty]$, are Banach spaces; if $p \in]1, +\infty[$ they are reflexive spaces.*

Proof. We will first show that the spaces $W^{k,p}(\mathbb{R}^n)$ are complete, i.e., if u_m is a Cauchy sequence in $W^{k,p}(\mathbb{R}^n)$ then there exists $u_0 \in W^{k,p}(\mathbb{R}^n)$ such that $u_m \to u_0$ in $W^{k,p}(\mathbb{R}^n)$. Now to say that u_m is a Cauchy sequence with respect to the norm (5.5) means that

$$\forall \varepsilon > 0 \quad \exists \bar{m} \in \mathbb{N} \quad \forall m, l \in \mathbb{N} \quad m > \bar{m} \Rightarrow \|u_m - u_{m+l}\|_{W^{k,p}(\mathbb{R}^n)}$$
$$= \left[\sum_{|\alpha| \leq k} \|D^\alpha(u_m - u_{m+l})\|_{L^p(\mathbb{R}^n)}^p \right]^{1/p} < \varepsilon, \quad (5.6)$$

and this implies that the sequences $D^\alpha u_m$, $|\alpha| \leq k$, are Cauchy sequences in $L^p(\mathbb{R}^n)$. Since $L^p(\mathbb{R}^n)$ is complete we can find, for every α, $|\alpha| \leq k$, a function $u_\alpha \in L^p(\mathbb{R}^n)$ such that $D^\alpha u_m \to u_\alpha$ in $L^p(\mathbb{R}^n)$. Putting $u_0 = u_{(0,\ldots,0)}$, to show that $W^{k,p}(\mathbb{R}^n)$ is complete it is sufficient to prove that $u_\alpha = D^\alpha u_0$, $|\alpha| \leq k$. Now since $u_m \to u_0$ in $L^p(\mathbb{R}^n)$ and $L^p(\mathbb{R}^n) \subsetneq \mathscr{D}'(\mathbb{R}^n)$ we conclude that $u_m \to u_0$ in $\mathscr{D}'(\mathbb{R}^n)$, and this, together with the continuity of the operators D^α in $\mathscr{D}'(\mathbb{R}^n)$, implies that $D^\alpha u_m \to D^\alpha u_0$ in $\mathscr{D}'(\mathbb{R}^n)$. In the same way we can show that $D^\alpha u_m \to u_\alpha$ in $\mathscr{D}'(\mathbb{R}^n)$, and since this is a separate space the limit is unique and therefore $u_\alpha = D^\alpha u_0$. For the proof of the reflexivity of $W^{k,p}(\mathbb{R}^n)$ when $p \in]1, +\infty[$ (note also that if $p = 1$ or $p = +\infty$ the spaces $L^p(\mathbb{R}^n)$ are themselves not reflexive ...) the reader is referred to the second part of the proof of theorem 5.3, where the more complicated result for spaces with fractional indices is proved. ∎

$W^{s,p}(\mathbb{R}^n)$, $s \in \mathbb{R}_+ \setminus \mathbb{N}$

We next consider the Sobolev spaces with s being a positive real non-integer:

DEFINITION 5.2. Let $s \in \mathbb{R}_+ \setminus \mathbb{N}$ and $p \in]1, +\infty[$. Putting $s = [s] + \sigma$, with $[s] = \max\{k \in \mathbb{N} : k \leq s\}$, and putting

$$d_{\sigma,\alpha}(u) = \frac{|D^\alpha u(x) - D^\alpha u(y)|}{\|x - y\|^{(n/p)+\sigma}}, \quad (5.7)$$

we denote by $W^{s,p}(\mathbb{R}^n)$ the vector space $\{u \in \mathscr{D}'(\mathbb{R}^n) : u \in W^{[s],p}(\mathbb{R}^n)$ and

$d_{\sigma,\alpha}(u) \in L^p(\mathbb{R}^n \times \mathbb{R}^n)$, $|\alpha| = [s]\}$ provided with the norm

$$\|u\|_{W^{s,p}(\mathbb{R}^n)} = \left[\|u\|^p_{W^{[s],p}(\mathbb{R}^n)} + \sum_{|\alpha|=[s]} \|d_{\sigma,\alpha}(u)\|^p_{L^p(\mathbb{R}^n \times \mathbb{R}^n)}\right]^{1/p}. \tag{5.8}$$

The norm of $d_{\sigma,\alpha}(u)$ in $L^p(\mathbb{R}^n \times \mathbb{R}^n)$, i.e., the pth root of the integral

$$\int_{\mathbb{R}^n} \int_{\mathbb{R}^n} |d_{\sigma,\alpha}(u)|^p \, dx \, dy, \tag{5.9}$$

is connected with the concept of *fractional derivative* and it is for this reason that the elements of $W^{s,p}(\mathbb{R}^n)$ are said to have derivatives of fractional order of summable pth power. We can also consider the spaces $W^{s,1}(\mathbb{R}^n)$ and $W^{s,\infty}(\mathbb{R}^n)$ with s not being an integer: the latter, with the natural interpretation of (5.8) for $p = +\infty$, can be identified with $C^{[s],\sigma}(\mathbb{R}^n)$ (in this context see Peetre, 1966, p. 286).

The closed graph theorem

Before proving a result analogous to theorem 5.1 for the spaces $W^{s,p}(\mathbb{R}^n)$, with s fractional, we will digress slightly and discuss the concept of closed operator:

DEFINITION 5.3. Let E and F be two topological spaces and $f: D \subset E \to F$ a function defined on a subset D of E. The set

$$G(f) = \{(x, f(x)) \in E \times F : x \in D\}, \tag{5.10}$$

will be called the *graph* of f and we will say that f is *closed* if $G(f)$ is a closed subset of $E \times F$ with the product topology.

Let us remark that this concept is of a different kind from that of the open function. We say that $f: D \subset E \to F$ is *open* if it transforms open sets of D (with the topology induced by E) in open sets of F (the two concepts are however closely connected; see, e.g., Husain, 1965, p. 58).

The most interesting case of the definition above is that in which E and F are t.v.s., D is a vector subspace of E and f a linear operator (in such conditions $G(f)$ is then a vector subspace of $E \times F$): in this case there are interesting connections between the properties 'f is closed' and 'f is continuous'. We will first show a basic result, in this direction, in the particular case in which E and F are Banach spaces:

THEOREM 5.2 (Closed graph theorem). Let E and F be two Banach spaces (norm $\|.\|_E$ and $\|.\|_F$ respectively) and $f: E \to F$ a linear operator. The propositions

(i) f is continuous
(ii) f is closed

are equivalent.

Proof. We first show that (i) $\Rightarrow$ (ii). It is sufficient to see that, since E and F are normed spaces, the fact that f is closed can be described in terms of the convergence of sequences: f is closed iff for every sequence u_m of elements of E the conditions $u_m \xrightarrow{E} u_0$ and $f(u_m) \xrightarrow{F} v_0$ imply that $v_0 = f(u_0)$—which is a necessary condition for the continuity of f. This part of the theorem is true in much more general conditions: it is in fact sufficient that E and F be two t.v.s. and F be separate—the proof is the same, taking nets or filters instead of sequences (see, e.g. Edwards, 1965, p. 68; for another proof see Husain, 1965, p. 34).

We next show that (ii) $\Rightarrow$ (i). In order to do this, we introduce in E a second norm, the so-called *graph norm*: for every u in E we put

$$\|\|u\|\|_E = \|u\|_E + \|f(u)\|_F. \tag{5.11}$$

If we can prove that

$$\exists M > 0 \quad \forall u \in E \quad \|\|u\|\|_E \leq M \|u\|_E \tag{5.12}$$

then we have completed the proof, since from (5.11) and (5.12) it follows that $\|f(u)\|_F \leq (M-1)\|u\|_E$ and hence that f is continuous. Now from (5.11) we always have

$$\forall u \in E \quad \|u\|_E \leq \|\|u\|\|_E, \tag{5.13}$$

and hence what we have to show is that the two norms $\|.\|_E$ and $\|\|.\|\|_E$ are equivalent. For this purpose we will use a theorem due to Banach which states that *two Banach norms on a vector space such that one is greater than the other, are necessarily equivalent* (see, e.g. Cotlar–Cignoli, 1974, p. 421). In order to satisfy the condition of this theorem, it is sufficient to show that $\|\|.\|\|_E$ is a Banach norm, i.e. that E with this norm—a space which we will denote, to avoid confusions, by E_*—is a Banach space. Let then u_m be a Cauchy sequence in E_*, i.e. such that

$$\forall \varepsilon > 0 \quad \exists \bar{m} \in \mathbb{N} \quad \forall m, l \in \mathbb{N}$$

$$m > \bar{m} \Rightarrow \|\|u_m - u_{m+l}\|\|_E = \|u_m - u_{m+l}\|_E + \|f(u_m) - f(u_{m+l})\|_F < \varepsilon, \tag{5.14}$$

and let us show that there exists $u_0 \in E$ such that $u_m \xrightarrow{E_*} u_0$. Now, from (5.14), u_m and $f(u_m)$ are Cauchy sequences in E and F, respectively, and since these are Banach spaces there exist $u \in E$ and $v \in F$ such that $u_m \xrightarrow{E} u$ and $f(u_m) \xrightarrow{F} v$: since f is closed $v = f(u)$ and hence $u_m \xrightarrow{E_*} u = u_0$. Also this second part of the theorem is still true in much more general conditions on E and F (see e.g. Edwards, 1965, ch. 6; Husain, 1965, ch. 3). ∎

Again in the context of structures and operators of the linear kind, the following concept, which is more general than that of a closed operator, is of interest: $f : D \subset E \to F$ is said to be a *linear preclosed operator* if $\overline{G(f)}$ is the graph of a linear operator, $\bar{f}$, which we will call the *strong closure* or *Friedrich's closure* of f.

Going from a linear operator f to its closure is in some cases an important way of generalization (comparable to and connected with transposition—the concepts of strong and weak derivatives, e.g., can be viewed in this context). Finally, let us remark that there are connections between measure theory and the closed graph theorem: see, L. Schwartz, 1973, p. 160.

More on $W^{s,p}(\mathbb{R}^n)$, $s \in \mathbb{R}_+ \setminus \mathbb{N}$

We now return to the study of Sobolev spaces, proving the following

THEOREM 5.3. $W^{s,p}(\mathbb{R}^n)$, $s \in \mathbb{R}_+ \setminus \mathbb{N}$ and $p \in]1, +\infty[$, are reflexive Banach spaces.

Proof. The operator

$$\delta^{s,p} : L^p(\mathbb{R}^n) \to [\mathscr{D}'(\mathbb{R}^n)]^{[s]^{\leq}-1} \times [\mathscr{D}'(\mathbb{R}^n \times \mathbb{R}^n - \Delta)]^{[s]^{=}}, \qquad (5.15)$$

where $\Delta = \{(x, x) \in \mathbb{R}^n \times \mathbb{R}^n : x \in \mathbb{R}^n\}$, $[s]^{\leq} = \text{card}\{\alpha \in \mathbb{N}^n : |\alpha| \leq [s]\}$ and $[s]^{=} = \text{card}\{\alpha \in \mathbb{N}^n : |\alpha| = [s]\}$ (if $[s]^{=} = 0$ put $\delta^{\sigma,p} : L^p(\mathbb{R}^n) \to \mathscr{D}'(\mathbb{R}^n \times \mathbb{R}^n - \Delta)$), which with every $u \in L^p(\mathbb{R}^n)$ associates $\delta^{s,p}(u) = (D^{(1,0,\ldots,0)}u, \ldots, D^\alpha u, \ldots, d_{\sigma,\alpha}(u), \ldots)$ is linear and continuous and therefore closed (note that the value space is obviously separate). Putting

$$D^{s,p} = \{u \in L^p(\mathbb{R}^n) : D^\alpha u \in L^p(\mathbb{R}^n), |\alpha| \leq [s]$$
$$\text{and } d_{\sigma,\alpha}(u) \in L^p(\mathbb{R}^n \times \mathbb{R}^n), |\alpha^2| = [s]\} \quad (5.16)$$

we can define, starting from $\delta^{s,p}$, the operator

$$d^{s,p} : D^{s,p} \to [L^p(\mathbb{R}^n)]^{[s]^{\leq}-1} \times [L^p(\mathbb{R}^n \times \mathbb{R}^n)]^{[s]^{=}}, \qquad (5.17)$$

also linear and closed as can be easily seen; it can also be immediately seen that $D^{s,p}$ provided with the graph norm associated with $d^{s,p}$ is a complete space, which coincides with $W^{s,p}(\mathbb{R}^n)$. Thus, we have proved that $W^{s,p}(\mathbb{R}^n)$ is a Banach space. The space $W^{s,p}(\mathbb{R}^n)$ can be considered as the graph $G(d^{s,p})$ with the topology induced by $[L^p(\mathbb{R}^n)]^{[s]^{\leq}} \times [L^p(\mathbb{R}^n \times \mathbb{R}^n)]^{[s]^{=}}$: we are then dealing with a closed subspace of a reflexive space, which is therefore reflexive (see, e.g., Taylor, 1958, p. 192; Dunford–Schwartz, 1958, p. 67). ∎

$W^{s,p}(\mathbb{R}^n)$, $s \in \mathbb{R}_- \setminus \{0\}$

We will next study the Sobolev spaces with s being real and strictly negative.

DEFINITION 5.4. Let $s \in \mathbb{R}_- \setminus \{0\}$ and $p \in]1, +\infty[$. We denote by $W^{s,p}(\mathbb{R}^n)$ the dual space of the space $W^{-s,p^*}(\mathbb{R}^n)$, where $p^* = p/(p-1)$.

For these spaces the proof of the theorem analogous to theorems 5.1 and 5.3 follow immediately: the duals of reflexive Banach spaces are reflexive

Banach spaces. However, the definition 5.4 raises a new problem: that of characterizing the elements of the spaces defined. In fact, if for $s \geq 0$ are definitions 5.1 and 5.2 themselves which tell us that we are dealing with distributions, this is not immediately true for $s < 0$. The following theorem, which shows that $W^{s,p}(\mathbb{R}^n)$, $s \geq 0$ and $p \in]1, +\infty[$, is a normal space of distributions, and hence that its dual is also a space of distributions, solves this problem:

THEOREM 5.4. *If s is a non-negative real number and $p \in]1, +\infty[$ then $\mathcal{D}(\mathbb{R}^n)$ is dense in $W^{s,p}(\mathbb{R}^n)$.*

Proof. Since $W^{s,p}(\mathbb{R}^n)$ is a normal space, to show that $\mathcal{D}(\mathbb{R}^n) \subset^{ds} W^{s,p}(\mathbb{R}^n)$ we need only to show that $\mathcal{D}(\mathbb{R}^n)$ is sequentially dense in $W^{s,p}(\mathbb{R}^n)$, i.e. that every element $u \in W^{s,p}(\mathbb{R}^n)$ can be approximated, in the topology of this space, by elements of $\mathcal{D}(\mathbb{R}^n)$.

In the proof we will use the *technique of truncation and regularization*, which is essentially a process for constructing approximating sequences:

(1) *Truncation*. Let us take $u \in W^{s,p}(\mathbb{R}^n)$ and, for every real number $r > 0$, let us consider the ball $B(0, r) = \{x \in \mathbb{R}^n : \|x\| < r\}$ and a function $\phi_r \in \mathcal{D}(\mathbb{R}^n)$ such that $0 \leq \phi_r(x) \leq 1$, $\phi_r(x) = 1$ if $x \in B(0, r)$ and $\phi_r(x) = 0$ if $x \in \mathbb{R}^n \setminus B(0, 2r)$: the product $u_r = u\phi_r \in W^{s,p}(\mathbb{R}^n)$ will be called a truncation of u. Let us remark that simply restricting u to $B(0, r)$ and then extending the restriction to zero outside $B(0, r)$ (i.e. considering the function $\bar{u}_r = u\chi_{B(0,r)}$ where $\chi_{B(0,r)}$ is the characteristic function of $B(0, r)$) does not maintain the regularity of u, which is on the other hand accomplished by the operation of truncation defined above.

The family u_r converges to u, when $r \to +\infty$, in the norm of $W^{s,p}(\mathbb{R}^n)$ since

$$\|u - u_r\|_{W^{s,p}(\mathbb{R}^n)} = \|u - u\phi_r\|_{W^{s,p}(\mathbb{R}^n)} = \|u(1 - \phi_r)\|_{W^{s,p}(\mathbb{R}^n)} \tag{5.18}$$

tends to zero as $r \to +\infty$ (since, for $\|x\| < r$, $1 - \phi_r \equiv 0$). We have thus proved that the set of functions of $W^{s,p}(\mathbb{R}^n)$ whose support is compact is dense in $W^{s,p}(\mathbb{R}^n)$. All we need to show now is that every function of $W^{s,p}(\mathbb{R}^n)$ with compact support can be approximated by functions of $\mathcal{D}(\mathbb{R}^n)$. For this we use:

(2) *Regularization*. Let now $u \in W^{s,p}(\mathbb{R}^n)$ be such that, for a fixed r, supp $u \subset B(0, r)$. Let u_ε be the sequence of ε-regularizates of u in the sense of definition 15.1: the proof of theorem 15.2 can be automatically adapted to the proof of the fact that $u_\varepsilon \to u$ in $W^{s,p}(\mathbb{R}^n)$. ∎

This proof is valid for the space $W^{k,1}(\mathbb{R}^n)$, $k \in \mathbb{N}$, while for the space $W^{k,\infty}(\mathbb{R}^n)$ the theorem is false (let us remember that $C^0(\mathbb{R}^n)$ is not dense in $L^\infty(\mathbb{R}^n) = W^{0,\infty}(\mathbb{R}^n)$...). This theorem shows that we can define the space $W^{s,p}(\mathbb{R}^n)$, $s \in [0, +\infty[$ and $p \in]1, +\infty[$, as the *completion* of $\mathcal{D}(\mathbb{R}^n)$ with respect to the norm (5.5) or to the norm (5.8), depending on whether $s \in \mathbb{N}$ or not (this generalizes the so-called definition by abstract completion of the spaces $L^p(\mathbb{R}^n)$).

The following theorem characterizes those distributions which are elements of $(W^{k,p}(\mathbb{R}^n))'$, $k \in \mathbb{N}$ and $p \in]1, +\infty[$, i.e. characterizes, in the particular case of an integer index, the elements of the spaces introduced in an abstract manner by means of the definition 5.4:

THEOREM 5.5. A distribution T belongs to $(W^{k,p}(\mathbb{R}^n))' = W^{-k,p^*}(\mathbb{R}^n)$, $k \in \mathbb{N}$ and $p \in]1, +\infty[$, iff T can be written as

$$T = \sum_{|\alpha| \leq k} D^\alpha g_\alpha \tag{5.19}$$

with $g_\alpha \in L^{p^*}(\mathbb{R}^n)$.

Proof. We will first show that if $T \in \mathscr{D}'(\mathbb{R}^n)$ is of the form (5.19) then $T \in W^{-k,p^*}(\mathbb{R}^n)$. By the definition of equality in $\mathscr{D}'(\mathbb{R}^n)$

$$\forall \phi \in \mathscr{D}(\mathbb{R}^n) \langle T, \phi \rangle = \sum_{|\alpha| \leq k} \langle D^\alpha g_\alpha, \phi \rangle = \sum_{|\alpha| \leq k} (-1)^{|\alpha|} \langle g_\alpha, D^\alpha \phi \rangle$$

$$= \sum_{|\alpha| \leq k} (-1)^{|\alpha|} \int_{\mathbb{R}^n} g_\alpha(x) D^\alpha \phi(x) \, dx, \tag{5.20}$$

and from this together with the Schwarz–Hölder inequality it follows that

$$\forall \phi \in \mathscr{D}(\mathbb{R}^n) \, |\langle T, \phi \rangle| \leq \sum_{|\alpha| \leq k} \|g_\alpha\|_{L^{p^*}(\mathbb{R}^n)} \|D^\alpha \phi\|_{L^p(\mathbb{R}^n)} \leq c \|\phi\|_{W^{k,p}(\mathbb{R}^n)}, \tag{5.21}$$

where c is a constant independent of ϕ. From the density of $\mathscr{D}(\mathbb{R}^n)$ in $W^{k,p}(\mathbb{R}^n)$ it follows that T is a continuous linear functional on $W^{k,p}(\mathbb{R}^n)$.

Conversely, we will show that if $T \in (W^{k,p}(\mathbb{R}^n))'$ then T can be written in the form (5.19), which in particular shows that T is a distribution of finite order $\leq k$ (the decomposition (5.19) being not generally unique). We saw in the proof of theorem 5.3 that $W^{k,p}(\mathbb{R}^n)$ can be considered as the graph $G(d^{k,p})$ of the operator $d^{k,p} : D^{k,p} \to [L^p(\mathbb{R}^n)]^{k^\leq - 1}$. On $G(d^{k,p})$ we define the functional L by the formula.

$$L((v, d^{k,p}(v))) = \langle T, v \rangle \quad \forall v \in D^{k,p}. \tag{5.22}$$

This functional is obviously linear, and from the continuity of T in $W^{k,p}(\mathbb{R}^n)$ it follows that is also continuous: from the Hahn–Banach theorem it then follows that we can extend it as a continuous linear functional on $[L^p(\mathbb{R}^n)]^{k^\leq}$. Given that the dual of $[L^p(\mathbb{R}^n)]^{k^\leq}$ can be identified with $[L^{p^*}(\mathbb{R}^n)]^{k^\leq}$, there exists $k^\leq$ functions $\bar{g}_\alpha \in L^{p^*}(\mathbb{R}^n)$ such that for every $v \in W^{k,p}(\mathbb{R}^n)$

$$\langle T, v \rangle = L((v, d^{k,p}(v))) = \int_{\mathbb{R}^n} \sum_{|\alpha| \leq k} \bar{g}_\alpha(x) D^\alpha v(x) \, dx, \tag{5.23}$$

and hence in particular for every $\phi \in \mathscr{D}(\mathbb{R}^n) \subset W^{k,p}(\mathbb{R}^n)$. From (5.23) we have then (5.19) with $g_\alpha = (-1)^{|\alpha|} \bar{g}_\alpha$. ∎

The Hilbertian case of Sobolev spaces

If $p = 2$ the spaces $W^{s,2}(\mathbb{R}^n)$, $s \in \mathbb{R}$, have a particular structure, reflection of the particular structure of $L^2(\mathbb{R}^n)$:

THEOREM 5.6. The spaces $W^{s,2}(\mathbb{R}^n)$, $s \in \mathbb{R}$, are Hilbert spaces: with the scalar product

$$(u, v)_{W^{k,2}(\mathbb{R}^n)} = \sum_{|\alpha| \leq k} (D^\alpha u, D^\alpha v)_{L^2(\mathbb{R}^n)}$$
$$= \sum_{|\alpha| \leq k} \int_{\mathbb{R}^n} D^\alpha u(x) \overline{D^\alpha v(x)} \, dx \qquad (5.24)$$

if $s = k \in \mathbb{N}$; with the scalar product

$$(u, v)_{W^{s,2}(\mathbb{R}^n)} = (u, v)_{W^{[s],2}(\mathbb{R}^n)} + \sum_{|\alpha| = [s]} (d_{\sigma, \alpha}(u), d_{\sigma, \alpha}(v))_{L^2(\mathbb{R}^n \times \mathbb{R}^n)} \qquad (5.25)$$

if $s \in \mathbb{R}_+ \setminus \mathbb{N}$; and with the scalar product associated with the duality if $s \in \mathbb{R}_- \setminus \{0\}$.

Proof. If $s \geq 0$, it is enough to note that the norms (5.5) and (5.8) satisfy the *parallelogram law*

$$\|u + v\|^2_{W^{s,2}(\mathbb{R}^n)} + \|u - v\|^2_{W^{s,2}(\mathbb{R}^n)} = 2(\|u\|^2_{W^{s,2}(\mathbb{R}^n)} + \|v\|^2_{W^{s,2}(\mathbb{R}^n)}) \qquad (5.26)$$

and that the products (5.24) and (5.25) satisfy

$$(u, v)_{W^{s,2}(\mathbb{R}^n)} = \tfrac{1}{4}(\|u + v\|^2_{W^{s,2}(\mathbb{R}^n)} - \|u - v\|^2_{W^{s,2}(\mathbb{R}^n)})$$
$$+ \frac{i}{4}(\|u + v\|^2_{W^{s,2}(\mathbb{R}^n)} - \|u - v\|^2_{W^{s,2}(\mathbb{R}^n)}) \qquad (5.27)$$

(see, e.g., Yosida, 1971, p. 39). If $s < 0$ it is enough to note that the dual of a Hilbert space is a Hilbert space. ∎

The functions of $W^{k,2}(\mathbb{R}^n)$, $k \in \mathbb{N}$, are said to be functions with *finite Dirichlet integral* since the norm (5.5) can be decomposed into a sum of k integrals of this kind, namely

$$\|u\|^2_{W^{k,2}(\mathbb{R}^n)} = \sum_{i=1}^{k} \sum_{|\alpha| = i} \int_{\mathbb{R}^n} D^\alpha u(x) \overline{D^\alpha u(x)} \, dx \qquad (5.28)$$

(see, e.g., Aronszajn–Smith, 1961, p. 401).

Spaces $H^{s,p}(\mathbb{R}^n)$

We will now present an interesting characterization of the spaces $W^{s,2}(\mathbb{R}^n)$, $s \geq 0$, based on the Fourier transform. We will first introduce

DEFINITION 5.5. Let $s \in \mathbb{R}$ and $p \in]1, +\infty[$. We will denote by $H^{s,p}(\mathbb{R}^n)$ the space of temperate distributions such that $\mathscr{F}^{-1}[(1+\|y\|^2)^{s/2}\mathscr{F}u] \in L^p(\mathbb{R}^n)$, provided with the norm

$$\|u\|_{H^{s,p}(\mathbb{R}^n)} = \|\mathscr{F}^{-1}[(1+\|y\|^2)^{s/2}\mathscr{F}u]\|_{L^p(\mathbb{R}^n)}. \tag{5.29}$$

These spaces are the so-called *Bessel potentials* (see Aronszajn–Smith, 1961; Adams–Aronszajn–Smith, 1967). They are reflexive Banach spaces, and if $p=2$ they are moreover Hilbert spaces.

The object of the following theorem is that of showing that the spaces $W^{s,2}(\mathbb{R}^n)$ and $H^{s,2}(\mathbb{R}^n)$, $s \geq 0$, are isomorphic (let us remark that if $p=2$ we can eliminate $\mathscr{F}^{-1}$ in the definition since $\mathscr{F}$ is an isometry in $L^2(\mathbb{R}^n)$):

THEOREM 5.7. *Let* $s \geq 0$; $u \in W^{s,2}(\mathbb{R}^n)$ *iff* $(1+\|y\|^2)^{s/2}\hat{v}(y) \in L^2(\mathbb{R}^n)$. *Further the norm*

$$\|u\|_{H^{s,2}(\mathbb{R}^n)} = \|(1+\|y\|^2)^{s/2}\hat{u}(y)\|_{L^2(\mathbb{R}^n)} \tag{5.30}$$

is equivalent to the norm (5.5) *(for* $p=2$) *or to the norm* (5.8) *(for* $p=2$) *depending on whether* $s \in \mathbb{N}$ *or not.*

Proof. We will first take the case $s = k \in \mathbb{N}$. If $u \in W^{k,2}(\mathbb{R}^n)$ then $D^\alpha u \in L^2(\mathbb{R}^n)$, $|\alpha| \leq k$, and therefore, by applying (4.84) iteratively:

$$\prod_{j=1}^n (2\pi i)^{\alpha_j} y_j^{\alpha_j} \hat{u}(y) \in L^2(\mathbb{R}^n) \tag{5.31}$$

and thus, putting $\prod_{j=1}^n x_j^{\alpha_j} = x^\alpha$,

$$\sum_{|\alpha| \leq k} (2\pi)^{2\alpha} y^{2\alpha} |\hat{u}(y)|^2 \in L^1(\mathbb{R}^n). \tag{5.32}$$

Keeping in mind the first part of the double inequality

$$(1+\|y\|^2)^k \leq \sum_{|\alpha| \leq k} y^{2\alpha} \leq c(1+\|y\|^2)^k, \tag{5.33}$$

where c is an appropriate constant, it follows that

$$\int_{\mathbb{R}^n} (1+\|y\|^2)^k |\hat{u}(y)|^2 \, dy \leq \int_{\mathbb{R}^n} \sum_{|\alpha| \leq k} (2\pi)^{2\alpha} y^{2\alpha} |\hat{u}(y)|^2 \, dy < +\infty \tag{5.34}$$

(remember that $(2\pi)^{2\alpha} \geq 1$), which means that $u \in H^{k,2}(\mathbb{R}^n)$ and shows that $\|u\|_{H^{k,2}(\mathbb{R}^n)} \leq \|u\|_{W^{k,2}(\mathbb{R}^n)}$. The converse is proved in an analogous manner. Let now $s \in \mathbb{R}_+ \setminus \mathbb{N}$: we need only consider the case $0 < \sigma = s < 1$. We will see that if $u \in W^{\sigma,2}(\mathbb{R}^n)$ then $u \in H^{\sigma,2}(\mathbb{R}^n)$. From the second part of the double inequality

$$c_1(1+\|y\|^{2\sigma}) \leq (1+\|y\|^2)^\sigma \leq c_2(1+\|y\|^{2\sigma}), \tag{5.35}$$

where c_1 and c_2 are appropriate constants, it follows that

$$\int_{\mathbb{R}^n} (1+\|y\|^2)^\sigma |\hat{u}(y)|^2 \, dy \leq c_2 \left(\int_{\mathbb{R}^n} |\hat{u}(y)|^2 \, dy + \int_{\mathbb{R}^n} \|y\|^{2\sigma} |\hat{u}(y)|^2 \, dy \right) \tag{5.36}$$

and, since by Plancherel's formula we can substitute $\hat{u}$ by u in the first term of the right-hand side, we need only to show (consider (5.8) with $[s]=0$ and $p=2$) that there exists c such that

$$\int_{\mathbb{R}^n}\int_{\mathbb{R}^n} |d_{\sigma,0}(u)|^2 \, dx \, dy = c\int_{\mathbb{R}^n} \|y\|^{2\sigma} |\hat{u}(y)|^2 \, dy \tag{5.37}$$

or, putting $y = x + z$ for convenience, that

$$\int_{\mathbb{R}^n}\int_{\mathbb{R}^n} \frac{|u(y)-u(x)|^2}{\|x-y\|^{n+2\sigma}} \, dy \, dx = \int_{\mathbb{R}^n}\int_{\mathbb{R}^n} \frac{|u(x+z)-u(x)|^2}{\|z\|^{n+2\sigma}} \, dx \, dz$$

$$= c\int_{\mathbb{R}^n} \|y\|^{2\sigma} |\hat{u}(y)|^2 \, dy. \tag{5.38}$$

From the formula for the inversion of the Fourier transform

$$u(x) = \int_{\mathbb{R}^n} e^{2\pi i(x,y)} \hat{u}(y) \, dy \tag{5.39}$$

and from (4.86) we have

$$u(x+z) = \tau_{-z} u(x) = \int_{\mathbb{R}^n} e^{2\pi i(x,y)} e^{2\pi i(z,y)} \hat{u}(y) \, dy, \tag{5.40}$$

and hence

$$u(x+z) - u(x) = \int_{\mathbb{R}^n} e^{2\pi i(x,y)} (e^{2\pi i(z,y)} - 1) \hat{u}(y) \, dy. \tag{5.41}$$

From (5.41) and (4.52) we have

$$\int_{\mathbb{R}^n} |u(x+z)-u(x)|^2 \, dx = \int_{\mathbb{R}^n} |\hat{u}(y)|^2 |e^{2\pi i(z,y)} - 1|^2 \, dy \tag{5.42}$$

and we can then write successively

$$\int_{\mathbb{R}^n}\int_{\mathbb{R}^n} \frac{|u(x+z)-u(x)|^2}{\|z\|^{n+2\sigma}} \, dx \, dz$$

$$= \int_{\mathbb{R}^n} \frac{1}{\|z\|^{n+2\sigma}} \left(\int_{\mathbb{R}^n} |u(x+z)-u(x)|^2 \, dx\right) dz$$

$$= \int_{\mathbb{R}^n} \frac{1}{\|z\|^{n+2\sigma}} \left(\int_{\mathbb{R}^n} |\hat{u}(y)|^2 |e^{2\pi i(z,y)} - 1|^2 \, dy\right) dz$$

$$= \int_{\mathbb{R}^n}\int_{\mathbb{R}^n} \frac{|\hat{u}(y)|^2 |e^{2\pi i(z,y)} - 1|^2}{\|z\|^{n+2\sigma}} \, dy \, dz$$

$$= \int_{\mathbb{R}^n}\int_{\mathbb{R}^n} \frac{|\hat{u}(y)|^2 |e^{2\pi i(x,y)} - 1|^2 \|y\|^{2\sigma}}{\|z\|^{2\sigma+n} \|y\|^{2\sigma}} \, dy \, dz$$

$$= \int_{\mathbb{R}^n} |\hat{u}(y)|^2 \|y\|^{2\sigma} \left(\int_{\mathbb{R}^n} \frac{|e^{2\pi i(z,y)} - 1|^2}{\|z\|^{n+2\sigma} \|y\|^{2\sigma}} \, dz\right) dy$$

$$= \int_{\mathbb{R}^n} |\hat{u}(y)|^2 \|y\|^{2\sigma} c(y) \, dy, \tag{5.43}$$

where we have put

$$c(y) = \int_{\mathbb{R}^n} \frac{|e^{2\pi i (z, y)} - 1|^2}{\|z\|^{n+2\sigma} \|y\|^{2\sigma}} \, dz. \tag{5.44}$$

Substituting

$$t = z \|y\| \tag{5.45}$$

in (5.44) we obtain (the modulus of the Jacobian of the transformation (5.45) being $\|y\|^{-n}$)

$$c(y) = \int_{\mathbb{R}^n} \frac{|e^{2\pi i (t/\|y\|, y)} - 1|^2}{\|t\|^{n+2\sigma}} \, dt \tag{5.46}$$

which, as can easily be seen, does not depend on y: we have therefore proved (5.37) with $c = c(y)$. We can thus say that $u \in H^{\sigma,2}(\mathbb{R}^n)$ and that the norm $\|u\|_{H^{\sigma,2}(\mathbb{R}^n)}$ is less than the norm $\|u\|_{W^{\sigma,2}(\mathbb{R}^n)}$. The converse can be shown in a completely analogous manner. ∎

This result is true also for $s < 0$ and we will henceforth indicate the spaces $W^{s,2}(\mathbb{R}^n)$ and $H^{s,2}(\mathbb{R}^n)$, $s \in \mathbb{R}$, simply as $H^s(\mathbb{R}^n)$. More generally one can show, using the Mihlin theorem regarding the multipliers in the space $\mathscr{F}(L^p(\mathbb{R}^n))$ (see e.g., Bergh–Löfström, 1976, p. 135; Hörmander, 1960, p. 120; Edwards–Gaudry, 1977, p. 152) that $W^{k,p}(\mathbb{R}^n)$ is isomorphic to $H^{k,p}(\mathbb{R}^n)$ for every $p \in]1, +\infty[$ and $k \in \mathbb{Z}$. On the other hand one can show that for every $s \in \mathbb{R}$ and every $p \in]1, +\infty[$

$$\forall \varepsilon > 0 \qquad H^{s+\varepsilon,p}(\mathbb{R}^n) \subsetneq W^{s,p}(\mathbb{R}^n) \subsetneq H^{s-\varepsilon,p}(\mathbb{R}^n) \tag{5.47}$$

(see Lions–Magenes, 1961b, p. 66), which means that *the class of Sobolev spaces is contiguous* (in Gagliardo's sense, see Gagliardo, 1961; 1963) *to the class of Bessel potentials*. The function $\pi(y) = (1 + \|y\|^2)^s$ present in (5.29) is the simplest case of a *weight function*: taking other weight functions (in well defined classes, see e.g. Hörmander, 1964, p. 34; Volevich–Paneyakh, 1965, p. 7) one obtains the so-called weighted spaces.

The Sobolev imdedding theorem

The characterization of $W^{s,2}(\mathbb{R}^n)$ from theorem 5.7 gives us an interesting result: if $s > n/2$ then $(1 + \|y\|^2)^{-s/2} \in L^2(\mathbb{R}^n)$ and therefore, for every $f \in W^{s,2}(\mathbb{R}^n) = H^s(\mathbb{R}^n)$, $\hat{f}(y) = (1 + \|y\|^2)^{-s/2}(1 + \|y\|^2)^{s/2}\hat{f}(y) \in L^1(\mathbb{R}^n)$, which implies that $f \in C^0(\mathbb{R}^n)$ (in fact if $f \in L^1(\mathbb{R}^n)$ then $\hat{f} \in C^0(\mathbb{R}^n)$—see, e.g. Rudin, 1973, p. 169). We have then shown that $W^{s,2}(\mathbb{R}^n) \subset C^0(\mathbb{R}^n)$ if $s > n/2$.

This is a particular case of a more general result, the so-called imbedding theorem of Sobolev, which we will soon discuss. Before that, however, we wish to briefly analyse the meaning of the inclusion $W^{s,2}(\mathbb{R}^n) \subset C^0(\mathbb{R}^n)$. If $f \in W^{s,2}(\mathbb{R}^n)$, $s \geq 0$, then f is a function defined everywhere except on a set of zero measure, or more precisely, f is the representative of a class $[f]$ of functions which are equal a.e.: hence $f \in C^0(\mathbb{R}^n)$ must be interpreted as meaning that there exists $\tilde{f} \in [f]$ such that $\tilde{f} \in C^0(\mathbb{R}^n)$. On the other hand this inclusion shows that the weak regularity ($f \in W^{s,2}(\mathbb{R}^n)$) and the classical regularity ($f \in C^0(\mathbb{R}^n)$) are much more related than one would think *a priori*: the increase in the 'quantity' of regularity is accompanied by an improvement in the 'quality' of the regularity itself.

THEOREM 5.8 (Sobolev imbedding theorem). *If $k \geq 1$ is an integer and $p \in [1, +\infty[$ the following inclusions hold:*

(I) *If $n > kp$ then $W^{k,p}(\mathbb{R}^n) \hookrightarrow L^q(\mathbb{R}^n)$ with $q = np/(n-kp)$; more generally, if $m \in \mathbb{N}$, $m \leq k$ and $n > (k-m)p$ then $W^{k,p}(\mathbb{R}^n) \hookrightarrow W^{m,q}(\mathbb{R}^n)$ with $q = np/(n-(k-m)p)$;*

(II) *If $n = kp$ then $W^{k,p}(\mathbb{R}^n) \hookrightarrow L^q_{\text{loc}}(\mathbb{R}^n)$ for every $q \in [1, +\infty[$;*

(III) *If $n < kp$ then $W^{k,p}(\mathbb{R}^n) \hookrightarrow C^{0,\mu}(\mathbb{R}^n)$ with $\mu = k - (n/p)$ if $k - (n/p) < 1$, μ arbitrarly with $\mu < 1$ if $k - (n/p) = 1$, and $\mu = 1$ if $k - (n/p) > 1$; more generally, if $m \in \mathbb{N}$, $m \leq k$ and $n < (k-m)p$ then $W^{k,p}(\mathbb{R}^n) \hookrightarrow C^{m,\mu}(\mathbb{R}^n)$ with $\mu = k - m - (n/p)$ if $k - m - (n/p) < 1$, μ arbitrary with $\mu < 1$ if $k - m - (n/p) = 1$ and $\mu = 1$ if $k - m - (n/p) > 1$.*

The proof of this result is rather long and presents some technical difficulties: see Nečas, 1967, p. 72; Gagliardo, 1958, pp. 120 and 136; Lions, 1965, p. 37; Adams, 1975, p. 97 (in this last book more general results concerning immersion of spaces of the type $W^{s,p}(\mathbb{R}^n)$ in spaces of the type $W^{t,q}(\mathbb{R}^k)$, $k \leq n$, are considered).

The theorem holds even when k and m are arbitrary positive real numbers (except for the second part of III, which will then be meaningless, unless $m \in \mathbb{N}$): see Lions–Magenes, 1961b, p. 61; Adams, 1975, p. 217 and Peetre, 1966, pp. 300 *et seq.*

Together with the above-mentioned inclusions, we can consider those in which one of the indices, k or p, is fixed: we have thus, obviously, $W^{k,p}(\mathbb{R}^n) \hookrightarrow W^{m,p}(\mathbb{R}^n)$, $k, m \in \mathbb{N}$, $m \leq k$ and $p \in [1, +\infty[$ (this is true even if k and m are any real numbers—see Nečas, 1967, p. 97 and, for the Hilbertian case, Lions–Magenes, 1972a, p. 35); instead, the inclusion $W^{k,p}(\mathbb{R}^n) \subset W^{k,q}(\mathbb{R}^n)$, $p \neq q$, is always false, which is natural since the inclusion $L^p(\mathbb{R}^n) \subset L^q(\mathbb{R}^n)$, $p \neq q$, is false (note that $\mathbb{R}^n$ does not have finite measure ...).

The one-dimensional case

Theorem 5.8 is optimal in the sense that for k, p, and n arbitrary, one cannot improve the inclusions by substituting the spaces on the right side with 'smaller' spaces (for counter-examples which demonstrate the truth of this statement see Adams, 1975, p. 117); this is, however, possible in some particular cases: thus if $p = 1$ part II can be written as $W^{n,1}(\mathbb{R}^n) \hookrightarrow C^0(\mathbb{R}^n)$, and if $p = n = 1$ we can even prove the following

THEOREM 5.9. $W^{1,1}(\mathbb{R}) \subset AC_{\text{loc}}(\mathbb{R})$, *i.e., if $u \in W^{1,1}(\mathbb{R})$ then u is absolutely* continuous in every compact $K \subset \subset \mathbb{R}$ (and further u is 'zero at infinity' in the sense that $u(x) \to 0$ as $|x| \to +\infty$). On the other hand, writing $u' = Du$ for simplicity, we have the inequality

$$\max_{x \in K} |u(x)| \leq c_K \left(\int_K |u(x)| \, dx + \int_K |u'(x)| \, dx \right), \tag{5.48}$$

where c_K is a constant which depends on K but not on u.

Proof. In this proof we will denote the functions and their restrictions by the same symbol. Let $u \in W^{1,1}(\mathbb{R})$: we want to prove that u is absolutely continuous in every compact $K \subset \subset \mathbb{R}$; obviously, we need consider only compact sets of the form $K = [a, b]$ with $a, b \in \mathbb{R}$, $a < b$, and show that in $[a, b]$ the function u is summable and can be written in the form

$$u(x) = \tilde{c} + \int_a^x \tilde{u}(t) \, dt, \qquad x \in [a, b], \tag{5.49}$$

where $\tilde{c}$ is a constant and $\tilde{u}$, which will then be a.e. the derivative of u in the usual sense, is summable in $[a, b]$. Now if $u \in W^{1,1}(\mathbb{R})$ then u is summable in $\mathbb{R}$ and u', the derivative of u in the sense of $\mathcal{D}'(\mathbb{R})$, is also summable in $\mathbb{R}$. We can then consider in $[a, b]$ the absolutely continuous function

$$v(x) = \int_a^x u'(t) \, dt, \tag{5.50}$$

whose derivative in the usual sense exists a.e. and is u'. If we show that in $[a, b]$ the difference $u - v$ is constant then we have completed the proof, since we can take $\tilde{c} = v - u$ and $\tilde{u} = u'$. For this purpose, we need only show that, in the sense of $\mathcal{D}'(]a, b[)$, $D(u - v) = 0$, and then use the result referred to in section 4.2.2 which states that if $Df = 0$ (in the sense of distributions) then f is constant. From (5.50) we can then write, for every $\phi \in \mathcal{D}(]a, b[)$,

$$\langle Dv, \phi \rangle = \left\langle D\left(\int_a^x u'(t) \, dt \right), \phi \right\rangle = -\left\langle \int_a^x u'(t) \, dt, \phi' \right\rangle$$

$$= -\int_a^b \left(\int_a^x u'(t) \, dt \right) \phi'(x) \, dx = -\int_a^b dt \int_t^b u'(t) \phi'(x) \, dx$$

$$= -\int_a^b u'(t) \, dt \int_t^b \phi'(x) \, dx = \int_a^b u'(t) \phi(t) \, dt = \langle u', \phi \rangle \tag{5.51}$$

and therefore $D(u - v) = u' - Dv = 0$. We have therefore shown that $u \in AC([a, b])$.

We will now prove that the inequality (5.48) holds, which shows that the immersion $W^{1,1}(\mathbb{R}) \hookrightarrow L^\infty_{\text{loc}}(\mathbb{R})$ is continuous (the inclusion $W^{1,1}(\mathbb{R}) \subset L^\infty_{\text{loc}}(\mathbb{R})$ follows immediately from the first part of the theorem—see also the discussion which follows the proof). One can easily see that $\tilde{c} = u(a)$ and hence we can write (5.49) in the form

$$u(x) = u(a) + \int_a^x u'(t)\,dt. \tag{5.52}$$

The latter expression is still true if instead of a we take an arbitrary $y \in [a, b]$, i.e., we have for every $y \in [a, b]$

$$u(x) = u(y) + \int_y^x u'(t)\,dt. \tag{5.53}$$

From (5.53) we have

$$|u(x)| \leq |u(y)| + \left|\int_y^x |u'(t)|\,dt\right| \leq |u(y)| + \int_a^b |u'(t)|\,dt \tag{5.54}$$

and integrating in $[a, b]$, over the variable y, we obtain

$$(b-a)|u(x)| \leq \int_a^b |u(y)|\,dy + (b-a)\int_a^b |u'(t)|\,dt \tag{5.55}$$

and hence (5.48) with $c_K = \max\{1, (b-a)^{-1}\}$. ∎

It is important to remark that this proof cannot be generalized to prove theorem 5.8: we have given it because it involves reasoning which is very common in the one-dimensional case. With analogous reasoning it can be shown that $H^1(\mathbb{R}) = W^{1,2}(\mathbb{R}) \subset AC_{\text{loc}}(\mathbb{R})$, which improves part III of theorem 5.8 which states only that $H^1(\mathbb{R}) \subset C^{0,1/2}(\mathbb{R})$. Finally, if $u \in W^{1,1}(\mathbb{R})$ then u is 'zero at infinity' since it is absolutely continuous on the compact sets and its absolute value has a finite mean: the functions of $W^{1,1}(\mathbb{R})$ are therefore bounded in $\mathbb{R}$ and we can write $W^{1,1}(\mathbb{R}) \subset L^\infty(\mathbb{R})$, which improves what we have stated in the proof of the theorem. Analogous considerations give us $H^1(\mathbb{R}) \subsetneq L^\infty(\mathbb{R})$.

A counter-example related to the properties of the transpose

To end this section let us present, as we promised, a counter-example which shows that, if in theorem 4.2(ii) the hypothesis of the reflexivity of E is dropped, it is not necessarily true that 'T is injective $\Rightarrow {}'T(F') \subset^{ds} E'$'.

All the spaces involved in the following reasoning are real, but we denote them by the same symbols which we use in the complex case. Let then $E = L^1(\mathbb{R})$, $F = H^{-1}(\mathbb{R})$, and $T: E \to F$ be the operator defined by

$$(T(u))(v) = \int_\mathbb{R} u(x)v(x)\,dx \quad \forall v \in H^1(\mathbb{R}). \tag{5.56}$$

We observe that T is a well defined operator from $L^1(\mathbb{R})$ into $H^{-1}(\mathbb{R})$ since with every $u \in L^1(\mathbb{R})$ it associates the functional $T(u): H^1(\mathbb{R}) \to \mathbb{R}$ which is obviously linear and continuous. On the other hand T is a linear continuous and injective operator (note that if $T(u) = 0$ then $\int_\mathbb{R} u(x)v(x)\,dx = 0$ for every $v \in H^1(\mathbb{R})$ and therefore $u = 0$ a.e.). We intend to prove that ${}^tT[(H^{-1}(\mathbb{R}))']$ is not dense in $(L^1(\mathbb{R}))'$. Let us consider the diagram below:

$$\begin{array}{ccc} H^1(\mathbb{R}) & \xrightarrow{\tilde{T}} & L^\infty(\mathbb{R}) \\ {\scriptstyle j_H}\downarrow & & \downarrow{\scriptstyle j_L} \\ (H^{-1}(\mathbb{R}))' & \xrightarrow{{}^tT} & (L^1(\mathbb{R}))' \end{array}$$

In the diagram we have:

(1) $j_H = j_{H^1(\mathbb{R})(H^{-1}(\mathbb{R}))'}$ is the canonical injection of $H^1(\mathbb{R})$ in $(H^{-1}(\mathbb{R}))'$—an isomorphism (note that $H^{-1}(\mathbb{R})$ is the dual of $H^1(\mathbb{R})$ which is a Hilbert space and consequently reflexive) characterized, for every $u \in H^1(\mathbb{R})$, by

$$_{(H^{-1}(\mathbb{R}))'}\langle j_H u, v\rangle_{H^{-1}(\mathbb{R})} = {}_{H^{-1}(\mathbb{R})}\langle v, u\rangle_{H^1(\mathbb{R})} \qquad \forall v \in H^{-1}(\mathbb{R}); \tag{5.57}$$

(2) $j_L = j_{L^\infty(\mathbb{R})(L^1(\mathbb{R}))'}$ is the canonical injection of $L^\infty(\mathbb{R})$ in $(L^1(\mathbb{R}))'$—an isomorphism (note that $L^\infty(\mathbb{R})$ is the dual of $L^1(\mathbb{R})$) characterized, for every $u \in L^\infty(\mathbb{R})$, by

$$_{(L^1(\mathbb{R}))'}\langle j_L u, v\rangle_{L^1(\mathbb{R})} = \int_\mathbb{R} u(x)v(x)\,dx \qquad \forall v \in L^1(\mathbb{R}); \tag{5.58}$$

(3) ${}^tT$ is the transpose of T;
(4) $\tilde{T}$ defined by

$$\tilde{T} = j_L^{-1} \circ {}^tT \circ j_H \tag{5.59}$$

is the unique application of $H^1(\mathbb{R})$ in $L^\infty(\mathbb{R})$ which makes the diagram commutative (the uniqueness follows from the fact that the commutativity of the diagram means that $j_L \circ \tilde{T} = {}^tT \circ j_H$, and j_H, j_L and ${}^tT$ are univocally defined).

From (5.59), since j_H and j_L are homeomorphisms, it follows that ${}^tT$ has a dense image if and only if $\tilde{T}$ has a dense image: if we can show that $\tilde{T} = i_{H^1(\mathbb{R})L^\infty(\mathbb{R})}$ (we use the letter i, which means set immersion, from the discussion which we have given after the proof of theorem 5.9...) we have finished, because $H^1(\mathbb{R}) \subset C^0(\mathbb{R})$ and $C^0(\mathbb{R})$ is not dense in $L^\infty(\mathbb{R})$. Let us prove now that the set immersion of $H^1(\mathbb{R})$ in $L^\infty(\mathbb{R})$ makes the diagram commutative. We can write, for every $u \in H^1(\mathbb{R})$,

$$_{(L^1(\mathbb{R}))'}\langle({}^tT \circ j_H)u, v\rangle_{L^1(\mathbb{R})} = {}_{(H^{-1}(\mathbb{R}))'}\langle j_H u, Tv\rangle_{H^{-1}(\mathbb{R})} = {}_{H^{-1}(\mathbb{R})}\langle Tv, u\rangle_{H^1(\mathbb{R})}$$

$$= \int_\mathbb{R} u(x)v(x)\,dx \qquad \forall v \in L^1(\mathbb{R}) \tag{5.60}$$

and hence, keeping in mind (5.58), for every $u \in H^1(\mathbb{R})$

$$_{(L^1(\mathbb{R}))'}\langle j_L u, v \rangle_{L^1(\mathbb{R})} = {}_{(L^1(\mathbb{R}))'}\langle ({}^t T \circ j_H) u, v \rangle_{L^1(\mathbb{R})} \qquad \forall v \in L^1(\mathbb{R}) \qquad (5.61)$$

and thus

$$j_L u = ({}^t T \circ j_H) u \qquad \forall u \in H^1(\mathbb{R}), \qquad (5.62)$$

which when compared with $j_L \circ \tilde{T} = {}^t T \circ j_H$ enables us to conclude that $\tilde{T} =$ identity.

5.3 SPACES $W^{s,p}(\Omega)$

We will now consider the Sobolev spaces relative to an arbitrary open set $\Omega \subset \mathbb{R}^n$.

$W^{k,p}(\Omega)$, $k \in \mathbb{N}$ **and** $W^{s,p}(\Omega)$, $s \in \mathbb{R}_+ \setminus \mathbb{N}$

DEFINITION 5.6. Let $\Omega \subset \mathbb{R}^n$ be an arbitrary open set. If $k \in \mathbb{N}$ and $p \in [1, +\infty]$, we denote by $W^{k,p}(\Omega)$ the vector space $\{u \in \mathscr{D}'(\Omega): D^\alpha u \in L^p(\Omega), |\alpha| \leq k\}$ provided with the norm

$$\|u\|_{W^{k,p}(\Omega)} = \left[\sum_{|\alpha| \leq k} \|D^\alpha u\|_{L^p(\Omega)}^p \right]^{1/p}. \qquad (5.63)$$

If $s \in \mathbb{R}_+ \setminus \mathbb{N}$ and $p \in]1, +\infty[$ we denote by $W^{s,p}(\Omega)$ the vector space $\{u \in \mathscr{D}'(\Omega): u \in W^{[s],p}(\Omega)$ and $d_{\sigma,\alpha}(u) \in L^p(\Omega \times \Omega), |\alpha| = [s]\}$ provided with the norm

$$\|u\|_{W^{s,p}(\Omega)} = \left[\|u\|_{W^{[s],p}(\Omega)}^p + \sum_{|\alpha|=[s]} \|d_{\sigma,\alpha}(u)\|_{L^p(\Omega \times \Omega)}^p \right]^{1/p}. \qquad (5.64)$$

The spaces $W^{s,2}(\Omega)$ will also be denoted by $H^s(\Omega)$, $s \geq 0$.

It follows immediately that if $\Omega = \mathbb{R}^n$ we have once again the definitions 5.1 and 5.2. On the other hand our discussion concerning these definitions can be repeated here. Let us remark that the property '$u \in W^{s,p}(\Omega)$' has a local character: *if $\{\Omega_i\}_{i=1,\ldots,m}$ is an open covering of Ω and for every $i=1,\ldots,m$, $u \in W^{s,p}(\Omega_i)$, $s \geq 0$, then $u \in W^{s,p}(\Omega)$ and there exists a constant $c > 0$ such that $\|u\|_{W^{s,p}(\Omega)} \leq c \sum_{i=1}^m \|u\|_{W^{s,p}(\Omega_i)}$* (see, for example, Nečas, 1967, p. 62)—we denote by the same symbol the function $u \in \mathscr{F}(\bigcup_{i=1}^m \Omega_i)$ and its restrictions to Ω and to Ω_i, $i=1,\ldots,m$. Let us note, in this context of the local structure of the spaces $W^{s,p}(\Omega)$, that the spaces $W^{s,p}_{\text{loc}}(\Omega)$ and $W^{s,p}_{\text{comp}}(\Omega)$ can be defined in analogy with the spaces $L^p_{\text{loc}}(\Omega)$ and $L^p_{\text{comp}}(\Omega)$.

The topology of the spaces $W^{s,p}(\Omega)$, $s \geq 0$, has the same fundamental characteristics as the topology of the spaces $W^{s,p}(\mathbb{R}^n)$, namely:

THEOREM 5.10. *The spaces $W^{k,p}(\Omega)$, $k \in \mathbb{N}$ and $p \in [1, +\infty]$ are Banach spaces; if $p \in]1, +\infty[$ they are reflexive spaces, and if $p = 2$ they are Hilbert* spaces with the scalar product

$$(u, v)_{H^k(\Omega)} = \sum_{|\alpha| \leq k} (D^\alpha u, D^\alpha v)_{L^2(\Omega)}. \tag{5.65}$$

The spaces $W^{s,p}(\Omega)$, $s \in \mathbb{R}_+ \setminus \mathbb{N}$ and $p \in]1, +\infty[$, are reflexive Banach spaces, and if $p = 2$ they are Hilbert spaces with the scalar product

$$(u, v)_{H^s(\Omega)} = (u, v)_{H^{[s]}(\Omega)} + \sum_{|\alpha|=[s]} (d_{\sigma,\alpha}(u), d_{\sigma,\alpha}(v))_{L^2(\Omega \times \Omega)}. \tag{5.66}$$

Proof. It is sufficient to adapt the proofs of the theorems 5.3 and 5.6. ∎

Not all the properties of the spaces $W^{s,p}(\mathbb{R}^n)$ can be generalized to the spaces $W^{s,p}(\Omega)$ with arbitrary Ω; thus in particular theorem 5.4 is not true, namely:

THEOREM 5.11. *If Ω is a bounded open set of class C^∞ and $k \geq 1$ is an integer then $\mathscr{D}(\Omega)$ is not dense in $W^{k,p}(\Omega)$, $p \in [1, +\infty]$.*

Proof. Before showing that this result holds, we remark that it holds in much more general conditions: it is enough e.g., that Ω be an open set such that the measure of $\mathbb{R}^n \setminus \Omega$ is positive; for more precise conditions, based on the concept of $(k-p)$-*polar set*, see Deny–Lions, 1954 and Lions, 1965, p. 21.

The proof will be set out in two stages: in the first part we will show that if u is an element of the closure $\overline{\mathscr{D}(\Omega)}^{W^{k,p}(\Omega)}$ of $\mathscr{D}(\Omega)$ in $W^{k,p}(\Omega)$ then its *trivial extension* to $\mathbb{R}^n$ (i.e. the function $\tilde{u}$ defined by $\tilde{u} \equiv u$ in Ω and $\tilde{u} \equiv 0$ in $\mathbb{R}^n \setminus \Omega$) belongs to $W^{k,p}(\mathbb{R}^n)$; in the second part we will show that there exists $v \in W^{k,p}(\Omega)$ whose trivial extension does not belong to $W^{k,p}(\mathbb{R}^n)$. We denote by $P: \mathscr{D}(\Omega) \to \mathscr{D}(\mathbb{R}^n)$ the operator which with $\phi \in \mathscr{D}(\Omega)$ associates $\tilde{\phi} = P(\phi) \in \mathscr{D}(\mathbb{R}^n)$, the trivial extension of ϕ to $\mathbb{R}^n$. P is obviously linear and, providing $\mathscr{D}(\Omega)$ and $\mathscr{D}(\mathbb{R}^n)$ with the topologies induced, respectively, by $W^{k,p}(\Omega)$ and $W^{k,p}(\mathbb{R}^n)$, it is also continuous since $\|\phi\|_{W^{k,p}(\Omega)} = \|P(\phi)\|_{W^{k,p}(\mathbb{R}^n)}$. Thus it is possible to extend P by continuity to an operator $\tilde{P}: \overline{\mathscr{D}(\Omega)}^{W^{k,p}(\Omega)} \to W^{k,p}(\mathbb{R}^n)$ which is also linear and continuous (besides $\tilde{P}(u|_\Omega) = u$ and $P(\tilde{u}|_{\mathbb{R}^n \setminus \Omega}) = 0$). We have shown then that if $u \in \overline{\mathscr{D}(\Omega)}^{W^{k,p}(\Omega)}$ then $\tilde{u} = \tilde{P}(u) \in W^{k,p}(\mathbb{R}^n)$.

We proceed now to the second part. Since Ω is bounded, the function $v \equiv 1$ in Ω is an element of $W^{k,p}(\Omega)$. Instead, its trivial extension, the function $\tilde{u} \equiv \chi_\Omega$ is not an element of $W^{k,p}(\mathbb{R}^n)$ since the derivatives $\partial \chi_\Omega / \partial x_i$ are distributions equal to zero outside Γ, the boundary of Ω, and thus, were they belong to $L^p(\mathbb{R}^n)$, from grad $\chi_\Omega \equiv 0$ it would follow that χ_Ω is a constant. ∎

$W^{s,p}_0(\Omega)$

The spaces $\overline{\mathscr{D}(\Omega)}^{W^{s,p}(\Omega)}$ are of great interest in the theory of Sobolev spaces: we did not explicitly mention them in section 5.2 because for $\Omega = \mathbb{R}^n$ they are not distinct from the $W^{s,p}(\mathbb{R}^n)$ spaces. We will see that some of the results of section 5.2 will now be interpreted in terms of $\overline{\mathcal{D}(\mathbb{R}^n)}^{W^{s,p}(\mathbb{R}^n)}$ since what plays a fundamental role is its approximability through regular functions.

DEFINITION 5.7. Let $\Omega \subset \mathbb{R}^n$ be an arbitrary open set. If $k \in \mathbb{N}$ and $p \in [1, +\infty]$ we denote by $W_0^{k,p}(\Omega)$ the closure of $\mathcal{D}(\Omega)$ in $W^{k,p}(\Omega)$, provided with the topology induced by the latter space. If $s \in \mathbb{R}_+ \setminus \mathbb{N}$ and $p \in]1, +\infty[$ we denote by $W_0^{s,p}(\Omega)$ the closure of $\mathcal{D}(\Omega)$ in $W^{s,p}(\Omega)$, provided with the topology induced by the latter space.

Intuitively we can say that if a function $u \in W^{s,p}(\Omega)$ is approximable through functions of $\mathcal{D}(\Omega)$ then u is 'null' on the boundary of Ω (this concept of annulment will be clarified in the next section)—this is the origin of the subscript 0 (zero) in $W_0^{s,p}(\Omega)$; in this context we can say that the functions of $W^{s,p}(\mathbb{R}^n)$ 'vanish at infinity'. We remark also that if $s \leq 1/p$ then $W_0^{s,p}(\Omega) = W^{s,p}(\Omega)$ (for the proof see Lions–Magenes, 1961c and, for the case $p = 2$, Lions–Magenes, 1972a, p. 60).

THEOREM 5.12. The spaces $W_0^{k,p}(\Omega)$, $k \in \mathbb{N}$ and $p \in [1, +\infty]$, are Banach spaces; if $p \in]1, +\infty[$ they are reflexive, and if $p = 2$ they are Hilbert spaces which we denote by $H_0^k(\Omega)$. By analogy, the spaces $W_0^{s,p}(\Omega)$, $s \in \mathbb{R}_+ \setminus \mathbb{N}$ and $p \in]1, +\infty[$, are reflexive Banach spaces, and if $p = 2$ they are Hilbert spaces which we denote by $H_0^s(\Omega)$.

Proof. It is enough to see that the spaces $W_0^{s,p}(\Omega)$, $s \geq 0$, are closed subspaces of $W^{s,p}(\Omega)$ and keep in mind theorem 5.10. ■

It is interesting to note that if Ω is bounded the topology of $H_0^k(\Omega)$ can be defined from the scalar product

$$(u, v)_{H_0^k(\Omega)} = \sum_{|\alpha|=k} (D^\alpha u, D^\alpha v)_{L^2(\Omega)}, \qquad (5.67)$$

namely, the following theorem holds:

THEOREM 5.13. Let $\Omega \subset \mathbb{R}^n$ be a bounded open set and $k \in \mathbb{N}$. There exists two constants, c_1 and c_2, such that for every $u \in H_0^k(\Omega)$ we have the inequality

$$c_1 \left(\sum_{|\alpha|=k} \int_\Omega |D^\alpha u|^2 \right)^{1/2} \leq \left(\sum_{|\alpha|\leq k} \int_\Omega |D^\alpha u|^2 \right)^{1/2} \leq c_2 \left(\sum_{|\alpha|=k} \int_\Omega |D^\alpha u|^2 \right)^{1/2}. \qquad (5.68)$$

Proof. For the inequality on the left-hand side we just put $c_1 = 1$. For the one on the right-hand side, we need only show that for every $\phi \in \mathcal{D}(\Omega)$ the so-called *Friedrich's inequality*

$$\int_\Omega |\phi|^2 \, dx \leq c \sum_{i=1}^n \int_\Omega |\partial \phi / \partial x_i|^2 \, dx \tag{5.69}$$

holds, and then to note that $H_0^k(\Omega)$ is the closure of $\mathcal{D}(\Omega)$ in $H^k(\Omega)$.

Let us prove (5.69): since Ω is bounded, we can assume that $\Omega \subset Q = \{x \in \mathbb{R}^n : |x_i| < a, \; i = 1, \ldots, n\}$, where $a \in \mathbb{R}_+$, and thus

$$\phi(x_1, \ldots, x_n) = -\int_{-a}^{x_1} \frac{\partial \phi}{\partial x_1}(\xi, x_2, \ldots, x_n) \, d\xi, \tag{5.70}$$

which, using the Schwarz–Hölder inequality, can be written as

$$|\phi(x_1, \ldots, x_n)|^2 \leq 2a \int_{-a}^{a} \left| \frac{\partial \phi}{\partial x_1}(\xi, x_2, \ldots, x_n) \right|^2 d\xi. \tag{5.71}$$

Integrating (5.71) with respect to the variable x_1, in the interval between $-a$ and a, we have

$$\int_{-a}^{a} |\phi(x_1, \ldots, x_n)|^2 \, dx_1 \leq 4a^2 \int_{-a}^{a} \left| \frac{\partial \phi}{\partial x_i}(x_1, \ldots, x_n) \right|^2 dx_1 \tag{5.72}$$

and integrating (5.72) with respect to the variables $x_2, \ldots, x_n$, between $-a$ and a, we obtain (5.69) with $c = 4a^2$. ∎

We have given this proof essentially because it involves Friedreich's inequality, which is interesting on its own; it is used in particular in the study of the homogeneous Dirichlet problem. Another very important inequality in the context of Sobolev spaces is the *Poincaré inequality*: if $\Omega \subset \mathbb{R}^n$ is a bounded open set of class C^0 then there exists $c \in \mathbb{R}$ such that for every $u \in H^k(\Omega)$

$$c \|u\|_{H^k(\Omega)} \leq \left[\sum_{|\alpha| < k} \left| \int_\Omega D^\alpha u \right|^2 + \sum_{|\alpha| = k} \int_\Omega |D^\alpha u|^2 \right]^{1/2} \tag{5.73}$$

(see, e.g., Nečas, 1967, p. 18). This inequality provides a norm equivalent to (5.63) for $p = 2$ (we need only to take as the norm of $u \in H^k(\Omega)$ the second term of (5.73)). The Poincaré inequality can be useful, among other things, in the study of problems which involve spaces of functions with zero mean. It is important to observe that the search for equivalent norms is of great practical importance: the efficacy of a methodology for the study of a problem formulated in Sobolev spaces often depends on a convenient choice of the norms of these spaces.

$W^{s,p}(\Omega)$, $s \in \mathbb{R}_- \setminus \{0\}$

As a consequence of the fact that $\mathcal{D}(\Omega)$ is in general not dense in $W^{s,p}(\Omega)$ (and the latter is therefore not in general a normal space of distributions) two serious problems arise: whether or not the elements of $W^{s,p}(\Omega)$ are approximable through regular functions and how to define the spaces $W^{s,p}(\Omega)$ with s strictly negative.

There are essentially two ways of defining the spaces $W^{s,p}(\Omega)$, $s<0$: either we simply carry over mechanically the definition 5.4 from the case '$\Omega=\mathbb{R}^n$' to the case 'general Ω', or we look for a definition which has as a consequence a characterization of the elements of $W^{s,p}(\Omega)$, $s<0$, analogous to that reported in theorem 5.5. The path we will follow is the latter, since the former might take us outside the context of the theory of distributions, and on the other hand the need for considering distributions which can be so characterized arises in a natural way in the treatment of many problems. Hence:

DEFINITION 5.8. Let $\Omega \subset \mathbb{R}^n$ be an arbitrary open set, $s \in \mathbb{R}_-\setminus\{0\}$ and $p \in]1, +\infty[$. We denote by $W^{s,p}(\Omega)$ the dual of the space $W_0^{-s,p^*}(\Omega)$, where $p^* = p/(p-1)$. The spaces $W^{s,2}(\Omega)$, $s<0$, will again be denoted by $H^s(\Omega)$.

It follows immediately that if $\Omega = \mathbb{R}^n$ this definition reduces to definition 5.4. It is also immediate that the spaces $W^{s,p}(\Omega)$, $s<0$, are reflexive Banach spaces. Further, as we required, the following result holds:

THEOREM 5.14. A distribution T belongs to $(W_0^{k,p}(\Omega))' = W^{-k,p^*}(\Omega)$, $k \in \mathbb{N}$ and $p \in]1, +\infty[$, iff T can be written in the form

$$T = \sum_{|\alpha| \leq k} D^\alpha g_\alpha \qquad (5.74)$$

with $g_\alpha \in L^{p^*}(\Omega)$.

Proof. The proof is analogous to that of theorem 5.5. ∎

Density results

We will now deal with the problem of the approximability of the elements of $W^{s,p}(\Omega)$, $s \geq 0$, through more general functions. This problem, which is of great theoretical and practical importance consists essentially on the search for a function space dense in $W^{s,p}(\Omega)$ whose elements satisfy certain requisites of regularity and, specially in numerical studies, the ease of representation in computers (for numerical problems several function spaces with special characteristics have been introduced: see, e.g., Aubin, 1972; Raviart, 1972; Temam, 1970; and in particular for the case of the variational inequalities see, e.g., Glowinski–Lions–Trémolières, 1976a; 1976b; Falk, 1974; Glowinski, 1980; Brezzi–Hager–Raviart, 1977; 1978).

We will only consider here the following important result:

THEOREM 5.15. If $\Omega \subset \mathbb{R}^n$ is a bounded open set of class C^0 then $\mathscr{D}(\bar{\Omega})$ is dense in $W^{s,p}(\Omega)$.

As an immediate consequence of this theorem, *if Ω is a bounded open set of class C^0, then the concepts of strong derivative and weak derivative coincide.* Another consequence is that if $1 \leq p < +\infty$ then $W^{m,p}(\Omega)$ can be defined as the abstract completion of $\{u \in C^m(\Omega) : \|u\|_{W^{m,p}(\Omega)} < +\infty\}$ with respect to the norm $\|.\|_{W^{m,p}(\Omega)}$.

For the proof of theorem 5.15 see Nečas, 1967, p. 67, Lions, 1965, p. 42; Lions–Magenes, 1961b, p. 52; let us remark also that the theorem can be proved under the weaker hypothesis that Ω has the segment property (see Adams, 1975, p. 54).

Besides, this result is natural: we need only to note that $\mathscr{D}(\mathbb{R}^n)$ is dense in $W^{s,p}(\mathbb{R}^n)$, that (Ω being sufficiently regular) the elements of $\mathscr{D}(\bar{\Omega})$ are restrictions of elements of $\mathscr{D}(\mathbb{R}^n)$, and that one expects that those of $W^{s,p}(\Omega)$ can be taken to be restrictions of elements of $W^{s,p}(\mathbb{R}^n)$.

The extension operators

The conjecture that the elements of $W^{s,p}(\Omega)$ can be taken to be restrictions of elements of $W^{s,p}(\mathbb{R}^n)$ brings us to the so-called problem of extension, which consists exactly of the search for conditions (on Ω) which ensure that the elements of $W^{s,p}(\Omega)$ are restrictions to Ω of elements of $W^{s,p}(\mathbb{R}^n)$, or, equivalently, for conditions which ensure that the elements of $W^{s,p}(\Omega)$ can be extended to $\mathbb{R}^n$ as elements of $W^{s,p}(\mathbb{R}^n)$. This question is of great interest since it is basic for the possibility of carrying over some of the results proved for $W^{s,p}(\Omega)$ to results relative to $W^{s,p}(\mathbb{R}^n)$ and vice versa.

DEFINITION 5.9. let $\Omega \subset \mathbb{R}^n$ be an arbitrary open set. We say that $v \in W^{s,p}(\mathbb{R}^n)$ is an *extension* of $u \in W^{s,p}(\Omega)$ if $v = u$ a.e. on Ω; and an operator $P \in \mathscr{L}(W^{s,p}(\Omega), W^{s,p}(\mathbb{R}^n))$ such that $P(u)$ is an extension of u, for every $u \in W^{s,p}(\Omega)$, is said to be an *extension operator* on $W^{s,p}(\Omega)$.

Let us remark that in the same way we can define extension operators on $W_0^{s,p}(\Omega)$: one example is the operator $\tilde{P}$ which we introduced in the proof of theorem 5.11.

The problem of extension can now be discussed in the following terms: what are the conditions on Ω which ensure the existence of at least one extension operator on $W^{s,p}(\Omega)$? As an answer we limit ourselves to the following:

THEOREM 5.16. *If $\Omega \subset \mathbb{R}^n$ is a bounded open set of class $C^{k-1,1}$, $k \geq 1$ is an integer and $p \in [1, +\infty[$, then there exists at least one extension operator on $W^{k,p}(\Omega)$.*

Let us remark that in the case of the spaces $W_0^{s,p}(\Omega)$ the answer is not as simple as it might seem to be: if $\Omega \subset \mathbb{R}^n$ is an arbitrary open set and $u \in \mathscr{D}(\Omega)$ (which is dense in $W_0^{s,p}(\Omega)$: it is in fact enough to construct a continuous extension on a dense set...) then the trivial extension $\tilde{u}$ of u is an extension of u in the sense of definition 5.9, but for the operator which associates $\tilde{u}$ with u to be an extension operator in the space $W_0^{s,p}(\Omega)$ we must have $s \neq \text{integer} + 1/p$, since otherwise it is not in general continuous (see e.g. Lions–Magenes, 1961c, p. 231, and 1972a, p. 66 for the case $p=2$).

For the proof and extensions of theorem 5.16 we refer the reader to Nečas, 1967, p. 75; Lions, 1965, p. 44; Calderón, 1961, p. 45; Lions–Magenes, 1961c, p. 319. Here we will merely explicitly construct an extension operator in the particularly simple case in which $\Omega =]0, +\infty[\subset \mathbb{R}$. The technique used, however, can be easily extended to the case in which $\Omega = \mathring{\mathbb{R}}_+^n$ and hence, using local maps, to the general case. We note that, in the case which we will consider, Ω is unbounded and hence we are not exactly in the conditions of theorem 5.16... : in this connection we must advise the reader that, in this and in the following sections of this chapter, we will not worry about how to present the minimal hypothesis on Ω which ensures us the validity of the results—thus many of them (like theorem 5.16) are true also when Ω is a half-space, even though it was stated only for bounded open sets.

So, we are faced with the following problem: '*given a function* $u \in W^{k,p}(\mathring{\mathbb{R}}_+)$, $k \in \mathbb{N}$ *and* $p \in [1, +\infty[$, *find* $\hat{u} \in W^{k,p}(\mathbb{R})$ *such that* $\hat{u} = u$ *a.e. in* $\mathring{\mathbb{R}}_+$.'
If $k=0$ we can take $\hat{u}$ to be the function

$$\hat{u}(x) = \begin{cases} u(x) & \text{if } x > 0 \\ 0 & \text{if } x < 0 \end{cases} \tag{5.75}$$

and arbitrarily defined for $x=0$. If we choose $\hat{u}(0) = 0$ then $\hat{u}$ is merely the trivial extension of u to $\mathbb{R}$: the case shown in fig. 5.1. This function is not, in general, continuous and hence if $k=1$ it does not solve the problem (note that $W^{1,p}(\mathbb{R}) \subset C^0(\mathbb{R})$—see theorems 5.8(III) and 5.9). In the case $k=1$ we can take $\hat{u}$ be be the function

$$\hat{u}(x) = \begin{cases} u(x) & \text{if } x > 0 \\ u(-x) & \text{if } x < 0. \end{cases} \tag{5.76}$$

If we wish to define it for $x=0$ also we do it in such a way so that it becomes continuous, which is always possible since $u(0^+) = u(0^-)$; this is the so-called *extension by symmetry or overturning*: see fig. 5.2. This function is not in

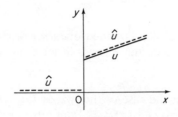

Figure 5.1

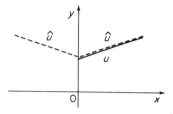

Figure 5.2

general differentiable, and hence if $k = 2$ it does not solve the problem...: following this line of reasoning leads us to take $\hat{u}$ to be a function of the type

$$\hat{u}(x) = \begin{cases} u(x) & \text{if } x > 0 \\ \sum_{j=1}^{k} \alpha_j u(-jx) & \text{if } x < 0 \end{cases} \tag{5.77}$$

and defined for $x = 0$ so that it is continuous, where α_j are coefficients to be determined in such a way that not only $\hat{u}$ is continuous but so are its derivatives up to order $k - 1$. We will now see that (5.77) does define effectively an extension, i.e., that there exist k numbers α_j which satisfy the required conditions. If $k = 0$ the problem is completely solved by the trivial extension, without further conditions. If $k = 1$ the only condition that we have to impose is

$$\hat{u}(0^+) = \hat{u}(0^-), \tag{5.78}$$

and it follows from (5.77) that we must then have (with $k = 1$)

$$\sum_{j=1}^{k} \alpha_j = 1 \tag{5.79}$$

or $\alpha_1 = 1$, which is in agreement with (5.76). If $k = 2$ we have not only the condition (5.78) of connection regarding the function but also one concerning the connection of the first derivative, namely

$$\hat{u}'(0^+) = \hat{u}'(0^-); \tag{5.80}$$

from (5.78) we have (5.79) (with $k = 2$), and from (5.80) together with (5.77) it follows that

$$\sum_{j=1}^{k} \alpha_j(-j) = 1 \tag{5.81}$$

(again with $k = 2$). The equations (5.79) and (5.81) together make up a linear system of two equations in two unknowns whose solutions are $\alpha_1 = 3$ and $\alpha_2 = -2$. In the general case we have the k conditions

$$\hat{u}^{(i)}(0^+) = \hat{u}^{(i)}(0^-), \quad i = 0, \ldots, k-1, \tag{5.82}$$

which, as can be easily seen, together with (5.77) allow us to obtain the linear system of k equations in k unknowns

$$\sum_{j=1}^{k} (-j)^i \alpha_j = 1, \quad i = 0, \ldots, k-1, \tag{5.83}$$

which always has a solution since its matrix is the Vandermonde matrix of the numbers $-1, -2, \ldots, -k$. We have thus shown that, for fixed $k > 0$, there exists at least one extension which works well from $H^i(\mathring{\mathbb{R}}_+)$ to $H^i(\mathbb{R})$, for $j = 0, 1, \ldots, k$. This extension is known as the *Babitch extension*. Taking a greater number of terms ($\sum_{j=0}^{2k} \ldots$) we could also construct extension operators acting from $H^i(\mathring{\mathbb{R}}_+)$ to $H^i(\mathbb{R})$ for $|j| \leq k$ (see Baiocchi, 1966).

The Sobolev imbedding theorem: the general case

Before proceeding to the Sobolev imbedding theorem analogous to theorem 5.8, we will introduce the concepts of compact operator and compact inclusion:

DEFINITION 5.10. Let E and F be two Banach spaces. $T \in \mathcal{L}(E, F)$ is said to be a *compact* (or *completely continuous*) *operator* if T transforms bounded sets of E in relatively compact sets of F (i.e., sets whose closure is compact). If $E \subset F$ and the set immersion $i_{EF}: E \to F$ is a compact operator, the inclusion of E in F is said to be a *compact inclusion* and is written $E \subset\subset F$.

It follows immediately from this definition that if f_m is a bounded sequence of elements of E and $T \in \mathcal{L}(E, F)$ is compact then the sequence $T(f_m)$ has a subsequence convergent in F. In general it is easier to show that a sequence is bounded than to show that it converges (even relative to different norms ...) and hence results which show that certain inclusions are compact are of interest—in Sobolev's theorem, which we will now state, there are results of this type.

THEOREM 5.17 (Sobolev imbedding theorem). If $\Omega \subset \mathbb{R}^n$ is a bounded open set of class $C^{0,1}$, $k \geq 1$ is an integer and $p \in [1, +\infty[$, then the following inclusions hold:

(I) If $n > kp$ then $W^{k,p}(\Omega) \subset L^q(\Omega)$ with $q \leq np/(n-kp)$, and $W^{k,p}(\Omega) \subset\subset L^q(\Omega)$ if $q < np/(n-kp)$; more generally if $m \in \mathbb{N}$, $m \leq k$ and $n > (k-m)p$ then $W^{k,p}(\Omega) \subset W^{m,q}(\Omega)$ with $q \leq np/(n-(k-m)p)$, and $W^{k,p}(\Omega) \subset\subset W^{m,q}(\Omega)$ if $q < np/(n-(k-m)p)$;

(II) If $n = kp$ then $W^{k,p}(\Omega) \subset\subset L^q(\Omega)$ for every $q \in [1, +\infty[$;

(III) If $n < kp$ then $W^{k,p}(\Omega) \subset\subset C^{0,\mu}(\bar{\Omega})$ with $\mu = k - (n/p)$ if $k - (n/p) < 1$, $\mu < 1$ arbitrary if $k - (n/p) = 1$ and $\mu = 1$ if $k - (n/p) > 1$; more generally if $m \in \mathbb{N}$, $m \leq k$ and $n < (k-m)p$ then $W^{k,p}(\Omega) \subset\subset C^{m,\mu}(\bar{\Omega})$ with $\mu = k - m - (n/p)$ if $k - m - (n/p) < 1$, $\mu < 1$ arbitrarly if $k - m - (n/p) = 1$ and $\mu = 1$ if $k - m - (n/p) > 1$.

The corresponding inclusions relative to the spaces $W_0^{k,p}(\Omega)$ are valid whatever the bounded open set Ω is.

For the proof see the references given for theorem 5.8 and also, in particular for the results concerning compactness, Nečas, 1967, p. 107; Lions, 1965, p. 53; Adams, 1975, p. 144; Gagliardo, 1958, p. 133.

This theorem can be proved under more general conditions on Ω. On the other hand, the discussion concerning theorem 5.8 can be repeated here—we should, however, note that $W^{k,p}(\Omega) \subset W^{k,q}(\Omega)$ if $p > q$ since Ω (being bounded) has finite measure. Let us report also the following very important result: *for every* $s \in \mathbb{R}$ *and* $\varepsilon > 0$, $H^s(\Omega) \subsetneq H^{s-\varepsilon}(\Omega)$ (Lions–Magenes, 1972a, p. 110).

Finally, we have the analogue of theorem 5.9: *if* $-\infty < a < b < +\infty$, *then* $W^{1,1}(]a, b[) = AC([a, b])$. And under the same conditions, $H^1(]a, b[) \subsetneq AC([a, b])$. However, for $n > 1$ this result is false: the function $f(x) = \log|\log|x||$ defined in the disc $D = \{x \in \mathbb{R}^2 : \|x\| < 1/2\}$ is an element of $H^1(D)$ but is not a continuous function. Again, if Ω is not an interval there are some pathological phenomena; thus, e.g., if $a < c < b$ then the step constant functions belong to $H^k(]a, c[\cup]c, b[)$ for every k....

To end this section we must mention that sometimes the spaces $H^{s,p}(\Omega)$, generalization of the spaces $H^{s,p}(\mathbb{R}^n)$ introduced in definition 5.5 and which can be defined as the spaces of restrictions to Ω of the elements of $H^{s,p}(\mathbb{R}^n)$, are used (see, e.g., Volevich–Paneyakh, 1965, p. 17; Lions–Magenes, 1961b, p. 67).

5.4 SPACES $W^{s,p}(\Gamma)$

The concept of trace

In the study of differential problems, and in particular when one is dealing with boundary conditions associated with partial differential equations, one needs to speak about the values which certain elements of $W^{s,p}(\Omega)$ take on the boundary of Ω. If $u \in W^{s,p}(\Omega)$ is continuous up to the boundary Γ of Ω then one can say that the value which u takes on Γ is the restriction to Γ (of the extension by continuity to $\bar{\Omega}$) of the function u. In general, however, the elements of $W^{s,p}(\Omega)$ are defined except for a set of zero measure and it is meaningless therefore to speak of their restrictions to Γ, which has a n-dimensional zero measure (except for the pathological case which does not interest us). We need therefore a new tool which can substitute, by generalizing it, that of restriction when the latter is inapplicable. Here we will be concerned exclusively with the value on the boundary—more generally it might be of interest to consider the value on any manifold of dimension $< n$ contained in $\bar{\Omega}$ (see, in this context, Nikol'skiĭ, 1961; 1975; Prodi, 1958).

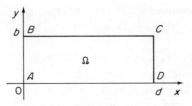

Figure 5.3

In order to motivate the results which we shall present here, we will first treat a simple case. Let Ω be the interior of the rectangle shown in fig. 5.3 and let u be an element of $H^1(\Omega)$; u is then a function of $L^2(\Omega)$ such that both the derivatives $u_x = D^{(1,0)}u$ and $u_y = D^{(0,1)}u$ also belong to $L^2(\Omega)$. We intend to make the statement '$g(y) = u(0, y)$' meaningful, i.e., we intend to make the phrase 'g is the value which u takes on the side AB' meaningful.

For this purpose let us consider first the linear application

$$\rho : C^1(\bar{\Omega}) \to C^1([0, b]) \tag{5.84}$$

which to every $u(x, y) \in C^1(\bar{\Omega})$ associates the function $\rho[u(x, y)] = u(0, y)$. This application is well defined since if u is continuous in $\bar{\Omega}$ it makes sense to consider its restriction to the segment AB, which is differentiable as long as u is. As we know, $C^1(\bar{\Omega}) \subset^{ds} H^1(\Omega)$ and $C^1([0, b]) \subset^{ds} L^2(]0, b[)$; hence, if we can show that, providing $C^1(\bar{\Omega})$ and $C^1([0, b])$ with the topologies induced respectively by $H^1(\Omega)$ and by $L^2(]0, b[)$, ρ is a continuous application, we can conclude then that it can be univocally extended in an application

$$\tilde{\rho} : H^1(\Omega) \to L^2(]0, b[) \tag{5.85}$$

which is linear and continuous.

If $u \in H^1(\Omega)$ it is natural to call $\tilde{\rho}u$ the *restriction of u to AB*—in order to avoid confusion and to emphasize that we are no longer dealing with a restriction in the classical sense of the term, we prefer to call $\tilde{\rho}u$ the *trace of u on AB*.

We will see now that ρ is in fact continuous. By the fundamental formula of integral calculus

$$u(x, y) = u(x_0, y) + \int_{x_0}^{x} u_x(\xi, y) \, d\xi, \quad x_0 \in [0, d], \tag{5.86}$$

we can write, with $x = 0$,

$$|u(0, y)| \leq |u(x_0, y)| + \int_{0}^{x_0} |u_x(\xi, y)| \, d\xi, \tag{5.87}$$

from which, keeping in mind in particular that $(\alpha + \beta)^2 \leq 2\alpha^2 + 2\beta^2$ and the

Schwarz–Hölder inequality, it follows that

$$|u(0, y)|^2 \leq 2 |u(x_0, y)|^2 + 2\left(\int_0^{x_0} |u_x(\xi, y)| \, d\xi\right)^2$$

$$\leq 2 |u(x_0, y)|^2 + 2d \int_0^{x_0} |u_x(\xi, y)|^2 \, d\xi$$

$$\leq 2 |u(x_0, y)|^2 + 2d \int_0^d |u_x(\xi, y)|^2 \, d\xi. \tag{5.88}$$

Integrating the first and the last terms of (5.88) with respect to x_0, between 0 and d, we have

$$d |u(x_0, y)|^2 \leq 2 \int_0^d |u(x_0, y)|^2 \, dx_0 + 2d^2 \int_0^d |u_x(\xi, y)|^2 \, d\xi, \tag{5.89}$$

and integrating both the terms of (5.89) with respect to y, between 0 and b, we obtain

$$d \int_0^b |u(0, y)|^2 \, dy \leq 2 \int_\Omega |u(x_0, y)|^2 \, dx_0 \, dy + 2d^2 \int_\Omega |u_x(\xi, y)|^2 \, d\xi \, dy, \tag{5.90}$$

or

$$d \|\rho(u)\|_{L^2(]0, b[)}^2 \leq 2 \|u\|_{L^2(\Omega)}^2 + 2d^2 \|u_x\|_{L^2(\Omega)}^2, \tag{5.91}$$

which proves that ρ is continuous. Naturally we can define the traces of u on BC, BD, and DA in an analogous manner, and hence the trace of u on $\Gamma = AB \cup BC \cup CD \cup DA$ which we will denote, for reasons which we will explain later, by the symbol $\gamma_0 u$.

The above proof is applicable, with simple modifications, to the case of a general open set Ω (as long as it is sufficiently regular) whose boundary has been previously 'flattened' by means of a system of local maps—hence the following theorem holds:

THEOREM 5.18. Let $\Omega \subset \mathbb{R}^n$ be a bounded open set with a Lipschitz continuous boundary Γ. There exists a unique application $\gamma_0 \in \mathscr{L}(H^1(\Omega), L^2(\Gamma))$ such that if $u \in \mathscr{D}(\bar{\Omega})$ then $\gamma_0 u = u|_\Gamma$. If $u \in H^1(\Omega)$ we will call $\gamma_0 u$ the *trace* (of order 0) of u on Γ. (For the definition of $L^p(\Gamma)$, see section 16.2.)

More generally one can prove the following result (see, e.g. Nečas, 1967, pp. 86 and 107):

THEOREM 5.19. Let $\Omega \subset \mathbb{R}^n$ be a bounded open set with a Lipschitz continuous boundary Γ, $k \geq 1$ an integer and $p \in [1, +\infty[$. Then the following propositions are true:

(I) If $kp < n$ and $1 \leq q \leq (n-1)p/(n-kp)$ then there exists a unique application $\gamma_0 \in \mathscr{L}(W^{k,p}(\Omega), L^q(\Gamma))$ such that if $u \in \mathscr{D}(\bar{\Omega})$ then

$\gamma_0 u = u|_\Gamma$—if $u \in W^{k,p}(\Omega)$ we call $\gamma_0 u$ the *trace* (of order 0) of u on Γ; if $p > 1$ then γ_0 is compact;
(II) If $kp = n$ then (I) holds for arbitrary $q \geqslant 1$;
(III) If $kp > n$ the trace $\gamma_0 u$ of $u \in W^{k,p}(\Omega) \subset C^0(\bar{\Omega})$ is the classical restriction.

One question which naturally arises is whether the trace operator γ_0 is surjective or not. The answer is no: the following reasoning (which we will develop under very particular conditions but which can easily be generalized) shows that γ_0 cannot be 'invertible' with continuity, which implies that it is not surjective. Let $p = 2$, $k = 1$, and $\Omega = S$ the unit open ball of $\mathbb{R}^2$ with boundary C; further let u_m be the sequence of elements of $L^2(C) \simeq L^1(]0, 2\pi[)$ defined by $u_1 = 1/\sqrt{(2\pi)}, \ldots, u_{2l} = \cos lx/\sqrt{(\pi)}$, $u_{2l+1} = \sin lx/\sqrt{(\pi)}, \ldots (l = 1, 2, \ldots)$: u_m is such that, for every m, $\|u_m\|_{L^2(C)} = 1$ and on the other hand $u_m \rightharpoonup 0$ in $L^2(C)$ (note that the functions u_m constitute a complete orthonormal system in $L^2(]0, 2\pi[)$). Were it possible to 'inverte' γ_0 with continuity, we could extend the u_m to functions $\tilde{u}_m \in H^1(S)$, with $\|\tilde{u}_m\|_{H^1(S)} \leqslant c$ (an appropriate constant): from the compactness of γ_0 it would then follow that there exists a subsequence $u_{m_i} = \gamma_0(\tilde{u}_{m_i})$ strongly convergent in $L^2(C)$, which is absurd.

However, in some particular cases γ_0 is surjective: thus, e.g., $\gamma_0(W^{1,1}(S)) = L^1(C)$. Besides $\overline{\gamma_0(H^1(S))}^{L^2(C)} = L^2(C)$; more generally, if Ω is a bounded open set with a Lipschitz continuous boundary Γ then $\overline{\gamma_0(W^{1,p}(\Omega))}^{L^p(\Gamma)} = L^p(\Gamma)$ (see, Nečas, 1967, p. 87).

The trace spaces

The non-surjectivity of γ_0 poses another problem: that of characterizing $\gamma_0(W^{s,p}(\Omega))$, i.e., of characterizing the elements of $L^q(\Gamma)$ which can be the trace of elements of $W^{s,p}(\Omega)$. In order to do this we need to introduce the so-called *trace spaces*:

DEFINITION 5.11. Let $s \geqslant 0$ be a real number, $p \in]1, +\infty[$ (or $p \in [1, +\infty[$ if $s \geqslant 1$ is an integer) and $\Omega \subset \mathbb{R}^n$ a bounded open set on whose boundary Γ of class $C^{[s],1}$ we assume defined a system of local maps under the conditions of definition 16.5. We denote by $W^{s,p}(\Gamma)$ the vector space $\{u \in L^p(\Gamma): u(x'_\alpha, \phi_\alpha(x'_\alpha)) \in W^{s,p}(\Delta\alpha), \alpha = 1, \ldots, m\}$ provided with the norm

$$\|u\|_{W^{s,p}(\Gamma)} = \left[\sum_{\alpha=1}^m \|u(x'_\alpha, \phi_\alpha(x'_\alpha))\|^p_{W^{s,p}(\Delta\alpha)} \right]^{1/p}. \quad (5.92)$$

If $s < 0$ and $p \in]1, +\infty[$ we denote by $W^{s,p}(\Gamma)$ the space $(W^{-s,p^*}(\Gamma))'$, where $p^* = p/(p-1)$. The spaces $W^{s,2}(\Gamma)$ will be denoted by $H^s(\Gamma)$.

The spaces $W^{s,p}(\Gamma)$ are Banach spaces, and they are reflexive if $p > 1$; further, many other results which have been mentioned for the spaces $W^{s,p}(\Omega)$ are also true for the spaces $W^{s,p}(\Gamma)$: among these, we will mention in particular theorem 5.17, regarding the Sobolev inclusions, and theorem

5.15, which in this case is stated as follows: $\mathscr{D}^b(\Gamma)$ *is dense in* $W^{s,p}(\Gamma)$, where $b \in \mathbb{N} \cup \{+\infty\}$ is the regularity of Γ (see Lions–Magenes, 1961b, p. 58 for the case in which Γ is indefinitely differentiable). In an analogous manner, one can define the spaces $H^{s,p}(\Gamma)$ (see, e.g., Lions–Magenes, 1961b, p. 68).

The normal derivative

Before proceeding to the study of the *theorems of traces*, which will enable us completely to characterize the images of the spaces $W^{s,p}(\Omega)$ through the operator γ_0, we will introduce the following concept:

DEFINITION 5.12. Let $\Omega \subset \mathbb{R}^n$ be a bounded open set with a boundary Γ of class $C^{0,1}$, let $\nu \equiv (\nu_1, \ldots, \nu_n)$ be the (external) normal to Γ, and let $u \in C^k(\bar{\Omega})$, with $k \geq 1$ being an integer. If $0 \leq j \leq k$ the function defined a.e. on Γ by the formula

$$\frac{\partial^j u}{\partial \nu^j} = \sum_{|\alpha|=j} \frac{j!}{\alpha!} (D^\alpha u|_\Gamma) \nu^j, \qquad (5.93)$$

where $\alpha! = \alpha_1! \alpha_2! \ldots \alpha_n!$ and $\nu^j = \nu_1^{\alpha_1} \nu_2^{\alpha_2} \ldots \nu_n^{\alpha_n}$ is said to be the *jth derivative of u with respect to ν*.

Note that this definition is meaningful since, on the one hand, the functions $D^\alpha u$ are defined on all the points of Γ and, on the other, Γ being Lipschitz continuous has a normal a.e. In the following theorem we will generalize this concept to the functions of $W^{s,p}(\Omega)$ by means of a tool analogous to the one we have introduced in order to generalize the concept of restriction (the operator γ_0); namely, we introduce a family of operators γ_j, $j = 1, 2, \ldots$, which we will call *trace operators* (of order $1, 2, \ldots$), defined on $W^{s,p}(\Omega)$ and which coincide with those defined by (5.93) when applied to a sufficiently regular function. The trace $\gamma_0 u$ (which we will often denote merely by u, which should not create any confusion) appears then as a *zero order derivative with respect to the normal* and γ_0 as the first element of the family of trace operators—this is why we have used the subscript 0 (zero) in γ_0.

Trace theorems

THEOREM 5.20 (The trace theorem). *Let* $p \in]1, +\infty[$, *let* $s \in]1/p, +\infty[$ *such that* $s - 1/p \notin \mathbb{N}$, *and let* $\Omega \subset \mathbb{R}^n$ *be a bounded open set with a boundary* Γ *of class* $C^{[s],1}$ *and (external) normal* $\nu \equiv (\nu_1, \ldots, \nu_n)$; *putting* $\{s\} = \max\{k \in \mathbb{N} : k < s\}$ *the following propositions are true*:

(I) *There exists an unique application*

$$\bar{\gamma} \equiv (\gamma_0, \gamma_1, \ldots, \gamma_{\{s-1/p\}}) \in \mathscr{L}\left(W^{s,p}(\Omega), \prod_{j=0}^{\{s-1/p\}} W^{s-j-1/p,p}(\Gamma)\right)$$

(5.94)

such that if $u \in \mathscr{D}(\bar{\Omega})$ then $\gamma_j u = \partial^j u/\partial \nu^j$, $j = 1, 2, \ldots, \{s-1/p\}$, $\gamma_0 u = u|_\Gamma$; $\bar{\gamma}u$ is called the *trace* of u on Γ. We remark that the continuity of $\bar{\gamma}$ can be expressed through the inequality

$$\sum_{j=0}^{\{s-1/p\}} \|\gamma_j u\|_{W^{s-j-1/p,p}(\Gamma)} \leq c \|u\|_{W^{s,p}(\Omega)}, \tag{5.95}$$

with c constant independent of u;

(II) There exists at least one operator $\mathscr{R} \in \mathscr{L}(\prod_{j=0}^{\{s-1/p\}} W^{s-j-1/p,p}(\Gamma), W^{s,p}(\Omega))$ such that given any $\{s-1/p\}+1$ functions $g_j \in W^{s-j-1/p,p}(\Gamma)$ then, putting $\bar{g} \equiv (g_0, \ldots, g_{\{s-1/p\}})$ and $u = \mathscr{R}(\bar{g})$, we have $\gamma_j u = g_j$ for $j = 0, \ldots, \{s-1/p\}$; u is called the 'inverse' of the system of traces $\bar{g}$ and $\mathscr{R}$ is called the operator of 'inversion'. The continuity of $\mathscr{R}$ is expressed through the inequality

$$\|u\|_{W^{s,p}(\Omega)} \leq c' \sum_{j=0}^{\{s-1/p\}} \|g_j\|_{W^{s-j-1/p,p}(\Gamma)}, \tag{5.96}$$

with c' independent of the g_j.

For a proof see Gagliardo, 1957; Nečas, 1967, pp. 103 and 104. It is important to note that if $s - 1/p \in \mathbb{N}$ we must consider the family of Besov spaces: the family of Sobolev spaces is not 'closed' for the trace operators—Besov spaces are in fact the 'good' trace spaces. Let us remark that part (II) of the theorem tells us that $\bar{\gamma}$ is surjective; on the other hand the 'inverse' of the system $\{g_j\}$ is not unique, as a consequence of the fact that $\bar{\gamma}$ is not injective. As a first application of this theorem we observe that the range of the application $\tilde{\rho}$ which we considered at the beginning of this section is the space $H^{1/2}(]0, b[)$.

Let us present now the following interesting characterization of the spaces $W_0^{s,p}(\Omega)$, based on the trace operators:

THEOREM 5.21. *If* $s > 0$, $p \in]1, +\infty[$ (*or* $p \in [1, +\infty[$ *if* $s \geq 1$ *is an integer*) *and* $\Omega \subset \mathbb{R}^n$ *is a bounded open set with a boundary* Γ *of class* $C^{[s],1}$ *then*

$$W_0^{s,p}(\Omega) = \ker(\bar{\gamma}) = \{u \in W^{s,p}(\Omega) : \gamma_j u = 0, j = 0, 1, \ldots, \{s-1/p\}\}. \tag{5.97}$$

For the proof see, e.g., Lions–Magenes, 1961b, p. 69; Nečas, 1967; p. 90; Lions, 1965; p. 87. In some cases the theorem can be proved under less restrictive conditions on Ω; thus, e.g., if $s = 1$ it is sufficient that Ω be of class $C^{0,1}$ (see Nečas, 1967, p. 87). This result enables us to justify the statement: *the functions of $W_0^{s,p}(\Omega)$ are null on the boundary of Ω*, and others of the same kind.

For the functions which satisfy appropriate regularity conditions, for example, partial derivatives in $L^2(\Omega)$, the definition of the trace as an extension by continuity is not the only one possible. For another point of view which provides a definition of the concept of trace, see Fichera, 1950; see also De Vito, 1958 and the relative bibliography. Under less regular conditions one can still define the trace: thus, for the functions u of $H^1(\Omega)$ such that $\Delta u \in L^2(\Omega)$, which are less regular than the functions of $H^2(\Omega)$, it is meaningful to speak of the trace of the normal derivative; for this type of problem we have developed an approach in Chapter 18, and other possible treatments have been mentioned at the end of that chapter.

5.5 NORMAL CONTRACTIONS AND DIRICHLET SPACES

The one-dimensional case

In this section we will present some properties of the spaces $H^1(\Omega)$ and $H_0^1(\Omega)$ which we will use later, in particular when we study the lattice structure of these spaces.

To introduce the subject we will be dealing with, let us first consider the following problem: '*if $u \in H_0^1(]a, b[)$, $-\infty < a < b < +\infty$, then $|u| \in H_0^1(]a, b[)$?*' The answer is yes. In fact, if $u \in H_0^1(]a, b[)$ then $u \in AC([a, b])$ and $|u| \in AC([a, b])$ (this follows from $u \in AC([a, b])$ and from the inequality $||u|(y) - |u|(x)| = ||u(y)| - |u(x)|| \leq |u(y) - u(x)|$, which holds for every $y, x \in [a, b]$): so there exists almost everywhere in $[a, b]$ the limits, as $y \to x$, of the difference ratios

$$\frac{||u|(y) - |u|(x)|}{|x - y|} \leq \frac{|u(y) - u(x)|}{|x - y|}, \qquad (5.98)$$

which leads to

$$||u|'(x)| \leq |u'(x)| \qquad \text{a.e. in } [a, b] \qquad (5.99)$$

and hence to $|u|' \in L^2(]a, b[)$. This result is true, with the same proof, if instead of $H_0^1(]a, b[)$ we consider the space $H^1(]a, b[)$ (and more generally we can consider the space $H^1(\Omega)$ by splitting Ω into its connected components...). This reasoning is not, however, applicable to the case in which Ω is an open set of $\mathbb{R}^n$, $n \geq 2$, since in general the elements of $H_0^1(\Omega)$ are not absolutely continuous functions. However, we have the following interesting characterization of the elements of $H^1(\Omega)$, where $\Omega \subset \mathbb{R}^n$ is an open set with a Lipschitz continuous boundary: *every function $u \in H^1(\Omega)$ conveniently modified in a set of measure zero is absolutely continuous on almost all the lines parallel to the axis of the Cartesian system to which $\mathbb{R}^n$ is referred*; *conversely, if $u \in L^2(\Omega)$ is absolutely continuous on almost all the lines parallel to the axis and its usual derivatives (which then exist almost everywhere) are square summable, then $u \in H^1(\Omega)$ and its distributional derivatives coincide with the usual ones*: for a proof see, e.g., Nečas, 1967, p. 61; see also Mizohata, 1973, p. 73.

Normal contractions and Dirichlet spaces

To study whether '$u \in H_0^1(\Omega) \Rightarrow |u| \in H_0^1(\Omega)$' and other questions of the same type in the case where $\Omega \subset \mathbb{R}^n$, $n \geq 2$, we will use some concepts widely used in potential theory (the 'natural' context is provided here by the functions of one complex variable ...). Thus:

DEFINITION 5.13. Let A be an arbitrary subset of $\mathbb{R}^n$; if $u, v \in \mathcal{F}(A) = \mathbb{C}^A$ we say that v is a *normal contraction of u* if

$$|v(x) - v(y)| \leq |u(x) - u(y)| \quad \forall x, y \in A \qquad (5.100)$$

$$|v(x)| \leq |u(x)| \quad \forall x \in A. \qquad (5.101)$$

DEFINITION 5.14. $T : \mathbb{C} \to \mathbb{C}$ is said to be a *normal contraction of the complex plane* if

$$|T(x) - T(y)| \leq |x - y| \quad \forall x, y \in \mathbb{C} \qquad (5.102)$$

$$T(0) = 0. \qquad (5.103)$$

DEFINITION 5.15. Let A be an arbitrary subset of $\mathbb{R}^n$ and H a Hilbert space such that $H \subset \mathcal{F}(A)$ setwise. H is said to be a *Dirichlet space* if the following condition is satisfied: 'if $u \in H$ and T is a normal contraction of $\mathbb{C}$ then $T(u) \in H$ and $\|T(u)\|_H \leq \|u\|_H$' (i.e., Dirichlet spaces are Hilbert spaces 'closed' for the normal contractions).

The following result shows that there is a close connection between the concepts introduced in definitions 5.13 and 5.14:

THEOREM 5.22. If T is a normal contraction of $\mathbb{C}$ and $u \in \mathcal{F}(A)$ then $T \circ u = T(u)$ is a normal contraction of u; conversely, if v is a normal contraction of u then there exists at least one normal contraction T of $\mathbb{C}$ such that $v = T(u)$.

Proof. The first part is obvious. In order to show the second part, we consider the application $t : u(A) \cup \{0\} \to \mathbb{C}$ defined by $t(u(x)) = v(x)$ and $t(0) = 0$. If $0 \in u(A)$ from (5.101) we have $t(0) = 0$ and on the other hand (5.100) shows that if $u(x) = u(y)$ then $t(u(x)) = t(u(y))$. The application t is then well defined; we now have to prove the existence of an extension T of t under the conditions laid down—this is ensured by the Kirszbraun–Valentine lemma which states that *if $C_1 \subset C_2$ are two subsets of $\mathbb{R}^n$ and $f : C_1 \to \mathbb{C}$ is a Lipschitz continuous map with constant L then there exists a Lipschitz continuous map with the same constant L, $F : C_2 \to \mathbb{C}$, such that $f = F|_{C_1}$.* The lemma is applied with $C_1 = u(A) \cup \{0\}$, $C_2 = \mathbb{R}^2 \simeq \mathbb{C}$, $L = 1$, $f = t$ and $F = T$. ∎

For a proof of the Kirszbraun–Valentine lemma see, e.g., Valentine, 1943, p. 105; Mickle, 1949, p. 164; in the latter paper a very general result has been proved which has as particular cases not only the Kirszbraun–Valentine lemma but also an analogous result with respect to Hölder continuous functions—for other general results see Wells–Williams, 1975; for an interesting geometric interpretation of the lemma see Minty, 1969, p. 69; Valentine, 1943, p. 101. Let us remark also that F can be constructed in such a way that $F(C_2)$ is an arbitrarily assigned subset of a closed convex set $K \supset F(C_2)$.

If $u \in \mathscr{F}(A)$ the following normal contractions of u are of particular importance: the contraction *modulus* ($|u|$), the contraction *real part* ($\mathscr{R}e\,(u) = (\bar{u} + u)/2$, $\bar{z}$ being the conjugate of the complex number z), the contraction *imaginary part* ($\mathscr{I}m\,(u) = (\bar{u} - u)/2i$, $i = \sqrt{(-1)}$) and, if u is real, the contraction *positive part* ($[u]^+ = (|u| + u)/2$) and the contraction *negative part* ($[u]^- = (|u| - u)/2$).

The n-dimensional case

The main purpose of this section is to show that if $u \in H_0^1(\Omega)$ (or: $u \in H^1(\Omega)$) then $|u|$, $\mathscr{R}e\,(u)$, $\mathscr{I}m\,(u)$ and, if u is real, $[u]^+$ and $[u]^-$ are still elements of $H_0^1(\Omega)$ (or: $H^1(\Omega)$). Now, theorem 5.22 ensures that with these normal contractions can be associated normal contractions of the complex plane—which we will again denote by $|\cdot|$, $\mathscr{R}e$, $\mathscr{I}m$, $[\cdot]^+$, $[\cdot]^-$—and hence the problem is solved by means of the following theorem:

THEOREM 5.23. If $\Omega \subset \mathbb{R}^n$ is a bounded open set then $H_0^1(\Omega)$ is a Dirichlet space; if, further, Ω is Lipschitz continuous then $H^1(\Omega)$ is also a Dirichlet space.

Proof. It is enough to show the first part of the theorem since the second part follows immediately as a corollary (since if Ω is Lipschitz continuous then, given any open set $\mathscr{O} \supset \bar{\Omega}$, the functions of $H^1(\Omega)$ can be extended to $\mathscr{O}$ as elements of $H_0^1(\mathscr{O})$; besides we can adapt for $H^1(\Omega)$ the discussion we will present for $H_0^1(\Omega)$—one would then exploit the fact that $\mathscr{D}(\bar{\Omega})$ is dense in $H^1(\Omega)$, which implies an assumption regarding the regularity of Ω, e.g. that it is Lipschitz continuous).

We will now prove that $H_0^1(\Omega)$ is a Dirichlet space by showing that if T is a normal contraction of $\mathbb{C}$ and $u \in H_0^1(\Omega)$ then $T(u) \in H_0^1(\Omega)$ and $\|T(u)\|_{H_0^1(\Omega)} \leq \|u\|_{H_0^1(\Omega)}$. The proof will be in two parts: in the first we will show that if $u \in \mathscr{D}(\Omega)$ then $T(u) \in H_0^1(\Omega)$ and $\|T(u)\|_{H_0^1(\Omega)} \leq \|u\|_{H_0^1(\Omega)}$; in the second we will consider the general case, exploiting the fact that $\mathscr{D}(\Omega)$ is dense in $H_0^1(\Omega)$ and the conclusions of the first part.

If $u \in \mathscr{D}(\Omega)$ then u and $T(u)$ are absolutely continuous functions on every line parallel to the axis and hence there exist almost everywhere the limits, as $h \to 0$ (we use the notations of Chapter 15), of the difference ratios

$$\left| \frac{T(u)(x) - T(u)(x + he_i)}{h} \right| \leq \left| \frac{u(x) - u(x + he_i)}{h} \right|, \quad i = 1, \ldots, n \quad (5.104)$$

which leads to

$$\left| \frac{\partial T(u)}{\partial x_i} \right| \leq \left| \frac{\partial u}{\partial x_i} \right|, \quad i = 1, \ldots, n; \quad (5.105)$$

from this together with

$$|T(u)(x)| = |T(u)(x) - T(0)| \leq |u(x) - 0| = |u(x)|, \qquad (5.106)$$

which implies in particular that $T(u) \in L^2(\Omega)$ (the measurability of $T(u)$ follows from (5.102) and from that of u), it follows that $T(u) \in H_0^1(\Omega)$. From (5.105) and (5.106) we have besides that

$$\|T(u)\|_{H_0^1(\Omega)} \leq \|u\|_{H_0^1(\Omega)}. \qquad (5.107)$$

We now proceed to the second part. Let u be a general element of $H_0^1(\Omega)$: by the definition of this space we can ensure the existence of a sequence u_m of elements of $\mathcal{D}(\Omega)$ such that $u_m \to u$ in $H_0^1(\Omega)$. From $u_m \to u$ in $H_0^1(\Omega)$ and from the first part we have

$$\exists c > 0 \quad \forall m \in \mathbb{N} \; \|T(u_m)\|_{H_0^1(\Omega)} \leq \|u_m\|_{H_0^1(\Omega)} < c, \qquad (5.108)$$

and hence from the sequence $T(u_m)$ we can extract a subsequence which is weakly convergent in $H_0^1(\Omega)$: let $T(u_{m_l})$ be this subsequence and $v = \lim{}^* T(u_{m_l})$ its weak limit. On the other hand, from $|T(u_m) - T(u)| \leq |u_m - u|$ we have that $T(u_m) \to T(u)$ in $L^2(\Omega)$ and hence, from the uniqueness of the weak limit, that $v = T(u)$: thus $T(u) \in H_0^1(\Omega)$. We can easily see also that the whole sequence $T(u_m)$ converges to $T(u)$ and hence

$$\|T(u)\|_{H_0^1(\Omega)} = \|\lim{}^* T(u_m)\|_{H_0^1(\Omega)} \leq \lim{}' \|T(u_m)\|_{H_0^1(\Omega)}$$
$$\leq \lim{}' \|u_m\|_{H_0^1(\Omega)} = \|u\|_{H_0^1(\Omega)}, \qquad (5.109)$$

which is the inequality that we wished to prove. Let us remark that if we know that $\|T(u)\|_{H_0^1(\Omega)} = \|u\|_{H_0^1(\Omega)}$, as for the contraction $|\cdot|$, then (5.109) ensures that the convergence of $T(u_m)$ to $T(u)$ is also strong in $H_0^1(\Omega)$. ∎

In the last part of the above proof we have incidentally shown for $H_0^1(\Omega)$ a property which is true in every Dirichlet space: *if u_m is a sequence of elements of a Dirichlet space H which is strongly convergent to u in H, and if T is a normal contraction of $\mathbb{C}$, then $T(u_m)$ converges weakly to $T(u)$ in H, the convergence being strong if $\|T(u)\| = \|u\|_H$* (indeed, this result is true independently of the latter assumption: see Ancona, 1976).

Among the other properties of Dirichlet spaces, and hence in particular of the spaces $H_0^1(\Omega)$ and $H^1(\Omega)$, we mention the following:

THEOREM 5.24. *Let H be a Dirichlet space which is setwise contained in $\mathcal{F}(A)$; then the following propositions hold:*

(I) If $u \in H$ is a function bounded by a constant c (i.e. if $\forall x \in A \; |u(x)| \leq c$) then $u^2 \in H$ and

$$\|u^2\|_H \leq 2d \|u\|_H, \qquad (5.110)$$

where $d = \max\{c, 1/2\}$;

(II) If u and v are functions bounded by c' and c'' respectively, then $uv \in H$ and

$$\|uv\|_H \leq d'(\|u\|_H + \|v\|_H), \tag{5.111}$$

where $d' = \max\{c' + c'', 1\}$;

(III) If $u, v \in H$ are real functions then (cf. definitions (5.115) and (5.116)) $\inf(u, v) \in H$, $\sup(u, v) \in H$ and

$$\|\inf(u, v)\|_H^2 + \|\sup(u, v)\|_H^2 \leq \|u\|_H^2 + \|v\|_H^2. \tag{5.112}$$

Proof. To prove (I) we need only to see that u^2 is a normal contraction of $2du$ since

$$|u^2(x) - u^2(y)| = |u(x) + u(y)||u(x) - u(y)| \leq 2d|u(x) - u(y)|. \tag{5.113}$$

For (II), it is enough to write the product uv in the form

$$uv = \tfrac{1}{4}[(u+v)^2 - (u-v)^2] \tag{5.114}$$

and use result (I). Finally for (III) recall that

$$\inf(u, v) \triangleq \tfrac{1}{2}(u+v) - \tfrac{1}{2}|u-v|, \tag{5.115}$$

$$\sup(u, v) \triangleq \tfrac{1}{2}(u+v) + \tfrac{1}{2}|u-v|, \tag{5.116}$$

and that for the contraction $|\cdot|$, as for any contraction, $\||u|\|_H \leq \|u\|_H$. ∎

If $u \in H_0^1(]a, b[)$ we can easily see that $|u|'(x) = u'(x) \cdot \text{sgn } u(x)$ a.e. in $[a, b]$ (the real function of a real variable sgn (for 'sign') is so defined: $\text{sgn } x = 1$ if $x > 0$, $\text{sgn } x = -1$ if $x < 0$ and $\text{sgn } 0 = 0$). It is natural then to ask whether there is an analogous result in several dimensions and whether, in general, there is one for other normal contractions of u. The answer is given by the following

THEOREM 5.25. *If $u \in H_0^1(\Omega)$ and T is a normal contraction of $\mathbb{C}$ such that T' has a finite number of points of discontinuity then*

$$\frac{\partial T(u)}{\partial x_i} = \frac{\partial u}{\partial x_i} T'(u), \quad i = 1, \ldots, n, \tag{5.117}$$

a.e. in the sense of $\mathscr{D}'(\Omega)$.

For the proof see, e.g., Stampacchia, 1966, pp. 15 and 321.

6
Examples of Variational Problems in One Dimension

6.1 THE OBSTACLE PROBLEM. GENERALITIES ABOUT REGULARITY RESULTS

Variational formulation of the obstacle problem

The Sobolev spaces introduced in the previous chapter provide a framework which is suitable for the formulation of a large number of problems. As the first example we will consider once again the problem introduced in section 1.1 of a string stretched below an obstacle. The formulation of the problem through the equations (1.1), (1.2), (1.3), and (1.4) or (1.5) (problem 1.0) leads us to the conclusion that it is of (differential) order 2, since in these equations the second derivative of the unknown function appears.

Now, the previous formulation is one of the simplest possible but is not the only one. Thus, a physicist would most probably prefer to formulate the obstacle problem by saying that the configuration that the string takes is such as to minimize the energy of the system. We will see that these two formulations are equivalent, and that the second can be expressed using only first derivatives; further this second formulation will be of the *variational type*.

In the system made up of the string and the obstacle, given that the only energy involved is that due to elastic deformation (the string being weightless), we can say, using the principle of minimum energy, that *the configuration u taken by the string is that which minimizes the energy of elastic deformation*

$$E(v) = \frac{1}{2} \int_0^l v'(x)^2 \, dx \qquad (6.1)$$

(note that the string is homogeneous) *among all the configurations which are consistent with the constraints imposed on the string*. In other words, we can reformulate problem 1.0 thus:

PROBLEM 1.0*. Find $u \in K$ such that

$$E(u) \leq E(v) \quad \forall v \in K, \tag{6.2}$$

where $K = \{v : v(0) = v(l) = 0 \text{ and } v \leq \psi\}$ is the set of possible configurations.

The problem 1.0* is incomplete because we have not specified the regularity of v and ψ and as such, in particular, K is not completely defined. We will try to complete it so that it becomes a well-posed problem.

Let us point out that in the present chapter, except when the contrary is stated, only real spaces are considered and so, in particular, we can say that a function is, for example, less or equal to another one.

The expression (6.1) leads us to require that v has a square summable first derivative in $[0, l]$, so that the integral is meaningful. It is natural, therefore, to impose $v \in H^1(]0, l[)$; in our particular case we can even consider the space $H^1_0(]0, l[)$, which we will do later (but if P_1 or P_2 (or both) are not points on the xx-axis the formulation in $H^1(]0, l[)$ is then necessary).

As for ψ we begin by assuming simply that it is continuous in $]0, l[$ since this is sufficient for defining K completely:

$$K = \{v \in H^1(]0, l[) : v(0) = v(l) = 0; \; v(x) \leq \psi(x) \quad \forall x \in]0, l[\}$$
$$= \{v \in H^1_0(]0, l[) : v(x) \leq \psi(x) \quad \forall x \in]0, l[\}. \tag{6.3}$$

This definition of K is meaningful since v is continuous in $[0, l]$ (or rather, absolutely continuous). One can easily see that K is a closed convex set, possibly empty; in figs 6.1 and 6.2 we present two situations in which $K = \emptyset$—in both of these $v \leq \psi \Rightarrow v \notin H^1_0(]0, l[)$ (let us note that $\psi(x) = -(xl - x^2)^{1/2} \notin H^1(]0, l[)$ and from the characterization $v \in H^1_0(]0, +\infty[)$ iff $v \in H^1(]0, +\infty[)$ and $x^{-1}v(x) \in L^2(]0, +\infty[)$ (see, e.g., Lions–Magenes, 1972a, p. 69) it follows more generally that if $v(x) \leq \psi(x)$ then $v \notin H^1_0(]0, l[)$).

To avoid K being empty, and the problem therefore becoming meaningless, we impose further restrictions on ψ; namely, ψ must satisfy at least one of the following conditions:

(ψ1) $\psi \in C^0(]0, l[), \psi(0^+) > 0$ and $\psi(l^-) > 0,$

(ψ2) $\psi \in H^1(]0, l[), \psi(0) \geq 0$ and $\psi(l) \geq 0$

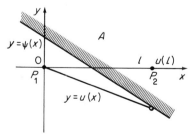

Figure 6.1

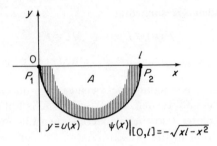

Figure 6.2

(if P_1 or P_2 do not belong to the x-axis we have to impose conditions of the type '$\psi(0^+)>$ordinate of P_1', etc.).

One can easily see that with these conditions, K is non-empty: if ψ satisfies (ψ1) then a function of the type shown in fig. 6.3 is one element of K (note that (ψ1) implies the existence of $\delta>0$ under the conditions shown in the figure); and if ψ satisfies (ψ2) then the function $v=-[\psi]^-$ belongs to K (see theorem 5.23).

We have now completed the formulation 1.0* of the obstacle problem:

PROBLEM 6.1. Find $u \in K$ such that

$$E(u) \leq E(v) \quad \forall v \in K \tag{6.4}$$

with E given by (6.1) and K by (6.3) where ψ satisfies (ψ1) or (ψ2).

Writing the functional E in the form

$$E(v) = \tfrac{1}{2} a(v, v), \tag{6.5}$$

where $a(u, v) : H_0^1(]0, l[) \times H_0^1(]0, l[) \to \mathbb{R}$ is the symmetric, coercive and continuous (on $H_0^1(]0, l[)$ and hence in particular on $K - K \subset H_0^1(]0, l[)$) bilinear form defined by

$$a(u, v) = \int_0^l u'(x) v'(x) \, dx \tag{6.6}$$

(one should keep in mind theorem 5.13 which tells us that $a(u, v)$ is a scalar product in $H_0^1(]0, l[)$), one can easily see that we are under the conditions of

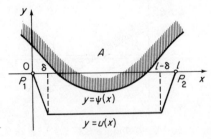

Figure 6.3

theorem 2.2 which ensure thus that there exists one and only one solution of problem 6.1.

On the other hand, from theorem 2.3 we can see that problem 6.1 is equivalent to the following variational problem:

PROBLEM 6.2. Find $u \in K$ such that

$$\int_0^l u'(x)(u'(x) - v'(x))\,dx \leq 0 \qquad \forall v \in K, \qquad (6.7)$$

where (6.7) are the Euler inequalities associated with the minimum problem 6.1.

Interpretation of the variational problem as a boundary value problem

We will now show that the formulation 6.2 is equivalent to the formulation 1.0 appropriately completed. We will first show that the solution of problem 6.2 is the solution of a 'good completion of problem 1.0' (which will be problem 6.3) and the reasoning is roughly as follows: we start from problem 6.2 and look for ways in which (1.1), (1.2), (1.3), (1.4), and (1.5) can be interpreted, knowing that we look for u in $H_0^1(]0, l[)$ and that ψ is in $\{\psi \in C^0(]0, l[) : (\psi 1)\}$ or in $\{\psi \in H^1(]0, l[) : (\psi 2)\}$ (in fact (1.5) will not be considered since it follows from the others).

From $u \in K$ it follows immediately that

$$u(0) = u(l) = 0 \quad \text{in the sense of } C^0([0, l]) \qquad (1.1)^*$$

and

$$u(x) \leq \psi(x) \quad \text{in the sense of } C^0(]0, l[). \qquad (1.2)^*$$

Also, if $u \in H^1(]0, l[)$ then $u' \in L^2(]0, l[)$ and, given that $L^2(]0, l[) \subset \mathcal{D}'(]0, l[)$, it is meaningful to consider $u'' = Du' \in \mathcal{D}'(]0, l[)$; we will now show that if u is a solution of problem 6.2 then

$$u'' \geq 0 \quad \text{in the sense of } \mathcal{D}'(]0, l[), \qquad (1.3)^*$$

i.e.,

$$_{\mathcal{D}'(]0,l[)}\langle u'', \phi \rangle_{\mathcal{D}(]0,l[)} \geq 0 \qquad \forall \phi \in \mathcal{D}_+(]0, l[). \qquad (6.8)$$

Let then $\phi \in \mathcal{D}_+(]0, l[)$ and consider $v_\phi = u - \phi$; one can easily see that, no matter what $\phi \in \mathcal{D}_+(]0, l[)$ is, $v_\phi \in K$ and hence, from (6.7),

$$\int_0^l u'(x)\phi'(x)\,dx \leq 0 \qquad \forall \phi \in \mathcal{D}_+(]0, l[). \qquad (6.9)$$

Now, from the definition of derivative in $\mathcal{D}'(]0, l[)$,

$$\int_0^l u'(x)\phi'(x)\,dx = -_{\mathcal{D}'(]0,l[)}\langle Du', \phi \rangle_{\mathcal{D}(]0,l[)}$$

$$= -_{\mathcal{D}'(]0,l[)}\langle u'', \phi \rangle_{\mathcal{D}(]0,l[)} \qquad (6.10)$$

and thus we have shown $(1.3)^*$ to be true.

Finally we consider (1.4). For this purpose, we decompose $]0, l[$ into the sets in which the (continuous) function $u - \psi$ is negative and where it is zero, i.e., we put $]0, l[= I^- \cup I^0$ where $I^- = \{x \in]0, l[: u(x) < \psi(x)\}$ and $I^0 = \{x \in]0, l[: u(x) = \psi(x)\}$ (the latter set is for obvious reasons known as the *contact set*). Given that u and ψ are continuous, I^- must be an open set and as such it makes sense to consider $\mathcal{D}'(I^-)$; further, since in $(1.3)^*$ we have considered u'' in the sense of distributions, it is natural to write, denoting also by u'' the restriction of the distribution $u'' \in \mathcal{D}'(]0, l[)$ to I^-,

$$u < \psi \Rightarrow u'' = 0 \quad \text{in the sense of } \mathcal{D}'(I^-). \qquad (1.4)^*$$

We will now prove $(1.4)^*$; in other words we will prove

$$_{\mathcal{D}'(I^-)}\langle u'', \phi \rangle_{\mathcal{D}(I^-)} = 0 \quad \forall \phi \in \mathcal{D}(I^-). \qquad (6.11)$$

One can easily see that if $\phi \in \mathcal{D}(I^-)$ then there exists $\lambda_\phi \in \mathbb{R}_+ \setminus \{0\}$ such that if $|\lambda| \leq \lambda_\phi$ then (denoting also by ϕ the trivial extension of ϕ to $]0, l[$) $u + \lambda \phi \leq \psi$ in $]0, l[$ and hence $v = u + \lambda \phi \in K$ (it is enough to see that $\min_{\text{supp } \phi}(\psi - u) = m > 0$ (m exists because $\psi - u$ is continuous in supp ϕ which is compact—on the other hand $m > 0$ because supp $\phi \subset I^-$) and hence in order that $\psi - u \geq \lambda \phi$ we can take $\lambda_\phi = m/\max_{I^-} |\phi|$). Introducing this function v in (6.7) we have

$$\int_0^l u'(x)(-\lambda \phi'(x)) \, dx \leq 0 \quad \forall \lambda \in \mathbb{R}, |\lambda| \leq \lambda_\phi, \qquad (6.12)$$

or

$$-\lambda \int_0^l u'(x) \phi'(x) \, dx \leq 0 \quad \forall \lambda \in \mathbb{R}, |\lambda| \leq \lambda_\phi \qquad (6.13)$$

from which, since λ can be negative or positive given that ϕ is arbitrary, (6.11) follows.

We thus come to the conclusion that if u is a solution of problem 6.2 then u is a solution of the following problem:

PROBLEM 6.3. Given ψ in $\{\psi \in C^0(]0, l[) : (\psi 1)\}$ or in $\{\psi \in H^1(]0, l[) : (\psi 2)\}$ find $u \in H_0^1(]0, l[)$ which satisfies $(1.1)^*$, $(1.2)^*$, $(1.3)^*$, $(1.4)^*$.

Equivalence of the two formulations of the obstacle problem

We will now show that problem 6.3 is well-posed and hence that it is a good completion of problem 1.0. To do this all we need to show is that if u is a solution of problem 6.3 then u is a solution of problem 6.2, since this shows that u is unique—its existence being ensured by the above reasoning and the continuous dependence from ψ being an immediate consequence of the fact that this is true for problem 6.2.

To show that problem 6.3 $\Rightarrow$ problem 6.2 we will assume that $u'' \in L^1(]0, l[)$ and hence we will take $(1.3)^*$ and $(1.4)^*$ to be true a.e. instead of in the sense of distributions (we will return later to this hypothesis on u''). The equations $(1.1)^*$ and $(1.2)^*$ imply that $u \in K$, and hence we have only to prove that

$$\int_0^l u'(x)(u'(x) - v'(x)) \, dx \le 0 \quad \forall v \in K. \tag{6.14}$$

We note in the first place that it is meaningful to consider this integral since $u', v' \in L^2(]0, l[)$ and hence $u'^2 \in L^1(]0, l[)$ and $u'v' = \frac{1}{2}[(u'+v')^2 - u'^2 - v'^2] \in L^1(]0, l[)$. Integrating the first term of (6.14) by parts, which we can do because of the assumption that $u'' \in L^1(]0, l[)$, we have

$$\int_0^l u'(x)(u'(x) - v'(x)) \, dx = \int_0^l (-u''(x))(u(x) - v(x)) \, dx$$
$$+ [u'(x)(u(x) - v(x))]_0^l \tag{6.15}$$

or, given that $u, v \in K$ and hence $[u'(x)(u(x) - v(x))]_0^l = 0$,

$$\int_0^l u'(x)(u'(x) - v'(x)) \, dx = \int_0^l (-u''(x))(u(x) - v(x)) \, dx$$
$$= \int_{I^-} (-u''(x))(u(x) - v(x)) \, dx$$
$$+ \int_{I^0} (-u''(x))(u(x) - v(x)) \, dx \tag{6.16}$$

or yet again, using $(1.4)^*$,

$$\int_0^l u'(x)(u'(x) - v'(x)) \, dx = \int_{I^0} (-u''(x))(u(x) - v(x)) \, dx; \tag{6.17}$$

now, given that $u \equiv \psi$ in I^0, we can write (6.17) as follows

$$\int_0^l u'(x)(u'(x) - v'(x)) \, dx = \int_{I^0} (-u''(x))(\psi(x) - v(x)) \, dx, \tag{6.18}$$

from which follows (6.14) since $-u'' \le 0$ and $\psi - v \ge 0$.

We have thus shown that the two problems are equivalent as long as $u'' \in L^1(]0, l[)$. The question which naturally arises now is: *does u'' effectively belong to $L^1(]0, l[)$?*, i.e. is $u'' \in L^1(]0, l[)$ a consequence of the fact that u is a solution of the problem? The answer is no, and the situation shown in fig. 6.4 provides a counter-example (note that u''', being a multiple of Dirac's δ concentrated in $x = l/2$, cannot belong to $L^1(]0, l[)$).

The proof of the equivalence of the two problems is not therefore complete. We will not, however, prove this result, which is true independently of the assumption $u'' \in L^1(]0, l[)$, since we will return to questions of this type in a much wider context and we prefer, for the present, to develop the problems raised by the above question.

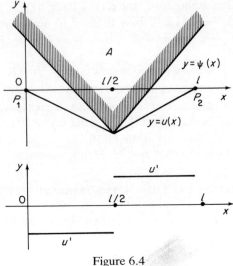

Figure 6.4

Basically, we are asking whether the solution of problem 6.3, which is to be looked for *a priori* in $H_0^1(]0, l[)$, belongs to a smaller set (namely to $\{u \in H_0^1(]0, l[) : u'' \in L^1(]0, l[)\}$), i.e.: we are asking whether the regularity of u is more than was necessary in order to formulate the problem itself.

The answer turns out to be no, but this question raises another: *what further assumptions need we make on ψ in order to have $u'' \in L^1(]0, l[)$?* It can

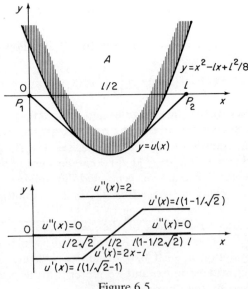

Figure 6.5

be shown (see, e.g., Mosco–Troianiello, 1973, p. 65) that it is sufficient that ψ be such that ψ'' is a Radon measure and its positive part can be represented by a function $f \in L^1(]0, l[)$. Another question, which is in a sense the inverse of the former, is the following: *what does increasing the regularity of ψ do to that of u?*. In our case we should not expect much: fig. 6.5 shows a case in which $\psi \in C^\infty(]0, l[)$ and on the other hand $u \notin C^2(]0, l[)$ (in fact u''' involves Dirac's δ in $x = l/2\sqrt{2}$ and in $x = l(1 - 1/2\sqrt{2})$). One can show (see, e.g., Mosco–Troianiello, 1973, p. 66) that if ψ satisfies (ψ1) or (ψ2) and ψ'' is a Radon measure such that $[\psi'']^+ \in L^p(]0, l[)$, $p \in \,]1, +\infty[$, then the solution u of the problem of the obstacle belongs to $W^{2,p}(]0, l[)$ and therefore, by theorem 5.17, to $C^{1,1-1/p}([0, l])$. As is evident, under particular conditions the regularity of u can be greater—thus, e.g., if $\psi(x) = x^2 - lx$ then $u \in C^\infty([0, l])$.

Results of regularity

Problems of the type considered above are called *regularity problems*.

Summarizing and generalizing, we can say that if the solution $u = \{i_m\}_{m=1,...,M}$ of a well-posed problem is to be looked for in a space $I = \prod \mathcal{T}_m$ and the data $d = \{d_n\}_{n=1,...,N}$ are taken from a space $D = \prod \mathcal{D}_n$ then we can essentially consider three types of *regularity results*:

(i) Find the effective regularity of u, i.e., find a subspace I' of I such that for every $d \in D$ then $u \in I'$ and on the other hand I' does not have a proper subspace under these conditions.

(ii) Given a subspace I' of I find a subspace D' of D such that if $d \in D'$ then $u \in I'$ and further D' does not have a proper overspace under these conditions.

(iii) Given a subspace D' of D find a subspace I' of I such that $u \in I'$ if $d \in D'$ and further I' does not have a proper subspace under these conditions.

The first type of result deserves some comment. Let us suppose that a problem $\mathcal{P}$ is well-posed with respect to a pair of spaces (I, D) and we prove that its solutions $u(d)$, with d varying in D, belong to a subspace I' of I. This means that the problem is well-posed with respect to (I', D). Why then did we not formulate the problem in (I', D) right from the beginning? Essentially there can be two reasons: either it is much more difficult to show that the problem $\mathcal{P}$ is well-posed in (I', D) than to show that it is so in (I, D), or the formulation in (I, D) is more 'natural' than that in (I', D). The term 'natural' is necessarily vague: we have seen that the spaces which are useful for the proper formulation of problem 1.0* follow naturally from the need to make the functional $E(v)$ meaningful (which itself arises somewhat mysteriously ...) and that these spaces allow an (equivalent) proper formulation of problem 1.0, but here it is not obvious that these are the 'natural' spaces ...

Let us remark also, always referring to the first type of results, that the determination of I' is in general an extremely difficult problem and often we have to content ourselves with a subspace which does not satisfy the second condition (i.e. that is not minimal)—and we have an analogous situation for results of types (ii) and (iii). On the other hand, faced with a concrete problem in which the data d is fixed, it is often interesting to determine the regularity of that particular solution u which corresponds to d.

The choice of spaces in which to pose a problem is one of the most important and difficult stages of its treatment. In principle the data spaces should be as large as possible, for reasons of generality; as for the spaces of the unknowns we can say roughly that they should be large enough to be able to prove results regarding existence and small enough to be able to prove results regarding uniqueness; finally both of them must be suitable for the problem in order to prove results regarding regularity of the solutions.

The membrane problem

The generalization of the problem of the string and the obstacle to two dimensions consists of finding the form which an elastic membrane takes when stretched under a three-dimensional obstacle (see, more generally for the n-dimensional case, Brézis–Stampacchia, 1968; Lewy–Stampacchia, 1969; Mosco–Troianiello, 1973; Murthy–Stampacchia, 1972 and what we will present in section 7.3).

A closely related problem is the so-called *Plateau's problem with obstacle*, which consists of finding surfaces of minimum area supported by an obstacle (see, e.g., Giusti, 1973; Nitsche, 1975 and, for a classical treatment, Radò, 1933). In the one-dimensional case the obstacle problem and Plateau's problem with obstacle (which is then a problem of minimum length) have the same solution, but this is not true in the general case.

6.2 SOME CONSIDERATIONS REGARDING SECOND ORDER LINEAR PROBLEMS

In this section we introduce, in the particularly simple case which is the one-dimensional one, the study of some problems which we will deal with in a much more general context in the following sections. At the same time, with the help of a few examples, we will try to clarify the meaning of the results regarding regularity which we considered at the end of the previous section.

A boundary value problem

We will start with the following boundary value problem:

PROBLEM 6.4. Find u such that

$$-u''(x) = f(x) \quad \text{in } [0, l]; \qquad u(l) = u(0) = 0. \tag{6.19}$$

We have considered the equation $-u''=f$ rather than $u''=f$ only for coherence with what we will do later; also, the choice of the minus sign will become clear later.

This problem is obviously incomplete since we have not indicated the regularity of u and f; we will try, however, to solve it formally and then to come back to the 'natural' spaces for a good formulation of the problem. Integrating formally the equation $-u''(x) = f(x)$ we write

$$u(x) = \alpha x + \beta - \int_0^x d\xi \int_0^\xi f(\theta)\, d\theta. \tag{6.20}$$

For this expression to be meaningful it is sufficient that $f \in L^1(]0, l[)$ —and under these conditions it follows immediately from the expression itself that $u \in C^1([0, l])$. Thus, putting in (6.20)

$$\alpha = \frac{1}{l}\int_0^l d\xi \int_0^\xi f(\theta)\, d\theta, \qquad \beta = 0, \tag{6.21}$$

we can say that

$$u(x) = -\int_0^x d\xi \int_0^\xi f(\theta)\, d\theta + \frac{x}{l}\int_0^l d\xi \int_0^\xi f(\theta)\, d\theta \tag{6.22}$$

is a solution of the following problem:

PROBLEM 6.5. Given $f \in L^1(]0, l[)$ find $u \in C^1([0, l])$, with $u' \in AC([0, l])$, such that

$$-u''(x) = f(x) \text{ a.e. in }]0, l[; \qquad u(l) = u(0) = 0 \tag{6.23}$$

(Let us observe that if $u \in C^1([0, l])$ and $u' \in AC([0, l])$ then u'' exists a.e. in $[0, l]$...).

This problem, as can easily be seen, has only one solution which depends continuously on f: the problem 6.5 is then well-posed.

There is an analogy between the above result regarding existence and that of Carathéodory for the Cauchy problem (see, e.g. Sansone–Conti, 1956, p. 14): under both cases there is a generalization of the concept of solution, and for the first as much as for the second the classical results are merely regularity results of the type (iii). Thus, in our case, if we assume that f satisfies the (classical) hypothesis $f \in C^0([0, l])$ then from (6.22) it follows that there exists $u \in C^2([0, l])$ such that $-u''(x) = f(x)$ in $[0, l]$ and $u(0) = u(l) = 0$, which is the classical result; more generally, again from (6.22), we can say that if $f \in C^k([0, l])$ then $u \in C^{k+2}([0, l])$.

It is worth noting the radical difference between this result concerning regularity and that already referred to for the obstacle problem: this difference in the behaviour of the regularity of the solution when the regularity of the data increase is essentially a consequence of the fact that problem 6.5 is linear while the obstacle problem is not. More generally we can say, roughly, that in most cases the regularity of the solution of a linear problem increases along with the regularity of the data, but for nonlinear problems there is often a *regularity threshold* which the solution cannot exceed (except, of course, for particular cases) whatever the increase in regularity of the data. Later we will see other examples of all this.

Variational formulation of a boundary value problem

In order to consider problems which are closer to those we will be dealing with in the following sections, we will now complete problem 6.4 in another way. Multiplying both sides of the equation $-u''=f$ by a generic function v, zero at 0 and at l, and formally integrating between 0 and l, we obtain

$$-\int_0^l u''(x)v(x)\,dx = \int_0^l f(x)v(x)\,dx \tag{6.24}$$

or, integrating by parts (again formally) the right-hand side and recalling that $v(l)=v(0)=0$,

$$\int_0^l u'(x)v'(x)\,dx = \int_0^l f(x)v(x)\,dx. \tag{6.25}$$

For this expression to be meaningful it is sufficient that $u \in H_0^1(]0, l[)$, $v \in H_0^1(]0, l[)$ and $f \in L^2(]0, l[)$, which leads us naturally to the problem:

PROBLEM 6.6. Given $f \in L^2(]0, l[)$ find $u \in H_0^1(]0, l[)$ such that

$$\int_0^l u'(x)v'(x)\,dx = \int_0^l f(x)v(x)\,dx \quad \forall v \in H_0^1(]0, l[) \tag{6.26}$$

(The choice of the minus sign already referred to has been made so as to eliminate it in variational expressions such as this—since we deal essentially with expressions of this type, and not properly with differential equations, this is somewhat convenient.)

The problem 6.6 is a variational problem, which can be seen, from the Lax–Milgram lemma, to be well-posed. In fact, on the left of (6.26) we have the scalar product between u and v in $H_0^1(]0, l[)$—see theorem 5.13—and on the right we have the value that the continuous linear functional in $H_0^1(]0, l[)$ represented by f takes on v—see theorem 5.14.

Since we wished to correctly complete problem 6.4, it is natural to ask what the relationship between problem 6.6 and problem 6.4 is. The answer derives from the fact (proved later) that problem 6.6 is equivalent to the problem:

PROBLEM 6.7. Given $f \in L^2(]0, l[)$ find $u \in H_0^1(]0, l[)$ such that

$$-u''=f \quad \text{in the sense of } \mathscr{D}'(]0, l[). \tag{6.27}$$

This boundary value problem obviously constitutes a completion of problem 6.4. On the other hand, problem 6.7 can undoubtedly be more readily interpreted than can problem 6.5: it is then natural to ask why we have 'complicated the question' by considering the variational formulation and thus leaving the 'framework of the boundary value problems'.

Now, except in particularly simple cases such as problems 6.5 and 6.7, we are not generally able to show directly that a boundary value problem is well-posed; thus, given a boundary value problem, $\mathcal{PL}$, we will systematically try to construct a variational problem, $\mathcal{PV}$, which is equivalent to it (which is not always possible!) and to apply to the latter the theory developed in Chapter 3, and in particular the powerful results of Lions–Stampacchia and Lax–Milgram. On the other hand, as we will see, the $\mathcal{PL}$ formulation is frequently more suitable for the proof of results concerning regularity than the $\mathcal{PV}$ formulation: thus, once a problem is posed in the variational form (as are the problems of physics obtained by applying the principle of minimum of the energy or the principle of virtual work) we will try to construct a boundary value problem which is equivalent to it and which, for obvious reasons, is called *the interpretation of the variational problem as a boundary value problem*.

Variational setting of the problem

The completions 6.5 and 6.7 of the problem 6.4 are different. More honestly we must say that not only is problem 6.7 different from problem 6.5 but it is also poorer, in that it requires more of the data ($f \in L^2(]0, l[)$ rather than $f \in L^1(]0, l[)$) and provides less information on the unknown ($u \in H_0^1(]0, l[)$ rather than $u \in C^1([0, l])$). This might lead us to think that variational techniques give us poorer formulations, but this is not so. In fact, the second defect (that the solution is less regular) is more apparent than real: if u is a solution of problem 6.7 then from (6.27) it follows immediately that $u \in H^2(]0, l[)$ (this is a regularity result of type (i)!). As for the first defect (that the data are less general), we can enrich the formulation of problem 6.6 by considering the

PROBLEM 6.8. Given $f \in H^{-1}(]0, l[)$ find $u \in H_0^1(]0, l[)$ such that

$$(u, v)_{H_0^1(]0,l[)} = {}_{H^{-1}(]0,l[)}\langle f, v\rangle_{H_0^1(]0,l[)} \quad \forall v \in H_0^1(]0, l[). \quad (6.28)$$

This is a variational problem which is still well-posed, and which will be shown to be equivalent to the problem:

PROBLEM 6.9. Given $f \in H^{-1}(]0, l[)$ find $u \in H_0^1(]0, l[)$ such that

$$-u'' = f \quad \text{in the sense of } \mathcal{D}'(]0, l[). \quad (6.29)$$

Finally, this is a boundary value problem which is not poorer than problem 6.5, only different.

THEOREM 6.1. Problems 6.8 and 6.9 are equivalent, i.e., u is a solution of problem 6.8 iff it is a solution of problem 6.9.

Proof. We will first show that problem 6.8 $\Rightarrow$ problem 6.9. If u is a solution of problem 6.8 then (6.28) is true and hence, given that $\mathscr{D}(]0, l[) \subset H_0^1(]0, l[)$ and $H^{-1}(]0, l[) \subset \mathscr{D}'(]0, l[)$, in particular we have

$$\int_0^l u'(x)v'(x)\,dx = {}_{\mathscr{D}'(]0,l[)}\langle f, v\rangle_{\mathscr{D}(]0,l[)} \quad \forall v \in \mathscr{D}(]0, l[). \tag{6.30}$$

Keeping in mind the definition of derivative in $\mathscr{D}'(]0, l[)$ we can write (6.30) in the form

$$-{}_{\mathscr{D}'(]0,l[)}\langle u'', v\rangle_{\mathscr{D}(]0,l[)} = {}_{\mathscr{D}'(]0,l[)}\langle f, v\rangle_{\mathscr{D}(]0,l[)} \quad \forall v \in \mathscr{D}(]0, l[), \tag{6.31}$$

which is the same as (6.29): this shows that u is a solution of problem 6.9. Now we will prove that problem 6.9 $\Rightarrow$ problem 6.8. If u is a solution of problem 6.9 then we have (6.31) and, keeping in mind that $u \in H_0^1(]0, l[)$, we have (6.30). Now, if w is any element of $H_0^1(]0, l[)$ and w_m is a sequence of elements of $\mathscr{D}(]0, l[)$ convergent to it in $H_0^1(]0, l[)$, we can write

$$\int_0^l u'(x)w_m'(x)\,dx = {}_{\mathscr{D}'(]0,l[)}\langle f, w_m\rangle_{\mathscr{D}(]0,l[)} = {}_{H^{-1}(]0,l[)}\langle f, w_m\rangle_{H_0^1(]0,l[)} \quad \forall m \in \mathbb{N} \tag{6.32}$$

and thus, since on the one hand f is a continuous linear functional on $H_0^1(]0, l[)$ and on the other $\int_0^l u'(x)w_m'(x) = (u, w_m)_{H_0^1(]0,l[)} \to (u, w)_{H_0^1(]0,l[)}$, we have

$$\int_0^l u'(x)w'(x)\,dx = {}_{H^{-1}(]0,l[)}\langle f, w\rangle_{H_0^1(]0,l[)} \tag{6.33}$$

i.e., (6.28): this shows that u is a solution of problem 6.8. ∎

It follows immediately as a corollary of this theorem that problems 6.6 and 6.7 are also equivalent. Let us remark that the fact that problem 6.7 (or problem 6.6) is well-posed together with the regularity result of type (i) which we have proved (and which ensures that the solution belongs to $C^1([0, l])$) constitutes a regularity result of type (iii) for problem 6.9 (or problem 6.7). Again, keeping in mind the discussion in the previous section immediately following the presentation of the different types of regularity results, it is possible to reformulate problem 6.7 on the pair of spaces $(L^2(]0, l[), C^1([0, l]))$ rather than on the pair $(L^2(]0, l[), H_0^1(]0, l[))$—this, however, is not very convenient since on the one hand we are going outside the Hilbert framework which we were using for convenience, and on the other we are no longer dealing with a problem which can be automatically formulated in the variational form.

More general operators

The operator $-d^2/dx^2$ which we have considered in the previous examples is the simplest example of a second order differential operator. In the following sections we will study, in the n-dimensional case, problems associated with a much more general operator which for $n=1$ can be written

$$L\square \equiv -\frac{d}{dx}\left[a(x)\frac{d}{dx}\square\right]+b(x)\frac{d}{dx}\square+\frac{d}{dx}[c(x)\square]+d(x)\square, \tag{6.34}$$

(by an expression of the type $-(d^2/dx^2)\square$ we mean that we are referring to the operator $-(d^2/dx^2)$ and not to the result of applying it to something) where the coefficients a, b, c, and d are taken in appropriate function spaces. If these coefficients are regular enough we can write the operator L in the equivalent form

$$\tilde{L}\square \equiv -a(x)\frac{d^2}{dx^2}\square+\tilde{b}(x)\frac{d}{dx}\square+\tilde{c}(x)\square, \tag{6.35}$$

where

$$\tilde{b}(x) = -\frac{da(x)}{dx}+b(x)+c(x) \tag{6.36}$$

and

$$\tilde{c}(x) = \frac{dc(x)}{dx}+d(x). \tag{6.37}$$

However, in general the operators L and $\tilde{L}$ are different, one not being more general than the other: here we will use the operator L because its structure (called *divergence structure*) is most suitable for the application of the formula of integration by parts (or Green's formulae in the n-dimensional case) which is a fundamental tool in the variational study of differential problems.

Eigenvalue problems

To end this section, we present an example of the difficulties which we encounter in proceeding with our study. Consider then the operator

$$L_\lambda \square \equiv -\frac{d^2}{dx^2}\square+\lambda\square, \quad \lambda \in \mathbb{R} \tag{6.38}$$

—a particularly simple case of the operator L (or $\tilde{L}$)—and the following boundary value problem associated with it:

PROBLEM 6.10. Find $u \in H_0^1(]0, l[)$ such that

$$L_\lambda u \equiv -u''(x)+\lambda u(x) = 0 \quad \text{in }]0, l[. \tag{6.39}$$

Simple calculations show that if λ is of the form

$$\lambda_{[k]} = -(k\pi/l)^2, \qquad k = 1, 2, \ldots, \tag{6.40}$$

then the function

$$u_{[k]}(x) = \sin \frac{k\pi}{l} x, \qquad k = 1, 2, \ldots, \tag{6.41}$$

is a solution of problem 6.10, which proves that this problem is not well-posed since (for any $\lambda \in \mathbb{R}$) the function $u(x) \equiv 0$ is also a solution.

In the face of this phenomenon it is natural to consider, together with problems such as problem 6.10, those such as

PROBLEM 6.11. Find the set $\sigma(d^2/dx^2)$ of the real numbers $\lambda \in \mathbb{R}$ for which problem 6.10 is not well-posed.

A problem of this type is called a *spectral problem* or an *eigenvalue problem* since the set $\sigma(d^2/dx^2)$ is called the *point spectrum* of the operator d^2/dx^2 and its elements, the $\lambda_{[k]}$ given by (6.40), are called the *eigenvalues* of problem 6.10—in this context the solutions $u_{[k]}(x)$ given by (6.41) are called *eigenfunctions* of the same problem. The study of problems of the type of problem 6.11 lies in the field of spectral analysis, one of the most highly developed branches of functional analysis (see among others, Yosida, 1971; Kato, 1966; Schechter, 1971; Weinberger, 1974; Berezanskiĭ, 1968; Fichera, 1965) to which we will return, although very briefly, in the following chapter.

To end, we take the opportunity provided by the above problem of presenting a typical example of an application of spectral analysis: thus, we remark that the square roots of the eigenvalues $\lambda_{[k]}$ of problem 6.10 can be interpreted as the natural frequencies (or proper modes of vibration) of a *vibrating string* of length l with its ends fixed, and that the eigenfunctions $u_{[k]}(x)$ represent the form that the string takes when it vibrates with the proper frequency associated with $\lambda_{[k]}$. Let us remark that the equation $-u''(x) + \lambda u(x) = 0$ is derived from the equation of vibrating strings through the well known procedure of *separation of variables*—see e.g., L. Schwartz, 1960, p. 34; Courant–Hilbert, 1953, p. 287; for an analogous analysis of the motion of a vibrating membrane, the natural generalization of the previous problem to two dimensions, see e.g. Sommerfeld, 1964, p. 166.

7

Examples of Variational Problems in Several Dimensions

7.1 GENERAL COMMENTS ON DIFFERENTIAL OPERATORS

We are interested exclusively in the second order elliptic linear operators; however we feel it will be useful to make some comments on differential operators in general so as to place our study in the proper framework.

Differential operators: quasi-linear, semi-linear, and linear

What do we mean by *differential operator*? A non-exhaustive answer, but sufficient for our purposes, is provided by the following considerations. Let Ω be an arbitrary open set of $\mathbb{R}^n$, m a positive integer and Φ a real function defined on the cylinder $\Omega \times \prod_{i=0}^{m} \mathbb{R}^{[i]^-}$, $[i]^- = \text{card}\{\alpha \in \mathbb{N}^n : |\alpha| = i\}$: the identity

$$\Lambda u \equiv \Phi\left(x_1, \ldots, x_n, u, \frac{\partial}{\partial x_1} u, \ldots, \frac{\partial}{\partial x_n} u, \ldots, \frac{\partial^m}{\partial x_1^m} u, \ldots, \frac{\partial^m}{\partial x_n^m} u\right) \quad (7.1)$$

defines Λ, *differential operator or order m* on Ω, whenever the right-hand side is meaningful (to be precise, the definition of Λ is incomplete without the datum of its domain $D(\Lambda)$—the set of the u to which it is to be applied—but here we will simply assume that the u's belong to an appropriate functional space $\mathcal{U}(\Omega)$, and come back to this subject later).

Henceforth, we will always consider $m = 2$ and therefore operators defined by identities of the type

$$\Lambda u \equiv \Phi\left(x_1, \ldots, x_n, u, \frac{\partial}{\partial x_1} u, \ldots, \frac{\partial}{\partial x_n} u, \frac{\partial^2}{\partial x_1^2} u, \ldots, \frac{\partial^2}{\partial x_n^2} u\right). \quad (7.2)$$

The operator Λ is said to be *quasi-linear*, *semi-linear*, or *linear* depending on whether the function Φ is such that (7.2) can be written in the form

$$\Lambda u \equiv \sum_{i,j=1}^{n} \phi_{ij}\left(x_1, \ldots, x_n, u, \frac{\partial}{\partial x_1} u, \ldots, \frac{\partial}{\partial x_n} u\right) \frac{\partial^2}{\partial x_i \, \partial x_j} u$$
$$+ \phi\left(x_1, \ldots, x_n, u, \frac{\partial}{\partial x_1} u, \ldots, \frac{\partial}{\partial x_n} u\right), \quad (7.3)$$

$$\Lambda u \equiv \sum_{i,j=1}^{n} \phi_{ij}(x_1, \ldots, x_n) \frac{\partial^2}{\partial x_i \, \partial x_j} u$$
$$+ \phi\left(x_1, \ldots, x_n, u, \frac{\partial}{\partial x_1} u, \ldots, \frac{\partial}{\partial x_n} u\right), \quad (7.4)$$

$$\Lambda u \equiv \sum_{i,j=1}^{n} \phi_{ij}(x_1, \ldots, x_n) \frac{\partial^2}{\partial x_i \, \partial x_j} u$$
$$+ \sum_{i=1}^{n} \phi_i(x_1, \ldots, x_n) \frac{\partial}{\partial x_i} u + \phi(x_1, \ldots, x_n) u, \quad (7.5)$$

respectively.

In the latter case, if in addition the functions ϕ_{ij}, ϕ_i and ϕ are constants, we say that Λ is a second order linear differential operator with constant coefficients. Naturally, these are the operators for which most results are known (see, e.g., Treves, 1966 and, for a more abstract treatment, Palamodov, 1970).

Differential equations: quasi-linear, semilinear and linear

A *differential equation* in Ω (partial differential equation if $n > 1$ and ordinary differential equation if $n = 1$) is an expression of the type

$$\Lambda u = f, \quad (7.6)$$

where Λ is a differential operator and u (the *unknown*) and f (the *data*) are in appropriate function spaces on Ω, and the equality is interpreted in a very precise manner (closely connected besides to these spaces . . .)—if $\bar{u}$ is such that $\Lambda \bar{u} \equiv f$ then $\bar{u}$ is said to be a *solution* of equation (7.6). The equation is said to be quasi-linear, semi-linear, etc., depending on Λ.

Fundamental solutions

It is well known that, in general, to a differential equation one associates further conditions to be satisfied by the solutions. It may, however, be useful to study certain differential equations independently of other conditions—in this context the following result (see, e.g., Hörmander, 1964, p. 64) is interesting:

THEOREM 7.1. For any linear differential operator Λ with constant coefficients there exists at least one distribution $\bar{u} \in \mathscr{D}'(\mathbb{R}^n)$ which is a solution of

the equation
$$\Delta u = \delta_0 \quad \text{in the sense of } \mathscr{D}'(\mathbb{R}^n). \tag{7.7}$$

The distribution $\bar{u}$ is called a *fundamental solution* of the operator Λ. The concept of a fundamental solution of an operator is closely connected to that of *Green's function* associated with the operator, a concept which has been extensively studied in the classical treatises on partial differential equations (see, e.g., Courant–Hilbert, 1953, pp. 351 *et seq.*). For example, the distribution

$$\bar{u} = \frac{\Gamma(n/2)}{(2-n)2\pi^{n/2}}(x_1^2 + \ldots + x_n^2)^{1-n/2} \quad \text{if } n > 2;$$

$$\bar{u} = \frac{1}{2\pi}\log(x_1^2 + x_2^2)^{1/2} \quad \text{if } n = 2; \quad \bar{u} = [x]^+ \quad \text{if } n = 1, \tag{7.8}$$

is a fundamental solution of the Laplace operator

$$\Delta \square \equiv \sum_{i=1}^{n} \frac{\partial^2}{\partial x_i^2} \square \tag{7.9}$$

in $\mathbb{R}^n$ (in (7.8) Γ is the 'gamma function'—for this, and in general for all the special functions which appear in this book, see Whittaker–Watson, 1965).

A classification of second order differential operators

We have said that we will deal with elliptic operators. What then is an elliptic operator? This question brings us to the classification of differential operators into types. We present here a classification of second order linear differential operators, but this is easily applicable also to the second order quasi-linear or semi-linear operators, and further there are classifications which extend this to operators of higher order.

The classification is based on the fact (which we will examine later for elliptic operators) that the properties of the linear differential operators depend essentially on that part which involves derivatives of higher order: we will call in fact the *principal part* of the operator defined by (7.5) the operator

$$^p\Lambda\square \equiv \sum_{i,j=1}^{n} \phi_{ij}(x_1, \ldots, x_n) \frac{\partial^2}{\partial x_i \partial x_j}\square. \tag{7.10}$$

It is obvious on the other hand that the properties of the operator $^p\Lambda$ depend exclusively on the functions $\phi_{ij}(x_1, \ldots, x_n)$, and as such on the values that they take in every point of Ω: what we need then is a classification of a local character. Let then $x^0 \equiv (x_1^0, \ldots, x_n^0)$ be a point of Ω and consider the operator

$$^p\Lambda_{x^0}\square \equiv \sum_{i,j=1}^{n} \phi_{ij}(x_1^0, \ldots, x_n^0) \frac{\partial^2}{\partial x_i \partial x_j}\square. \tag{7.11}$$

One can easily see that the formula

$$^P\Lambda_{x^0}(e^{(x,\xi)}) = Q_{x^0}(\xi)e^{(x,\xi)}, \tag{7.12}$$

where $\xi = (\xi_1, \ldots, \xi_n) \in \mathbb{R}^n$, $(x, \xi) = \sum_{i=1}^n x_i \xi_i$ and Q_{x^0} is the quadratic form

$$Q_{x^0}(\xi) \equiv \sum_{i,j=1}^n \phi_{ij}(x_1^0, \ldots, x_n^0)\xi_i\xi_j, \tag{7.13}$$

provides an isomorphism between the set of operators (7.11) and that of the forms (7.13): we have then converted the problem of classifying differential operators into that of classifying quadratic forms.

Thus, the operator defined by (7.5) is said to be *elliptic, hyperbolic* or *parabolic* in x^0 depending on whether the quadratic form associated with its principal part in x^0 is definite, semi-definite, or indefinite. In general we will be dealing with function spaces in which successive derivatives do not depend on the order in which they are carried out and hence we can assume that in (7.5) $\phi_{ij} = \phi_{ji}$ in Ω: under these conditions we can provide a finer classification of differential operators. In fact, in this case the matrix $[\phi_{ij}(x^0)]$ associated with the form (7.13) is symmetric and hence it can be diagonalized by means of a suitable change of variables $\theta_i = \sum_{j=1}^m c_{ij}(x^0)\,\xi_j$: the quadratic form (7.13) can then be written as an algebraic sum of squares

$$Q_{x^0}(\theta) = \sum_{i=1}^n k_i(x^0)\theta_i^2, \tag{7.14}$$

where the $k_i(x^0)$ are constants depending on the matrix $[c_{ij}(x^0)]$ which transforms the variables. It is known however (Sylvester's law of inertia of quadratic forms) that the so-called invariants of the quadratic form, the *inertia* $I = \text{card } \{k_i(x^0) = k_i(x^0) > 0\}$ and the *deficiency* $D = \text{card }\{k_i(x^0): k_i(x^0) = 0\}$, do not depend on the matrix $[c_{ij}(x^0)]$. Now, with $T = n - (I+D)$, we say that in x^0 the operator Λ defined by (7.5), with $\phi_{ij} = \phi_{ji}$, is

elliptic	when $I = n, D = 0, T = 0$ or $I = 0, D = 0\ T = n$
hyperbolic	when $I > 0, D = 0, T > 0$
normal hyperbolic	when $I = n-1, D = 0, T = 1$ or
	$I = 1, D = 0, T = n-1$
ultra hyperbolic	when $I > 1, D = 0, T > 1$
parabolic	when $D > 0$
normal parabolic	when $I = n-1, D = 1, T = 0$ or
	$I = 0, D = 1, T = n-1$.

Naturally, Λ is said to be elliptic, hyperbolic, etc., in Ω if Λ is elliptic, hyperbolic, etc., at every point of Ω. Various combinations of these terms are also used: for example, if Λ is elliptic in $\Omega_1 \subset \Omega$, parabolic in $\Omega_2 \subset \Omega$ ($\Omega_1 \cap \Omega_2 = \emptyset$) and hyperbolic in $\Omega\setminus(\Omega_1 \cup \Omega_2)$, Λ is said to be of *mixed type* elliptic-parabolic-hyperbolic in Ω. Further, an equation associated with an operator which is elliptic, parabolic, etc., in Ω, is also said to be elliptic, parabolic, etc., in Ω.

The 'geometric' terminology used in classifying differential operators is derived from the fact that, for two variables, the operator

$$\Lambda \square \equiv \phi_{11}(x_1, x_2) \frac{\partial^2}{\partial x_1^2} \square + 2\phi_{12}(x_1, x_2) \frac{\partial^2}{\partial x_1 \partial x_2} \square + \phi_{22}(x_1, x_2) \frac{\partial^2}{\partial x_2^2} \square$$
$$+ \phi_1(x_1, x_2) \frac{\partial}{\partial x_1} \square + \phi_2(x_1, x_2) \frac{\partial}{\partial x_2} \square + \phi(x_1, x_2) \square \quad (7.15)$$

is *elliptic*, *parabolic*, or *hyperbolic* in $x = x^0 \in \Omega$ depending on whether

$$d(x^0) \equiv \phi_{21}^2(x_1^0, x_2^0) - \phi_{11}(x_1^0, x_2^0)\phi_{22}(x_1^0, x_2^0) \quad (7.16)$$

is $<0, =0$ or >0 (see, e.g., M. Smirnov, 1966, p. 31), i.e., whether the equation

$$\phi_{11}(x_1^0, x_2^0)\xi_1^2 + 2\phi_{12}(x_1^0, x_2^0)\xi_1\xi_2 + \phi_{22}(x_1^0, x_2^0)\xi_2^2$$
$$+ \phi_1(x_1^0, x_2^0)\xi_1 + \phi_2(x_1^0, x_2^0)\xi_2 + \phi(x_1^0, x_2^0) = 0 \quad (7.17)$$

is that of an *ellipse*, a *parabola*, or a *hyperbola* (remark that $d(x^0)$ is the discriminant of the quadric (7.17)).

We will not deal here with a question which is particularly important for parabolic and hyperbolic equations: the theory of *characteristics*, which are, for an operator such as (7.15), the curves defined by the equation

$$\phi_{11} dx_1^2 + \phi_{22} dx_2^2 - 2\phi_{12} dx_1 dx_2 = 0 \quad (7.18)$$

—in this context see M. Smirnov, 1966, p. 33; Cibrario–Cinquini, 1956.

Some classical operators

Classical examples of elliptic, (normal) hyperbolic and (normal) parabolic operators in $\mathbb{R}^3$ are, respectively, the *Laplace* operator or potential operator

$$\Delta \square \equiv \frac{\partial^2}{\partial x_1^2} \square + \frac{\partial^2}{\partial x_2^2} \square + \frac{\partial^2}{\partial x_3^2} \square \quad (7.19)$$

the *D'Alembert* operator or wave operator

$$\square \square \equiv \frac{\partial^2}{\partial t^2} \square - \frac{\partial^2}{\partial x_1^2} \square - \frac{\partial^2}{\partial x_2^2} \square \quad (7.20)$$

and the *Fourier* operator or heat operator

$$\star \square \equiv \frac{\partial}{\partial t} \square - \frac{\partial^2}{\partial x_1^2} \square - \frac{\partial^2}{\partial x_2^2} \square \quad (7.21)$$

(in (7.20) and (7.21) we have written t in place of x_3 since this variable ('time') has a physical significance which is quite different from those of the ('space') variables x_1 and x_2). An example of a mixed operator in $\mathbb{R}^2$ is the

Tricomi operator

$$T\square \equiv y\frac{\partial^2}{\partial x^2}\square + \frac{\partial^2}{\partial y^2}\square, \tag{7.22}$$

which is of great importance in aerodynamics, and which is hyperbolic for $y<0$, elliptic for $y>0$ and parabolic along the line $y=0$.

Initial value problems and boundary value problems

The classification of operators into types which we have presented here may seem artificial (if not completely useless!) in that we have characterized entities which have little to do with algebra by means of algebraic invariants; however, the following considerations will show that such a classification is of profound significance.

The phenomena of radiation (or electromagnetic energy, for example) are generally described fairly well by hyperbolic equations, and diffusion phenomena (of heat, for example) are generally described quite clearly by parabolic equations. Both irradiation and diffusion are phenomena of evolution, i.e. phenomena which depend explicitly on time: in their mathematical modelling it is natural to formulate the so-called *Cauchy problems*, in which we associate with the differential equations certain conditions—called *initial conditions*—which the solution (and, in the hyperbolic case, its derivative 'with respect to time') must satisfy at a given moment. Later we will give an example of such problems; let us remark, however, that these are not the only 'natural' problems in modelling phenomena of evolution: of particular importance are the *mixed problems in Hadamard's sense*, in which initial conditions and *boundary conditions* are simultaneously imposed. We will discuss now this second kind of conditions.

For elliptic equations Cauchy problems are usually ill-posed: these equations describe fairly well stationary phenomena, i.e. phenomena which do not explicitly depend on time (such as those concerned with electric potential, for example), and for those it is natural to formulate problems of a completely different kind. Given that our main concern is centred on elliptic operators, it is worth recalling, through some examples, the classical formulation of the most important problems of this kind: we will take them up again in the next section in a more modern context.

Let then $\Omega \subset \mathbb{R}^2$ be a bounded open set of class C^1, $f \in C^0(\Omega)$ a function given in Ω, $f_1, f_2, f_3 \in C^0(\Gamma)$ three functions given on the boundary Γ of Ω and consider the differential equation

$$-\Delta u + \lambda u = f \quad \text{on } \Omega, \tag{7.23}$$

where Δ is the Laplace operator and λ is a real constant. Depending on what further conditions we associate with this equation we obtain different problems: they are called *boundary value problems* or *problems with conditions on the boundary* for reasons which will soon become obvious.

The following problem is known as a *Dirichlet problem*:

PROBLEM 7.1. Find $u \in C^2(\Omega) \cap C^0(\bar{\Omega})$ such that

$$-\Delta u + \lambda u = f \quad \text{in } \Omega, \tag{7.24}$$

$$u = f_1 \quad \text{on } \Gamma \text{ (Dirichlet condition).} \tag{7.25}$$

The following problem, where by $\partial/\partial \nu$ we denote the derivative operator with respect to the (external) normal to Γ, is known as a *Neumann problem*:

PROBLEM 7.2. Find $u \in C^2(\Omega) \cap C^1(\bar{\Omega})$ such that

$$-\Delta u + \lambda u = f \quad \text{in } \Omega, \tag{7.26}$$

$$\frac{\partial u}{\partial \nu} = f_2 \quad \text{on } \Gamma \text{ (Neumann condition).} \tag{7.27}$$

Note that this is not a Cauchy problem!—the derivative with respect to the normal has only 'geometrical' characteristics, and not 'temporal' characteristics. A natural generalization of this problem is the following *oblique derivative problem*, where $\partial/\partial l$ denotes the derivative with respect to a direction l which is associated with continuity with every point of Γ:

PROBLEM 7.3. Find $u \in C^2(\Omega) \cap C^1(\bar{\Omega})$ such that

$$-\Delta u + \lambda u = f \quad \text{in } \Omega, \tag{7.28}$$

$$\frac{\partial u}{\partial l} = f_3 \quad \text{on } \Gamma \text{ (oblique derivative condition).} \tag{7.29}$$

If there exists a constant α such that, for every point of Γ, $\cos(\nu, l) \geq \alpha > 0$ then the oblique derivative problem is said to be *regular* (let us remark that if at some point $\cos(\nu, l) = 0$ then the problem degenerates into a problem with *tangential derivatives*, which is difficult to treat).

Generalizations and combinations of these problems are possible; of particular interest is the *mixed Dirichlet–Neumann problem*:

PROBLEM 7.4. Find $u \in C^2(\Omega) \cap C^1(\bar{\Omega})$ such that, Γ_1 being a part of Γ and $\Gamma_2 = \Gamma \backslash \Gamma_1$,

$$-\Delta u + \lambda u = f \quad \text{in } \Omega, \tag{7.30}$$

$$u = f_1 \quad \text{on } \Gamma_1 \text{ (Dirichlet condition)} \tag{7.31}$$

$$\frac{\partial u}{\partial \nu} = f_2 \quad \text{on } \Gamma_2 \text{ (Neumann condition).} \tag{7.32}$$

Let us note the following important fact regarding nomenclature: if $f \equiv 0$ the equation (7.23) is said to be a *homogeneous equation*; further if $f_1 = f_2 =$

$f_3 \equiv 0$ (even if $f \not\equiv 0$) the previous problems are said to be *homogeneous problems*.

The operator $\Delta \square + \lambda \square$ which we considered in the presentation of the different kinds of elliptic problems is sometimes called the *metaharmonic operator* because Δ is sometimes called the *harmonic operator*. It is curious that every elliptic second order linear differential operator with constant coefficients can be written, by means of an appropriate change of variable, as a metaharmonic operator. Further, still in connection with a change of variables, let us remark that whether a differential operator has constant coefficients or not is not an invariant with respect to these changes. Thus, e.g., the parabolic operator with variable coefficients $\Lambda \square \equiv (\partial^2/\partial x_2^2)\square + 2x_2(\partial^2/\partial x_1 \partial x_2)\square + x_2^2(\partial^2/\partial x_1^2)\square$ becomes the parabolic operator with constant coefficients $\Lambda \square \equiv (\partial^2/\partial y_2^2)\square - (\partial/\partial y_1)\square$ by the (nonlinear) change of variables $y_2 = x_2$ and $y_1 = x_1 - (x_2^2/2)$.

In the same line of thought as that which gave us theorem 7.1 (the study of a differential equation independently of further conditions) the following result is of interest: *if $\Omega \subset \mathbb{R}^2$ is a simply connected bounded open set containing the origin then every regular real solution* (i.e. of class C^2 in Ω) *of the equation $\Delta u + \lambda u = 0$ can be written in the form*

$$u(x_1, x_2) = \alpha J_0(\lambda^{1/2}|z|) + \mathcal{R}e\left(\int_0^z \phi(\zeta) J_0((\lambda \bar{z}(z-\zeta))^{1/2})\,d\zeta\right),$$

where $z = x_1 + ix_2$, J_0 is the Bessel function of first kind and order zero, α is a real constant and $\phi(\zeta)$ is a holomorphic function in Ω (Ω is here taken to be a subset of $\mathbb{C}$ in agreement with the correspondence $(x_1, x_2) \leftrightarrow z = x_1 + ix_2$)— further conditions on the solution will enable us to find in each case the constant α and the function ϕ (see, e.g., Vekua, 1968, p. 58).

A Cauchy problem

We will now give, as promised, an example of a *Cauchy problem*; this is a famous example due to Hadamard which confirms our assertion that Cauchy problems for elliptic operators are in general ill-posed (see in this context Hadamard, 1968a; 1968b; 1968c; John, 1955; Lavrentiev, 1967).

PROBLEM 7.5. 'Find $u \in C^2(\mathring{\mathbb{R}}_+^2) \cap C^1(\overline{\mathbb{R}}_+^2)$ such that

$$\Delta u \equiv \frac{\partial^2 u}{\partial x_1^2} + \frac{\partial^2 u}{\partial x_2^2} = 0 \quad \text{in } \mathring{\mathbb{R}}_+^2, \tag{7.33}$$

$$u(x_1, 0) = 0, \tag{7.34}$$

$$\frac{\partial u}{\partial x_2}(x_1, 0) = \frac{1}{n}\sin(nx_1). \tag{7.35}$$

Conditions (7.34) and (7.35) are the so-called *initial conditions* and together they are often known as the *Cauchy data* (the variable x_2 replaces the 'time variable'...). This problem, which has as its unique solution the analytic function

$$u(x_1, x_2) = \frac{1}{n^2} \sinh(nx_2) \sin(nx_1), \tag{7.36}$$

is ill-posed since as $n \to +\infty$ the condition (7.35) tends (uniformly) to the homogeneous condition

$$\frac{\partial u}{\partial x_2}(x_1, 0) = 0, \tag{7.37}$$

to which corresponds the solution $u = 0$: on the other hand (7.36) does not converge (at least in spaces of classical type...) to the identically zero function as $n \to +\infty$ (there is then no continuous dependence on the data and the problem as such is ill-posed).

For another version of this example see Mizohata, 1973, p. 256.

The operator L

The differential operator which we will deal with next is the n-dimensional analogue of the operator defined by (6.34):

$$L\square \equiv -\sum_{i,j=1}^{n} \left(\frac{\partial}{\partial x_i} a_{ij}(x) \frac{\partial}{\partial x_j} \square \right) + \sum_{i=1}^{n} b_i(x) \frac{\partial}{\partial x_i} \square$$
$$+ \sum_{i=1}^{n} \frac{\partial}{\partial x_i}(c_i(x)\square) + d(x)\square. \tag{7.38}$$

This operator, which is said to have a *principal part in divergence form*, cannot be written in the form (7.5) (unless we require that the coefficients be differentiable, which we will not do). Operators of this type appear in a natural way in problems of physics—in Ladyženskaja–Ural'ceva, 1968, pp. 190 et seq.) we find an example taken from diffraction theory. This is anyway a second order differential operator which is evidently linear and which we can certainly classify. We will restrict ourselves to the case in which L is elliptic, or more precisely to the case in which satisfies the conditions of the following definition:

DEFINITION 7.1. The operator (7.38) is said to be *uniformly strongly elliptic* in Ω if

$$\exists c > 0 \quad \frac{1}{c} \|\xi\|^2 \leq \left| \sum_{i,j=1}^{n} a_{ij}(x)\xi_i\xi_j \right| \leq c \|\xi\|^2 \quad \text{a.e. in } \Omega, \xi \in \mathbb{R}^n. \tag{7.39}$$

It follows immediately that *if L is uniformly strongly elliptic in Ω then L is elliptic in Ω*. Since we will exclusively consider operators under these conditions we will use 'elliptic' synonymously with 'uniformly strongly elliptic'. For a weaker form of the theory, which does not require the operator to be uniformly elliptic (elliptic-parabolic operators) see, e.g., Oleĭnik–Radkevič, 1973 and Fichera, 1956.

We have stated previously that in order to define an operator completely it is not enough to write a formula such as (7.1)—we also need to specify its domain. In the present case we do not yet have to define (7.38) completely: indeed we must also specify the regularity of the coefficients a_{ij}, b_i, c_i, d ($i, j = 1, \ldots, n$). At this point, with the aim of providing the greatest generality, we can follow two lines: either we specify the regularity of the coefficients and find 'the space of least regular functions' to which it is meaningful to apply (7.38), or we fix the domain of the operator and find 'the least regularity that the coefficients can have' so as to be able to apply it to the elements of the chosen domain. Here we will follow the second possibility: we will fix as the domain of L the space $H^1_{\text{loc}}(\Omega)$ and we will find the least regularity of the coefficients (in the framework of the families of spaces which we have considered up till now) so as to be able to make sense of an equality of the type

$$Lu = f \quad \text{in } \mathscr{D}'(\Omega), \tag{7.40}$$

with $u \in H^1_{\text{loc}}(\Omega)$ and $f \in \mathscr{D}'(\Omega)$.

So, we will develop our theory in the Hilbertian framework; for a treatment of elliptic operators in the framework of Banach spaces see, e.g., Nečas, 1967; Agmon, 1959; Lions–Magenes, 1960–1963b.

We will, for the present, impose on u and on the coefficients of L a local regularity so that among the operators L are also those with constant coefficients; later we will restrict ourselves to considering bounded sets and then we will impose conditions of global regularity.

Analysing term by term the expression

$$-\sum_{i,j=1}^{n} \frac{\partial}{\partial x_i}\left(a_{ij}\frac{\partial u}{\partial x_j}\right) + \sum_{i=1}^{n} b_i \frac{\partial u}{\partial x_i} + \sum_{i=1}^{n} \frac{\partial}{\partial x_i}(c_i u) + du \tag{7.41}$$

(i.e.: Lu) we can see that, for this to be an element of $\mathscr{D}'(\Omega)$, it is enough that the following hypotheses be satisfied:

(h1) $\quad a_{ij} \in L^2_{\text{loc}}(\Omega) \qquad (i, j = 1, \ldots, n)$

—since in that case, given that $u \in H^1_{\text{loc}}(\Omega)$ implies that $\partial u/(\partial x_j) \in L^2_{\text{loc}}(\Omega)$, it follows that $a_{ij}(\partial u/\partial x_j) \in L^1_{\text{loc}}(\Omega)$ and therefore it is meaningful to consider its derivative in the sense of $\mathscr{D}'(\Omega)$;

(h2) $\quad b_i \in L^p_{\text{loc}}(\Omega), \qquad$ with $p \geq 2$ arbitrary $(i = 1, \ldots, n)$

—since in that case $b_i(\partial u/\partial x_i) \in L^1_{\text{loc}}(\Omega)$ and therefore is an element of $\mathscr{D}'(\Omega)$;

(h3) $\quad c_i \in L^q_{\text{loc}}(\Omega), \qquad$ with $q \geq 2$ arbitrary $(i = 1, \ldots, n)$

—since in that case $c_i u \in L^1_{\text{loc}}(\Omega)$ and as such it is meaningful to consider its derivative in the sense of $\mathcal{D}'(\Omega)$;

(h4) $d \in L^r_{loc}(\Omega)$, with $r \geq 2$ arbitrary

—since in that case $du \in L^1_{loc}(\Omega)$ and hence is an element of $\mathcal{D}'(\Omega)$.

Within hypotheses (h1), (h2), (h3), and (h4) we have $L \in \mathcal{L}(H^1_{loc}(\Omega), \mathcal{D}'(\Omega))$: linearity is obvious and continuity follows from the fact that differentiation, addition, and multiplication, which (7.41) involves, are continuous. This is a very important conclusion since, in the modern theory of differential equations, the general tendency is to reduce the study of differential equations to that of differential operators between appropriate function spaces.

Hypotheses (h1), (h2), (h3), and (h4) can be further weakened—this weakening, however, is not very interesting since they are already too weak for an effective study of the operator L (in the sense of being able to show that certain problems associated with it are well-posed). In the next section we will, with this aim, replace them with assumptions of greater regularity, through further restrictions on the powers of summability (and imposing in addition that the regularity be global in Ω).

7.2 LINEAR PROBLEMS

7.2.1 IINTRODUCTION. THE HOMOGENEOUS DIRICHLET PROBLEM

We will first consider the following boundary value problem for the operator L defined by (7.38):

PROBLEM 7.6. Given $f \in \mathcal{D}'(\Omega)$, $\Omega \subset \mathbb{R}^n$ being a bounded open set, find $u \in H^1_0(\Omega)$ such that

$$Lu = f \quad \text{in } \mathcal{D}'(\Omega). \tag{7.42}$$

First, we observe that problem 7.6 is in effect a boundary value problem, the data on the boundary being included in the fact that u belongs to $H^1_0(\Omega)$. These data have a purely formal interpretation if Ω is not regular, but if the boundary Γ of Ω is, e.g., of class $C^{0,1}$ then we can interpret it as $u|_\Gamma = 0$ (in the sense of traces of the functions of $H^1(\Omega)$—recall theorem 5.21). It is then appropriate to call the problem 7.6 the *homogeneous Dirichlet problem*.

Variational formulation of the homogeneous Dirichlet problem

We will now try to present problem 7.6 in variational form and therefore as one that can be treated in terms of the theory developed in chapter 3. With this aim we will write (7.42) fully, keeping in mind (7.38) and the definition of equality in the sense of distributions:

$$-\sum_{i,j=1}^{n} {}_{\mathscr{D}'(\Omega)}\left\langle \frac{\partial}{\partial x_i}\left(a_{ij}\frac{\partial u}{\partial x_j}\right), \phi \right\rangle_{\mathscr{D}(\Omega)} + \sum_{i=1}^{n} {}_{\mathscr{D}'(\Omega)}\left\langle b_i \frac{\partial u}{\partial x_i}, \phi \right\rangle_{\mathscr{D}(\Omega)}$$
$$+ \sum_{i=1}^{n} {}_{\mathscr{D}'(\Omega)}\left\langle \frac{\partial}{\partial x_i}(c_i u), \phi \right\rangle_{\mathscr{D}(\Omega)} + {}_{\mathscr{D}'(\Omega)}\langle du, \phi \rangle_{\mathscr{D}(\Omega)} = {}_{\mathscr{D}'(\Omega)}\langle f, \phi \rangle_{\mathscr{D}(\Omega)}$$
$$\forall \phi \in \mathscr{D}(\Omega) \quad (7.43)$$

or, from the definition of derivative in the sense of distributions,

$$\sum_{i,j=1}^{n} {}_{\mathscr{D}'(\Omega)}\left\langle a_{ij}\frac{\partial u}{\partial x_j}, \frac{\partial \phi}{\partial x_i} \right\rangle_{\mathscr{D}(\Omega)} + \sum_{i=1}^{n} {}_{\mathscr{D}'(\Omega)}\left\langle b_i \frac{\partial u}{\partial x_i}, \phi \right\rangle_{\mathscr{D}(\Omega)}$$
$$- \sum_{i=1}^{n} {}_{\mathscr{D}'(\Omega)}\left\langle c_i u, \frac{\partial \phi}{\partial x_i} \right\rangle_{\mathscr{D}(\Omega)} + {}_{\mathscr{D}'(\Omega)}\langle du, \phi \rangle_{\mathscr{D}(\Omega)} = {}_{\mathscr{D}'(\Omega)}\langle f, \phi \rangle_{\mathscr{D}(\Omega)}$$
$$\forall \phi \in \mathscr{D}(\Omega). \quad (7.44)$$

Now, $u \in H_0^1(\Omega)$ and this together with hypotheses (h1), (h2), (h3), and (h4) concerning the coefficients of L allows us to write (7.44) in the form

$$\sum_{i,j=1}^{n} \int_\Omega a_{ij} \frac{\partial u}{\partial x_j}\frac{\partial \phi}{\partial x_i} dx + \sum_{i=1}^{n} \int_\Omega b_i \frac{\partial u}{\partial x_i} \phi \, dx$$
$$- \sum_{i=1}^{n} \int_\Omega c_i u \frac{\partial \phi}{\partial x_i} dx + \int_\Omega du\phi \, dx = {}_{\mathscr{D}'(\Omega)}\langle f, \phi \rangle_{\mathscr{D}(\Omega)} \quad \forall \phi \in \mathscr{D}(\Omega), \quad (7.45)$$

which leads us to the consideration of the *bilinear form* $a(u, \phi): H_0^1(\Omega) \times \mathscr{D}(\Omega) \to \mathbb{R}$ defined by

$$a(u, \phi) = \int_\Omega \left(\sum_{i,j=1}^{n} a_{ij}\frac{\partial u}{\partial x_j}\frac{\partial \phi}{\partial x_i} + \sum_{i=1}^{n} b_i \frac{\partial u}{\partial x_i} \phi - \sum_{i=1}^{n} c_i u \frac{\partial \phi}{\partial x_i} + du\phi \right) dx \quad (7.46)$$

and hence to rewrite (7.45) in the more compact form

$$a(u, \phi) = {}_{\mathscr{D}'(\Omega)}\langle f, \phi \rangle_{\mathscr{D}(\Omega)} \quad \forall \phi \in \mathscr{D}(\Omega) \quad (7.47)$$

(let us note that, as in Chapter 6, all the function spaces considered in this chapter are assumed to be real: so the bilinear form a is supposed to take real values).

The equation (7.47) is not a variational equation in the sense of Chapter 3 in that, on the one hand, $a(u, \phi)$ is not a bilinear form on $H_0^1(\Omega) \times H_0^1(\Omega)$ but only on $H_0^1(\Omega) \times \mathscr{D}(\Omega)$ and, on the other, that f is not necessarily a continuous linear functional on $H_0^1(\Omega)$.

The second objection can be immediately overcome by imposing the condition that henceforth $f \in H^{-1}(\Omega)$ instead of $f \in \mathscr{D}'(\Omega)$.

The first objection is the object of the following discussion. It is our purpose to prove that $a(u, \phi)$ can be extended from $H_0^1(\Omega) \times \mathscr{D}(\Omega)$ to $H_0^1(\Omega) \times H_0^1(\Omega)$ as long as the coefficients of L are sufficiently regular. With this aim, we need essentially that the coefficients a_{ij}, b_i, c_i, d ($i, j = 1, \ldots, n$) be such that the products $a_{ij}(\partial u/\partial x_j)(\partial \phi/\partial x_i)$, $b_i(\partial u/\partial x_i)\phi$, $c_i u(\partial \phi/\partial x_i)$, $du\phi$

($i, j = 1, \ldots, n$) are summable in Ω if $u, \phi \in H_0^1(\Omega)$. Making use of the inclusions

$$H_0^1(\Omega) \subset L^{2n/n-2}(\Omega) \quad \text{if } n > 2; \; H_0^1(\Omega) \subset L^r(\Omega), \text{ for every } r \geq 1,$$
$$\text{if } n = 2; \; H_0^1(\Omega) \subset C^0(\bar{\Omega}) \quad \text{if } n = 1, \quad (7.48)$$

it follows immediately that, to make $a(u, \phi): H_0^1(\Omega) \times H_0^1(\Omega) \to \mathbb{R}$ meaningful, it is sufficient that the coefficients of L satisfy the following conditions (with $i, j = 1, \ldots, n$):

(H1) $a_{ij} \in L^\infty(\Omega)$;

(H2) $b_i \in L^n(\Omega) + L^\infty(\Omega)$, if $n > 2$;
$b_i \in L^{2+\varepsilon}(\Omega) + L^\infty(\Omega)$, if $n = 2$ (with $\varepsilon > 0$);
$b_i \in L^2(\Omega) + L^\infty(\Omega)$, if $n = 1$;

(H3) $c_i \in L^n(\Omega) + L^\infty(\Omega)$, if $n > 2$;
$c_i \in L^{2+\varepsilon}(\Omega) + L^\infty(\Omega)$, if $n = 2$ (with $\varepsilon > 0$);
$c_i \in L^2(\Omega) + L^\infty(\Omega)$, if $n = 1$;

(H4) $d \in L^{n/2}(\Omega) + L^\infty(\Omega)$, if $n > 2$;
$d \in L^{1+\varepsilon}(\Omega) + L^\infty(\Omega)$, if $n = 2$ (with $\varepsilon > 0$);
$d \in L^1(\Omega) + L^\infty(\Omega)$, if $n = 1$

(of course, for bounded Ω the space $L^\infty(\Omega)$ can be suppressed in (H2), (H3), (H4)).

We denote once more by a the extension of the form (7.46) from $H_0^1(\Omega) \times \mathcal{D}(\Omega)$ to $H_0^1(\Omega) \times H_0^1(\Omega)$. We will subsequently prove that the form a is continuous on $H_0^1(\Omega) \times H_0^1(\Omega)$—it follows then, in particular, that it is continuous on $H_0^1(\Omega) \times \mathcal{D}(\Omega)$ with the topology of $H_0^1(\Omega) \times H_0^1(\Omega)$ and hence, since $\mathcal{D}(\Omega)$ is dense in $H_0^1(\Omega)$, that (7.46) has a unique continuous extension with respect to this topology. Under the conditions (H1), (H2), (H3), (H4), which are stronger than (h1), (h2), (h3), (h4), we can in fact prove the following theorem:

THEOREM 7.2. *If the coefficients a_{ij}, b_i, c_i, d ($i, j = 1, \ldots, n$) satisfy the conditions (H1), (H2), (H3), (H4) then the form $a(u, \phi): H_0^1(\Omega) \times H_0^1(\Omega) \to \mathbb{R}$ is continuous, i.e.:*

$$\exists k \in \mathbb{R} \, |a(u, \phi)| \leq k \, \|u\|_{H_0^1(\Omega)} \|\phi\|_{H_0^1(\Omega)} \quad \forall u, \phi \in H_0^1(\Omega). \quad (7.49)$$

Proof. We will show the calculations for $n > 2$ only, using many times the inequality

$$\|u\|_{L^{2n/n-2}(\Omega)} \leq c_0 \|u\|_{H_0^1(\Omega)} \quad \forall u \in H_0^1(\Omega), \text{ with } c_0 \in \mathbb{R}, \quad (7.50)$$

which expresses the continuity of the inclusion of $H_0^1(\Omega)$ in $L^{2n/n-2}(\Omega)$. The cases $n = 1$ and $n = 2$ are analogous. In the proof we will make use of the equivalence of the usual norm of $H_0^1(\Omega)$

$$\|u\|_{H_0^1(\Omega)} = \left(\sum_{i=1}^n \left\|\frac{\partial u}{\partial x_i}\right\|_{L^2(\Omega)}^2\right)^{1/2} \tag{7.51}$$

and of the (non-Hilbertian!) norm

$$\|\|u\|\|_{H_0^1(\Omega)} = \sum_{i=1}^n \left\|\frac{\partial u}{\partial x_i}\right\|_{L^2(\Omega)}, \tag{7.52}$$

an equivalence which is expressed by the double inequality

$$c_1 \|u\|_{H_0^1(\Omega)} \leq \|\|u\|\|_{H_0^1(\Omega)} \leq c_2 \|u\|_{H_0^1(\Omega)}, \text{ with } c_1, c_2 \in \mathbb{R}. \tag{7.53}$$

Given that $u, \phi \in H_0^1(\Omega)$ and that the conditions (H1), (H2), (H3), (H4) are satisfied by hypothesis, it follows from (7.46) that

$$|a(u, \phi)| \leq \sum_{i,j=1}^n \|a_{ij}\|_{L^\infty(\Omega)} \|\partial u/\partial x_j\|_{L^2(\Omega)} \|\partial \phi/\partial x_i\|_{L^2(\Omega)}$$

$$+ \sum_{i=1}^n \|b_i\|_{L^n(\Omega)} \|\partial u/\partial x_i\|_{L^2(\Omega)} \|\phi\|_{L^{2n/n-2}(\Omega)}$$

$$+ \sum_{i=1}^n \|c_i\|_{L^n(\Omega)} \|u\|_{L^{2n/n-2}(\Omega)} \|\partial \phi/\partial x_i\|_{L^2(\Omega)}$$

$$+ \|d\|_{L^{n/2}(\Omega)} \|u\|_{L^{2n/n-2}(\Omega)} \|\phi\|_{L^{2n/n-2}(\Omega)} \tag{7.54}$$

and hence, given (7.50) and (7.53) (and recalling that $\|\partial u/\partial x_i\|_{L^2(\Omega)} \leq \|\|u\|\|_{H_0^1(\Omega)}$), the theorem is shown to be true with

$$k = An^2 c_2^2 + Bnc_0 c_2 + Cnc_0 c_2 + Dc_0^2, \tag{7.55}$$

where $A = \max_{i,j} \|a_{ij}\|_{L^\infty(\Omega)}$, $B = \max_i \|b_i\|_{L^n(\Omega)}$, $C = \max_i \|c_i\|_{L^n(\Omega)}$ and $D = \|d\|_{L^{n/2}(\Omega)}$. ∎

We can now formulate the

PROBLEM 7.7. Given $f \in H^{-1}(\Omega)$ find $u \in H_0^1(\Omega)$ such that

$$a(u, v) = {}_{H^{-1}(\Omega)}\langle f, v\rangle_{H_0^1(\Omega)} \quad \forall v \in H_0^1(\Omega). \tag{7.56}$$

This is a true *variational problem* and is equivalent to problem 7.6 (with $f \in H^{-1}(\Omega)$ instead of $f \in \mathscr{D}'(\Omega)$). We have thus achieved our aim, i.e. that of presenting problem 7.6 in the form of a variational problem. We will now study the problem 7.7.

On the coerciveness of the bilinear form

Unfortunately the form a is, in general, not coercive and thus we cannot apply the results of Chapter 3 to problem 7.7. We have two alternatives: either we impose further restrictions on the coefficients a_{ij}, b_i, c_i, d ($i, j =$

1,..., n) which ensure that the form a is coercive, or we perturb the problem by changing the form itself into a coercive one (we will then have to translate the information about the perturbed problem into information about the original one!...). We will present here both these alternatives but essentially we will follow the second, which gives very interesting results.

With respect to the first alternative suggested above we present the following result:

THEOREM 7.3. *If the coefficients a_{ij}, b_i, c_i, d ($i, j = 1, \ldots, n$) are such that there exists $\alpha > 0$ such that*

$$\sum_{i,j=1}^{n} a_{ij}(x)\xi_i\xi_j \geq \alpha \sum_{i=1}^{n} \xi_i^2 \quad \forall \xi \in \mathbb{R}^n, \text{ a.e. in } \Omega, \quad (7.57)$$

and

$$c_0 c_2 nB + c_0 c_2 nC + c_0^2 D < \alpha/2 \quad (7.58)$$

then

$$a(v, v) \geq \frac{\alpha}{2}\|v\|^2_{H_0^1(\Omega)} \quad \forall v \in H_0^1(\Omega), \quad (7.59)$$

i.e. the form a is coercive on $H_0^1(\Omega)$.

For a proof see, e.g., Stampacchia, 1965, p. 200. Essentially, the result states that if the operator L is elliptic in Ω (i.e. (7.57)) and the coefficients of its non-principal part are small enough (i.e. (7.58)) then the form a is coercive on $H_0^1(\Omega)$.

The hypothesis (7.58), which is easier to satisfy for $\mu(\Omega)$ smaller, is of an entirely different kind from the hypotheses (H1), (H2), (H3), and (H4): the latter are hypotheses of regularity, while (7.57) is an algebraic hypothesis and (7.58) is an hypothesis which we can call metric. As we have previously stated (section 7.1), we assume that (7.57) is always true, and deal therefore with elliptic operators only.

With respect to the second alternative, we will first prove the following result:

THEOREM 7.4. *There exists $\bar{\lambda} \in \mathbb{R}$ such that for every $\lambda \geq \bar{\lambda}$ the bilinear form*

$$a_\lambda(u, v) = a(u, v) + \lambda(u, v)_{L^2(\Omega)} \quad (7.60)$$

is coercive on $H_0^1(\Omega)$, and more precisely

$$\exists \bar{\lambda} > 0 \quad \forall \lambda \geq \bar{\lambda} \quad a_\lambda(u, v) \geq \frac{\alpha}{4}\|v\|^2_{H_0^1(\Omega)}, \quad (7.61)$$

with α as in (7.57).

Proof. We will first note that *given $f \in L^p(\Omega)$ and $\varepsilon > 0$ we can always find*

$f^{[2]} \in L^p(\Omega)$ and $f^{[1]} \in L^\infty(\Omega)$ such that $\|f^{[2]}\|_{L^p(\Omega)} < \varepsilon$ and

$$f = f^{[1]} + f^{[2]}. \tag{7.62}$$

Effectively, let k be a positive real number and consider the sets $\Omega_{k^+} = \{x \in \Omega : f(x) \geq k\}$, $\Omega_{k^-} = \{x \in \Omega : f(x) \leq -k\}$ and the functions

$$f^{[1]}(x) = \begin{cases} k & \text{if } x \in \Omega_{k^+} \\ f(x) & \text{if } x \in \Omega \setminus (\Omega_{k^+} \cup \Omega_{k^-}) \\ -k & \text{if } x \in \Omega_{k^-} \end{cases} \tag{7.63}$$

$$f^{[2]} = f - f^{[1]}. \tag{7.64}$$

To see that there exists $k > 0$ such that $f^{[1]}$ and $f^{[2]}$ satisfy the above conditions, we need only note that

$$\|f^{[2]}\|_{L^p(\Omega)}^p = \int_\Omega |f^{[2]}|^p \, dx = \int_{\Omega \setminus (\Omega_{k^+} \cup \Omega_{k^-})} |f^{[2]}|^p \, dx$$
$$+ \int_{\Omega_{k^+} \cup \Omega_{k^-}} |f^{[2]}|^p \, dx = \int_{\Omega_{k^+}} |f - k|^p \, dx + \int_{\Omega_{k^-}} |f + k|^p \, dx, \tag{7.65}$$

which shows that as k increases $\|f^{[2]}\|_{L^p(\Omega)}$ tends to zero (since $\mu(\Omega_{k^+})$ and $\mu(\Omega_{k^-})$ tend to zero) and hence there certainly exists k (dependent on ε) such that $\|f^{[2]}\|_{L^p(\Omega)} < \varepsilon$ and, at the same time, $\|f^{[1]}\|_{L^\infty(\Omega)} = \operatorname{supess}_\Omega |f^{[1]}| \leq k$.

We proceed now to proving the theorem proper. Let us decompose the functions b_i, c_i, d $(i = 1, \ldots, n)$ according to the previous comments and at the same time let us consider the decomposition of the linear form a into the sum of the two forms

$$a^{[1]}(u, v) = \int_\Omega \left(\sum_{i,j=1}^n a_{ij} \frac{\partial u}{\partial x_j} \frac{\partial v}{\partial x_i} + \sum_{i=1}^n b_i^{[1]} \frac{\partial u}{\partial x_i} v - \sum_{i=1}^n c_i^{[1]} u \frac{\partial v}{\partial x_i} + d^{[1]} uv \right) dx \tag{7.66}$$

and

$$a^{[2]}(u, v) = \int_\Omega \left(\sum_{i=1}^n b_i^{[2]} \frac{\partial u}{\partial x_i} v - \sum_{i=1}^n c_i^{[2]} u \frac{\partial v}{\partial x_i} + d^{[2]} uv \right) dx, \tag{7.67}$$

which we will deal with separately.

For every $v \in H_0^1(\Omega)$ we have

$$a^{[2]}(v, v) \leq \sum_{i=1}^n \varepsilon \left\| \frac{\partial v}{\partial x_i} \right\|_{L^2(\Omega)} \|v\|_{L^{2n/n-2}(\Omega)}$$
$$+ \sum_{i=1}^n \varepsilon \left\| \frac{\partial v}{\partial x_i} \right\|_{L^2(\Omega)} \|v\|_{L^{2n/n-2}(\Omega)} + \varepsilon \|v\|_{L^{2n/n-2}(\Omega)} \|v\|_{L^{2n/n-2}(\Omega)} \tag{7.68}$$

or, given (7.50) and (7.53),

$$a^{[2]}(v, v) \leq (2\varepsilon c_0 c_2 + \varepsilon c_0^2) \|v\|_{H_0^1(\Omega)}^2. \tag{7.69}$$

If we fix ε (which was till now free of any restraint) in such a way that

$$2\varepsilon c_0 c_2 + \varepsilon c_0^2 \leq \alpha/4 \tag{7.70}$$

we can say that

$$a^{[2]}(v, v) \leq \frac{\alpha}{4} \|v\|^2_{H_0^1(\Omega)} \qquad \forall v \in H_0^1(\Omega), \qquad (7.71)$$

and hence proving (7.61) is equivalent to proving that

$$\exists \bar\lambda \in \mathbb{R} \quad \forall \lambda \geq \bar\lambda \ a^{[1]}(v, v) + \lambda \|v\|^2_{L^2(\Omega)} \geq \frac{\alpha}{2} \|v\|^2_{H_0^1(\Omega)} \qquad \forall v \in H_0^1(\Omega). \qquad (7.72)$$

Now, given that L is elliptic, we can write (putting $\xi = \operatorname{grad} v(x)$ in (7.57))

$$\int_\Omega \sum_{i,j=1}^n a_{ij} \frac{\partial v}{\partial x_i} \frac{\partial v}{\partial x_j} dx \geq \alpha \int_\Omega \sum_{i=1}^n \left|\frac{\partial v}{\partial x_i}\right|^2 dx = \alpha \|v\|^2_{H_0^1(\Omega)} \qquad (7.73)$$

and, further, putting $\beta = \max_i \|b_i^{[1]}\|_{L^\infty(\Omega)}$, $\gamma = \max_i \|c_i^{[1]}\|_{L^\infty(\Omega)}$ and $\delta = \|d^{[1]}\|_{L^\infty(\Omega)}$,

$$\left| \sum_{i=1}^n \int_\Omega b_i^{[1]} \frac{\partial v}{\partial x_i} v \, dx \right| \leq \beta c_2 \|v\|_{H_0^1(\Omega)} \|v\|_{L^2(\Omega)}, \qquad (7.74)$$

$$\left| \sum_{i=1}^n \int_\Omega c_i^{[1]} \frac{\partial v}{\partial x_i} v \, dx \right| \leq \gamma c_2 \|v\|_{H_0^1(\Omega)} \|v\|_{L^2(\Omega)} \qquad (7.75)$$

and

$$\left| \int_\Omega d^{[1]} v v \, dx \right| \leq \delta \|v\|_{L^2(\Omega)} \|v\|_{L^2(\Omega)}. \qquad (7.76)$$

We have thus the estimate (rough, but enough for our purposes)

$$a^{[1]}(v, v) \geq \alpha \|v\|^2_{H_0^1(\Omega)} - c_2(\beta + \gamma) \|v\|_{H_0^1(\Omega)} \|v\|_{L^2(\Omega)} - \delta \|v\|^2_{L^2(\Omega)}. \qquad (7.77)$$

Applying the formula (valid for every $a \in \mathbb{R}$, $b \in \mathbb{R}$ and $\sigma > 0$)

$$2 |a| |b| \leq \sigma a^2 + \frac{1}{\sigma} b^2 \qquad (7.78)$$

(from $(|A| - |B|)^2 = |A|^2 + |B|^2 - 2 |A| |B| \geq 0$ we get $2 |A| |B| \leq |A|^2 + |B|^2$, and putting $A = \sqrt{(\sigma)} a$ and $B = b/\sqrt{(\sigma)}$ we get (7.78)) to the case $a = \|v\|_{H_0^1(\Omega)}$, $b = c_2(\beta + \gamma) \|v\|_{L^2(\Omega)}$ and $\sigma = \alpha$, we get

$$c_2(\beta + \gamma) \|v\|_{H_0^1(\Omega)} \|v\|_{L^2(\Omega)} \leq \frac{\alpha}{2} \|v\|^2_{H_0^1(\Omega)} + \frac{c_2^2(\beta + \gamma)^2}{2\alpha} \|v\|^2_{L^2(\Omega)} \qquad (7.79)$$

which, substituted in (7.77), gives us

$$a^{[1]}(v, v) \geq \alpha \|v\|^2_{H_0^1(\Omega)} - \frac{\alpha}{2} \|v\|^2_{H_0^1(\Omega)} - \frac{c_2^2(\beta + \gamma)^2}{2\alpha} \|v\|^2_{L^2(\Omega)} - \delta \|v\|^2_{L^2(\Omega)}$$

$$= \frac{\alpha}{2} \|v\|^2_{H_0^1(\Omega)} - \left(\frac{c_2^2(\beta + \gamma)^2}{2\alpha} + \delta\right) \|v\|^2_{L^2(\Omega)}. \qquad (7.80)$$

Taking

$$\bar\lambda = \frac{c_2^2(\beta+\gamma)^2}{2\alpha}+\delta \qquad (7.81)$$

gives us the result we want. ∎

REMARK. The nomenclature used in the works mentioned in the bibliography is not uniform. In many of them, if (7.61) is true the form a is said to be *coercive* (or, for reasons which will become clearer in what follows, *coercive with respect to* $L^2(\Omega)$), and the form a is said to be *elliptic* (or *strongly coercive*) if (7.59) is true.

The form a_λ is obviously continuous on $H_0^1(\Omega) \times H_0^1(\Omega)$ and, as such, if $\lambda \geq \bar\lambda$, we can say that the following problem is well-posed:

PROBLEM 7.8. Given $f \in H^{-1}(\Omega)$ find $u \in H_0^1(\Omega)$ such that

$$a_\lambda(u, v) = {}_{H^{-1}(\Omega)}\langle f, v \rangle_{H_0^1(\Omega)} \qquad \forall v \in H_0^1(\Omega). \qquad (7.82)$$

Now, problem 7.8 is equivalent to the following problem 7.9 which therefore is also well-posed:

PROBLEM 7.9. Given $f \in H^{-1}(\Omega)$ find $u \in H_0^1(\Omega)$ such that

$$Lu + \lambda u = f \quad \text{in the sense of } \mathcal{D}'(\Omega). \qquad (7.83)$$

The Riesz–Fredholm theorem

Now the original problem is problem 7.6—which is nothing but problem 7.9 with $\lambda = 0$ (except for the question of the regularity of the data f)—and naturally we want to know as much as we can about this problem. Now, given that in general the form a is not coercive, we cannot be sure that problem 7.6 is well-posed; however, the fact that problem 7.9 is well-posed for some values of λ (and we know that it is so for every $\lambda \geq \bar\lambda$) enables us to obtain useful information regarding problem 7.6—and in general about all the problems 7.9 with λ not necessarily $\geq \bar\lambda$. This information is contained in the statement of the following theorem, which follows from the application of the Riesz–Fredholm theory to the pair formed by the problem 7.9 and by the following problem 7.10:

PROBLEM 7.10. Find $u \in H_0^1(\Omega)$ such that

$$Lu + \lambda u = 0 \quad \text{in the sense of } \mathcal{D}'(\Omega). \qquad (7.84)$$

THEOREM 7.5 (Theorem of the alternative or Riesz–Fredholm theorem). The following results are true:

(i) Necessary and sufficient condition for problem 7.9 to have one and only one solution is that problem 7.10 has only the null solution.

(ii) Problem 7.10 has only the null solution for all values of λ which do not belong to a countable set $\sigma(L)$, the *spectrum* of the problem.

(iii) If $\lambda \in \sigma(L)$ then problem 7.10 has at most a finite number of linearly independent solutions, $u_1, \ldots, u_s$.

Let us remark that part (iii) can be improved (better: completed) by considering the homogeneous problem adjoint to problem 7.9, which under the conditions of (iii) has s solutions $u'_1, \ldots, u'_s$: *necessary and sufficient condition for problem 7.9 to have a solution is that f be orthogonal to $u'_1, \ldots, u'_s$*. For a proof see Lions, 1955, p. 30; see also Lions, 1965, p. 108 and Magenes–Stampacchia, 1958; p. 293 for this theorem and Riesz–Sz.–Nagy, 1955; chs. IV and V for the classical theory of the alternative.

Problem 7.6 has finally been studied completely as regards existence, uniqueness, and continuous dependence of the solution. For a study of the regularity and other properties of the solutions of the Dirichlet problems see, e.g., Stampacchia, 1965 (on the subject of regularity we will say something later (section 7.3.1)).

The success of the variational method, also known as the *finite Dirichlet integral method*, in studying the homogeneous Dirichlet problem suggests that we consider other boundary value problems. This is what we intend doing in the following sections where, however, contrary to what we have done so far, we will start with a problem already posed in the variational form, which we will then interpret (if possible) as a boundary value problem. It is true that, in most cases, physics and technology gives us the problems under the form of boundary value problems and not under that of variational problems, but this is not always the case, and, roughly, we can say that the problems which arise from applying the principle of the minimum of the energy or the theorem of virtual work are more naturally formulated variationally (one example, for the case of equalities, is problem 6.2). Anyway, what we are doing is not useless since, on the one hand, it enables us to frame a large portion of the set of results on partial differential equations within a single theory, and, on the other, it teaches us (by just inverting what we have done) how to formulate variationally some of the most important and frequent boundary value problems (for a further discussion on subjects of this kind, see Lions, 1955, pp. 38 *et seq.*).

7.2.2 GENERAL FORMULATION IN VARIATIONAL TERMS

We will present here a very simplified version of the formulation which Lions (1955) gave to the theory of boundary value problems. For a discussion of this formulation in a much more general framework see Lions, 1955 and Magenes–Stampacchia, 1958; in L. Schwartz, 1957, p. 218 and in Nečas, 1967, p. 131 the reader will find shorter versions, and in Lions, 1961, p. 9 and Lions, 1965, p. 93 a slightly different one.

The data and the variational problem

Here we start with the following data:

(L1) an open set $\Omega \subset \mathbb{R}^n$,
(L2) a Hilbert space Q such that $\mathscr{D}(\Omega) \hookrightarrow^{ds} Q \hookrightarrow \mathscr{D}'(\Omega)$,
(L3) a Hilbert space V such that $\mathscr{D}(\Omega) \hookrightarrow V \hookrightarrow Q \hookrightarrow \mathscr{D}'(\Omega)$,
(L4) a bilinear form $\mathfrak{A}(u, v): V \times V \to \mathbb{R}$ continuous and coercive on V,

and consider the following variational problem:

PROBLEM 7.11. Given $f \in Q'$ find $u \in V$ such that

$$\mathfrak{A}(u, v) = {}_{Q'}\langle f, v \rangle_Q \qquad \forall v \in V. \tag{7.85}$$

First, the problem 7.11 is effectively a variational problem, even though in (7.85) we have the duality ${}_{Q'}\langle\,,\,\rangle_Q$ instead of the duality ${}_{V'}\langle\,,\,\rangle_V$, since $V \hookrightarrow^{ds} Q$ and v varies in V. The space Q was not introduced when we studied (in section 3.1) the abstract theory of variational problems: we introduce it here so as to avoid considering the space V' which is not, in general, a space of distributions (if V is normal we can of course take $Q = V$, and thus consider problem 7.11 in its most general form).

The Lions–Stampacchia theorem—in the form of its corollary which is the Lax–Milgram lemma—ensures us that problem 7.11 is well-posed (let us remark that if $\mathfrak{A}$ is coercive on V then it is so also on $V - V = V$ (cf. (3.27)).

Interpretation of the variational problem

We will now interpret problem 7.11, when possible, as a boundary value problem.

With this aim, we will start by associating with the form $\mathfrak{A}$ an operator $\Lambda \in \mathscr{L}(V, \mathscr{D}'(\Omega))$. Given $u \in V$ fixed, the functional '$\phi \mapsto \mathfrak{A}(u, \phi)$' is linear and continuous on $\mathscr{D}(\Omega)$ (since $\mathfrak{A}$ is continuous and $\mathscr{D}(\Omega) \hookrightarrow V$); denoting this functional by $\Lambda(u)$ we can write

$$\mathfrak{A}(u, \phi) = {}_{\mathscr{D}'(\Omega)}\langle \Lambda u, \phi \rangle_{\mathscr{D}(\Omega)} \qquad \forall \phi \in \mathscr{D}(\Omega). \tag{7.86}$$

(7.86) defines, with u varying in V, an operator $\Lambda: V \to \mathscr{D}'(\Omega)$ which is obviously linear and continuous: $\Lambda \in \mathscr{L}(V, \mathscr{D}'(\Omega))$ (let us remark that if $u_n \xrightarrow{V} 0$ then, since $\mathfrak{A}$ is continuous, $\mathfrak{A}(u_n, \phi) \to 0$ uniformly in ϕ, as long as ϕ remains in a bounded set of $\mathscr{D}(\Omega)$; this means that $\Lambda u_n \to 0$ in the (strong) topology of $\mathscr{D}'(\Omega)$).

The association of an operator with a form is not new to us—in section 3.1 (whose H and a are now V and $\mathfrak{A}$) we associated with $\mathfrak{A}$ two operators,

$A: V \to V'$ and $\mathscr{A}: V \to V$, by the formulae

$$\mathfrak{A}(u, v) = {}_{V'}\langle Au, v\rangle_V \qquad \forall v \in V \tag{7.87}$$

and

$$\mathfrak{A}(u, v) = (\mathscr{A}u, v)_V \qquad \forall v \in V, \tag{7.88}$$

respectively. The operator Λ, which we have now defined, is essentially a restriction of the operator A which, when V is normal, actually coincides with A (in other words we substitute in (7.86) the duality ${}_{\mathscr{D}'(\Omega)}\langle\,,\rangle_{\mathscr{D}(\Omega)}$ by the duality ${}_{V'}\langle\,,\rangle_V$ and let ϕ vary in V, which is possible since $\mathscr{D}(\Omega)$ is dense in V). We will soon introduce the analogue of the operator $\mathscr{A}$.

In (7.85) we have considered the duality ${}_{Q'}\langle\,,\rangle_Q$ and not the duality ${}_{\mathscr{D}'(\Omega)}\langle\,,\rangle_{\mathscr{D}(\Omega)}$, which instead we have used to define Λ. It is then natural to think that, to interpret problem 7.11 as a boundary value problem, it is sufficient to consider the restriction of Λ to the set

$$D = \{u \in V : \Lambda u \in Q'\} \tag{7.89}$$

(it is interesting to observe that D is dense in V and hence in Q (see Lions, 1961, p. 12; Nečas, 1967, p. 132).

This restriction, which we will continue to denote by Λ for convenience, can be defined directly by means of the formula

$$u \in D, \quad \mathfrak{A}(u, \phi) = {}_{Q'}\langle \Lambda u, \phi\rangle_Q \qquad \forall \phi \in \mathscr{D}(\Omega), \tag{7.90}$$

since to say that Λu belongs to Q' means that Λu is a continuous linear functional on $\mathscr{D}(\Omega)$ in the topology of Q (i.e., $|\mathfrak{A}(u, \phi)| = |{}_{\mathscr{D}'(\Omega)}\langle \Lambda u, \phi\rangle_{\mathscr{D}(\Omega)}| \leq$ constant $\cdot \|\phi\|_Q \quad \forall \phi \in \mathscr{D}(\Omega)$) and hence, since $\mathscr{D}(\Omega)$ is dense in Q, we can substitute in (7.86) the duality ${}_{\mathscr{D}'(\Omega)}\langle\,,\rangle_{\mathscr{D}(\Omega)}$ by the duality ${}_{Q'}\langle\,,\rangle_Q$ (but not $\forall \phi \in \mathscr{D}(\Omega)$ by $\forall \phi \in Q$!). Since Q is a Hilbert space we can, by Riesz's theorem, represent the functional $\Lambda u \in Q'$ by an element $\mathscr{L}(u) \in Q$ and hence write the formula

$$\mathfrak{A}(u, \phi) = (\mathscr{L}u, \phi)_Q \qquad \forall \phi \in \mathscr{D}(\Omega), \tag{7.91}$$

which defines the operator $\mathscr{L}: Q \to Q$ analogous to the operator $\mathscr{A}$ (and which coincides with $\mathscr{A}$ if V is normal and we take $Q = V$, since then $V = D$).

In what follows we will provide D with the (Hilbertian) topology given by the norm of the graph

$$\|u\|_D^2 = \|u\|_V^2 + \|\Lambda u\|_{Q'}^2, \tag{7.92}$$

unless, of course, $D = V$ (this is the least fine of the topologies on D which enables us to say that $D \hookrightarrow V$ and $\Lambda \in \mathscr{L}(D, Q')$).

The formulae (7.85) and (7.90) lead us to consider the linear variety

$$N = \{u \in D : \mathfrak{A}(u, \phi) = {}_{Q'}\langle \Lambda u, \phi\rangle_Q \quad \forall \phi \in V\}, \tag{7.93}$$

which of course we will provide with the topology induced by D.

The space N, which will play a very important role, is in general a proper subset of D, but coincides with D if V is normal and we take $Q = V$—in this limit case we will then have $V = Q = N = D$. A curious characterization of the space N, the proof of which can be seen in Lions, 1955, p. 27, is the following: N is the *supplementary* of $D_\Lambda = \{u \in D : \Lambda u = 0\} =$ kernal (Λ), i.e., $D_\Lambda \oplus N = D$. In terms of N, the fact that problem 7.11 is well-posed means that (a further restriction of) Λ is a topological isomorphism of N on Q'; for a direct proof of this fact see Lions, 1955, p. 26; Magenes–Stampacchia, 1958, pp. 273 and 281.

We can now interpret problem 7.11 as a boundary value problem. It is evident that if u is a solution of the problem 7.11 then in particular

$$\mathfrak{A}(u, v) = {}_{Q'}\langle f, v \rangle_Q \qquad \forall v \in \mathscr{D}(\Omega) \tag{7.94}$$

and therefore, comparing it to (7.90) (observe that the solution, u, is fixed), we have $f = \Lambda u \in Q'$; this together with (7.85) shows also that $u \in N$. On the other hand, it is also evident that if we put $f = \Lambda u \in Q'$ and $u \in N$ then u is a solution of the problem 7.11. We have thus proved that problem 7.11 is equivalent to the following one:

PROBLEM 7.12. Given $f \in Q'$ find $u \in V$ such that

$$\Lambda u = f \tag{7.95}$$

and

$$u \in N. \tag{7.96}$$

We will call this problem the *homogeneous boundary value problem* and (7.96) the *homogeneous boundary condition*. Condition (7.96) has a very unusual form and must be appropriately interpreted in every concrete case—it is essentially to these interpretations that we will address the next section.

Natural and forced conditions

One question that arises naturally is the following: where are the boundary value conditions in the variational formulation 7.11? They are in part contained in the fact that u belongs to V and in part in the variational equation (7.85) itself. The former are known as *stable, forced,* or *rigid conditions* and the latter as *natural* or *transversality conditions*. For a discussion of the 'physical significance' of these two (very different!) types of conditions see Courant, 1943, p. 4.

Non-homogeneous boundary value problems

Another question that can be asked, given the name which we have associated to problem 7.12, is the following: does the variational formulation enable us to study only homogeneous boundary value problems? Not at all! Variational techniques can be applied also to non-homogeneous boundary value problems. However we have to complicate the picture we have presented somewhat by introducing another functional space, *the space of data on the boundary*. Let then $K \supseteq D$ be a Hilbert space such that $\Lambda \in \mathcal{L}(D, Q')$ can be considered as the restriction to D of a continuous linear operator from K to Q' (which we will again denote by Λ for convenience). Under these conditions it is meaningful to consider the following.

PROBLEM 7.13. Given $f \in Q'$ and $h \in K$ find $u \in K$ such that

$$\Lambda u = f \qquad (7.97)$$

and

$$u - h \in N. \qquad (7.98)$$

We will call this problem a *non-homogeneous boundary value problem*. More precisely, the problem is non-homogeneous if $h \notin N$ (and hence in particular $h \neq 0$). In this context note that, if $\eta \in N$, the condition (7.98) is equivalent to the condition $u - (h + \eta) \in N$—this has consequences for the uniqueness of some problems.

It can be easily seen that problem 7.13 is equivalent to the following:

PROBLEM 7.14. Given $f \in Q'$ and $h \in K$ find $u \in K$ such that $w = u - h \in V$ satisfies the variational equation

$$\mathfrak{A}(w, v) = {}_{Q'}\langle f - \Lambda h, v\rangle_Q \qquad \forall v \in V. \qquad (7.99)$$

This is a variational problem completely analogous to problem 7.11. Essentially we have used the well-known technique which changes a non-homogeneous problem into a homogeneous one.

More on the Dirichlet problem

Let us now return to the problems 7.8 and 7.9 considered in the previous section and try to analyse them in the light of the abstract theory just presented. There we already had the spaces $V = H_0^1(\Omega)$, $Q = H_0^1(\Omega)$ ($H_0^1(\Omega)$ being normal) and the form $\mathfrak{A} = a_\lambda$, which we have seen to be continuous and (as long as λ is large enough) coercive on $H_0^1(\Omega) \times H_0^1(\Omega)$: the problem 7.8 is then a concrete case of problem 7.11. Problem 7.9, with $L + \lambda = \Lambda \in \mathcal{L}(H_0^1(\Omega), H^{-1}(\Omega))$, is instead a concrete case of problem 7.11: the condition (7.96) has not been explicitly stated in problem 7.9 because, with $N = D = H_0^1(\Omega)$, it is no more restrictive than the condition $u \in H_0^1(\Omega)$ which has already been imposed once (there are no natural conditions here). The theory confirms then what we knew already: problem 7.9 is an interpretation of problem 7.8 as a boundary value problem.

There is still one question to be cleared up: if we take, as we have done, $\mathfrak{A}(u, v) = a_\lambda(u, v)$ with appropriate λ (i.e. $\lambda \geq \bar{\lambda}$) then we are completely within the framework (L1), (L2), (L3), (L4); but we can just take the form a and then find $\bar{\lambda}$ such that $\mathfrak{A}(u, v) = a(u, v) + \lambda(u, v)_{L^2(\Omega)}$ is coercive if $\lambda \geq \bar{\lambda}$: the form $\mathfrak{A}$ is essentially the same in both approaches, but in the second case we are not within the framework (L1), (L2), (L3), (L4)! In fact, when in the second case we go from form $a(u, v)$ to the form $\mathfrak{A}(u, v) = a_\lambda(u, v) = a(u, v) + \lambda(u, v)_{L^2(\Omega)}$ we are introducing a 'parasitic' space—the space $L^2(\Omega)$ whose presence is not foreseen by the theory. This is a particular aspect of a more general situation which we will now briefly discuss.

Suppose, on the one hand, that we eliminate in (L4) the assumption regarding the coercivity of the form $\mathfrak{A}$ on V—which implies, among other things, the collapse of the whole theory so far exposed—and, on the other, that H is a Hilbert space which satisfies the two following conditions

(p1) $V \hookrightarrow^{ds} H$,
(p2)   there exists $\bar{\lambda} \in \mathbb{R}$ such that if $\lambda \geq \bar{\lambda}$ then the form $\mathfrak{A}_\lambda(u, v) = \mathfrak{A}(u, v) + \lambda(u, v)_H$ is coercive on V.

Under these conditions, since it is impossible to study the problem with respect to the form $\mathfrak{A}$, we naturally study it with respect to the forms $\mathfrak{A}_\lambda$ with $\lambda \geq \bar{\lambda}$ and then, if possible, apply the Riesz–Fredholm theory—this is exactly what we have done in the previous section.

In most of the interesting applications we can take as H, which we call the *pivot space*, the space $L^2(\Omega)$—there are, however, exceptions, as one can see in Lions, 1969b, p. 127; Lions–Magenes, 1972a, p. 214. We will systematically identify H with its dual H' (even when $H \neq L^2(\Omega)$!) and hence write

$$V \hookrightarrow^{ds} H \equiv H' \hookrightarrow V'. \tag{7.100}$$

This chain is sometimes known as the *Hilbertian triplet* since it contains essentially three Hilbert spaces, V, H, and V'.

In this context note that *if we identify H and H' we cannot identify also V and V'*! The reason for this is that the diagram

$$\begin{array}{ccc} V & \xrightarrow{j_{VH}} & H \\ {\scriptstyle J_V}\downarrow & & \downarrow{\scriptstyle J_H} \\ V' & \xrightarrow{{}^t j_{VH}} & H' \end{array}$$

is commutative iff V and H are algebraically and topologically isomorphic (in short: they are the same space—$V = H$).

Now, to write equation (7.100) signifies that the injections j_{VH}, ${}^t j_{VH}$ and the Riesz operator J_H are identified, respectively, to the set immersions i_{VH}, ${}^t i_{VH}$ and $i_{HH'}$—which in turn implies that the set immersion $i_{VV'}$ is identified to ${}^t j_{VH} \circ J_H \circ j_{VH}$; on the other hand, to identify V and V' means to identify the Riesz operator J_V with $i_{VV'}$, and this is not possible, if $V \neq H$, since the non-commutativity of the diagram can be interpreted through $J_V \neq {}^t j_{VH} \circ J_H \circ j_{VH}$. As a concrete example of this situation take the case $V = H_0^1(\Omega)$ and

$H = L^2(\Omega)$: in this case $i_{VV'} = i_{H_0^1(\Omega)H^{-1}(\Omega)}$ is the canonical immersion of $H_0^1(\Omega)$ in $H^{-1}(\Omega)$, while $J_V = J_{H_0^1(\Omega)}$ is the operator $-\Delta$ (if we provide $H_0^1(\Omega)$ with its usual norm; if instead we provide $H_0^1(\Omega)$ with the norm induced on it by $H_0^1(\Omega)$ we get $J_{H_0^1(\Omega)} = -\Delta + I$, where I is the identity operator).

7.2.3 EXAMPLES OF BOUNDARY VALUE PROBLEMS

The data

The examples of boundary value problems which we present here can be obtained by considering particular cases of the following set of data:

(d1) $\Omega \subset \mathbb{R}^n$ is a bounded connected open set with boundary Γ of class $C^{0,1}$; we take $V = H_0^1(\Omega)$, or $V = H^1(\Omega)$, or $V = V_{\Gamma_0}$, V_{Γ_0} being the
(d2) V is a closed subspace of $H^1(\Omega)$ such that $H_0^1(\Omega) \subsetneq V \subsetneq H^1(\Omega)$ (more precisely: we take $V = H_0^1(\Omega)$, or $V = H^1(\Omega)$, or $V = V_{\Gamma_0}$, V_{Γ_0} being the space $\{u \in H^1(\Omega) : \gamma_0 u = 0 \text{ on } \Gamma_0\}$ with the topology induced by $H^1(\Omega)$, where $\Gamma_0 \subset \Gamma$ is a regular subset of Γ with an $(n-1)$-dimensional positive measure);
(d3) $Q = H_0^1(\Omega)$ if $V = H_0^1(\Omega)$, and $Q = L^2(\Omega)$ otherwise;
(d4) $H = L^2(\Omega)$;
(d5) $\mathfrak{A}_\lambda : V \times V \to \mathbb{R}$ is the restriction to $V \times V$ of the bilinear form (which we denote by the same symbol) $\mathfrak{A}_\lambda : H^1(\Omega) \times H^1(\Omega) \to \mathbb{R}$ defined by

$$\mathfrak{A}_\lambda(u, v) = \int_\Omega \left[\sum_{i,j=1}^n a_{ij} \frac{\partial u}{\partial x_j} \frac{\partial v}{\partial x_i} + \sum_{i=1}^n b_i \frac{\partial u}{\partial x_i} v - \sum_{i=1}^n c_i u \frac{\partial v}{\partial x_i} + duv \right] dx$$

$$+ \lambda \int_\Omega uv \, dx +{}_{H^{-1/2}(\Gamma)}\langle Tu, \gamma_0 v \rangle_{H^{1/2}(\Gamma)}, \quad (7.101)$$

where a_{ij}, b_i, c_i, d $(i, j = 1, \ldots, n)$ satisfy the conditions (H1), (H2), (H3), (H4), and (7.57), λ is an appropriate real number and $T \in \mathcal{L}(H^1(\Omega), H^{-1/2}(\Gamma))$ (more precisely: we take $T \equiv 0$, or $T = t\gamma_0$, or $T = t(\partial/\partial\tau)$, where $t \in C^1(\bar{\Omega})$, and $\partial/\partial\tau$ is an operator of tangential differentiation on Γ);
(d6) $K = H^1(\Omega)$ if $V = H_0^1(\Omega)$ and $K = H_L^1(\Omega) = \{u \in H^1(\Omega) : Lu + \lambda u \in L^2(\Omega)\}$ otherwise, where $L + \lambda$ is the operator

$$L\square + \lambda\square \equiv -\sum_{i,j=1}^n \frac{\partial}{\partial x_i}\left(a_{ij}\frac{\partial}{\partial x_j}\square\right) + \sum_{i=1}^n b_i \frac{\partial}{\partial x_i}\square + \sum_{i=1}^n \frac{\partial}{\partial x_i}(c_i \square) + d\square + \lambda\square$$

(7.102)

with a_{ij}, b_i, c_i, d, and λ as in (d5).

Before going on to consider particular cases of these data in concrete examples, we should analyse them more closely.

Relative to (d1), we impose this regularity to Γ so as to be able to give a precise meaning to the boundary conditions which appear in the problems considered. The problems themselves are meaningful in conditions of less regularity on Ω (all we need is $V \hookrightarrow L^{2n/(n-2)}(\Omega)$ if $n > 2 \ldots$) but the boundary conditions should then be interpreted purely formally.

Relative to (d2), let us observe that Lions, 1961, p. 17, conjectures that the closed subspaces $V \subset H^1(\Omega)$ which are modules over the algebra $W^{1,\infty}(\Omega)$ (i.e., such that $\forall u \in V \ \forall v \in W^{1,\infty}(\Omega) \ uv \in V$) are essentially the spaces which we will consider here. The correct statement of the conjecture (which in particular does not involve any regularity of Γ_0) introduces concepts which we have not defined. The spaces V_{Γ_0} are defined by means of *local conditions*; an example of a closed subspace of $H^1(\Omega)$ defined by means of a *global condition* is given by $\{u \in H^1(\Omega) : \int_\Gamma \gamma_0 u \, d\sigma = 0\}$ (see Lions, 1961, p. 21).

Relative to (d5), the operator $\partial/\partial \tau$ can be defined by interpreting appropriately (according to the trace theory) the identity

$$\frac{\partial u}{\partial \tau} \equiv \sum_{i=1}^{n} \frac{\partial u}{\partial x_i} \tau_i$$

(with $\sum_{i=1}^{n} \tau_i \nu_i = 0$)—in this context see Nečas, 1967, pp. 124 and 138. On the two-dimensional case the tangential derivative, $\partial u/\partial \tau$, of a function $u \in H^1(\Omega)$ can be interpreted as the derivative with respect to the arc length (s, which we use as the parameter for Γ) of the trace of u:

$$\frac{\partial u}{\partial \tau} \triangleq \frac{d(\gamma_0 u)}{ds}$$ (this being meaningful since $\gamma_0 u \in H^{1/2}(\Gamma)$).

Relative to (d6), let us remark that these choices of K are in agreement with the theory developed, since, on the one hand, $\{u \in H^1(\Omega) : Lu + \lambda u \in H^{-1}(\Omega)\} = H^1(\Omega)$ (recall the hypothesis (H1), (H2), (H3), (H4)) and, on the other, $H_L^1(\Omega)$ is nothing but the space D introduced in (7.89); let us remark also that the notation $H_L^1(\Omega)$ is consistent because, given that $u \in H^1(\Omega)$, $Lu + \lambda u \in L^2(u)$ iff $Lu \in L^2(\Omega)$.

Finally, let us return to (d5) and analyse more closely the bilinear form $\mathfrak{A}_\lambda$. It is important to note that we already have a form which is perturbed with the part in λ, which is not fixed—as are for example the functions a_{ij}—but which must be chosen so that the form $\mathfrak{A}_\lambda$ is coercive (and we will see that under our conditions this is always possible): this is the significance of the word 'appropriate' in (d5) (let us remark that in these circumstances it is not absolutely necessary to specify the datum (d4) regarding the pivot space, as we have done). So, we must show that there exists λ such that $\mathfrak{A}_\lambda$ is continuous and coercive on $H^1(\Omega) \times H^1(\Omega)$ and hence on every closed subspace $V \times V \hookrightarrow H^1(\Omega) \times H^1(\Omega)$. Besides, as is obvious, if the form $\mathfrak{A}_\lambda$ is continuous and coercive on $H^1(\Omega) \times H^1(\Omega)$ or, in short, on $H^1(\Omega)$, then it is evidently continuous and coercive on $K - K$, K being a closed convex set in $H^1(\Omega)$; this is of fundamental importance for the problems connected with variational inequalities which we will consider in section 7.3.

The proof becomes easier if we consider that the form $\mathfrak{A}_\lambda(u, v)$ can be decomposed into the sum of the form $a_\lambda(u, v)$, already considered in section

7.2.1, and the form $_{H^{-1/2}(\Gamma)}\langle Tu, \gamma_0 v\rangle_{H^{1/2}(\Gamma)}$, which for obvious reasons we call a *boundary bilinear form*. In section 7.2.2 we have assumed a_λ to be defined on $H_0^1(\Omega) \times H_0^1(\Omega)$ and not on $H^1(\Omega) \times H^1(\Omega)$; the extension from the first space to the second, however, follows immediately from the fact that, Ω being regular (as in (d1)), we have for $H^1(\Omega)$ the inclusions which are written in terms of $H_0^1(\Omega)$ in (7.48).

Now, in order to show that (for every λ) the form $\mathfrak{A}_\lambda$ is continuous, it is enough to see that, on the one hand, we have the inequality

$$|_{H^{-1/2}(\Gamma)}\langle Tu, \gamma_0 v\rangle_{H^{1/2}(\Gamma)}| \leq \|T\|_{\mathscr{L}(H^1(\Omega), H^{-1/2}(\Gamma))} \|\gamma_0\|_{\mathscr{L}(H^1(\Omega), H^{1/2}(\Gamma))} \|u\|_{H^1(\Omega)} \|v\|_{H^1(\Omega)} \tag{7.103}$$

and that, on the other, the proof of theorem 7.2 can be easily modified to prove that the form $a_\lambda : H^1(\Omega) \times H^1(\Omega) \to \mathbb{R}$ is continuous. All we need to do is to use the equivalence of the norms

$$\|u\|_{H^1(\Omega)} = \left(\|u\|_{L^2(\Omega)}^2 + \sum_{i=1}^n \|\partial u/\partial x_i\|_{L^2(\Omega)}^2\right)^{1/2} \tag{7.104}$$

and

$$\|\|u\|\|_{H^1(\Omega)} = \|u\|_{L^2(\Omega)} + \sum_{i=1}^n \|\partial u/\partial x_i\|_{L^2(\Omega)} \tag{7.105}$$

instead of that of the norms (7.51) and (7.52).

We now go on to the question of the coercivity of $\mathfrak{A}_\lambda$. The proof of theorem 7.4 can be modified, like that of theorem 7.2, to prove that there exists $\bar\lambda$ such that the form a_λ is coercive on $H^1(\Omega)$ as long as $\lambda \geq \bar\lambda$. We must now study the effect on the coercivity of $\mathfrak{A}_\lambda$ of the addition of the boundary bilinear form to a_λ. To do this, we will consider the three types of bilinear boundary forms which are of interest to us separately; i.e. those corresponding to the operators $T \equiv 0$, $T = t\gamma_0$ and $T = t(\partial/\partial\tau)$. For a more general treatment of the boundary bilinear forms see, e.g., Lions, 1955, pp. 44 *et seq.*; Magenes–Stampacchia, 1958; p. 272; Nečas, 1967, pp. 29 and 123; Lions, 1956; p. 225.

The first case is trivial: if $T \equiv 0$ then $\mathfrak{A}_\lambda = a_\lambda$ and hence $\mathfrak{A}_\lambda$ is coercive on $H^1(\Omega)$ as long as $\lambda \geq \bar\lambda$.

For the second case we see that the duality $_{H^{-1/2}(\Gamma)}\langle\,,\,\rangle_{H^{1/2}(\Gamma)}$ can be written as

$$_{H^{-1/2}(\Gamma)}\langle t\gamma_0 u, \gamma_0 v\rangle_{H^{1/2}(\Gamma)} = \int_\Gamma t\gamma_0 u \gamma_0 v \, d\sigma, \tag{7.106}$$

since $\gamma_0 u \in L^2(\Gamma)$, $\gamma_0 v \in L^2(\Gamma)$ and $t|_\Gamma \in C^1(\Gamma)$; from (7.106) we have then

$$|_{H^{-1/2}(\Gamma)}\langle t\gamma_0 v, \gamma_0 v\rangle_{H^{1/2}(\Gamma)}| = \left|\int_\Gamma t(\gamma_0 v)^2 \, d\sigma\right| \leq k \|\gamma_0 v\|_{L^2(\Gamma)}^2, \tag{7.107}$$

and such term will be handled as we will do for (7.110).

Let us come to the third case. Let us suppose that $\Omega \subset \mathbb{R}^2$ and interpret $\partial u/\partial \tau$ as $d(\gamma_0 u)/ds$, in accordance with what we have already said. Since, in general, $(d(\gamma_0 u)/ds) \notin L^2(\Gamma)$ we cannot write the duality ${}_{H^{-1/2}(\Gamma)}\langle\,,\,\rangle_{H^{1/2}(\Gamma)}$ in the form of an integral as we did in the previous case—let us nevertheless write it formally as

$$ {}_{H^{-1/2}(\Gamma)}\left\langle t\frac{d(\gamma_0 u)}{ds}, \gamma_0 v \right\rangle_{H^{1/2}(\Gamma)} = \int_\Gamma t\frac{d(\gamma_0 u)}{ds} \gamma_0 v \, ds. \qquad (7.108) $$

Once again let us write formally

$$ \left| \int_\Gamma t\frac{d(\gamma_0 v)}{ds} \gamma_0 v \, ds \right| = \left| \frac{1}{2}\int_\Gamma t\frac{d(\gamma_0 v)^2}{ds} ds \right| $$

$$ = \left| -\int_\Gamma \frac{1}{2}\frac{dt}{ds}(\gamma_0 v)^2 \, ds \right| \leq k \|\gamma_0 v\|^2_{L^2(\Gamma)} \qquad (7.109) $$

(with $k = \frac{1}{2}\max_\Gamma |dt/ds|$) and therefore, from this and from (7.108),

$$ \left| {}_{H^{-1/2}(\Gamma)}\left\langle t\frac{d(\gamma_0 v)}{ds}, \gamma_0 v \right\rangle_{H^{1/2}(\Gamma)} \right| \leq k \|\gamma_0 v\|^2_{L^2(\Gamma)}. \qquad (7.110) $$

A rigorous (non-formal...) proof of this inequality can be found in Lions, 1956, p. 221; here we will take it as a valid one and transform it into another which will be useful to us later. For this purpose we require the following abstract result:

THEOREM 7.6. *Let E, F, and G be three Banach spaces with the respective norms $\|\cdot\|_E$, $\|\cdot\|_F$, and $\|\cdot\|_G$ and suppose that $F \subset E$. If $\phi \in \mathcal{L}(F, G)$ is compact then*

$$ \forall \varepsilon > 0 \quad \exists k_\varepsilon \quad \forall u \in F \quad \|\phi(u)\|_G \leq \varepsilon \|u\|_F + k_\varepsilon \|u\|_E. \qquad (7.111) $$

Proof. By *reductio ad absurdum*. If (7.111) is false then

$$ \exists \varepsilon > 0 \quad \forall n \quad \exists u_n \in F \quad \|\phi(u_n)\|_G > \varepsilon \|u_n\|_F + k_\varepsilon \|u_n\|_E \qquad (7.112) $$

and hence there exists a sequence u_n of elements of F such that

$$ \|\phi(u_n)\|_G = 1 > \varepsilon \|u_n\|_F + n \|u_n\|_E \qquad (7.113) $$

In fact, putting successively $n = 1, 2, 3, \ldots$, in (7.112) and denoting by $\tilde{u}_1, \tilde{u}_2, \tilde{u}_3, \ldots$, the corresponding u of whose existence we are sure from (7.112) we can see that the sequence $u_n = \tilde{u}_n/\|\phi(\tilde{u}_n)\|_G$ satisfies the required conditions. It follows from (7.113) that

$$ u_n \underset{E}{\rightrightarrows} 0 \qquad (7.114) $$

and that

$$ \exists c \quad \forall n \in \mathbb{N} \quad \|u_n\|_F \leq c. \qquad (7.115) $$

From this and from the fact that ϕ is compact we can be sure that there exists a subsequence u_{n_k} of u_n such that $\phi(u_{n_k})$ converges in G. Now from (7.114) and the continuity of ϕ we must have $\phi(u_{n_k}) \xrightarrow{G} 0$, which contradicts (7.113) and hence proves the theorem. ∎

Putting $E = L^2(\Omega)$, $F = H^1(\Omega)$, $G = L^2(\Gamma)$, and $\phi = \gamma_0$, it follows from the previous theorem that

$$\forall \varepsilon > 0 \quad \exists k_\varepsilon > 0 \quad \forall u \in H^1(\Omega) \qquad \|\gamma_0 u\|^2_{L^2(\Gamma)} \leq \varepsilon \|u\|^2_{H^1(\Omega)} + k_\varepsilon \|u\|^2_{L^2(\Omega)} \qquad (7.116)$$

(the presence of the squares is of secondary importance—remember (7.78)...). From this together with (7.110) we have

$$\forall \varepsilon > 0 \quad \exists k_\varepsilon > 0 \quad \forall u \in H^1(\Omega)$$

$$_{H^{-1/2}(\Gamma)}\left\langle t\frac{d(\gamma_0 u)}{ds}, \gamma_0 u\right\rangle_{H^{1/2}(\Gamma)} \leq \varepsilon \|u\|^2_{H^1(\Omega)} + k_\varepsilon \|u\|^2_{L^2(\Omega)}, \qquad (7.117)$$

an inequality which enables us to study the coercivity of $\mathfrak{A}_\lambda$.

We will state the result in a form which is more general than we need, but which has the same simple proof. For other results of a general nature concerning coercivity of forms with 'boundary pieces' see Nečas, 1967, pp. 136 et seq., Magenes–Stampacchia, 1958, pp. 285–6, and Lions, 1956. So:

THEOREM 7.7. If the form a_λ is coercive on $H^1(\Omega)$ for $\lambda \geq \bar{\lambda}$ and if $T \in \mathscr{L}(H^1(\Omega), H^{-1/2}(\Gamma))$ is such that

$$\forall \varepsilon > 0 \quad \exists k_\varepsilon > 0 \quad \forall v \in H^1(\Omega)$$

$$|_{H^{-1/2}(\Gamma)}\langle Tv, \gamma_0 v\rangle_{H^{1/2}(\Gamma)}| \leq \varepsilon \|v\|^2_{H^1(\Omega)} + k_\varepsilon \|v\|^2_{L^2(\Omega)} \qquad (7.118)$$

then the form $\mathfrak{A}_\lambda$ is coercive on $H^1(\Omega)$ as long as $\lambda \geq \bar{\lambda} + k_{\alpha/8}$.

Proof. This follows from (7.61) (with $H^1(\Omega)$ in place of $H^1_0(\Omega)$—we have not carried out the calculations for the space $H^1(\Omega)$ but the formula remains nevertheless true, changing only the value of $\bar{\lambda}$...) and putting $\varepsilon = \alpha/8$ in (7.118). ∎

We have finally solved completely the problem which we had set ourselves: we know that, under the conditions of (d5), the form $\mathfrak{A}_\lambda$ is continuous and that, under the same conditions, it is coercive if λ is large enough. In the examples in which a_{ij}, b_i, c_i, d, and t $(i, j = 1, \ldots, n)$ are not specified (but are fixed and, by hypothesis, satisfy all the conditions in (d5)) we will always assume that λ is in these conditions and we will not mention this again. In concrete cases, in which a_{ij}, b_i, c_i, d, and t $(i, j = 1, \ldots, n)$ are given functions, we must examine whether the hypotheses in (d5) are satisfied; further, if we are given a fixed value of λ then we also have to examine whether it is suitable, a task which, except for some particular cases, is generally very difficult.

The differential operator

We now have to associate with the form $\mathfrak{A}_\lambda$ an operator, in agreement with (7.86). One can easily see that the operator associated with $\mathfrak{A}_\lambda$ is the operator $L+\lambda$ defined by (7.102)—the same operator which in section 7.21 we associated with the form a_λ! We have come then to the conclusion that *two distinct forms can define the same operator*. This shows us the need to discuss the relationship between forms and operators, which we will now do.

Formula (7.86) enables us to associate a single operator $\Lambda \in \mathcal{L}(V, \mathcal{D}'(\Omega))$ with a (continuous) bilinear form $\mathfrak{A}: V \times V \to \mathbb{R}$ which was previously given. Suppose now that, instead of the form $\mathfrak{A}$, we are given the operator Λ: what then does (7.86) tell us? It associates with Λ a form $\mathfrak{A}: V \times \mathcal{D}(\Omega) \to \mathbb{R}$, but does not define $\mathfrak{A}$ on all of $V \times V$! Hence, in order to associate with Λ a form $\mathfrak{A}: V \times V \to \mathbb{R}$ we need to extend the form (defined (by (7.86)) on $V \times \mathcal{D}(\Omega)$) to $V \times V$, and this extension will not in general be unique. The extension will be unique if V is normal, since in this case we will be dealing with an extension by continuity of $V \times \mathcal{D}(\Omega)$ to $\overline{V \times \mathcal{D}(\Omega)}^{V \times V} = V \times V$. In general we can say, roughly, that the difference between two extensions is determined by the 'boundary pieces' which vanishes on $V \times \mathcal{D}(\Omega)$. This explains why two distinct forms can be associated with a single operator. In terms of boundary value problems this means, essentially, that with a single differential equation we can associate several boundary conditions—we will see some examples of this.

But the diversification of extensions is not the only way in which we can associate two distinct bilinear forms with a single operator. In fact, what does the phrase 'given the operator $L \in \mathcal{L}(H^1(\Omega), \mathcal{D}'(\Omega))$' mean? L is the name of a well defined part of the Cartesian product $H^1(\Omega) \times \mathcal{D}'(\Omega)$, but we will of course not present it in this way; by the phrase 'given the operator L' we mean that the coefficients a_{ij}, b_i, c_i, d $(i, j = 1, \ldots, n)$ in (7.38) are specified. Now, two different choices of coefficients can correspond to with the same part L of the Cartesian product $H^1(\Omega) \times \mathcal{D}'(\Omega)$: i.e., *the same operator L can be decomposed in different ways with respect to the elementary operators* $\partial/\partial x_i$. On the important subject of decomposition of operators, which we mention just briefly here, see e.g. Lions, 1955, p. 43; Magenes–Stampacchia, 1958, p. 265; Lions–Magenes, 1972a, p. 208; Nečas, 1967, p. 25. With every decomposition of L we will of course associate a distinct bilinear form, and this, in terms of boundary value problems, means once again an association of different boundary conditions with a single differential equation.

Decomposition of the Laplace operator

To show that a single operator can indeed be decomposed in different ways with respect to the elementary operators, we present the following example which we will take up again later. Suppose $\Omega \subset \mathbb{R}^2$ for convenience in the writing of the expressions. That part of $H^1(\Omega) \times \mathcal{D}'(\Omega)$ which we call the

Laplace operator and denote by Δ can be represented, in the form (7.38), in different ways: (1) By the coefficients $a_{11} = a_{22} = -1$, $a_{21} = a_{12} = b_1 = b_2 = c_1 = c_2 = d = 0$ (the *standard decomposition* of Δ), corresponding to which we have the way of writing

$$\Delta \square \equiv \frac{\partial}{\partial x_1}\left(\frac{\partial}{\partial x_1}\square\right) + \frac{\partial}{\partial x_2}\left(\frac{\partial}{\partial x_2}\square\right) \tag{7.119}$$

and, e.g., the bilinear form

$$a_{\Delta 1}(u, v) = -\int_\Omega \left(\frac{\partial u}{\partial x_1}\frac{\partial v}{\partial x_1} + \frac{\partial u}{\partial x_2}\frac{\partial v}{\partial x_2}\right) dx. \tag{7.120}$$

(2) By the coefficients $a_{11} = a_{22} = -1$, $a_{12} = -t$, $a_{21} = t$, $b_1 = \partial t/\partial x_2$, $b_2 = -(\partial t/\partial x_1)$, $c_1 = c_2 = d = 0$, where $t \in C^1(\bar{\Omega})$ (the *asymmetric decomposition* of Δ), corresponding to which we have the way of writing

$$\Delta \square \equiv \frac{\partial}{\partial x_1}\left(\frac{\partial}{\partial x_1}\square\right) + \frac{\partial}{\partial x_2}\left(\frac{\partial}{\partial x_2}\square\right) + \frac{\partial}{\partial x_1}\left(t\frac{\partial}{\partial x_2}\square\right) - \frac{\partial}{\partial x_2}\left(t\frac{\partial}{\partial x_1}\square\right)$$
$$+ \frac{\partial t}{\partial x_1}\frac{\partial}{\partial x_1}\square - \frac{\partial t}{\partial x_1}\frac{\partial}{\partial x_2}\square \tag{7.121}$$

and, e.g., the bilinear form

$$a_{\Delta 2}(u, v) = -\int_\Omega \left(\frac{\partial u}{\partial x_1}\frac{\partial v}{\partial x_1} + \frac{\partial u}{\partial x_2}\frac{\partial v}{\partial x_2} + t\frac{\partial u}{\partial x_1}\frac{\partial v}{\partial x_2} - t\frac{\partial u}{\partial x_2}\frac{\partial v}{\partial x_1}\right.$$
$$\left. - \frac{\partial t}{\partial x_2}v\frac{\partial u}{\partial x_1} + \frac{\partial t}{\partial x_1}v\frac{\partial u}{\partial x_2}\right) dx. \tag{7.122}$$

(3) By the coefficients $a_{11} = a_{22} = -1$, $a_{12} = a_{21} = 0$, $b_1 = b_2 = -c_1 = -c_2 = t$, $d = (\partial t/\partial x_1) + (\partial t/\partial x_2)$, where $t \in C^1(\bar{\Omega})$, corresponding to which we have the way of writing

$$\Delta \square \equiv \frac{\partial}{\partial x_1}\left(\frac{\partial}{\partial x_1}\square\right) + \frac{\partial}{\partial x_2}\left(\frac{\partial}{\partial x_2}\square\right) + t\frac{\partial}{\partial x_1}\square + t\frac{\partial}{\partial x_2}\square$$
$$- \frac{\partial}{\partial x_1}(t\square) - \frac{\partial}{\partial x_2}(t\square) + \left(\frac{\partial t}{\partial x_1} + \frac{\partial t}{\partial x_2}\right)\square \tag{7.123}$$

and, e.g., the bilinear form

$$a_{\Delta 3}(u, v) = -\int_\Omega \left(\frac{\partial u}{\partial x_1}\frac{\partial v}{\partial x_1} + \frac{\partial u}{\partial x_2}\frac{\partial v}{\partial x_2} - t\frac{\partial u}{\partial x_1}v - t\frac{\partial u}{\partial x_2}v\right.$$
$$\left. - tu\frac{\partial v}{\partial x_1} - tu\frac{\partial v}{\partial x_2} - \left(\frac{\partial t}{\partial x_1} + \frac{\partial t}{\partial x_2}\right)uv\right) dx. \tag{7.124}$$

The examples

Finally, we will present some examples, which have been chosen on the basis of the following criteria: they seem to us to be typical of situations which are frequently encountered in practice and to be suitable for clarifying some aspects of the above theory (for other examples see Magenes–Stampacchia, 1958; Lions, 1955, 1961, 1965, etc.).

As we have stated earlier, all the examples which we present can be obtained by particularizing the data (d1), (d2), (d3), (d4), (d5), (d6) (in the way in which this set of data can be particularized, that is in the choice of the space V and in the choice of the form $\mathfrak{A}_\lambda$). We will deal with two kinds of examples. In that of the first type (7.1 to 7.3) we will take $\mathfrak{A} = a_\lambda$ (i.e., $\mathfrak{A}_\lambda$ with $T \equiv 0$) as the bilinear form and for the space V we will take in turn $H_0^1(\Omega)$, $H^1(\Omega)$ and V_{Γ_0}: for each of these choices of V we will write the problems 7.12 and 7.13 and try to rewrite (7.96) and (7.98) in a more 'legible' form. In the second type of examples (7.4 to 7.8) we will take $V = H^1(\Omega)$ (with $\Omega \subset \mathbb{R}^2$) and for the bilinear form, $\mathfrak{A}$, we will take in turn five different forms associated with the operator $-\Delta + \lambda$: for each of these we will write problem 7.12, trying once again to interpret (7.96) in a convenient manner.

We choose the operator $-\Delta + \lambda$ for the same reason that we take $n = 2$: convenience in writing the expressions we need; what we will do, however, can be perfectly generalized, both with respect to the number of variables and with respect to the differential operator.

Due to identities of the type of the 'formula of integration by parts', the same problem can be formulated in different ways by adding integrals on the boundary and subtracting the corresponding integrals on the domain: see the pairs of problems (7.22, 7.24) and (7.23, 7.25).

EXAMPLE 7.1 (Dirichlet problems). Data: (d1), (d3), (d4), (d5) (with $T \equiv 0$), (d6) and $V = H_0^1(\Omega)$.

In this case, as we know, $N = H_0^1(\Omega)$. The homogeneous case, which corresponds to problem 7.12, has already been extensively dealt with in section 7.2.1 (see also the analysis carried out in section 7.2.2): in fact we are dealing with problem 7.9 which we will rewrite, with the boundary conditions written explicitly:

PROBLEM 7.15 (The homogeneous Dirichlet problem). Given $f \in H^{-1}(\Omega)$ find $u \in H^1(\Omega)$ (or $u \in H_0^1(\Omega)$) such that

$$Lu + \lambda u = f \quad \text{in } \Omega, \qquad (7.125)$$

$$\gamma_0 u = 0 \quad \text{in } \Gamma. \qquad (7.126)$$

As we have already mentioned, there are no natural conditions here: there is only a boundary condition (7.126) and this is forced.

Next we will consider the non-homogeneous case. The condition (7.98) can be written in the form

$$u - h \in H_0^1(\Omega), \qquad (7.127)$$

which in turn can be written in the form

$$\gamma_0 u = \gamma_0 h \quad \text{in } \Gamma. \qquad (7.128)$$

The only thing about h that is of interest to us is therefore its trace on Γ, which is a function of $H^{1/2}(\Gamma)$: $g_0 = \gamma_0 h \in H^{1/2}(\Gamma)$. This is, on the other hand, a natural conclusion, since in the classical formulation of boundary value problems the boundary conditions are given by functions defined on Γ and not on Ω. Now, since for every $g_0 \in H^{1/2}(\Gamma)$ we can find a 'inverse' $h \in H^1(\Omega)$ of g_0 (i.e. $h \in H^1(\Omega)$ such that $\gamma_0 h = g_0 \in H^{1/2}(\Gamma)$), we come to the conclusion that, in this case, we can write problem 7.13 in the following form:

PROBLEM 7.16 (The non-homogeneous Dirichlet problem). Given $f \in H^1(\Omega)$ and $g_0 \in H^{1/2}(\Gamma)$ find $u \in H^1(\Omega)$ such that

$$Lu + \lambda u = f \quad \text{in } \Omega, \tag{7.129}$$

$$\gamma_0 u = g_0 \quad \text{in } \Gamma. \tag{7.130}$$

Here too there is only one boundary condition and that is forced. Further, the absence of natural conditions can be used in order to characterize the Dirichlet problems, namely: *a boundary value problem is said to be a Dirichlet problem if all its boundary conditions are forced.*

EXAMPLE 7.2 (Neumann problems). Data: (d1), (d3), (d4), (d5) (with $T \equiv 0$), and $V = H^1(\Omega)$.

We will first determine the space N. From (7.93) and from the above data we have

$$N = \{u \in H^1(\Omega) : Lu + \lambda u \in L^2 \text{ and } a_\lambda(u, v)$$
$$= {}_{L^2(\Omega)}\langle Lu + \lambda u, v\rangle_{L^2(\Omega)} \quad \forall v \in H^1(\Omega)\}. \tag{7.131}$$

Now, from this together with Green's formula (18.32) (see in particular Remark 18.1) we have

$$N = \{u \in H^1_L(\Omega) : {}_{H^{-1/2}(\Gamma)}\langle \gamma_\mathfrak{A} u, \gamma_0 v\rangle_{H^{1/2}(\Gamma)} = 0 \quad \forall v \in H^1(\Omega)\}$$
$$= \{u \in H^1_L(\Omega) : \gamma_\mathfrak{A} u = 0 \text{ in } \Gamma\}, \tag{7.132}$$

and therefore in this case problem 7.12 can be written in the form

PROBLEM 7.17 (The homogeneous Neumann problem). Given $f \in L^2(\Omega)$ find $u \in H^1_L(\Omega)$ such that

$$Lu + \lambda u = f \quad \text{in } \Omega, \tag{7.133}$$

$$\gamma_\mathfrak{A} u = 0 \quad \text{in } \Gamma. \tag{7.134}$$

Here, contrary to the Dirichlet problems, there are no forced conditions: the only boundary condition, (7.134), is natural.

Next, we consider the non-homogeneous case. Under our conditions, (7.98) can be written as

$$\gamma_\mathfrak{A}(u - h) = 0 \quad \text{in } \Gamma \tag{7.135}$$

or, since $u \in H_L^1(\Omega)$ and $h \in H_L^1(\Omega)$,

$$\gamma_\mathfrak{A} u = \gamma_\mathfrak{A} h \quad \text{in } \Gamma. \tag{7.136}$$

From theorem 18.8—and in particular that part of it which refers to the possibility of 'inverting' the traces $g \in H^{-1/2}(\Gamma)$ relative to $\gamma_\mathfrak{A}$—it can be easily seen that we can write problem 7.13 in the following form:

PROBLEM 7.18 (Non-homogeneous Neumann problem). Given $f \in L^2(\Omega)$ and $g_1 \in H^{-1/2}(\Gamma)$ find $u \in H_L^1(\Omega)$ such that

$$Lu + \lambda u = f \quad \text{in } \Omega, \tag{7.137}$$

$$\gamma_\mathfrak{A} u = g_1 \quad \text{in } \Gamma. \tag{7.138}$$

Here too we have just one boundary condition, and this is a natural one. The absence of forced conditions are characteristic of Neumann problems: *a boundary value problem is said to be a Neumann problem if all its boundary conditions are natural.* Let us remark that our nomenclature here is in contrast with that presented in section 7.1 for the classical formulation of boundary value problems: the problems 7.17 and 7.18 have, as particular cases, not only those problems with which the name of Neumann is associated (and which correspond to a condition on the normal derivative or on a direct extension of it) but also many others, among which we will mention those which correspond to a condition on the oblique derivative—the so-called *oblique derivative problems*. This question concerning nomenclature will be clarified with the problems treated in examples 7.4 and 7.8.

The non-homogeneous Neumann problems, and in general all non-homogeneous problems, can be dealt with directly, without first reducing it to the homogeneous case: this has been done, e.g., for problem 18.1. We do not develop further this approach because, as for problem 18.1, it goes out of the distributional setting. For a treatment of these subjects see Lions–Magenes, 1961a, p. 162; Magenes–Stampacchia, 1958, p. 300.

EXAMPLE 7.3 (Mixed problems). Data: (d1), (d3), (d4), (d5), (with $T \equiv 0$), (d6) and $V = V_{\Gamma_0}$.

In this case we have

$$N = \{u \in V_{\Gamma_0} : Lu + \lambda u \in L^2(\Omega) \quad \text{and} \quad a_\lambda(u, v) = {}_{L^2(\Omega)}\langle Lu + \lambda u, v\rangle_{L^2(\Omega)}$$
$$\forall v \in V_{\Gamma_0}\} \tag{7.139}$$

and thus, putting $\Gamma_1 = \Gamma \setminus \bar{\Gamma}_0$ and using (18.36),

$$N = \{u \in V_{\Gamma_0} \cap H_L^1(\Omega) : {}_{(H_{00}^{1/2}(\Gamma_1))'}\langle \gamma_\mathfrak{A} u |_{\Gamma_1}, \gamma_0 v |_{\Gamma_1}\rangle_{H_{00}^{1/2}(\Gamma_1)} = 0 \quad \forall v \in V_{\Gamma_0}\}$$
$$= \{u \in V_{\Gamma_0} \cap H_L^1(\Omega) : \gamma_\mathfrak{A} u |_{\Gamma_1} = 0\}. \tag{7.140}$$

The problem 7.12 can then be written in the following form:

PROBLEM 7.19 (Homogeneous mixed problem). Given $f \in L^2(\Omega)$ find $u \in H_L^1(\Omega)$ such that

$$Lu + \lambda u = f \quad \text{in } \Omega, \tag{7.141}$$

$$\gamma_0 u|_{\Gamma_0} = 0, \tag{7.142}$$

$$\gamma_\mathfrak{A} u|_{\Gamma_1} = 0. \tag{7.143}$$

Problem 7.13 can in turn be written, modulus the proof (which we will not give) of the existence of a 'inversion' $\mathcal{R} \in \mathcal{L}(H_{00}^{1/2}(\Gamma_0) \times (H_{00}^{1/2}(\Gamma_1))', H_L^1(\Omega))$, in the following form:

PROBLEM 7.20 (Non-homogeneous mixed problem). Given $f \in L^2(\Omega)$, $g_0 \in H^{1/2}(\Gamma_0)$, $g_1 \in (H_{00}^{1/2}(\Gamma_1))'$ find $u \in H_L^1(\Omega)$ such that

$$Lu + \lambda u = f \quad \text{in } \Omega, \tag{7.144}$$

$$\gamma_0 u|_{\Gamma_0} = g_0, \tag{7.145}$$

$$\gamma_\mathfrak{A} u|_{\Gamma_1} = g_1. \tag{7.146}$$

In both problems 7.19 and 7.20 we have two types of boundary conditions: conditions (7.142) and (7.145) are forced while conditions (7.143) and (7.146) are natural. For this reason, we call such problems 'mixed problems' (*problèmes mêlés* in French).

Problems for the Laplace operator

EXAMPLE 7.4. Data: (d1) (with $n = 2$), (d2) (with $V = H^1(\Omega)$), (d3), (d4), and, in (d5) and in (d6), $a_{11} = a_{22} = 1$, $a_{12} = a_{21} = b_1 = b_2 = c_1 = c_2 = d = 0$ and $T \equiv 0$.

We have then the bilinear form $\mathfrak{A}(u, v) = -a_{\Delta 1}(u, v) + \lambda(u, v)_{L^2(\Omega)}$, with $a_{\Delta 1}$ as given in (7.120), and the differential operator $-\Delta + \lambda$, where Δ is taken with its decomposition (7.119).

The hypotheses (H1), (H2), (H3), (H4), and (7.57) are obviously satisfied, so that the form $\mathfrak{A}$ is continuous and coercive on $H^1(\Omega)$ for λ large enough. This situation is a particular case of the one treated in example 7.2 and hence we can write problem 7.12 in the form:

PROBLEM 7.21. Given $f \in L^2(\Omega)$ find $u \in H_\Delta^1(\Omega)$ such that

$$-\Delta u + \lambda u = f \quad \text{in } \Omega, \tag{7.147}$$

$$\gamma_{a_{\Delta 1}} u = 0 \quad \text{in } \Gamma. \tag{7.148}$$

Remark that this problem is a particular case of problem 7.17.

We will now analyse the conditions (7.148) under the hypothesis that the solution u of problem 7.21 be an element of $H^3(\Omega)$. Under these conditions,

$u \in C^1(\bar{\Omega})$ (see theorem 5.19, with $n = 2$) and thus we can write

$$\gamma_{a_{\Delta 1}}(u) = \frac{\partial u}{\partial x_1}\bigg|_\Gamma \cos(\nu, x_1) + \frac{\partial u}{\partial x_2}\bigg|_\Gamma \cos(\nu, x_2) = \frac{\partial u}{\partial \nu} = 0 \quad \text{in } \Gamma. \tag{7.149}$$

Problem 7.21 is then a Neumann problem associated with the normal derivative: we can say then that we are dealing with a *Neumann problem proper*. Let us remark that the hypothesis $u \in H^3(\Omega)$ is quite natural since it can be proved that if $f \in H^1(\Omega)$ then $u \in H^3(\Omega)$ (this is a type (iii) regularity result which we will come back to); however, at present, we are merely interested in a formal discussion and will not impose further regularity on the data f.

We have previously remarked that it is in general very difficult to find the set of values λ which make a bilinear form of the type $\mathfrak{A}_\lambda$ coercive. In this case, however, it is very easy: the form $\mathfrak{A}(u, v) = -a_{\Delta 1}(u, v) + \lambda(u, v)_{L^2(\Omega)}$ is coercive on $H^1(\Omega)$ iff $\lambda \in]0, +\infty[$; we can prove, more generally, the following result:

THEOREM 7.8. *Let $\Omega \subset \mathbb{R}^n$ be an open set, let $a_{ij} \in L^\infty(\Omega)$ $(i, j = 1, \ldots, n)$ be n^2 functions which satisfy (7.57) and let $p \in L^\infty(\Omega)$ such that $p(x) \geq \alpha' > 0$ a.e. in Ω. Then the form*

$$a(u, v) = \int_\Omega \left(\sum_{i,j=1}^n a_{ij} \frac{\partial u}{\partial x_j} \frac{\partial v}{\partial x_i} + puv \right) dx \tag{7.150}$$

is coercive on $H^1(\Omega)$.

Proof. It is enough to note that

$$a(v, v) \geq \alpha \sum_{i=1}^n \|\partial v/\partial v_i\|^2_{L^2(\Omega)} + \alpha' \|v\|^2_{L^2(\Omega)} \geq \min\{\alpha, \alpha'\} \|v\|^2_{H^1(\Omega)}. \tag{7.151}$$

From (7.151) it follows that, for a to be coercive on $H^1_0(\Omega)$, with Ω bounded, it is enough that $p(x) \geq 0$ a.e. in Ω, which implies, in particular, that *the Dirichlet problems for the operator $-\Delta$ (or Δ) are well-posed in $H^1_0(\Omega)$, with Ω bounded*. ∎

EXAMPLE 7.5. Data: (d1) (with $n = 2$), (d2) (with $V = H^1(\Omega)$), (d3), (d4), and, in (d5) and (d6), $a_{11} = a_{22} = 1$, $a_{12} = t$, $a_{21} = -t$, $b_1 = -(\partial t/\partial x_2)$, $b_2 = \partial t/\partial x_1$, $c_1 = c_2 = d = 0$ and $T \equiv 0$ (where $t \in C^1(\bar{\Omega})$).

Here we have the bilinear form $\mathfrak{A}(u, v) = -a_{\Delta 2}(u, v) + \lambda(u, v)_{L^2(\Omega)}$, with $a_{\Delta 2}$ given by (7.122), and the operator $-\Delta + \lambda$, where Δ is taken along with its decomposition (7.121). It is easy to show that the conditions (H1), (H2), (H3), (H4), and (7.57) are satisfied, and therefore that the form $\mathfrak{A}$ is continuous and coercive on $H^1(\Omega)$ if λ is large enough (in general it is not enough that $\lambda > 0 \ldots$). We are still under the conditions of example 7.2, and hence problem 7.12 can be written in the form:

PROBLEM 7.22. Given $f \in L^2(\Omega)$ (and $t \in C^1(\bar{\Omega}) \ldots$) find $u \in H^1_\Delta(\Omega)$ such that

$$-\Delta u + \lambda u = f \quad \text{in } \Omega, \tag{7.152}$$
$$\gamma_{a_{\Delta 2}} u = 0 \quad \text{in } \Gamma. \tag{7.153}$$

This problem is another particular case of problem 7.17. Assuming, as before, that $u \in H^3(\Omega)$ we can write

$$\gamma_{a_{\Delta 2}} u = \frac{\partial u}{\partial x_1}\bigg|_\Gamma \cos(\nu, x_1) + \frac{\partial u}{\partial x_2}\bigg|_\Gamma \cos(\nu, x_2)$$
$$+ t|_\Gamma \frac{\partial u}{\partial x_2}\bigg|_\Gamma \cos(\nu, x_1) - t|_\Gamma \frac{\partial u}{\partial x_1}\bigg|_\Gamma \cos(\nu, x_2)$$
$$= \frac{\partial u}{\partial \nu} - t|_\Gamma \frac{\partial u}{\partial \tau} = 0 \quad \text{in } \Gamma, \tag{7.154}$$

where $\partial u / \partial \tau$ is the derivative along the tangent to the boundary of Ω (we will assume that the versor of the tangent is oriented in agreement with the 'direct direction' on $\Gamma \ldots$). The expression $(\partial u / \partial \nu) - t|_\Gamma \partial u / \partial \tau$ defines a derivative with respect to the vector $\nu - t|_\Gamma \tau$, which is an ('oblique') vector coming out of Ω. The problem 7.22 is then a Neumann problem associated with a condition on an oblique derivative: in the literature this problem is often known as the *oblique derivative problem*.

The problems 7.21 and 7.22 show that, depending on the decomposition of an operator, very different problems can be associated with it. The next problem gives another example of this.

EXAMPLE 7.6. Data: (d1) (with $n = 2$), (d2) (with $V = H^1(\Omega)$), (d3), (d4), and, in (d5) and in (d6), $a_{11} = a_{22} = 1$, $a_{12} = a_{21} = 0$, $b_1 = b_2 = -c_1 = -c_2 = -t$ and $d = -(\partial t / \partial x_1 + \partial t / \partial x_2)$ (where $t \in C^1(\bar{\Omega})$).

We now have the bilinear form $\mathfrak{A}(u, v) = -a_{\Delta 3}(u, v) + \lambda(u, v)_{L^2(\Omega)}$, with $a_{\Delta 3}$ given by (7.124), and the operator $-\Delta + \lambda$, where Δ is taken with its decomposition (7.123). The hypotheses (H1), (H2), (H3), (H4), and (7.57) are obviously satisfied and therefore the form $\mathfrak{A}$ is continuous and coercive on $H^1(\Omega)$ for large enough λ. We are again under the conditions of example 7.2 and can therefore write problem 7.12 in the form:

PROBLEM 7.23. Given $f \in L^2(\Omega)$ (and $t \in C^1(\bar{\Omega}) \ldots$) find $u \in H^1_\Delta(\Omega)$ such that

$$-\Delta u + \lambda u = f \quad \text{in } \Omega, \tag{7.155}$$
$$\gamma_{a_{\Delta 3}} u = 0 \quad \text{in } \Gamma. \tag{7.156}$$

This problem is another particular case of problem 7.17.

Again assuming $u \in H^3(\Omega)$ we can write

$$\gamma_{a_{\Delta 3}}(u) = \frac{\partial u}{\partial x_1}\bigg|_\Gamma \cos(\nu, x_1) + \frac{\partial u}{\partial x_2}\bigg|_\Gamma \cos(\nu, x_2)$$
$$- t|_\Gamma u|_\Gamma \cos(\nu, x_1) - t|_\Gamma u|_\Gamma \cos(\nu, x_2)$$
$$= \frac{\partial u}{\partial \nu} - t|_\Gamma u|_\Gamma (\cos(\nu, x_1) + \cos(\nu, x_2)) = 0. \quad (7.157)$$

Problem 7.23 is therefore a Neumann problem of a kind different from the previous ones: some authors call it the *Newton problem*.

EXAMPLE 7.7. Data: (d1) (with $n=2$), (d2) (with $V = H^1(\Omega)$), (d3), (d4), and, in (d5) and in (d6), $a_{11} = a_{22} = 1$, $a_{21} = a_{12} = b_1 = b_2 = c_1 = c_2 = d = 0$ and $T^\square \equiv t(\partial/\partial \tau)^\square$.

Here, we have the bilinear form

$$\mathfrak{A}(u,v) = -a_{\Delta 1}(u,v) + \lambda(u,v)_{L^2(\Omega)} + {}_{H^{-1/2}(\Gamma)}\langle t(\partial u/\partial \tau), \gamma_0 v\rangle_{H^{1/2}(\Gamma)}$$

and the operator $-\Delta + \lambda$, with $a_{\Delta 1}$ given by (7.120) and Δ given by (7.119). We know that the form $\mathfrak{A}$ is continuous and coercive on $H^1(\Omega)$ as long as λ is large enough.

As can be easily seen, from (7.93) and (18.32) follows

$$N = \bigg\{ u \in H^1_\Delta(\Omega): {}_{H^{-1/2}(\Gamma)}\langle \gamma_{a_{\Delta 1}} u, \gamma_0 v\rangle_{H^{1/2}(\Gamma)}$$
$$- {}_{H^{-1/2}(\Gamma)}\bigg\langle t|_\Gamma \frac{\partial u}{\partial \tau}, \gamma_0 v\bigg\rangle_{H^{1/2}(\Gamma)} = 0 \quad \forall v \in H^1(\Omega) \bigg\}$$
$$= \bigg\{ u \in H^1_\Delta(\Omega): \gamma_{a_{\Delta 1}} u - t|_\Gamma \frac{\partial u}{\partial \tau} = 0 \quad \text{in } \Gamma \bigg\} \quad (7.158)$$

and thus we can write problem 7.12 in the form:

PROBLEM 7.24. Given $f \in L^2(\Omega)$ and $t \in C^1(\bar\Omega)$ find $u \in H^1_\Delta(\Omega)$ such that

$$-\Delta u + \lambda u = f \quad \text{in } \Omega, \quad (7.159)$$

$$\gamma_{a_{\Delta 1}} u - t|_\Gamma \frac{\partial u}{\partial \tau} = 0 \quad \text{in } \Gamma. \quad (7.160)$$

The condition (7.160) can be written, if $u \in H^3(\Omega)$, in the form

$$\gamma_{a_{\Delta 1}} u - t|_\Gamma \frac{\partial u}{\partial \tau} = \frac{\partial u}{\partial \nu} - t|_\Gamma \frac{\partial u}{\partial \tau} = 0 \quad \text{in } \Gamma, \quad (7.161)$$

and therefore we have once more an *oblique derivative problem*.

Problems 7.22 and 7.24 show that the same boundary value problem can be obtained in very different ways: in the case of problem 7.22 we have 'desymmetrized' the Laplace operator and in the case of problem 7.24 we have, instead, added to the bilinear form $a_{\Delta 1}$, associated with the standard decomposition of the operator, an appropriate boundary bilinear form.

EXAMPLE 7.8. Data: (d1) (with $n = 2$), (d2) (with $V = H^1(\Omega)$), (d3), (d4), and, in (d5) and in (d6), $a_{11} = a_{22} = 1$, $a_{12} = a_{21} = b_1 = b_2 = c_1 = c_2 = d = 0$ and $T\square \equiv t\gamma_0\square$.

The bilinear form we have now is

$$\mathfrak{A}(u, v) = -a_{\Delta 1}(u, v) + \lambda(u, v)_{L^2(\Omega)} + {}_{H^{-1/2}(\Gamma)}\langle t\gamma_0 u, \gamma_0 v\rangle_{H^{1/2}(\Gamma)},$$

with $a_{\Delta 1}$ given by (7.120), and the differential operator is again $-\Delta + \lambda$, with Δ given by (7.119). We know that $\mathfrak{A}$ is continuous and coercive on $H^1(\Omega)$ if λ is large enough (namely, if $\lambda > 0$). In this case, we have

$$N = \{u \in H^1_\Delta(\Omega): {}_{H^{-1/2}(\Gamma)}\langle \gamma_{a_{\Delta 1}} u, \gamma_0 v\rangle_{H^{1/2}(\Gamma)}$$
$$= {}_{H^{-1/2}(\Gamma)}\langle t\gamma_0 u, \gamma_0 v\rangle_{H^{1/2}(\Gamma)} \quad \forall v \in H^1(\Omega)\}$$
$$= \{u \in H^1_\Delta(\Omega): \gamma_{a_{\Delta 1}} u - t\gamma_0 u = 0 \quad \text{in } \Gamma\} \tag{7.162}$$

and therefore we can write problem 7.12 in the form:

PROBLEM 7.25. Given $f \in L^2(\Omega)$ and $t \in C^1(\bar{\Omega})$ find $u \in H^1_\Delta(\Omega)$ such that

$$-\Delta u + \lambda u = f \quad \text{in } \Omega, \tag{7.163}$$

$$\gamma_{a_{\Delta 1}} u - t|_\Gamma \gamma_0 u = 0 \quad \text{in } \Gamma. \tag{7.164}$$

If $u \in H^3(\Omega)$ we can write (7.164) in the form

$$\gamma_{a_{\Delta 1}} u - t|_\Gamma \gamma_0 u = \frac{\partial u}{\partial \nu} - t|_\Gamma u|_\Gamma = 0 \quad \text{in } \Gamma, \tag{7.165}$$

which shows that problems 7.25 and 7.23 are essentially the same problem. We have, once again, the phenomenon mentioned in the previous example: the same boundary value problem can be associated with a differential operator in several different ways.

7.3 NONLINEAR PROBLEMS

We have already treated a nonlinear one-dimensional problem extensively: the problem of the string and the obstacle (sections 1.1 and 6.1). In this section we propose studying some multi-dimensional nonlinear problems, one of which is a generalization of the above problem (example 7.10).

Non-existence of a general theory

For nonlinear variational problems, there is no theory analogous to that which we have presented (following Lions (1955)) for the linear case. This is due, essentially, to the wide choice of factors which determine the non-linearity of a problem; i.e., of the nonlinear functionals which can be present in the variational inequalities and of the convex sets on which these are posed. This makes the already difficult task of interpreting variational problems as differential problems much more difficult in the nonlinear case, since every problem is a case by itself (except, of course, when analogies can be appropriately exploited . . .).

We will study three problems which seem to us to be typical. For many other examples of problems connected with variational inequalities, see the references cited in the bibliography, and in particular Duvaut–Lions, 1976; Mosco, 1973; Stampacchia, 1969; Lions, 1969a; 1971; Duvaut, 1974—also, later in this book, we present other examples connected with free boundary problems.

In all the examples we consider we will denote by Ω a bounded connected open set of $\mathbb{R}^n$ and by Γ the boundary of Ω, which we will assume to be Lipschitz continuous. Further, we will denote simply by $\mathfrak{A}$ the bilinear form $\mathfrak{A}_\lambda$ defined in (7.101), where we assume $T \equiv 0$ and λ to be such that the form is coercive on $H^1(\Omega)$. By $L + \lambda$ we denote, naturally, the differential operator (7.102) associated with $\mathfrak{A}$.

EXAMPLE 7.9 (Signorini's problem). Let us denote by K_1 the set

$$K_1 = \{v \in H^1(\Omega) : \gamma_0 v \geq 0 \quad \text{a.e. on } \Gamma\}. \tag{7.166}$$

This set K_1 is well-defined since $\gamma_0 v \in H^{1/2}(\Gamma) \subset L^2(\Gamma)$ and thus the statement '$\gamma_0 v \geq 0$ a.e. in Γ' is meaningful; this will be abbreviated, as we have done earlier in analogous circumstances, as '$\gamma_0 v \geq 0$ in Γ'. It can also be seen that K_1 is a convex set (or rather, a convex cone), is non-empty (in particular $H_0^1(\Omega) \subset K_1$), and is closed in $H^1(\Omega)$ (to show this, we only need to take in K_1 a sequence v_n such that $v_n \to v$ in $H^1(\Omega)$ and note that the continuity of the operator γ_0 implies that $\gamma_0 v \geq 0$ in Γ and hence $v \in K_1$).

The bilinear form $\mathfrak{A}$, being continuous and coercive on $H^1(\Omega)$, is continuous and coercive on $K_1 - K_1$ and hence we can use the Lions–Stampacchia theorem to ensure the existence and the uniqueness of the solution of the following problem:

PROBLEM 7.26. Given $f \in L^2(\Omega)$ find $u \in H^1(\Omega)$ such that

$$\mathfrak{A}(u, u-v) \leq \int_\Omega f(u-v)\, dx \quad \forall v \in K_1. \tag{7.167}$$

Problem 7.26 is a nonlinear variational problem. Observe that the right-hand side of (7.167) can be written in the form $_{L^2(\Omega)}\langle f, u-v\rangle_{L^2(\Omega)}$, or in the form $_{(H^1(\Omega))'}\langle f, u-v\rangle_{H^1(\Omega)}$, and hence (7.167) is in effect a variational inequality to which the Lions–Stampacchia theorem can be applied; we have not taken the datum f in $(H^1(\Omega))'$ since this would take us out of the distributional framework—besides, the boundary conditions and the equation in Ω would be 'mixed' in problem 7.26.

Let us consider now the following:

PROBLEM 7.27. Given $f \in L^2(\Omega)$ find $u \in H^1(\Omega)$ such that

$$Lu + \lambda u = f \quad \text{in } \Omega, \tag{7.168}$$

$$\gamma_0 u \geq 0 \quad \text{in } \Gamma, \tag{7.169}$$

$$\gamma_{\mathfrak{A}} u \geq 0 \quad \text{in } \Gamma, \tag{7.170}$$

$$\gamma_0 u \gamma_{\mathfrak{A}} u = 0 \quad \text{in } \Gamma. \tag{7.171}$$

This is a nonlinear boundary value problem associated with the linear differential equation (7.168) (the nonlinearity of the problem only regards the boundary conditions...).

As is natural, equation (7.168) and conditions (7.169) and (7.170) are to be taken, respectively, in the sense of $\mathcal{D}'(\Omega), H^{1/2}(\Gamma)$ (i.e., a.e. on Γ) and $H^{-1/2}(\Gamma)$ (or in the sense of the measures on Γ); the meaning of (7.171) is not obvious under the hypothesis $u \in H^1(\Omega)$ alone: if, in addition, u were to belong to $C^0(\bar{\Omega})$ then we could substitute (7.171) by

$$\gamma_{\mathfrak{A}} u = 0 \quad \text{in } \Gamma^+(u), \tag{7.172}$$

which is meaningful in the framework of the measures on $\Gamma^+(u)$, i.e.:

$$_{\mathcal{D}^0(\Gamma^+(u))}\langle \gamma_{\mathfrak{A}} u, \phi \rangle_{\mathcal{D}^0(\Gamma^+(u))} = 0 \quad \forall \phi \in \mathcal{D}^0(\Gamma^+(u)). \tag{7.173}$$

[A remark on notations: if A is an arbitrary set of $\mathbb{R}^n$ and $f \in C^0(A)$ we denote by $A^+(f), A^0(f)$ and $A^-(f)$, respectively, the sets where f is positive, zero or negative, i.e.: $A^+(f) = \{x \in A : f(x) > 0\}$, $A^0(f) = \{x \in A : f(x) = 0\}$ and $A^-(f) = \{x \in A : f(x) < 0\}$ (see the notation used in section 6.1); one can easily see that $A^+(f)$ and $A^-(f)$ are open sets (in A) and supp $f = \overline{A^+(f) \cup A^-(f)} = A \setminus \text{int}(A^0(f))$.]

Now we will show the equivalence of problems 7.26 and 7.27 under the assumption that $u \in C^0(\bar{\Omega}) \cap H^1(\Omega)$ (and in problem 7.27 we will then substitute (7.171) with (7.172)—we will not repeat this). It is important to observe that this assumption is not indispensable either for (7.171) to have a precise interpretation or in order to prove the equivalence of the two problems. In its absence, however, everything becomes more difficult, and what we lose in generality we gain in clarity of exposition. Also, such an assumption is not a serious limitation for most problems which have a physical interpretation. It can in fact be shown that the solution of problem 7.26 belongs to $H^2(\Omega)$ (this is a type (i) regularity result) and hence, from theorem 5.17, it is continuous if $n \leq 3$. If the continuity assumption is not true (and there are counter-examples which show that this is possible for values of n large enough—see Brézis–Stampacchia, 1968, p. 176—then our discussion is purely formal.

We will now show that if u is a solution of problem 7.26 then u is a solution of problem 7.27 (i.e. problem 7.26 $\Rightarrow$ problem 7.27). More precisely, we assume that we do not know problem 7.27 and try to interpret problem 7.26 as a differential problem. This will, naturally, give problem 7.27...

Let then ϕ be an arbitrary function of $\mathscr{D}(\Omega)$ and let us consider the functions $v_1 = u - \phi$ and $v_2 = u + \phi$ which are, as is evident, elements of K_1: substituting v in (7.167) by v_1 and v_2 in turn we get

$$\mathfrak{A}(u, u - v_1) \leq \int_\Omega f(u - v_1)\, dx \quad \text{and} \quad \mathfrak{A}(u, u - v_2) \leq \int_\Omega f(u - v_2)\, dx, \tag{7.174}$$

or

$$\mathfrak{A}(u, \phi) \leq \int_\Omega f\phi\, dx \quad \text{and} \quad \mathfrak{A}(u, -\phi) \leq \int_\Omega f(-\phi)\, dx \tag{7.175}$$

and thus

$$\mathfrak{A}(u, \phi) = \int_\Omega f\phi\, dx. \tag{7.176}$$

Since ϕ is arbitrary in $\mathscr{D}(\Omega)$ we can say (recalling section 7.2) that u satisfies, in the sense of $\mathscr{D}'(\Omega)$, the equation

$$Lu + \lambda u = f \quad \text{in } \Omega. \tag{7.177}$$

We will now look for the boundary conditions which in problem 7.26 are implicitly associated with (7.177). As in the linear case, they are in part 'contained' in (7.167) and in part 'contained' in K_1 (which now takes the place of the former $V\ldots$). A first condition (forced) is obvious: since $u \in K_1$ we can write

$$\gamma_0 u \geq 0 \quad \text{in } \Gamma, \tag{7.178}$$

a condition to be interpreted in the sense a.e. on Γ (or, since u is continuous on $\bar{\Omega}$, in the classical sense). Further, (7.167) and Green's formula (18.32), which can be written in the form

$$\mathfrak{A}(u, u - v) = \int_\Omega (Lu + \lambda u)(u - v)\, dx + {}_{H^{-1/2}(\Gamma)}\langle \gamma_\mathfrak{A} u, \gamma_0(u - v)\rangle_{H^{1/2}(\Gamma)} \tag{7.179}$$

(and which we can apply since, from (7.177), $u \in H_L^1(\Omega)$), imply that

$${}_{H^{-1/2}(\Gamma)}\langle \gamma_\mathfrak{A} u, \gamma_0(u - v)\rangle_{H^{1/2}(\Gamma)} \leq 0 \quad \forall v \in K_1, \tag{7.180}$$

a condition of which we will now give the interpretation. Let then $\phi \in H^{1/2}(\Gamma)$, $\phi \geq 0$ in Γ and let us put $v = u + \mathscr{R}\phi$ (where $\mathscr{R} \in \mathscr{L}(H^{1/2}(\Gamma), H^1(\Omega))$ is an 'inversion' operator): $v \in K_1$ and hence, from (7.180), we have that

$${}_{H^{-1/2}(\Gamma)}\langle \gamma_\mathfrak{A} u, \gamma_0(-\mathscr{R}\phi)\rangle_{H^{1/2}(\Gamma)} \leq 0 \quad \forall \phi \in H^{1/2}(\Gamma), \phi \geq 0 \quad \text{in } \Gamma, \tag{7.181}$$

or
$$\gamma_\mathfrak{A} u \geq 0 \quad \text{in } \Gamma \tag{7.182}$$

in the sense of $H^{-1/2}(\Gamma)$ (or in the sense of measures on Γ, given that in (7.181) we can substitute '$\phi \in H^{1/2}(\Gamma)$, $\phi \geq 0$ in Γ' by '$\phi \in H^{1/2}(\Gamma) \cap C^0(\Gamma)$, $\phi \geq 0$ in Γ' and that $C^0(\Gamma) \cap H^{1/2}(\Gamma)$ is dense in $C^0(\Gamma)$)—(7.182) is a second (now natural) condition to be associated with (7.177).

Recalling the linear case, one would naturally think that one has finished and therefore that one can take as an interpretation of problem 7.26 the problem 'given $f \in L^2(\Omega)$ find $u \in H^1(\Omega)$ such that (7.177), (7.178) and (7.182) are satisfied'. But this problem is ill-posed and hence we cannot consider it to be an interpretation of problem 7.26 which is on the contrary well-posed. This shows that problem 7.26 is richer than one would expect: it contains conditions which we have not yet explicitly stated—we must therefore go on. There are two conditions on Γ, (7.178) and (7.182), and one naturally suspects that there is some connection between them, i.e. some compatibility—in other words: that they are not completely independent of each other. There is, indeed, a connection: we will now show that '$\gamma_\mathfrak{A} u = 0$ wherever $\gamma_0 u > 0$', a statement whose meaning will become clear as we prove it.

From the continuity of u in $\bar{\Omega}$ it follows that $\gamma_0 u = u|_\Gamma$ is continuous on Γ and therefore that it is meaningful to consider the set $\Gamma^+(\gamma_0 u)$ (this is the first time we are using effectively the assumption that u is continuous!). Let us denote by X the space of $\phi \in \mathcal{D}^0(\Gamma^+(\gamma_0 u))$ such that $\tilde{\phi} \in H^{1/2}(\Gamma)$, where $\tilde{\phi}$ denotes the trivial extension of ϕ to Γ; it is easy to see that X is dense in $\mathcal{D}^0(\Gamma^+(\gamma_0 u))$. Let now ϕ be an arbitrary function in X and let $\mu = \min_{\text{supp } \phi} \gamma_0 u / \max_{\Gamma^+(\gamma_0(u))} |\phi|$ if $\phi \not\equiv 0$, $\mu = 0$ if $\phi \equiv 0$: it can be easily seen that the functions $v_1 = u - \mu \mathcal{R}(\tilde{\phi})$ and $v_2 = u + \mu \mathcal{R}(\tilde{\phi})$ are elements of K_1 and therefore, from (7.180), it follows that

$$_{H^{-1/2}(\Gamma)}\langle \gamma_\mathfrak{A} u, \gamma_0(\mu \mathcal{R}(\tilde{\phi}))\rangle_{H^{1/2}(\Gamma)} \leq 0 \quad \text{and} \quad _{H^{-1/2}(\Gamma)}\langle \gamma_\mathfrak{A} u, \gamma_0(-\mu \mathcal{R}(\tilde{\phi}))\rangle_{H^{1/2}(\Gamma)} \leq 0, \tag{7.183}$$

or
$$_{H^{-1/2}(\Gamma)}\langle \gamma_\mathfrak{A} u, \tilde{\phi}\rangle_{H^{1/2}(\Gamma)} = 0. \tag{7.184}$$

Since ϕ is arbitrary in X and X is dense in $\mathcal{D}^0(\Gamma^+(\gamma_0 u))$ we can write

$$_{\mathcal{D}'^0(\Gamma^+(\gamma_0 u))}\langle \gamma_\mathfrak{A} u, \phi\rangle_{\mathcal{D}^0(\Gamma^+(\gamma_0 u))} = 0 \quad \forall \phi \in \mathcal{D}^0(\Gamma^+(\gamma_0 u)), \tag{7.185}$$

which tells us that

$$\gamma_\mathfrak{A} u = 0 \quad \text{in } \Gamma^+(\gamma_0 u) \tag{7.186}$$

in the sense of the measures on $\Gamma^+(\gamma_0 u)$—(7.186) is a third condition to be associated with (7.177).

All this brings us to the conclusion that the solution of problem 7.26 is a solution of problem 7.27. To confirm that we have interpreted problem 7.26 properly we must now show that the problem 7.27 is well-posed, which we will do by proving that problem 7.27 $\Rightarrow$ problem 7.26, i.e., by completing the proof of the (formal) equivalence of the two problems.

Let then $u \in H^1(\Omega) \cap C^0(\bar{\Omega})$ be a solution of problem 7.27. (Observe that we do not yet have a uniqueness result for problem 7.27: this will follow from the equivalence of problems 7.26 and 7.27; we have, however, an existence result.) From (7.169) it follows that $u \in K_1$ and from (7.168) that

$$\int_\Omega (Lu + \lambda u)(u-v) = \int_\Omega f(u-v) \qquad \forall v \in K_1, \qquad (7.187)$$

an expression which, in accordance with Green's formula (7.179), we can change to

$$\mathfrak{A}(u, u-v) = \int_\Omega f(u-v)\, dx + {}_{H^{-1/2}(\Gamma)}\langle \gamma_\mathfrak{A} u, \gamma_0(u-v)\rangle_{H^{1/2}(\Gamma)} \qquad \forall v \in K_1. \qquad (7.188)$$

To complete the proof, we need only show that if u satisfies (7.188) then u satisfies (7.167), which we can do by proving that

$$_{H^{-1/2}(\Gamma)}\langle \gamma_\mathfrak{A} u, \gamma_0(u-v)\rangle_{H^{1/2}(\Gamma)} \leq 0 \qquad \forall v \in K_1. \qquad (7.189)$$

Now

$$_{H^{-1/2}(\Gamma)}\langle \gamma_\mathfrak{A} u, \gamma_0(u-v)\rangle_{H^{1/2}(\Gamma)} = {}_{H^{-1/2}(\Gamma)}\langle \gamma_\mathfrak{A} u, \gamma_0 u\rangle_{H^{1/2}(\Gamma)}$$
$$- {}_{H^{-1/2}(\Gamma)}\langle \gamma_\mathfrak{A} u, \gamma_0 v\rangle_{H^{1/2}(\Gamma)} \qquad (7.190)$$

and this, along with (7.170), gives

$$_{H^{-1/2}(\Gamma)}\langle \gamma_\mathfrak{A} u, \gamma_0(u-v)\rangle_{H^{1/2}(\Gamma)} \leq {}_{H^{-1/2}(\Gamma)}\langle \gamma_\mathfrak{A} u, \gamma_0 u\rangle_{H^{1/2}(\Gamma)} \qquad \forall v \in K_1. \qquad (7.191)$$

Thus, we have to show that

$$_{H^{-1/2}(\Gamma)}\langle \gamma_\mathfrak{A} u, \gamma_0 u\rangle_{H^{1/2}(\Gamma)} = 0 \qquad (7.192)$$

or, given that $\gamma_0 u \in C^0(\Gamma)$, that

$$_{\mathscr{D}^0(\Gamma)}\langle \gamma_\mathfrak{A} u, \gamma_0 u\rangle_{\mathscr{D}^0(\Gamma)} = 0, \qquad (7.193)$$

which follows from (7.172) and from the fact (easily proved) that $\gamma_0 u$ can be approximated, in the topology of $\mathscr{D}^0(\Gamma)$, by functions of $\mathscr{D}^0(\Gamma^+(\gamma_0 u))$. The problems 7.26 and 7.27 are therefore equivalent.

Problem 7.27 (and consequently problem 7.26) is called *Signorini's problem*. It belongs to the family of the so-called *problems with unilateral conditions*, due to the presence of the sign $\geq$ in (7.169) and (7.170). Problem 7.27 is also called a *mixed problem with ambiguous conditions* (see Signorini, 1959, p. 128, since a condition of the Neumann type is imposed on $\Gamma^+(\gamma_0 u)$ and a condition of the Dirichlet type on $\Gamma^0(\gamma_0 u) = \Gamma \setminus \Gamma^+(\gamma_0 u)$, but, since $\Gamma^+(\gamma_0 u)$ (for example) is itself an unknown of the problem, we do not know what condition is in fact imposed on any given point of Γ. In a sense, problem 7.27 is then a *free boundary problem*, the free boundary being that of $\Gamma^+(\gamma_0 u)$. The Signorini problems arise naturally in some problems of the mechanics of continuous media—in this connection see Signorini, 1959 and Fichera, 1966.

Problem 7.27 is a homogeneous problem. In the corresponding non-homogeneous problem the conditions (7.169), (7.170), and (7.171) are replaced, respectively, by the conditions $\gamma_0 u \geq g_0$ in Γ, $\gamma_\mathfrak{A} u \geq g_1$ in Γ and $(\gamma_0 u - g_0)(\gamma_\mathfrak{A} u - g_1) = 0$ in Γ, where g_0 and g_1 are given, for example, respectively in $H^{1/2}(\Gamma)$ and $H^{-1/2}(\Gamma)$ (the third condition has to be interpreted appropriately....—if $u \in C^0(\bar{\Omega}) \cap H^1(\Omega)$ and $g_0 \in C^0(\Gamma)$ the sense is obvious: $\gamma_\mathfrak{A} u = g_1$ in $\Gamma^+(\gamma_0 u - g_0)$ in the sense of the measures on $\Gamma^+(\gamma_0 u - g_0)$). This non-homogeneous problem is the interpretation of the variational problem obtained from problem 7.26 by substituting in (7.166) '$\gamma_0 v \geq 0$ in Γ' by '$\gamma_0 v \geq g_0$ in Γ' and adding to the right-hand side of the inequality (7.167) the expression ${}_{H^{-1/2}(\Gamma)}\langle g_1, \gamma_0(u-v)\rangle_{H^{1/2}(\Gamma)}$.

Considerations of this kind hold also for problems which we will deal with later and we will not mention them again.

EXAMPLE 7.10 (The obstacle problem). Let $\psi \in H^1(\Omega)$ be a function such that $\gamma_0 \psi \geq 0$ in Γ and let us consider the set

$$K_2 = \{v \in H_0^1(\Omega) : v \leq \psi \text{ in } \Omega\}. \qquad (7.194)$$

It can easily be seen that K_2 is a closed convex subset of $H_0^1(\Omega)$ which is well-defined and non-empty $(-[\psi]^- \in K_2 \ldots)$. The bilinear form $\mathfrak{A}$ is then continuous and coercive on $K_2 - K_2$, and from the Lions–Stampacchia theorem we know there is one and only one solution of the following

PROBLEM 7.28. Given $f \in L^2(\Omega)$ find $u \in K_2$ such that

$$\mathfrak{A}(u, u-v) \leq \int_\Omega f(u-v) \, dx \qquad \forall v \in K_2. \qquad (7.195)$$

In order to interpret this problem we assume, as in example 7.9 and for analogous reasons, that its solution is a continuous function in $\bar{\Omega}$; we assume further that ψ is also a continuous function in $\bar{\Omega}$. Here too these assumptions are not too restrictive since it can be shown (see, e.g. Brézis–Stampacchia, 1968, p. 164, that if $f \in H^{-1}(\Omega) \cap L^p(\Omega)$ and $\psi \in H^1(\Omega) \cap W^{2,p}(\Omega) (1 < p < +\infty)$ then $u \in W^{2,p}(\Omega)$, and therefore if $p > n/2$ both u and ψ are continuous functions on $\bar{\Omega}$.

Now let $u \in C^0(\bar{\Omega}) \cap H^1(\Omega)$ be the solution of problem 7.28 and let ϕ be an arbitrary function of $\mathcal{D}_+(\Omega)$ (i.e., $\phi \in \mathcal{D}(\Omega)$ and $\phi \geq 0$ in Ω). The function $v = u - \phi$ is an element of K_2 and hence, from (7.195),

$$\mathfrak{A}(u, \phi) \leq \int_\Omega f\phi \, dx \qquad \forall \phi \in \mathcal{D}_+(\Omega), \qquad (7.196)$$

or

$$Lu + \lambda u \leq f \quad \text{in } \Omega \qquad (7.197)$$

in the sense of $\mathscr{D}'(\Omega)$. Further, since $u \in K_2$,

$$u \le \psi \quad \text{in } \Omega. \tag{7.198}$$

We have then a pair of unilateral conditions which coexist on a single set, Ω: (7.197) and (7.198). From these conditions we can obtain a third by reasoning analogous to that which, in example 7.9, from conditions (7.178) and (7.182) gave us (7.186). Let then $\phi \in \mathscr{D}(\Omega^-(u-\psi))$ and let $\mu = \min_{\text{supp } \phi} (\psi - u)/\max_{\Omega^-} |\phi|$; it follows immediately that $v_1 = u - \mu\tilde{\phi}$ and $v_2 = u + \mu\tilde{\phi}$ are elements of K_2 and thus

$$\mathfrak{A}(u, \mu\tilde{\phi}) \le \int_\Omega f\mu\tilde{\phi}\, dx \text{ e} \quad \text{and} \quad \mathfrak{A}(u, -\mu\tilde{\phi}) \le \int_\Omega f(-\mu\tilde{\phi})\, dx, \tag{7.199}$$

which implies

$$\mathfrak{A}(u, \tilde{\phi}) = \int_\Omega f\tilde{\phi}\, dx \quad \forall \phi \in \mathscr{D}(\Omega^-(u-\psi)), \tag{7.200}$$

from which we can conclude that

$$Lu + \lambda u = f \quad \text{in } \Omega^-(u-\psi) \tag{7.201}$$

in the sense of $\mathscr{D}'(\Omega^-(u-\phi))$. Equation (7.201) can be replaced by

$$(Lu + \lambda u - f)(u - \psi) = 0 \quad \text{in } \Omega, \tag{7.202}$$

which must be interpreted as if it were (7.201). Without the condition of continuity of u and ψ, (7.202) is the condition we must retain and interpret in the appropriate sense, which is no longer that of (7.201): note the parallels between the pairs of conditions (7.171), (7.172) and (7.201), (7.202).

The conditions (7.197), (7.198) and (7.201) have to be true in Ω; to these we can add the boundary condition

$$\gamma_0 u = 0 \quad \text{in } \Gamma, \tag{7.203}$$

in the sense of $H^{1/2}(\Gamma)$ (or in the classical sense since $\gamma_0 u = u|_\Gamma$), which follows from $u \in K_2$.

Summarizing, we can say that if u is a solution of problem 7.28 then u is (formally...) a solution of the following:

PROBLEM 7.29. Given $f \in L^2(\Omega)$ and $\psi \in H^1(\Omega)$, $\gamma_0 \psi \ge 0$ in Γ find $u \in H^1(\Omega)$ such that conditions (7.197), (7.198), (7.201) (or (7.202)), and (7.203) are satisfied.

To show that this is a proper interpretation of problem 7.28 we will show that problem 7.29 $\Rightarrow$ problem 7.28 under the assumptions of continuity of u and of ψ.

Expressions (7.198) and (7.203) imply that if u is a solution of problem 7.29 then $u \in K_2$; all we have to prove then is that u satisfies the variational inequality (7.195). We have to use for this a technique which is very different from the one we used in the previous example since we cannot multiply both sides of (7.197) by $u-v$. ... Since $\gamma_0(u-v) = \gamma_0 u - \gamma_0 v = 0$, for $u, v \in K_2$, we can write Green's formula (7.179) in the form

$$\mathfrak{A}(u, u-v) = \int_\Omega (Lu + \lambda u)(u-v)\, dx. \tag{7.204}$$

Now the right-hand side of (7.204) can be broken up into the sum

$$\int_\Omega (Lu + \lambda u)(u-v)\, dx = \int_{\Omega^-(u-\psi)} (Lu + \lambda u)(u-v)\, dx$$
$$+ \int_{\Omega^0(u-\psi)} (Lu + \lambda u)(u-v)\, dx, \tag{7.205}$$

in which the first part can be replaced, using (7.201), by

$$\int_{\Omega^-(u-\psi)} (Lu + \lambda u)(u-v)\, dx = \int_{\Omega^-(u-\psi)} f(u-v)\, dx \tag{7.206}$$

and for the second we have the inequality

$$\int_{\Omega^0(u-\psi)} (Lu + \lambda u)(u-v)\, dx = \int_{\Omega^0(u-\psi)} (Lu + \lambda u)(\psi - v)\, dx$$
$$\leq \int_{\Omega^0(u-\psi)} f(\psi - v)\, dx$$
$$\leq \int_{\Omega^0(u-\psi)} f(u-v)\, dx, \tag{7.207}$$

which we can write since in $\Omega^0((u-\psi))$ we have $u = \psi$, $Lu + \lambda u = f$ and $\psi - v \geq 0$. From (7.204) and (7.206) we have

$$\int_\Omega (Lu + \lambda u)(u-v)\, dx \leq \int_\Omega f(u-v)\, dx, \tag{7.208}$$

and this together with (7.204) gives (7.195). We have therefore proved that problems 7.28 and 7.29 are equivalent.

Like problem 7.27, problem 7.29 is a *problem with unilateral and ambiguous conditions* which can be considered as a *free boundary problem* (the free boundary being the boundary of $\Omega^-(u-\psi)$), but, while in problem 7.27 the conditions were imposed on Γ, this time they are imposed on Ω. The boundary condition in problem 7.29 is a Dirichlet condition and, by analogy with the linear case, one naturally thinks that if one replaces the phrase '$v \in H_0^1(\Omega)$' in the definition of K_2 by the phrase '$v \in H^1(\Omega)$' one would obtain a problem which differs from problem 7.29 only in that the Dirichlet condition is replaced by a Neumann condition. This is not so: the problem so obtained is ambiguous not only on the internal part but also on the boundary (see, e.g., Lions, 1969a, p. 265).

The problem 7.29 is called an *obstacle problem* since it is, as is evident, a generalization of the problem of the string and the obstacle considered in sections 1.1 and 6.1.

We will now present another particular case of problem 7.29. Let $n=2$, $f \equiv 0$, $\psi \in H^2(\Omega)$, $a_{11} = a_{22} = 1$ and $a_{12} = a_{21} = b_1 = b_2 = c_1 = c_2 = d = \lambda = 0$. Under these conditions problem 7.29 can then be written in the form:

PROBLEM 7.30. Given $\psi \in H^2(\Omega)$, $\psi|_\Gamma \geq 0$, find $u \in H^1(\Omega)$ such that

$$-\Delta u \leq 0 \quad \text{in } \Omega \tag{7.209}$$

$$u \leq \psi \quad \text{in } \Omega \tag{7.210}$$

$$\Delta u = 0 \quad \text{where } u < \psi \text{ (alias in } \Omega^-(u-\psi)) \tag{7.211}$$

$$u|_\Gamma = 0. \tag{7.212}$$

The solution of this problem represents the form which, in the absence of external forces ($f \equiv 0$), a weightless elastic membrane would take if stretched under the obstacle $A = \{x \in \mathbb{R}^3 : (x_1, x_2) \in \Omega \text{ and } x_3 = \psi(x_1, x_2)\}$ and fixed on the boundary Γ of Ω. The continuity of this solution, which follows from the regularity theorem we have already referred to, can be interpreted to mean that the membrane 'does not break'. The sets $\Omega^0(u-\psi)$ and $\Omega^-(u-\psi)$ are called, respectively, the *contact set* and the *non-contact set*—they are, as we have already mentioned, among the unknowns of the problem. It can easily be seen that, in analogy with the one-dimensional case, the configuration the membrane takes is that which minimizes the energy integral

$$E(v) = \tfrac{1}{2} \int_\Omega (\text{grad } v)^2 \, dx \tag{7.213}$$

among all admissible configurations, i.e. in K_2.

EXAMPLE 7.11 (Another nonlinear problem). In examples 7.9 and 7.10 we dealt with nonlinear problems in which the nonlinearity was due to a convex restraint (K_1 or K_2). We will now present a problem of a different kind: a nonlinear problem in which the nonlinearity is due to the presence of a non-differentiable functional. Let then $j : H_0^1(\Omega) \to \mathbb{R}$ be the functional defined by

$$j(v) = -\int_\Omega fv \, dx + \int_\Omega [v]^+ \, dx, \tag{7.214}$$

where $f \in L^2(\Omega)$, and let us consider the following:

PROBLEM 7.31. Given $f \in L^2(\Omega)$ find $u \in H_0^1(\Omega)$ such that

$$\mathfrak{A}(u, u-v) + j(u) \leq j(v) \quad \forall v \in H_0^1(\Omega). \tag{7.215}$$

Here j is clearly a proper convex functional, l.s.c. on $H_0^1(\Omega)$, and we can therefore apply theorem 3.5 which ensures the existence and the uniqueness of the solution of problem 7.31, which is a variational problem (obviously nonlinear).

In order to interpret problem 7.31 as a differential problem, we make the usual assumption regarding the continuity of the solution: $u \in H_0^1(\Omega) \cap C^0(\bar{\Omega})$.

Let now $\phi \in \mathcal{D}(\Omega)$ and put $v = u + \phi$; v is obviously a function of $H_0^1(\Omega)$ and hence, from (7.215),

$$\mathfrak{A}(u, -\phi) + \int_\Omega [u]^+ \, dx \leq \int_\Omega [u+\phi]^+ \, dx + \int_\Omega f(-\phi) \, dx, \qquad (7.216)$$

which can be written in the form

$$_{\mathcal{D}'(\Omega)}\langle Lu + \lambda u - f, -\phi \rangle_{\mathcal{D}(\Omega)} + \int_\Omega [u]^+ \, dx \leq \int_\Omega [u+\phi]^+ \, dx. \qquad (7.217)$$

Now, assuming that $\phi \in \mathcal{D}_+(\Omega)$ (and hence that $[\phi]^+ \equiv \phi$) (7.217) becomes

$$_{\mathcal{D}'(\Omega)}\langle Lu + \lambda u - f, -\phi \rangle_{\mathcal{D}(\Omega)} + \int_\Omega [u]^+ \, dx \leq \int_\Omega [u]^+ \, dx + \int_\Omega \phi \, dx \qquad (7.218)$$

or, given that $\int_\Omega \phi \, dx = {}_{\mathcal{D}'(\Omega)}\langle 1, \phi \rangle_{\mathcal{D}(\Omega)}$ and that ϕ is arbitrary in $\mathcal{D}_+(\Omega)$,

$$Lu + \lambda u - f + 1 \geq 0 \quad \text{in } \Omega \qquad (7.219)$$

in the sense of $\mathcal{D}'(\Omega)$. If we assume instead that $\phi \in \mathcal{D}_-(\Omega)$ (and hence that $[\phi]^+ \equiv 0$) we get

$$Lu + \lambda u - f \leq 0 \quad \text{in } \Omega \qquad (7.220)$$

in the sense of $\mathcal{D}'(\Omega)$. From (7.219) and (7.220) we get the double differential inequality

$$-1 + f \leq Lu + \lambda u \leq f \quad \text{in } \Omega, \qquad (7.221)$$

which is to be verified in the sense of $\mathcal{D}'(\Omega)$, with which we associate the boundary condition

$$\gamma_0 u = 0 \quad \text{in } \Gamma, \qquad (7.222)$$

which is to be verified a.e. in Γ, and which follows from the fact that $u \in H_0^1(\Omega)$. Up to this point we have not yet used the assumption that u is continuous (but, from (7.221) and from $f \in L^2(\Omega)$, it follows in fact that $u \in H^2(\Omega)$ and hence that $u \in C^0(\Omega)$ if $n = 1, 2, 3$ (cf. the following theorem 7.12)). We will use such a hypothesis now in order to obtain a finer result than (7.221), namely: in order to state in explicit form the ambiguous conditions which will complete the double unilateral condition (7.221). Let then $\phi \in \mathcal{D}(\Omega^+(u))$ and put $v = u + \mu\tilde{\phi}$, with $\mu = \min_{\text{supp } \phi} u/\max_{\Omega^+(u)} |\phi|$, where $\tilde{\phi}$ is the trivial extension of ϕ to Ω. Instead of (7.219) we now obtain

$$_{\mathcal{D}'(\Omega^+(u))}\langle Lu + \lambda u - f, -\mu\phi \rangle_{\mathcal{D}(\Omega^+(u))} + \int_{\Omega^+(u)} [u]^+ \, dx \leq \int_{\Omega^+(u)} [u + \mu\phi]^+ \, dx, \qquad (7.223)$$

which can also be written, given that $[u]^+ \equiv u$ and $[u+\mu\phi]^+ = u+\mu\phi$ on $\Omega^+(u)$ (since $u+\mu\phi \geq 0$ on it),

$$_{\mathscr{D}'(\Omega^+(u))}\langle Lu+\lambda u-f, -\mu\phi\rangle_{\mathscr{D}(\Omega^+)} + \int_{\Omega^+(u)} u\, dx \leq \int_{\Omega^+(u)} u\, dx + \int_{\Omega^+(u)} \mu\phi\, dx, \tag{7.224}$$

which gives us

$$Lu+\lambda u-f = -1 \quad \text{in } \Omega^+(u) \tag{7.225}$$

in the sense of $\mathscr{D}'(\Omega^+(u))$. In an analogous manner we have

$$Lu+\lambda u-f = 0 \quad \text{in } \Omega^-(u), \tag{7.226}$$

in th sense of $\mathscr{D}'(\Omega^-(u))$. Equations (7.225) and (7.226) are the ambiguous conditions mentioned above.

Summarizing, we can say that the solution of problem 7.31 is (formally) a solution of the following problem:

PROBLEM 7.32. Given $f \in L^2(\Omega)$ find $u \in H_0^1(\Omega)$ such that conditions (7.221), (7.225), and (7.226) are satisfied.

We will now show that, conversely, if u is a (continuous) solution of problem 7.32 then u is a solution of problem 7.31. Now, from Green's formula (note that $Lu+\lambda u-f \in L^2(\Omega)$ and hence that $u \in H_L^1(\Omega)$), u satisfies (7.215) iff it satisfies

$$\int_\Omega [(Lu+\lambda u-f)(u-v) + [u]^+]\, dx \leq \int_\Omega [v]^+\, dx \quad \forall v \in H_0^1(\Omega). \tag{7.227}$$

We will show that u in fact satisfies (7.227). With this aim we will break up the integral in the left-hand side of (7.227) into the sum

$$\int_\Omega [(Lu+\lambda u-f)(u-v) + [u]^+]\, dx = \int_{\Omega^+(u)} [(Lu+\lambda u-f)(u-v) + [u]^+]\, dx$$
$$+ \int_{\Omega^0(u)} [(Lu+\lambda u-f)(u-v) + [u]^+]\, dx$$
$$+ \int_{\Omega^-(u)} [(Lu+\lambda u-f)(u-v) + [u]^+]\, dx \tag{7.228}$$

and then treat each of the parts thus obtained separately.

We have:

$$\int_{\Omega^+(u)} [(Lu+\lambda u-f)(u-v) + [u]^+]\, dx = \int_{\Omega^+(u)} [(-1)(u-v) + u]\, dx$$
$$= \int_{\Omega^+(u)} v\, dx \tag{7.229}$$

since on $\Omega^+(u)$ is $[u]^+ \equiv u$ and (from (7.225)) $Lu + \lambda u - f = -1$,

$$\int_{\Omega^0(u)} [(Lu + \lambda u - f)(u - v) + [u]^+] \, dx \leq \int_{\Omega^0(u)} [v]^+ \, dx \qquad (7.230)$$

since on $\Omega^0(u)$ is $-1 \leq Lu + \lambda u - f \leq 0$ (from 7.221)) and $u \equiv [u]^+ \equiv 0$,

$$\int_{\Omega^-(u)} [(Lu + \lambda u - f)(u - v) + [u]^+] \, dx = 0 \qquad (7.231)$$

since on $\Omega^-(u)$ is $Lu + \lambda u - f = 0$ (from (7.226)) and $[u]^+ \equiv 0$. Now, since

$$\int_{\Omega^+(u)} v \, dx + \int_{\Omega^0(u)} v \, dx \leq \int_{\Omega} [v]^+ \, dx, \qquad (7.232)$$

we see that (7.227) is satisfied. We have thus completed the proof of the (formal) equivalence of problems 7.31 and 7.32.

REMARK 7.1. From what we have seen in section 3.2.3 we can say that conditions (7.221), (7.225), and (7.226) can be written in the compact form $Lu + \lambda u \in f - H(u)$, where H is the subdifferential of the positive-part function: $H = \partial [\ ]^+$ (we denote it by H since $\partial [\ ]^+$ can be represented as the monotonic maximal graph associated with the Heaviside function $H(x) = 1$ if $x > 0$ and $H(x) = 0$ if $x < 0$). Analogous notations can be used in order to write the unilateral conditions which appear in examples 7.9 and 7.10.

7.4 REGULARITY RESULTS

7.4.1 REGULARITY OF THE SOLUTIONS OF DIFFERENTIAL EQUATIONS

It is in general difficult to prove results regarding regularity, not so much because they are conceptually complicated but because they are very technical. We will deal with a particularly simple case in which these difficulties can still be noticed. The simplest possible case is that in which Ω does not have a boundary, i.e. $\Omega = \mathbb{R}^n$: in this case from the theory of Fourier transform we have, e.g., that

$-\Delta u + u : H^{s+2}(\mathbb{R}^n) \to H^s(\mathbb{R}^n)$
$\qquad\qquad$ is a surjective isomorphism for every $s \in \mathbb{R}$. (7.233)

We will now deal with a type (iii) regularity result (see section 6.1) with respect to the homogeneous Dirichlet problem for the equation $-\Delta u + u = f$ in the semispace. More precisely we consider the following:

PROBLEM 7.33. Given $f \in H^{-1}(\mathring{\mathbb{R}}^n_+)$ find $u \in H^1_0(\mathring{\mathbb{R}}^n_+)$ such that

$$-\Delta u + u = f \quad \text{in } \mathring{\mathbb{R}}^n_+. \qquad (7.234)$$

This problem is obviously well-posed. In fact, problem 7.33 is equivalent to the variational problem 'given $f \in H^{-1}(\mathring{\mathbb{R}}^n_+)$ find $u \in H^1_0(\mathring{\mathbb{R}}^n_+)$ such that $\int_{\mathring{\mathbb{R}}^n_+} uv \, dx + \int_{\mathring{\mathbb{R}}^n_+} (\text{grad } u, \text{grad } v) \, dx =\ _{H^{-1}(\mathring{\mathbb{R}}^n_+)}\langle f, v \rangle_{H^1_0(\mathring{\mathbb{R}}^n_+)}\ \forall v \in H^1_0(\mathring{\mathbb{R}}^n_+)$', which is a well-posed problem since the bilinear form in the first part of the variational equation is continuous and coercive on $H^1_0(\mathring{\mathbb{R}}^n_+)$; since this form is the same as $(u, v)_{H^1(\mathring{\mathbb{R}}^n_+)}$ the result can be obtained from Riesz's theorem, which states also that $\|u\|_{H^1(\mathring{\mathbb{R}}^n_+)} = \|f\|_{H^{-1}(\mathring{\mathbb{R}}^n_+)}$, an *a priori* estimate to which we will return later.

Our aim now is to prove:

THEOREM 7.9. If $f \in L^2(\mathring{\mathbb{R}}^n_+)$ then $u \in H^1_0(\mathring{\mathbb{R}}^n_+) \cap H^2(\mathring{\mathbb{R}}^n_+)$.

To prove this theorem we have essentially to show that $(\partial^2 u / \partial x_j \, \partial x_i) \in L^2(\mathring{\mathbb{R}}^n_+)$, $i, j = 1, \ldots, n$, i.e. that $(\partial u / \partial x_i) \in H^1(\mathring{\mathbb{R}}^n_+)$, $i = 1, \ldots, n$.

It is obvious that the *'tangential'* variables $x_1, \ldots, x_{n-1}$ and the *'normal'* variable x_n have to be considered in different ways. Let then x_i, $1 \le i \le n-1$, be a tangential variable and put $u_i = \partial u / \partial x_i$. We have assumed that u is a solution of:

PROBLEM 7.34. Given $f \in L^2(\mathring{\mathbb{R}}^n_+)$ find $u \in H^1_0(\mathring{\mathbb{R}}^n_+)$ such that

$$-\Delta u + u = f \quad \text{in } \mathring{\mathbb{R}}^n_+. \tag{7.235}$$

So, u_i solves formally the

PROBLEM 7.35. Given $f \in L^2(\mathring{\mathbb{R}}^n_+)$ find $u_i \in H^1_0(\mathring{\mathbb{R}}^n_+)$ such that

$$-\Delta u_i + u_i = \frac{\partial f}{\partial x_i} \quad \text{in } \mathring{\mathbb{R}}^n_+. \tag{7.236}$$

This is a well-posed problem (note that $(\partial f / \partial x_i) \in H^{-1}(\mathring{\mathbb{R}}^n_+)$). Under these conditions we can write

$$\frac{\partial u_i}{\partial x_j} = \frac{\partial^2 u}{\partial x_j \, \partial x_i} \in L^2(\mathring{\mathbb{R}}^n_+), \qquad j = 1, \ldots, n, \tag{7.237}$$

and hence, to show that $u \in H^2(\mathring{\mathbb{R}}^n_+)$, all we need to do is to prove that $(\partial^2 u / \partial x_n^2) \in L^2(\mathring{\mathbb{R}}^n_+)$, which is obvious since

$$\frac{\partial^2 u}{\partial x_n^2} = -f - \sum_{i=1}^{n-1} \frac{\partial^2 u}{\partial x_i^2} + u \in L^2(\mathring{\mathbb{R}}^n_+). \tag{7.238}$$

This is then a very simple proof of theorem 7.9, for which we had foreseen a fairly complicated one. Do we then conclude that we had overestimated the difficulties involved? Unfortunately, this is not so: *the above proof is incorrect!* It contains in fact a *petitio principio*, i.e. 'eating one's own tail'. We cannot ensure that u_i is a solution of problem 7.35 unless we know that $u_i \in H^1_0(\mathring{\mathbb{R}}^n_+)$, whereas this is exactly what we wished to prove.

We have presented this pseudo-proof as a warning: in proving regularity results we have to be specially careful in not accepting, as good ones, reasonings which are only formally true. We now give a correct proof.

Proof of theorem 7.9. We will show that $(\partial u/\partial x_1) \in H^1(\mathring{\mathbb{R}}^n_+)$: the reasoning is exactly the same for the other tangential variables, whereas for the normal variable the previous argument holds. For this purpose we use the *technique of (tangential) difference quotients* and make use of an *a priori estimate*. This is a very commonly used method for proving regularity results (see, Nirenberg, 1959; Nečas, 1967; Lions–Magenes, 1972a etc.).

We will use the following notation: if $u \in L^2(\mathring{\mathbb{R}}^n_+)$ and $h > 0$ we denote by $\rho_{1h} u$ the *difference ratio* of u with respect to the variable x_1 relative to the parameter h, i.e.:

$$\rho_{1h} u(x_1, \ldots, x_n) = \frac{1}{h}[u(x_1 + h, x_2, \ldots, x_n) - u(x_1, x_2, \ldots, x_n)]. \tag{7.239}$$

It is obvious that (7.239) defines a transformation of $L^2(\mathring{\mathbb{R}}^n_+)$ in itself, with respect to which $H^1(\mathring{\mathbb{R}}^n_+)$ and $H^1_0(\mathring{\mathbb{R}}^n_+)$ are closed (i.e., $\rho_{1h}(H^1(\mathring{\mathbb{R}}^n_+)) \subset H^1(\mathring{\mathbb{R}}^n_+)$ and $\rho_{1h}(H^1_0(\mathring{\mathbb{R}}^n_+)) \subset H^1_0(\mathring{\mathbb{R}}^n_+)$). It is less obvious that this transformation is *non-expansive* on $H^1(\mathring{\mathbb{R}}^n_+)$:

$$\|\rho_{1h} u\|^2_{L^2(\mathring{\mathbb{R}}^n_+)} \leq \|u\|_{H^1(\mathring{\mathbb{R}}^n_+)} \quad \forall u \in H^1(\mathring{\mathbb{R}}^n_+), \tag{7.240}$$

a fundamental result which we will now prove. It is easily seen that, whatever $u \in H^1(\mathring{\mathbb{R}}^n_+)$,

$$\rho_{1h} u(x_1, x_2, \ldots, x_n) = \frac{1}{h} \int_0^h \frac{\partial}{\partial x_1} u(x_1 + \tau, x_2, \ldots, x_n) \, d\tau \tag{7.241}$$

and hence

$$\|\rho_{1h} u\|^2_{L^2(\mathring{\mathbb{R}}^n_+)} = \int_{\mathring{\mathbb{R}}^n_+} \left(\frac{1}{h} \int_0^h \frac{\partial}{\partial x_1} u(x_1 + \tau, x_2, \ldots, x_n) \, d\tau\right)^2 dx. \tag{7.242}$$

Applying the Schwarz–Hölder inequality and the Fubini theorem to the right-hand side of (7.242) we can write successively

$$\|\rho_{1h} u\|^2_{L^2(\mathring{\mathbb{R}}^n_+)} \leq \frac{1}{h} \int_{\mathring{\mathbb{R}}^n_+} \left(\int_0^h \left(\frac{\partial}{\partial x_1} u(x_1 + \tau, x_2, \ldots, x_n)\right)^2 d\tau\right) dx$$

$$= \frac{1}{h} \int_0^h \left(\int_{\mathring{\mathbb{R}}^n_+} \left(\frac{\partial}{\partial x_1} u(x_1 + \tau, x_2, \ldots, x_n)\right)^2 dx\right) d\tau$$

$$= \left\|\frac{\partial u}{\partial x_1}\right\|^2_{L^2(\mathring{\mathbb{R}}^n_+)} \leq \|u\|^2_{H^1(\mathring{\mathbb{R}}^n_+)}, \tag{7.243}$$

which proves (7.240).

Since ρ_{1h} is obviously linear, we can say that $\rho_{1h} \in \mathscr{L}(H^1(\mathring{\mathbb{R}}^n_+), L^2(\mathring{\mathbb{R}}^n_+))$.

The above argument shows also that $\rho_{1h} \in \mathcal{L}(H_0^1(\mathring{\mathbb{R}}_+^n), L^2(\mathring{\mathbb{R}}_+^n))$, the inequality analogous to (7.240) being true. To conclude this digression regarding the properties of the operator ρ_{1h}, we note that if $\Delta u \in L^2(\mathring{\mathbb{R}}_+^n)$ then

$$\rho_{1h}(\Delta u) = \Delta(\rho_{1h} u). \tag{7.244}$$

We will now go on to the proof of the theorem. Consider then the

PROBLEM 7.36. Given $f \in L^2(\mathring{\mathbb{R}}_+^n)$ find $u \in H_0^1(\mathring{\mathbb{R}}_+^n)$ such that

$$-\Delta(\rho_{1h}u) + \rho_{1h}u = \rho_{1h}f \quad \text{in } \mathring{\mathbb{R}}_+^n. \tag{7.245}$$

This problem is well-posed since problem 7.34 is well-posed and (7.245) can be written in the form

$$\rho_{1h}(-\Delta u + u) = \rho_{1h}f \quad \text{in } \mathring{\mathbb{R}}_+^n, \tag{7.246}$$

which follows from (7.235). Problem 7.36 replaces here problem 7.35 in the same way that $\rho_{1h}u$ replaces u_1: we now intend to prove that $\rho_{1h}u \to u_1$ in $H^1(\mathring{\mathbb{R}}_+^n)$ as $h \to 0$, which will prove the theorem. To do this we make use of the *a priori* estimate

$$\|\rho_{1h}u\|_{H^1(\mathring{\mathbb{R}}_+^n)} \leq \|\rho_{1h}f\|_{H^{-1}(\mathring{\mathbb{R}}_+^n)}, \tag{7.247}$$

which follows from Riesz's theorem. We can easily see that

$$_{H^{-1}(\mathring{\mathbb{R}}_+^n)}\langle \rho_{1h}f, v \rangle_{H_0^1(\mathring{\mathbb{R}}_+^n)} = {}_{H^{-1}(\mathring{\mathbb{R}}_+^n)}\langle f, \rho_{1(-h)}v \rangle_{H_0^1(\mathring{\mathbb{R}}_+^n)} \quad \forall v \in H_0^1(\mathring{\mathbb{R}}_+^n) \tag{7.248}$$

and hence, given that $f \in L^2(\mathring{\mathbb{R}}_+^n)$,

$$\left|{}_{H^{-1}(\mathring{\mathbb{R}}_+^n)}\langle \rho_{1h}f, v \rangle_{H_0^1(\mathring{\mathbb{R}}_+^n)}\right| = \left|\int_{\mathring{\mathbb{R}}_+^n} f\rho_{1(-h)}v \, dx\right|$$

$$\leq \|f\|_{L^2(\mathring{\mathbb{R}}_+^n)} \|\rho_{1(-h)}v\|_{L^2(\mathring{\mathbb{R}}_+^n)}. \tag{7.249}$$

From this together with (7.240) we have then

$$\|\rho_{1h}f\|_{H^{-1}(\mathring{\mathbb{R}}_+^n)} \leq c, \tag{7.250}$$

where c is a constant (which depends on f but not on v or h). From (7.247) and (7.250) we have

$$\|\rho_{1h}u\|_{H^1(\mathring{\mathbb{R}}_+^n)} \leq c; \rho_{1h}u \in H_0^1(\mathring{\mathbb{R}}_+^n), \tag{7.251}$$

which tells us that from the family $\rho_{1h}u$ we can extract a subsequence weakly convergent in $H_0^1(\mathring{\mathbb{R}}_+^n)$: let v_1 be such a weak limit. If on the other hand we can show that

$$\rho_{1h}u \to u_1 \quad \text{in } \mathcal{D}'(\mathring{\mathbb{R}}_+^n), \text{ as } h \to 0, \tag{7.252}$$

then we will have finished, since from the uniqueness of the weak limit it follows that $u_1 = v_1 \in H_0^1(\mathring{\mathbb{R}}_+^n)$ (further, it is the whole family $\rho_{1h}u$ which converges weakly to $v_1 = u_1$ in $H_0^1(\mathring{\mathbb{R}}_+^n)$). In order to prove (7.252) we observe that, in analogy with (7.248),

$$_{\mathscr{D}'(\mathring{\mathbb{R}}_+^n)}\langle \rho_{1h}u, \phi\rangle_{\mathscr{D}(\mathring{\mathbb{R}}_+^n)} = {}_{\mathscr{D}'(\mathring{\mathbb{R}}_+^n)}\langle u, \rho_{1(-h)}\phi\rangle_{\mathscr{D}(\mathring{\mathbb{R}}_+^n)} \qquad \forall \phi \in \mathscr{D}(\mathring{\mathbb{R}}_+^n) \quad (7.253)$$

and that $\rho_{1(-h)}\phi$ converges uniformly to $-\partial\phi/\partial x_1$ in $\mathring{\mathbb{R}}_+^n$ as $h \to 0$, which e.g. follows from

$$\rho_{1(-h)}\phi = -\frac{\partial}{\partial x_1}\phi(x_1, x_2, \ldots, x_n) - \frac{h}{2!}\frac{\partial^2}{\partial x_1^2}\phi(x_1 - \xi, x_2, \ldots, x_n)$$
(7.254)

(where $0 \leq \xi \leq h$) and from the fact that the second derivative is limited. Equation (7.252) follows then from the definition of the derivative in $\mathscr{D}'(\mathring{\mathbb{R}}_+^n)$. ∎

By a finite induction procedure one can easily show that *if $f \in H^k(\mathring{\mathbb{R}}_+^n)$ then the solution of problem 7.33 belongs to $H_0^1(\mathring{\mathbb{R}}_+^n) \cap H^{k+2}(\mathring{\mathbb{R}}_+^n)$*. Further, by means of local maps, one can show (see Lions–Magenes, 1972a, p. 149), that it is possible to eliminate the restriction $\Omega = \mathring{\mathbb{R}}_+^n$. The two following results can also be proved (where the open set $\Omega \subset \mathbb{R}^n$ has a boundary Γ sufficiently regular for them to be meaningful):

THEOREM 7.10. If $f \in H^s(\Omega)$ and $g_0 \in H^{s+3/2}(\Gamma)$, $s \geq 0$, then the solution of problem 7.16 (non-homogeneous Dirichlet problem) belongs to $H^{s+2}(\Omega)$.

THEOREM 7.11. If $f \in H^s(\Omega)$ and $g_1 \in H^{s+1/2}(\Gamma)$, $s \geq 0$, then the solution of problem 7.18 (non-homogeneous Neumann problem) belongs to $H^{s+2}(\Omega)$.

For the problem 7.20 (mixed non-homogeneous problem), on the other hand, a general result of this type does not hold, unless f, g_0, and g_1 satisfy certain *compatibility conditions*. For many other results concerning the regularity of the solutions of elliptic differential equations see Nečas, 1967; Lions–Magenes, 1960–1963b; 1972a; Agmon, 1965; Nirenberg, 1959; Miranda, 1970; Ladyženskaja–Ural'ceva, 1968; Shamir, 1968; Pucci, 1964; Lopatinski, 1953; Shapiro, 1953; Stampacchia, 1965; Magenes–Stampacchia, 1958; Agmon–Douglis–Nirenberg, 1959; 1964; etc.

The theorem we have proved is a '*global*' or '*up to the boundary*' *regularity result*: it tells us about the regularity of the solution in the closure of the open set with respect to which the problem has been formulated. When one speaks of the regularity of the solution of a given problem one usually intends this kind of information: theorems 7.8 and 7.11, e.g., refer to such a concept. Now, analysing the statements of theorems 7.10 and 7.11 and keeping in mind what we said with respect to the mixed problem, we conclude that *the (global) regularity of the solutions of different problems associated with a given operator depends on the type (and not only on the regularity) of the data on the boundary*. These considerations lead us to examine, along with the results already discussed, the so-called '*local*' or '*on the interior' regularity results* which do not depend on the boundary conditions but only on the differential operators involved.

We will give an example of this:

THEOREM 7.12. Let $\Omega \subset \mathbb{R}^n$ be an arbitrary open set and $f \in L^2_{\text{loc}}(\Omega)$. If $u \in H^1_{\text{loc}}(\Omega)$ is such that

$$-\Delta u + u = f \quad \text{in } \Omega \tag{7.255}$$

then $u \in H^2_{\text{loc}}(\Omega)$.

Proof. Let $x_0 \in \Omega$ and let $B(x_0, r) = \{x \in \Omega : \|x - x_0\| < r\}$, with $r < \frac{1}{2} \cdot \text{dist}(x_0, \mathbb{R}^n \setminus \Omega)$. We wish to prove that if $-\Delta u + u = f \in L^2_{\text{loc}}(\Omega)$ and $u \in H^1_{\text{loc}}(\Omega)$ then $u|_{B(x_0, r)} \in H^2(B(x_0, r))$. For this purpose we need only show that $v = \tilde{u}\tilde{\phi} \in H^2(\mathbb{R}^n)$, where $\tilde{u}$ is the trivial extension of u to $\mathbb{R}^n$ and $\tilde{\phi}$ is the trivial extension to $\mathbb{R}^n$ of a function $\phi \in C_0^\infty(B(x_0, 2r))$ such that $\phi \equiv 1$ in $B(x_0, r)$ (recall the lemma which is used in proving the theorem of the partition of the unit (see Theorem 15.4)). Now denoting by $\tilde{f}$ the trivial extension of f to $\mathbb{R}^n$,

$$\begin{aligned}-\Delta v + v &= -\Delta(\tilde{u}\tilde{\phi}) + \tilde{u}\tilde{\phi} = -\tilde{u}\Delta\tilde{\phi} - \tilde{\phi}\Delta\tilde{u} - 2(\text{grad } \tilde{u}, \text{grad } \tilde{\phi}) + \tilde{u}\tilde{\phi} \\ &= -\tilde{u}\Delta\tilde{\phi} + \tilde{\phi}(-\Delta\tilde{u} + \tilde{u}) - 2(\text{grad } \tilde{u}, \text{grad } \tilde{\phi}) \\ &= -\tilde{u}\Delta\tilde{\phi} + \tilde{\phi}\tilde{f} - 2(\text{grad } \tilde{u}, \text{grad } \tilde{\phi}) \in L^2(\mathbb{R}^n),\end{aligned} \tag{7.256}$$

and thus, from (7.233), $v \in H^2(\mathbb{R}^n)$. ∎

This result is true, with a different proof, for much more general operators than $-\Delta + I$. Associated with these results is the concept of *hypoelliptic operator*: see, e.g., Hörmander, 1964.

7.4.2 REGULARITY OF SOLUTIONS OF VARIATIONAL INEQUALITIES

The study of the regularity of the solutions of problems associated with variational inequalities is, in general, much more complicated than the study of the regularity of the solutions of problems associated with equations. The reason for this is, naturally, the nonlinear nature of these problems. Besides, nonlinearity implies that a technique or a method of studying these problems which works well for some problems will not be applicable to other not too different problems. Further, direct methods of studying regularity, as for example the method of tangential difference ratios used in the previous section, are seldom applicable here. A notable exception is the case of the Signorini problem (see Lions, 1969a, p. 256), which is, however, a problem which is 'linear on the interior' (since in the open set Ω a differential equation ($Au = f$ in Ω) is satisfied, and not merely a differential inequality ($Au \leq f$ in Ω) as, for example, in the obstacle problem); the first step of the method for such problem works.

We will present here an indirect method—a fairly general one, even though it is not always applicable. The method consists of transferring the problem of the regularity for inequalities into one of regularity for equations. The basic idea of this method is that of eliminating the lack of information due to the fact that we have a differential inequality ($Au \leq f$ in Ω) rather than an equation ($Au = f$ in Ω), and thus showing that under certain conditions (i.e. for certain problems) the regularity of Au is anyway that of f. We will prove this result, which hinges essentially on the Lewy–Stampacchia inequality (see Lewy–Stampacchia, 1969) under fairly general conditions on the operator A, which can in particular be nonlinear (in contrast to those associated with the bilinear forms, the only ones we have considered so far in this chapter: to underline this fact we will denote the operator by A rather than by L). In Chapter 3 we have already considered variational inequalities associated with nonlinear operators (see theorem 3.2)—here we will use a still more general framework.

Coerciveness, monotonicity, hemicontinuity

In order to describe this framework we need certain definitions, and, for ease of reading, we will repeat here some of the definitions we have already included in previous chapters:

DEFINITION 7.2. Let V be a Banach space, with dual V', and $K \subset V$ a non-empty set. An operator $A : K \to V'$ is said to be
(i) *coercive*, if

$$\exists v_0 \in K \frac{{}_{V'}\langle Au, u - v_0 \rangle_V}{\|u\|_V} \to +\infty \quad \text{when } \|u\|_V \to +\infty, u \in K \quad (7.257)$$

(this definition is of course meaningless if K is bounded);
(ii) *monotone*, if

$$\forall u, v \in K \; {}_{V'}\langle Au - Av, u - v \rangle_V \geq 0; \quad (7.258)$$

(iii) *strictly monotone*, if it is monotone and

$${}_{V'}\langle Au - Av, u - v \rangle_V = 0 \Rightarrow u = v; \quad (7.259)$$

(iv) *strongly monotone*, if

$$\exists \alpha > 0 \; \forall u, v \in K \; {}_{V'}\langle Au - Av, u - v \rangle_V \geq \alpha \|u - v\|_V^2; \quad (7.260)$$

(v) *hemicontinuous*, if for every $u, v \in K$ and every sequence $t_m \to 0$ of non-negative real numbers such that $u + t_m v \in K$, the sequence $A(u + t_m v)$ converges weakly* to Au in V'.

In order to study these and other concepts closely related to them (such as *pseudo-monotonicity, maximal monotonicity, vague continuity,* etc.) see, e.g., Lions, 1969a, ch. 2; Kato, 1964; 1967; Biroli, 1972; Vainberg, 1973, ch. 1; Tartar, 1975; Brézis, 1968; 1973; Browder, 1964. Here we will merely point out the fairly evident fact that a *strongly monotone operator is strictly monotone and coercive* and make some brief comments on the concept of hemicontinuity. If K is convex, which is the most frequent case in applications, the definition of hemicontinuity can be reformulated as follows: $A: K \to V'$ *is hemicontinuous if for every* u, $v \in K$, $w \in V$, *the function* $\lambda \mapsto f(\lambda) = {}_{V'}\langle A(\lambda u + (1+\lambda)v), w\rangle_V$ *is continuous from* $[0, 1]$ *in* $\mathbb{R}$ (i.e., the restriction of A to the segments contained in K is continuous for the weak* topology in V'); written in this manner, the definition can be generalized very naturally: thus, $A: K \to V'$ *is said to be upper hemi-semicontinuous (resp.: lower hemi-semi-continuous) if* $f(\lambda)$ *is u.s.c. (resp.: l.s.c.)*.

Let us now consider the

PROBLEM 7.37. Let V be a reflexive Banach space with dual V', $K \subset V$ a non-empty closed convex set and $A: V \to V'$ a monotone operator, hemicontinuous and coercive (on K): given $f \in V'$ find

$$u \in K; \quad \forall v \in K \quad {}_{V'}\langle Au - f, u - v\rangle_V \le 0. \tag{7.261}$$

We are not yet in position to say whether this problem is well-posed or not. In Chapter 10 we will see that it has in effect a solution, which is unique if A is strictly monotone (see also Lions, 1969a, p. 247; Browder, 1965b; Hartman–Stampacchia, 1966, p. 273; Mosco, 1967a). We will, however, prove a regularity result for the solutions of problem 7.37. The problem of the regularity of the solution of variational inequalities, first studied in Lewy–Stampacchia, 1969, has been extensively treated by several authors (see, e.g., Brézis–Stampacchia, 1968; Mosco–Troianiello, 1973, ...) —here we will follow the approach of Brézis–Stampacchia, 1968.

The Brézis–Stampacchia theorem and Minty's lemma

THEOREM 7.13. Let us suppose that the conditions of problem 7.37 are satisfied and that K and A are *compatible* in the sense that

$$\forall u \in K \quad \forall \varepsilon > 0 \quad \exists u_\varepsilon \in K \quad u_\varepsilon + \varepsilon A u_\varepsilon = u. \tag{7.262}$$

Then, if H is a Hilbert space such that $V \hookrightarrow^{ds} H \simeq H' \hookrightarrow^{ds} V'$, we have the following regularity result: if $f \in H$ and u is a solution of problem 7.37 then $Au \in H$.

This is not a regularity result proper in the sense described in Chapter 6, since we are talking about the regularity of Au and not directly about that of u. It is nevertheless a result similar to the type (iii) results. This theorem can be proved in more general conditions on H, which can be a reflexive Banach space—in that case (7.262) must, however, be modified to a condition involving the so-called *duality mappings* (see Brézis–Stampacchia, 1968, p. 156; Lions, 1969a, p. 259).

To prove the theorem we need another result which is as important as the theorem itself (and which also has several applications of a different kind—see Minty, 1967, p. 318):

THEOREM 7.14 (Minty's lemma). Let V be a Banach space with dual V', $K \subset V$ a non-empty closed convex set, $f \in V'$ and $A: K \to V'$ a hemicontinuous monotone operator: u is a solution of

$$u \in K; \quad \forall v \in K \quad {}_{V'}\langle Au - f, u - v \rangle_V \le 0 \qquad (7.263)$$

iff it is a solution of

$$u \in K; \quad \forall v \in K \quad {}_{V'}\langle Av - f, u - v \rangle_V \le 0, \qquad (7.264)$$

i.e.:

$$\{u \in K : \forall v \in K \; {}_{V'}\langle Au - f, u - v \rangle_V \le 0\}$$
$$= \{u \in K : \forall v \in K \; {}_{V'}\langle Av - f, u - v \rangle_V \le 0\}. \qquad (7.265)$$

Further, the sets considered in (7.265) are closed convex sets (possibly empty).

Proof. If u satisfies (7.263) then it also satisfies (7.264) since we can write

$$\begin{aligned}{}_{V'}\langle Av - f, u - v \rangle_V &= {}_{V'}\langle Av - f + Au - Au, u - v \rangle_V \\ &= {}_{V'}\langle Av - Au, u - v \rangle_V + {}_{V'}\langle Au - f, u - v \rangle_V \end{aligned} \qquad (7.266)$$

and from (7.263) and the fact that A is monotone (note that ${}_{V'}\langle Av - Au, u - v \rangle_V = -{}_{V'}\langle Au - Av, u - v \rangle_V \le 0$) we obtain (7.264).

Conversely, if u satisfies (7.264) we fix $w \in K$ and put $v = \lambda w + (1 - \lambda)u$, $\lambda \in \,]0, 1[$ (note that $v \in K$ since K is convex). We can then write, from (7.264),

$${}_{V'}\langle A(\lambda w + (1-\lambda)u) - f, u - (\lambda w + (1-\lambda)u) \rangle_V$$
$$= \lambda \, {}_{V'}\langle A(\lambda w + (1-\lambda)u), u - w \rangle_V \le 0 \qquad (7.267)$$

and dividing by $\lambda (>0)$ and making $\lambda \to 0$ we obtain, given that A is hemicontinuous,

$${}_{V'}\langle Au - f, u - w \rangle_V \le 0 \qquad (7.268)$$

which, since w is arbitrary in K, is (7.263). The inequalities (7.263) and (7.264) are therefore equivalent. To complete the proof we observe that K is a closed convex set, that the intersection of convex sets is a convex set, that

$$\{u \in K : \forall v \in K \; {}_{V'}\langle Av - f, u - v \rangle_V \le 0\}$$
$$= \bigcap_{v \in K} \{K \cap \{u \in V : {}_{V'}\langle Av - f, u - v \rangle_V \le 0\}\} \qquad (7.269)$$

and, finally, that $\{u \in V : {}_{V'}\langle Av - f, u - v \rangle_V \le 0\}$ is a closed semi-space determined by the hyperplane with equation ${}_{V'}\langle Av - f, u - v \rangle_V = 0$. ∎

We will now go on to the

Proof of theorem 7.13. The main idea of this proof is that of approximating u by a family u_ε under the conditions of (7.262) and then showing that the family Au_ε converges to Au in H, which proves that $Au \in H$. From theorem 7.14, which can obviously be applied, we can substitute (7.261) by

$$u \in K; \quad \forall v \in K \quad {}_{V'}\langle Av - f, u - v \rangle_V \leq 0. \tag{7.270}$$

Now let $\varepsilon > 0$ and put, in (7.270), $v = u_\varepsilon$ with u_ε given by (7.262) (which is meaningful since $u_\varepsilon \in K$):

$$ {}_{V'}\langle Au_\varepsilon - f, u - u_\varepsilon \rangle_V \leq 0. \tag{7.271}$$

Given that

$$u = u_\varepsilon + \varepsilon Au_\varepsilon \tag{7.272}$$

we get, from (7.271)

$$\varepsilon \, {}_{V'}\langle Au_\varepsilon - f, Au_\varepsilon \rangle_V \leq 0 \tag{7.273}$$

or, since $\varepsilon > 0$,

$$ {}_{V'}\langle Au_\varepsilon - f, Au_\varepsilon \rangle_V \leq 0. \tag{7.274}$$

From (7.272) we have $Au_\varepsilon \in V \subset H$ and, since we have assumed that $f \in H$, we can write (7.274) in the form

$$ {}_{H'}\langle Au_\varepsilon - f, Au_\varepsilon \rangle_H = (Au_\varepsilon - f, Au_\varepsilon)_H \leq 0, \tag{7.275}$$

from which we obtain

$$\|Au_\varepsilon\|_H \leq \|f\|_H \quad \forall \varepsilon > 0. \tag{7.276}$$

From this inequality and from (7.272) it follows that

$$u_\varepsilon \underset{H}{\to} u \quad \text{as } \varepsilon \to 0 \tag{7.277}$$

and that

$$(Au_\varepsilon, u_\varepsilon - u) \underset{H}{\to} 0 \quad \text{as } \varepsilon \to 0. \tag{7.278}$$

Now let $v_0 \in K$ be one of the elements of which, from (7.257), we know the existence, and write

$$\begin{aligned} {}_{V'}\langle Au_\varepsilon, u_\varepsilon - v_0 \rangle_V &= {}_{V'}\langle Au_\varepsilon, u_\varepsilon - u + u - v_0 \rangle_V \\ &= {}_{V'}\langle Au_\varepsilon, u_\varepsilon - u \rangle_V + {}_{V'}\langle Au_\varepsilon, u - v_0 \rangle_V \\ &= {}_{V'}\langle Au_\varepsilon, u - v_0 \rangle_V + (Au_\varepsilon, u_\varepsilon - u)_H. \end{aligned} \tag{7.279}$$

From (7.278) and from the fact that ${}_{V'}\langle Au_\varepsilon, u - v_0 \rangle_V$ is bounded (since Au_ε is bounded and $u - v_0$ is fixed) we can see that ${}_{V'}\langle Au_\varepsilon, u_\varepsilon - v_0 \rangle_V$ is bounded. From this and from the coercivity of A it follows then that

$$\exists c > 0 \quad \forall \varepsilon > 0 \quad \|u_\varepsilon\|_V < c. \tag{7.280}$$

From (7.280) together with the fact that V is reflexive we conclude that from u_ε one can extract a subsequence weakly convergent in V:

$$u_{\varepsilon_h} \underset{V}{\rightharpoonup} \bar{u} \quad \text{as } \varepsilon_h \to 0. \tag{7.281}$$

From (7.277) together with the uniqueness of the weak limit we have $u = \bar{u}$ and thus we can write that

$$u_{\varepsilon_h} \underset{V}{\rightharpoonup} u \quad \text{as } \varepsilon_h \to 0. \tag{7.282}$$

Further, writing (7.276) for Au_{ε_h} and recalling that H is a Hilbert space, and therefore reflexive, we conclude that from Au_{ε_h} one can extract a subsequence weakly convergent in H:

$$Au_{\varepsilon_{h_m}} \underset{H}{\rightharpoonup} \chi \quad \text{as } \varepsilon_{h_m} \to 0. \tag{7.283}$$

To prove the theorem all we have to do is to show that $\chi = Au$. From the fact that A is monotone it follows that

$$\forall \varepsilon > 0 \quad \forall w \in V \quad {}_{V'}\langle Au_{\varepsilon_{h_m}} - Aw, u_{\varepsilon_{h_m}} - w \rangle_V \geq 0, \tag{7.284}$$

and taking

$$w = (1-\theta)u + \theta v, \, \theta \in \,]0, 1[, \tag{7.285}$$

with v fixed in V, we can write, from (7.284),

$${}_{V'}\langle Au_{\varepsilon_{h_m}} - A[(1-\theta)u + \theta v], u_{\varepsilon_{h_m}} - [(1-\theta)u + \theta v]\rangle_V$$
$$= (Au_{\varepsilon_{h_m}}, u_{\varepsilon_{h_m}} - u)_H + \theta(Au_{\varepsilon_{h_m}}, u - v)_H - {}_{V'}\langle A[(1-\theta)u + \theta v], u_{\varepsilon_{h_m}} - u\rangle_V$$
$$- \theta \, {}_{V'}\langle A[(1-\theta)u + \theta v], u - v\rangle_V \geq 0. \tag{7.286}$$

Making $\varepsilon_{h_m} \to 0$ we obtain (recalling (7.278), (7.283), (7.282) for the first three parts, respectively)

$$\theta \, {}_{H'}\langle \chi, u - v\rangle_H - \theta \, {}_{V'}\langle A[(1-\theta)u + \theta v], u - v\rangle_V \geq 0, \tag{7.287}$$

or, dividing by $\theta(>0)$ and making $\theta \to 0$, given that A is hemicontinuous

$${}_{V'}\langle \chi - Au, u - v\rangle_V \geq 0. \tag{7.288}$$

Since v is arbitrary in V (and consequently $u - v$ is so too) we obtain the result which we wished to prove: $\chi = Au$. ∎

A concrete example

We will present now an example in which we apply the abstract result just proved. Let us consider the

PROBLEM 7.38. Let $\Omega \subset \mathbb{R}^n$ be an open set with Lipschitz continuous boundary Γ, $V = H_0^1(\Omega)$, $K = \{v \in H_0^1(\Omega) : v \leq \psi\}$, where $\psi \in H^1(\Omega)$ and $\gamma_0 \psi \geq 0$ in Γ, and $a : V \times V \to \mathbb{R}$ a bilinear form continuous and coercive (on $K - K$): given $f \in H^{-1}(\Omega)$ find

$$u \in H_0^1(\Omega); \quad \forall v \in K \quad a(u, u - v) \leq {}_{H^{-1}(\Omega)}\langle f, v\rangle_{H_0^1(\Omega)}. \tag{7.289}$$

This is merely a generalization of problem 7.13, a generalization in that the form a is not specified and in the regularity of the datum f. Now let $H = L^2(\Omega)$: we wish to find out what further assumptions we must make on a (or on A, the operator associated with a) and on ψ (or on K) so that theorem 7.13 is applicable and we can say that if $f \in L^2(\Omega)$ then $Au \in L^2(\Omega)$ (and hence, at least in the case in which A is an elliptic differential linear operator—as in problem 7.13)—$u \in H_0^1(\Omega) \cap H^2(\Omega)$). As is evident, if a is bilinear, continuous and coercive, then the operator A associated with it is monotone (or rather, linear), hemicontinuous (or rather, continuous), and coercive: we have to find, therefore, what conditions must be imposed on a and ψ so that A and K are compatible in the sense of (7.262), i.e., so that the *problem*

$$u_\varepsilon \in K; (u_\varepsilon, v)_{L^2(\Omega)} + \varepsilon a(u_\varepsilon, v) = (u, v)_{L^2(\Omega)} \quad \forall v \in H_0^1(\Omega) \quad (7.290)$$

has a solution for every u given in K and for every $\varepsilon > 0$. Problem (7.290) obviously has a solution, u_ε, in $H_0^1(\Omega) = V$: we have to prove then that, under suitable assumptions which we have to discover, such a solution belongs to K, i.e. $u_\varepsilon \leq \psi$ or, equivalently (see section 5.5), $[u_\varepsilon - \psi]^+ \equiv 0$.

From (7.290) we can write

$$(u_\varepsilon - \psi, v)_{L^2(\Omega)} + \varepsilon a(u_\varepsilon - \psi, v) = (u, v)_{L^2(\Omega)} - (\psi, v)_{L^2(\Omega)} - \varepsilon a(\psi, v), \quad (7.291)$$

and choosing $v = [u_\varepsilon - \psi]^+$ (which is legitimate since $\gamma_0 u_\varepsilon = 0$ and $\gamma_0 \psi \geq 0$ imply that $[u_\varepsilon - \psi]^+ \in H_0^1(\Omega)$) we can write

$$\int_\Omega (u_\varepsilon - \psi)[u_\varepsilon - \psi]^- dx + \varepsilon a(u_\varepsilon - \psi, [u_\varepsilon - \psi]^+)$$
$$= \int_\Omega (u - \psi)[u_\varepsilon - \psi]^+ - \varepsilon a(\psi, [u_\varepsilon - \psi]^+). \quad (7.292)$$

Now $u - \psi \leq 0$ (since we have assumed that $u \in K$) and if we assume that

$$a(\psi, [u_\varepsilon - \psi]^+) \geq 0 \quad (7.293)$$

(*the first of the hypotheses we are looking for!*) we can write

$$\int_\Omega (u_\varepsilon - \psi)[u_\varepsilon - \psi]^+ dx + \varepsilon a(u_\varepsilon - \psi, [u_\varepsilon - \psi]^+) \leq 0. \quad (7.294)$$

Now, since

$$(u_\varepsilon - \psi) = [u_\varepsilon - \psi]^+ - [u_\varepsilon - \psi]^- \quad (7.295)$$

and

$$\mu(\operatorname{supp}[u_\varepsilon - \psi]^+ \cap \operatorname{supp}[u_\varepsilon - \psi]^-) = 0, \quad (7.296)$$

(7.294) becomes

$$\int_\Omega ([u_\varepsilon - \psi]^+)^2 dx + \varepsilon a([u_\varepsilon - \psi]^+, [u_\varepsilon - \psi]^+) - \varepsilon a([u_\varepsilon - \psi]^-, [u_\varepsilon - \psi]^+) \leq 0.$$
$$(7.297)$$

Assuming that
$$a([u_\varepsilon - \psi]^-, [u_\varepsilon - \psi]^+) = 0 \qquad (7.298)$$
(*the second of the hypotheses we are looking for!*) we can write (7.297) in the form
$$\int_\Omega ([u_\varepsilon - \psi]^+)^2 \, dx + \varepsilon a([u_\varepsilon - \psi]^+, [u_\varepsilon - \psi]^+) \leq 0, \qquad (7.299)$$
from which we obtain
$$a([u_\varepsilon - \psi]^+, [u_\varepsilon - \psi]^+) \leq 0. \qquad (7.300)$$
From (7.300) and the coercivity of a it follows that $[u_\varepsilon - \psi]^+ \equiv 0$, which is what we set out to prove.

In conclusion, we can say that if (7.293) and (7.298) are satisfied then we can apply theorem 7.13. These assumptions are difficult to verify, since we need to assume first that we know the solution u of the problem; hence we will rewrite them in a more restrictive form, namely: we will require that the form a and the obstacle ψ satisfy the conditions
$$a(\psi, v) \geq 0 \qquad \forall v \in H_0^1(\Omega), v \geq 0 \qquad (7.301)$$
and
$$a([v]^-, [v]^+) = 0 \qquad \forall v \in H_0^1(\Omega). \qquad (7.302)$$

The condition (7.302) is an assumption regarding a which is fulfilled, in particular, by the form (7.101)—it is not then excessively restrictive for most applications. The condition (7.301) is a *compatibility condition* between a and ψ which, in terms of the operator A, can be written
$$A\psi \geq 0 \qquad \text{in } \mathscr{D}'(\Omega) \qquad (7.303)$$
(or in the sense of the measures on Ω).

A faster proof that $u_\varepsilon \leq \psi$ can be obtained by applying a maximum principle which we have, to some extent, kept 'hidden' in the previous proof. We will now prove it, using this principle explicitly, in the particularly simple case of the form
$$a(u, v) = \int_\Omega (\operatorname{grad} u, \operatorname{grad} v) \, dx, \qquad (7.304)$$
which has the corresponding operator $A = -\Delta$. Under these conditions we can write, in place of (7.291),
$$u_\varepsilon - \psi - \varepsilon \Delta(u_\varepsilon - \psi) = u - \psi + \varepsilon \Delta \psi \qquad (7.305)$$
or further, putting $v = u_\varepsilon - \psi$ and recalling that $u - \psi \leq 0$ (since we have assumed that u is a solution of the problem) and that $\Delta \psi \leq 0$ (since, from (7.303), $-\Delta \psi \geq 0$),
$$v - \varepsilon \Delta v \leq 0. \qquad (7.306)$$

From this, since $\varepsilon > 0$ and $\gamma_0 v \leq 0$ by applying the maximum principle (see section 17.5), we have $v \leq 0$, i.e. $u_\varepsilon \leq \psi$.

In general, *the regularity of the solution of a problem associated with an inequality does not necessarily increase with the increase in the regularity of the data*: we have here a phenomenon which can be interpreted suggestively by stating that, for such problems, there are *regularity thresholds* which the solutions cannot cross (or cannot reach) however regular the data may be (see the comments made in Chapter 6). The exact determination of such thresholds is in general impossible and we must therefore be satisfied with *estimating* them. These estimates (underestimates...) are made usually with the help of counter-examples, among which the one invented by Shamir for the discussion of the regularity of the solutions of equations is very useful (for a treatment of this problem see Brézis–Stampacchia, 1968, p. 176 or Lions, 1969a, p. 255). The regularity thresholds depend naturally on the family (or scale) of functional spaces considered. Thus, if we work in the framework of the spaces $H^s(\Omega)$, $\bar{s}$ is said to be the regularity threshold achieved (respectively: not achieved) by the solution of a problem $\mathscr{P}$ if, when the data are as regular as possible (in a given framework...), in the general case $u \notin H^{\bar{s}+\varepsilon}(\Omega)$ ($\varepsilon > 0$) but $u \in H^{\bar{s}}(\Omega)$ (respectively: $u \in H^{\bar{s}-\varepsilon}(\Omega)$ (for every $\varepsilon > 0$) but $u \notin H^{\bar{s}}(\Omega)$).

For a study of other regularity results see, among others, Beirão da Veiga, 1969; Brézis–Kinderlehrer, 1974; Brézis–Stampacchia, 1968; Dias, 1973; Friedman, 1968; Lewy–Stampacchia, 1969; Lions, 1969a; Mosco–Troianiello, 1973; Murthy–Stampacchia, 1972; Stampacchia, 1969; Brézis, 1971; 1972. For the solutions of the inequalities with obstacle there also arises the problem of the regularity of the 'contact set'; see, in this connection, Kinderlehrer, 1975 and the bibliography of this paper.

8
Variational Formulation of a Free-Boundary Problem

8.1 GENERALITIES. THE PHYSICAL PROBLEM

Variational inequalities and free-boundary problems

The connection between variational inequalities and free-boundary problems has been pointed out by Lewy and Stampacchia (see Lewy–Stampacchia, 1969); many free-boundary problems which arise in applications, however, cannot be *directly* written in terms of variational inequalities. For one such problem (which arises in hydraulics, namely problem 8.1 which follows) it has been shown in Baiocchi, 1971; 1972 that it is possible to put it in a variational inequality form if one makes an appropriate change of the unknown function. This technique has also proved useful in many other problems: besides several hydraulic problems more general than that dealt with in Baiocchi, 1971; 1972 and for which we will provide some references later, analogous methods have been used for free boundary problems which arise in fluid dynamics (subsonic flow around a wing profile—see Brézis–Stampacchia, 1973a for the case without wake and Brézis–Duvaut, 1973 for the case with wake), for problems of a parabolic nature in thermodynamics (Stefan's problem—see Duvaut, 1973; 1974), and for problems of evolution which involve an elliptic equation on the inside and conditions of evolution on the free boundary (see Torelli, 1977d).

The dam problem

Here we will describe only free-boundary problems which arise in the study of the motion of fluids across porous materials. More precisely, we will deal with the case of two water basins, with different levels, separated by an earth dam. Water will filter by gravity from the higher to the lower level, and the problem is to determine some physical quantities (such as the distributions of velocity and pressure, carrying capacity, etc.) as well as some geometrical unknowns (such as the current-lines, the wet portion of the dam, etc.) associated with the motion.

We will study the problem within the hypothesis of bidimensional stationary flow in the absence of evaporation and capillary phenomena, assuming that the fluid is incompressible and that the porous material is homogeneous, isotropic, and incompressible.

For the treatment of other or more general problems see: Torelli, 1977d and relative bibliography for the evolution problems; Gilardi, 1977 and Stampacchia, 1974 for three-dimensional problems; Friedman, 1976d and Pozzi, 1974a; 1974b for evaporation phenomena; Comincioli–Guerri, 1976 for capillarity phenomena; Baiocchi–Comincioli–Magenes–Pozzi, 1973 and Benci, 1974 for problems relative to variable permeability. For problems of a physically different kind, such as the drainage of a channel and for the opening towards the sea of underground faults see, respectively, Torelli, 1974 and Baiocchi–Comincioli–Magenes–Pozzi, 1973.

A further restriction which we will impose concerns the geometry of the dam, which will be assumed to have vertical walls and a horizontal base; in Part II we will deal with a problem including a more complicated geometry.

Let us introduce some notations (for others, see figs 8.1 and 8.2):

y_1, y_2, c denote respectively the heights of the basins and the thickness of the dam,
Ω denotes the region of the flow, i.e. the wet portion of the dam,
$p(x, y)$ denotes the pressure at any given point of the dam (atmospheric pressure being assumed to be the zero pressure),
$\phi(x)$ denotes the 'free boundary', i.e. 'the upper boundary' of Ω,
q denotes the carrying capacity of the dam (it will be $q = -\int_0^{\phi(x)} p_x(x, t)\,dt$ a.e. in $]0, c[$ if we choose as 1 some physical coefficients).

In addition to assuming that the base of the dam is impermeable we will assume that the left wall, above a height d, is also impermeable; this does not, of course, influence the phenomenon if $d \geqslant y_1$ (as in fig. 8.1, in which the impermeable part is not shown) but plays an important part if $d < y_1$ (as

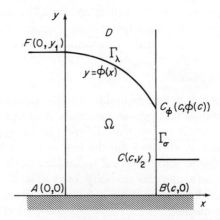

Figure 8.1

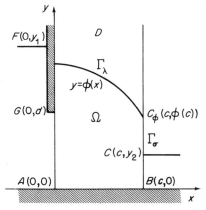

Figure 8.2

in fig. 8.2). This second case will be dealt with in section 8.3; here we will treat the case in fig. 8.1.

Without entering into the physical details of the problems (for which we refer the reader to the second part and to books on hydraulic engineering, e.g., Bear, 1972; Harr, 1962; Polubarinova-Kochina, 1962) we will merely say that we have to deal with the following mathematical problem:

PROBLEM 8.1. Given three numbers, c, y_1, y_2, with

$$c > 0, \quad y_1 > y_2 \geq 0, \tag{8.1}$$

find a curve, $y = \phi(x)$, such that

ϕ is defined and 'regular' on $[0, c]$, (8.2)

ϕ is non-decreasing, $\phi(0) = y_1$, $\phi(c) > y_2$, (8.3)

and such that, putting

$$\Omega = \{(x, y): 0 < x < c \text{ and } 0 < y < \phi(x)\}, \tag{8.4}$$

it is possible to solve in Ω the following boundary value problem in the unknown function $p(x, y)$:

p is defined and 'regular' in $\bar{\Omega}$, (8.5)

$\Delta p = 0 \quad$ in Ω, (8.6)

$p(0, y) = y_1 - y \quad$ for $0 \leq y \leq y_1$, (8.7)

$p(c, y) = y_2 - y \quad$ for $0 \leq y \leq y_2$, (8.8)

$p(c, y) = 0 \quad$ for $y_2 \leq y \leq \phi(c)$, (8.9)

$p(x, \phi(x)) = 0 \quad$ for $0 \leq x \leq c$, (8.10)

$\left(\dfrac{\partial(p+y)}{\partial \nu}\right)(x, \phi(x)) = 0 \quad$ for $0 < x < c$, (8.11)

$\left(\dfrac{\partial(p+y)}{\partial y}\right)(x, 0) = 0 \quad$ for $0 < x < c$. (8.12)

Let us remark that if only one of the conditions (8.10) and (8.11) is imposed, for every fixed ϕ the problem in p would be a mixed Dirichlet–Neumann problem; as we will see it is the imposition of both (8.10) and (8.11) along the curve ϕ that will make it possible to find this curve univocally.

Problem 8.1 is (obviously) not 'complete'. To complete it let us begin with the following considerations: under minimal regularity conditions on ϕ and p, the maximum principle gives us

$$p(x, y) > 0 \quad \text{in } \Omega. \tag{8.13}$$

To complete problem 8.1 we will utilize (8.13) and work, instead of on p, on the trivial extension of p to the set D defined by

$$D =]0, c[\times]0, +\infty[\tag{8.14}$$

—we must therefore to put $\tilde{p}(x, y) = p(x, y)$ for $(x, y) \in \bar{\Omega}$, and $\tilde{p}(x, y) = 0$ for $(x, y) \in \bar{D} \setminus \bar{\Omega}$: to simplify the notation we will continue to write p rather than $\tilde{p}$.

We consider now the problem in the following (complete) form:

PROBLEM 8.2. Given c, y_1, y_2 such that (8.1) holds, and D being defined as in (8.14), find p such that

$$p \in C^0(\bar{D}) \cap H^1(D), \tag{8.15}$$

$$p(x, y) \geq 0 \quad \text{in } \bar{D}, \tag{8.16}$$

$$p(0, y) = [y_1 - y]^+ \quad \text{for } y \in [0, +\infty[, \tag{8.17}$$

$$p(c, y) = [y_2 - y]^+ \quad \text{for } y \in [0, +\infty[, \tag{8.18}$$

$$(x_0, y_0) \in D, p(x_0, y_0) > 0, y \in]0, y_0[\Rightarrow p(x_0, y) > 0, \tag{8.19}$$

and such that, putting

$$\Omega = \{(x, y) \in D : p(x, y) > 0\}, \tag{8.20}$$

it follows that

$$\bar{\Omega} \text{ is compact}, \tag{8.21}$$

$$\int_\Omega \text{grad } (p + y) \text{ grad } \psi \, dx \, dy = 0 \quad \forall \psi \in C_0^1(]0, c[\times]-\infty, +\infty[). \tag{8.22}$$

The expressions (8.16), (8.17), (8.18), and (8.19) are meaningful because $p \in C^0(\bar{D})$, which also implies that Ω defined by (8.20) is open; (8.22) is meaningful because $p \in H^1(D)$. The expression (8.19) once again expresses the fact that Ω has the form (8.4), without, however, imposing any property on ϕ (actually $p \in C^0(\bar{D})$ implies that $x \mapsto \phi(x)$ is l.s.c.). The equation (8.21) says that ϕ is bounded; in the following we will fix a real number y_* such that

$$\bar{\Omega} \subset [0, c] \times [0, y_*[\tag{8.23}$$

and we will limit ourselves to work in the closure $\bar{D}_*$ of the rectangle D_* defined by
$$D_* = \,]0,c[\,\times\,]0, y_*[\tag{8.24}$$
—as regards the uniqueness of y_*, if we have two solutions p_1, p_2 and the corresponding Ω_i satisfy $\bar{\Omega}_i \subset [0, c]\times[0, y_{*i}]$ we will choose $y_* = \max(y_{*1}, y_{*2})$.

Formula (8.22) is the usual weak formulation of (8.6), (8.11), and (8.12); (8.10) is contained in (8.15); (8.7) is translated by (8.17); (8.8) and (8.9) are translated by (8.18). As far as the 'regularity' is concerned, see the discussion following theorem 8.1.

For problem 8.2 we have the following results, from which we will prove here only some of them (others will be proved, in a more general context, in Part II—however, for all the results of theorem 8.1 see Baiocchi, 1972; Baiocchi–Comincioli–Magenes–Pozzi, 1973 and Friedman–Jensen, 1978):

THEOREM 8.1. Problem 8.2 has one and only one solution; further

$$\forall r \in [1, +\infty[\ p \in W^{1,r}(D_*), \tag{8.25}$$

$$p_x \notin L^\infty(D_*), \tag{8.26}$$

$$-\int_0^{+\infty} p_x(x, t)\,\mathrm{d}t = \frac{y_1^2 - y_2^2}{2c} \quad \text{a.e. in }\,]0, c[\tag{8.27}$$

and setting:
$$\phi(x) = \max\{y : (x, y) \in \bar{\Omega}\} \tag{8.28}$$
(8.3) holds and, moreover,
$$\phi \in C^0([0, c]),\ \phi \text{ is analytic in }]0, c[, \tag{8.29}$$
$$\phi \text{ is concave};\ \phi'(0) = 0,\ \phi'(c) = -\infty. \tag{8.30}$$

The (complete) formulation given in problem 8.2 of the (incomplete) problem 8.1 is a very 'weak' formulation as far as the regularity imposed on ϕ is concerned, while it is very 'strong' as concerns the regularity imposed on p ((8.25) and (8.26) show that (8.15) is almost optimal).

8.2 TRANSFORMATION OF THE PROBLEM

In order to state problem 8.2 in a 'more convenient' form, we first note that (8.22) is equivalent to

$$\int_\Omega \operatorname{grad} p \operatorname{grad} \psi\,\mathrm{d}x\,\mathrm{d}y = -\int_\Omega \chi_\Omega \psi_y\,\mathrm{d}x\,\mathrm{d}y \quad \forall \psi \in C_0^\infty(\,]0, c[\,\times\,]-\infty, +\infty[\,) \tag{8.31}$$

(χ_Ω denotes the characteristic function of Ω in D). Indeed, the left-hand side is the same as $\int_\Omega \operatorname{grad} p \operatorname{grad} \psi\,\mathrm{d}x\,\mathrm{d}y$, from the definition of Ω, and the right-hand side is the same that $\int_\Omega \operatorname{grad} y \operatorname{grad} \psi\,\mathrm{d}x\,\mathrm{d}y$, from the definition of χ_Ω.

We will transform (8.31) still further, by showing that it implies the pair of relationships

$$\Delta p = D_y \chi_\Omega \text{ in the sense of } \mathcal{D}'(D), \tag{8.32}$$

$$x \mapsto \int_0^{+\infty} p_x(x, t) \, dt \text{ is linear on } [0, c] \tag{8.33}$$

(in fact (8.31) is equivalent to (8.32) and (8.33); for our purposes, however, it is enough to prove only what we stated). Now, relation (8.32) follows from (8.31) by choosing in it supp $\psi \subset D$; while to obtain (8.33) we choose in (8.31) $\psi(x, y) = \psi_1(x)\psi_2(y)$, with $\psi_1 \in \mathcal{D}(]0, c[)$ and $\psi_2 \in \mathcal{D}(\mathbb{R})$ with $\psi_2 \equiv 1$ on $[0, y_*]$: we have then $\int_0^c (\int_0^{y_*} p_x(x, t) \, dt) \psi_1'(x) \, dx = 0$, i.e. $D_x \int_0^{y_*} p_x(x, t) \, dt = 0$.

The new unknown

The operator D_y on the right-hand side of (8.32) prevents us from directly showing that the solutions of problem 8.2 are solutions of inequalities with obstacle (the obstacle is identically zero; see (8.16)). In order to 'destroy' this operator we consider a new unknown, a primitive of p with respect to y. Namely, we put:

$$w(x, y) = \int_y^{+\infty} p(x, t) \, dt \quad \forall (x, y) \in \bar{D}. \tag{8.34}$$

This change in the unknown, introduced in Baiocchi, 1971; 1972, is the main idea of this method. For other cases in which such a change is useful see Baiocchi, 1975a. A general discussion of free boundary problems which, after an analogous change in the unknown, become variational inequalities is presented in Baiocchi, 1976b and also in Chapter 13 of this book.

One has then

THEOREM 8.2. If p is a solution of problem 8.2, defining w by (8.34), one has

$$w \in H^2(D), \quad \text{supp } w \text{ is compact}, \tag{8.35}$$

$$w \geq 0 \text{ in } \bar{D}, \quad \Omega = \{(x, y) \in D : w(x, y) > 0\}, \tag{8.36}$$

$$\Delta w = \chi_\Omega \quad \text{in } D, \tag{8.37}$$

$$w(c, y) = \frac{1}{2} \int_y^{+\infty} [y_2 - t]^+ \, dt \quad \text{for } 0 \leq y < +\infty, \tag{8.38}$$

$$w(0, y) = \frac{1}{2} \int_y^{+\infty} [y_1 - t]^+ \, dt \quad \text{for } 0 \leq y < +\infty, \tag{8.39}$$

$$w(x, 0) = \frac{y_1^2 - y_2^2}{2c}(c - x) + \frac{y_2^2}{2} \quad \text{for } 0 \leq x \leq c. \tag{8.40}$$

Proof. χ_Ω and w (and hence Δw) obviously have compact supports; from this and from (8.32) (which is equivalent to $D_y(\Delta w - \chi_\Omega) = 0$) (8.37) follows; from (8.15) it follows that $w_y, w_{xy}, w_{yy} \in L^2(D)$; since w has a compact support, to show (8.35) we need only prove that $w_{xx} \in L^2(D)$: this follows from the fact that $w_{yy} \in L^2(D)$ together with (8.37).

Relation (8.36) obviously follows from (8.16), (8.19), (8.20); (8.38) and (8.39) follows from (8.17), (8.18), respectively; finally (8.40) follows from (8.38), (8.39) and from (8.33) (let us remark that (8.38) and (8.39) give the values of $w(0, 0)$ and $w(c, 0)$—remark that from (8.35), e.g., it follows that $w \in C^0(\bar{D})$). ∎

Expressions (8.36) and (8.37) have several consequences. We can for example deduce from them:

$$w \geq 0 \text{ in } D; \quad \Delta w \leq 1 \text{ in } D; \quad w > 0 \Rightarrow \Delta w = 1 \tag{8.41}$$

(which is the 'interior part' of the variational inequality (8.47) which follows); also:

$$\Delta w \in H(w) \tag{8.42}$$

(which is the 'interior part' of (8.45) which follows). (Recall that H denotes the monotonic maximal graph associated with the Heaviside function. Equation (8.42) is an abbreviated expression for the family of relationships: $\Delta w \leq 1$, $w > 0 \Rightarrow \Delta w = 1$, $w < 0 \Rightarrow \Delta w = 0$. See observation 7.1 at the end of section 7.3.)

In particular, if we fix D_*, then w satisfies the two following problems:

PROBLEM 8.3. Let $g(x, y)$ be defined on ∂D_* by

$$\left.\begin{aligned} g(x, y) &= 0, \quad g(x, 0) = \frac{y_1^2 - y_2^2}{2c}(x - c) + \frac{y_2^2}{2} \quad \text{for } 0 \leq x \leq c, \\ g(0, y) &= \int_y^{y_*} [y_1 - t]^+ \, dt, \\ g(c, y) &= \int_y^{y_*} [y_2 - t]^+ \, dt \quad \text{for } 0 \leq y \leq y_* \end{aligned}\right\} \tag{8.43}$$

and put:

$$K = \{v \in H^1(D_*) : v|_{\partial D_*} = g\}; \tag{8.44}$$

find w such that

$w \in K; \forall v \in K$

$$\int_{D_*} \text{grad } w \text{ grad } (z - w) \, dx \, dy + \int_{D_*} [z]^+ \, dx \, dy \geq \int_{D_*} [w]^+ \, dx \, dy. \tag{8.45}$$

PROBLEM 8.4. Let $g(x, y)$ and K be defined by (8.43) and (8.44), and put:

$$K_+ = \{v \in K : v \geq 0 \text{ in } D_*\}; \tag{8.46}$$

find w such that

$w \in K_+; \forall v \in K_+$

$$\int_{D_*} \text{grad } w \text{ grad } (z-w) \, dx \, dy \geq \int_{D_*} (-1)(z-w) \, dx \, dy. \quad (8.47)$$

Since both problem 8.3 (which is of the same type as problem 7.31) and problem 8.4 (which is of the same type as problem 7.28) have one and only one solution, we can conclude with the following theorem:

THEOREM 8.3. Problem 8.2 has at most one solution. If this solution exists it can be obtained by solving problem 8.3 and putting

$$p = -D_y w \quad (8.48)$$

$$\Omega = \{(x, y) \in D_* : w(x, y) > 0\}. \quad (8.49)$$

Let us observe that we can solve problem 8.3 by putting for example $y_* = y_1$, since the curve ϕ will be decreasing. One can of course use problem 8.4, but theoretically certain properties are simpler using problem 8.3. Conversely, from the numerical point of view, problem 8.4 is 'easier' to discretise (see e.g., Baiocchi–Comincioli–Guerri–Volpi, 1973).

Conversely, let w be the solution of problem 8.3 with $y_* = y_1$. From (8.42) together with the fact that $w|_{\partial D} = g$, one has

$$w \in W^{2,r}(D_*) \quad \forall r \in [1, +\infty[\quad (8.50)$$

(For the regularity of the solutions of boundary value problems in open sets with corners see Grisvard, 1975. From (8.42) it follows that $\Delta w \in L^\infty(D_*)$ and hence we apply only the regularity results for solutions of equations and not of inequalities.)

From (8.50), through (8.48), one concludes that (8.25) is true and further that $w, w_x, w_y \in C^0(\bar{D}_*)$. The maximum principle in the Hopf form then gives us (for the details, see Baiocchi–Comincioli–Magenes–Pozzi, 1973):

$$w \geq 0; \quad w_x \leq 0; \quad w_y \leq 0 \quad \text{in } \bar{D}_*, \quad (8.51)$$

and of these expressions the first expresses the fact that it was in fact equivalent whether we solve problem 8.4 or problem 8.3, while the latter two imply that Ω defined by (8.49) is bounded above by a Lipschitz continuous curve with respect to the axes $X = x + y$, $Y = x - y$. For the further properties mentioned in theorem 8.1, see the references cited there.

8.3 QUASIVARIATIONAL PROBLEMS

As we will see in Part II, in studying an analogous problem in the case of a dam with a 'more complicated geometry', such as that shown in fig. 8.3, it is

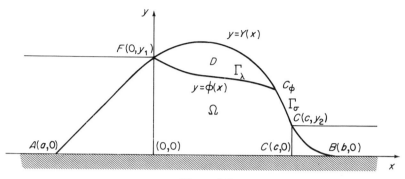

Figure 8.3

convenient to make the following change in the unknown:

$$z(x, y) = -\int_0^y p(x, t) \, dt, \qquad (8.52)$$

which is obviously related to the one introduced in (8.34) by the relations

$$w(x, y) = z(x, y) - z(x, Y(x)); \qquad z(x, y) = w(x, y) - w(x, 0). \qquad (8.53)$$

(Let us remark that fig. 8.1, e.g. with $D_* =]0, c[\times]0, y_1[$, is a particular case of fig. 8.3 in which $a = 0$, $b = c$, $Y(x) \equiv y_1$.)

The problem we will deal with has, as an equivalent of (8.41), an obstacle problem:

$$z \geq \psi \quad \text{in } D_*; \qquad \Delta z \leq 1 \quad \text{in } D_*; \qquad z > \psi \Rightarrow \Delta z = 1, \qquad (8.54)$$

where, however, the 'obstacle' ψ depends itself on the solution. It is, in fact:

$$\psi(x, y) = z(x, Y(x)). \qquad (8.55)$$

Problems of this type, i.e. in which the obstacle depends on the solution, are called *quasivariational inequalities* and will be treated in detail in Part II. Here we will merely observe that the fortunate circumstance which has made it possible for us to study problem 8.2 in a variational form (problem 8.3 or problem 8.4) is that we have been able to explicitly evaluate $z(x, Y(x))$ (which, from (8.53), is given by $w(x, y_*) - w(x, 0)$ and $w(x, y_*) \equiv 0$, while the values of $w(x, 0)$ are given by (8.40)). (A different technique for dealing with the problem with a general geometry (and in n variables), has been proposed in Alt, 1977a; this uses 'locally' the result of theorem 8.1 and the transformation (8.34), and subsequently a 'balayage' method.)

As we will see, the case shown in fig. 8.2 is an intermediate case—the values of $z(x, Y(x))$ are not known, but it is known that $z(x, Y(x))$ has the form $\lambda - qx$ with λ known and q unknown (see (8.40 *bis*) which follows); this suggests that we study this problem by expressing it as a family of variational inequalities with a real parameter q (rather than the functional parameter $z(x, Y(x))$.

This is the method followed in Baiocchi–Comincioli–Magenes–Pozzi, 1973 which we will outline briefly. Let us first formulate the analogues of problems 8.1 and 8.2:

PROBLEM 8.1 *bis*. Problem 8.1, but with (8.1), (8.3), and (8.7) substituted respectively by

$$c>0; \quad y_1>y_2\geq 0; \quad d\in\,]0, y_1[, \qquad (8.1\ bis)$$

$$\phi \text{ is non-decreasing}, \quad \phi(c)>y_2, \quad \phi(0)\in\,]d, y_1[, \qquad (8.3\ bis)$$

$$p(0, y) = y_1 - y \quad \text{for } 0\leq y\leq d; \quad p_x(0, y) = 0 \quad \text{for } d<y<\phi(0). \qquad (8.7\ bis)$$

PROBLEM 8.2 *bis*. Problem 8.2, but with (8.17), (8.22) substituted respectively by

$$p(0, y) = y_1 - y \quad \text{for } 0\leq y\leq d, \qquad (8.17\ bis)$$

$$\left. \begin{array}{l} \displaystyle\int_\Omega \text{grad}\,(p+y)\,\text{grad}\,\psi\,\mathrm{d}x\,\mathrm{d}y = 0 \quad \forall \psi \in C_0^1([0, c[\times\,]-\infty, +\infty[) \\ \text{with } \psi \text{ zero in a neighbourhood of } \{(0, y): 0\leq y\leq d\}. \end{array} \right\} \qquad (8.22\ bis)$$

We will now state the analogue of theorem 8.1:

THEOREM 8.1 *bis*. Problem 8.2 *bis* has one and only one solution. Further

$$\forall r \in [1, 4[\quad p \in W^{1,r}(D_*), \qquad (8.25\ bis)$$

$$p_x, p_y \notin L^\infty(D_*), \qquad (8.26\ bis)$$

$$\exists \bar{q} \in \mathbb{R} \int_0^\infty p_x(x, t)\,\mathrm{d}t = -\bar{q} \quad \text{a.e. in }\,]0, c[. \qquad (8.27\ bis)$$

Further, with ϕ defined as in (8.28), (8.3 *bis*), and (8.29) are true.

This theorem is much less 'precise' than theorem 8.1; nevertheless, though in some respects this lack of precision can probably be avoided (for example to prove (8.30) it should be possible to adapt the reasoning of Friedman–Jensen (1978)), in others the problem itself imposes a 'loss in precision'; the relation $p \in W^{1,r}(D_*)$ is probably false for $r>4$; nor do we know of an *a priori* formula for the explicit calculation of the carrying capacity $\bar{q}$ in terms of the parameters c, y_1, y_2, d of the type (8.27) which ensures the validity of the formula $\bar{q} = (y_1^2 - y_2^2)/2c$ for $d \geq y_1$).

THEOREM 8.2 *bis*. If p is a solution of problem 8.2 *bis*, and w is defined as in (8.34), then (8.35), (8.36), (8.37), (8.38) holds, and further there exists $\bar{q} \in \mathbb{R}$

such that

$$w(0, y) = \bar{q}c + \frac{y_2^2 + y^2}{2} - y_1 y \quad \text{for } y \in [0, d]$$

$$w_x(0, y) = 0 \quad \text{for } y \in \,]d, +\infty[\quad (8.39 \text{ bis})$$

$$w(x, 0) = \bar{q}(c - x) + \frac{y_2^2}{2}. \quad (8.40 \text{ bis})$$

Proof. The proof of theorem 8.2 holds unchanged up to the point where (8.38) is considered; from (8.33) and $w(c, 0) = y_2^2/2$ it follows that there exists $\bar{q} \in \mathbb{R}$ such that (8.40 bis) is true; from this (which gives $w(0, 0)$ in terms of $\bar{q}$) together with (8.17 bis), (8.39 bis) follows. ∎

With D_* fixed as in (8.24), we put

$$\Gamma_n = \{(0, y) : d < y < y_*\} \quad \text{and} \quad \Gamma_d = \partial D_* \setminus \Gamma_n, \quad (8.56)$$

and for every $q \in \mathbb{R}$ we defined the function $g_q(x, y)$ on Γ_d as

$$g_q(x, y_*) = 0; \quad g_q(x, 0) = q(x - c) + \frac{y_2^2}{2} \quad \text{for } 0 \leq x \leq c$$

$$g_q(c, y) = \int_y^{y_*} [y_2 - t]^+ \, dt; \quad g_q(0, y) = -\bar{q}c + \frac{y_2^2}{2} + y y_1 - \frac{y^2}{2} \quad \text{for } 0 \leq y \leq d$$

$$(8.43 \text{ bis})$$

and the set

$$K_q = \{v \in H^1(D_*) : v|_{\Gamma_d} = g_q\}. \quad (8.44 \text{ bis})$$

We will next consider the following problem:

PROBLEM 8.3 *bis.* Let $q \in \mathbb{R}$; with (8.56), (8.43 bis), (8.44 bis), find $w_q(x, y)$ such that

$$w_q \in K_q; \quad \forall v \in K_q \int_{D_*} \text{grad } w_q \text{ grad } (z - w_q) \, dx \, dy$$

$$+ \int_{D_*} [z]^+ \, dx \, dy \geq \int_{D_*} [w_q]^+ \, dx \, dy. \quad (8.45 \text{ bis})$$

To solve problem 8.3 *bis* we will have to study in particular the problem:

PROBLEM 8.5. Find $\bar{q} \in \mathbb{R}$ such that the corresponding solution $w_{\bar{q}}$ of problem 8.3 *bis* satisfies $w_{\bar{q}} \in H^2(D_*)$.

Let us remark that for every $q \in \mathbb{R}$ there is one and only one w_q which is a solution of problem 8.3 *bis*. So, denoting, for every $q \in \mathbb{R}$, by w_q the solution of problem 8.3 *bis*, the following results can be proved (for details, see Baiocchi–Comincioli–Magenes–Pozzi, 1973):

THEOREM 8.4. There exist $\alpha, \beta \in \mathbb{R}$, $F: \mathbb{R} \to \mathbb{R}$ continuous and bounded, such that

$q \in \mathbb{R}$ is such that $w_q \in H^2(D_*)$ if and only if $F(q) + \alpha q + \beta = 0$;

(8.57)

further $\alpha \neq 0$, and hence problem 8.5 has at least one solution.

THEOREM 8.5. Every $\bar{q}$ which is a solution of problem 8.5 is such that

$$w_{\bar{q}} \in W^{2,r}(\bar{D}) \qquad \text{for } r \in [1, 4[. \qquad (8.58)$$

THEOREM 8.6. There exists at most one $\bar{q} \in \mathbb{R}$ solution of problem 8.5.

Here we will give only a brief outline of the proofs of (8.57) and (8.58).

To prove (8.57) note that, as far as the solution of problem 8.3 *bis* is concerned, it can easily be shown that the application $q \mapsto w_q$ is continuous from $\mathbb{R}$ in $H^1(D_*)$; hence $q \mapsto \Delta w_q$ is continuous from $\mathbb{R}$ in $H^{-1}(D_*)$. Since further (in analogy with (8.42)) $0 \leq \Delta w_q \leq 1 \ \forall q \in \mathbb{R}$, we have:

the application $q \mapsto \Delta w_q$ is continuous from $\mathbb{R}$ in $L^2(D_*)$ provided with the weak topology; further it has a bounded image on $L^2(D_*)$. (8.59)

Let us now denote by X the space of traces on Γ_d of the elements of $H^2(D)$ (with the notation (8.56)). X turns out to be a Hilbert space with respect to the quotient norm (let us remark that one cannot work with $H^{3/2}(\Gamma_d)$ because of the corners of ∂D_*). Using the results of Peetre (see Peetre, 1961; 1963) regarding the mixed boundary value problems it follows that the following problem 8.6 is a *problem with index*:

PROBLEM 8.6. Fixed f, g with

$$f \in L^2(D_*), \qquad g \in X, \qquad (8.60)$$

find u such that

$$u \in H^2(D_*) \qquad (8.61)$$

$$\Delta u = f \quad \text{in } D_* \qquad (8.62)$$

$$u|_{\Gamma_d} = g; \qquad u_x|_{\Gamma_n} = 0. \qquad (8.63)$$

In general, the problem (8.62), (8.63) has one and only one variational solution (i.e. in $H^1(D_*)$); by the regularity results we have already referred to, this solution will have H^2 regularity in the neighbourhood of every point of $\bar{D}$ except at most the points $(0, d)$ and $(0, y_*)$. Because problem 8.6 is a problem with index, if g is zero in a set of the type $\{(x, y_*) \in \Gamma_d : 0 \leq x \leq \varepsilon\}$

for some $\varepsilon > 0$ (which corresponds to automatically satisfying the compatibility condition already mentioned in the neighbourhood of the point $(0, y_*))$, there exist h, l with

$$h \in L^2(D_*); \quad l \in X' \tag{8.64}$$

such that problem 8.6 has a solution iff

$$\int_{D_*} h(x, y) f(x, y) \, dx \, dy + {}_{X'}\langle l, g \rangle_X = 0. \tag{8.65}$$

The solution w_q of problem 8.36 satisfies (8.62), (8.63) with $f = \Delta w_q$, $g = g_q$; using (8.59) and the affinity of the application $q \mapsto g_q$ one obtains (8.57) with F given by

$$F(q) = \int_{D_*} h(x, y)(\Delta w_q)(x, y) \, dx \, dy \quad \text{and with} \quad \alpha q + \beta = \langle l, g_q \rangle.$$

Finally note that (8.58) is related to the fact that the problem (8.62) and (8.63) is an index problem even in the spaces $W^{2,r}(D_*)$ (see Shamir, 1968); while the uniqueness theorem 8.6 follows from the monotonic nature of the applications $q \mapsto w_q(x, y)$ and $q \mapsto w_q(x, 0) - w_q(x, y)$ (for further details see Baiocchi–Comincioli–Magenes–Pozzi, 1973). The theoretical procedure described here, though apparently very elaborate, is very amenable to the numerical treatment of the free boundary problem 8.1 *bis*; see Baiocchi–Comincioli–Guerri–Volpi, 1973 for a description of the numerical approach and Comincioli, 1975 and the relevant bibliography for a complete justification of this approach.

PART II

Quasivariational Problems

9
Fixed Point Theorems

9.1 INTRODUCTION

The different kinds of fixed point theorems

Many problems that can be treated in the framework of functional analysis, like differential problems, variational problems, integro-differential problems, and so on, can be reduced to the search for fixed points for transformations appropriately associated with the problems themselves. This justifies the interest in theorems which ensure that certain transformations have fixed points (theorems which are interpreted as theorems of existence of solutions...) and at the same time the great number of results of this kind.

Here we will develop only a small part of the theory of fixed points; namely that dealing with the following three types of problems:

(a) Fixed points for Lipschitz continuous operators in the framework of metric structures, both for the single-valued case (section 9.2.1) and for the multi-valued case (section 9.2.2). For extensions of this kind of questions, connected with the notion of α-contraction introduced by Darbo (cf. Darbo, 1955) see e.g., Martelli, 1975; Pejsachowicz, 1977; Furi–Martelli, 1972; 1974; Furi–Martelli–Vignoli, 1978 and their references.

(b) Fixed points for continuous applications in the framework of vector-topological structures, both in the single-valued case (section 9.3.2) and in the multi-valued case (section 9.3.3); here too our treatment will be very short: e.g. we will not deal with extensions connected with homological theory or questions connected with the theory of topological degree (cf. Eilenberg–Montgomery, 1946; Leray–Schauder, 1934). The literature in this field is vast; see e.g. Hamilton, 1947; Darbo, 1950; 1961; Martelli–Vignoli, 1947b and their references.

(c) Fixed points for monotone operators in the framework of ordered structures (section 9.4).

Concerning the proofs of results relative to case (b), we must point out that one can follow many alternative pathways, since the results are closely connected with each other. The pathway we follow does not coincide with the chronological order in which these results were obtained. For example, theorem 9.17, proved by Schauder in 1930 (cf. Schauder, 1930) and independently by Caccioppoli in 1931 (cf. Caccioppoli, 1931) has been obtained by us as a consequence of theorem 9.16 proved by Tychonov in 1935 (cf. Tychonov, 1935). For a clearer explanation of these connections see the table at the end of section 9.3.3.

We will develop the treatment of the single-valued and multi-valued cases in a parallel way. The treatment in the form of multi-valued applications brings together, in a certain sense, the whole treatment, but it requires a knowledge of concepts and notations which makes it very laborious; we have for this reason preferred to deal with these preliminaries in Chapter 20 and develop a double treatment in the present chapter.

Sections 9.2.2 and 9.3.3 (which are the generalizations of sections 9.2.1 and 9.3.2 to the case of multi-valued applications) can be skipped by the reader interested in the treatment of only single-valued applications: in fact, before reading sections 9.2.2 and 9.3.3 one is advised to read Chapter 20 in which we have put together certain fundamental notions and notations with respect to multi-valued applications.

For the use of fixed point theorems in fields other than that considered in this book (such as probability, numerical analysis, and so on), see, among many others, Istrățescu, 1973; Ortega–Rheinboldt, 1970; Swaminathan, 1976; Todd, 1976; Antosiewicz, 1977; Aubin, 1979b).

Fixed points and fixed point property

Let us introduce a few terms (restricting ourselves for the present to the single-valued case—for the multi-valued case see section 9.2.2) which will be used more or less frequently in the following.

DEFINITION 9.1. Let M be a non-empty set and $f: M \to M$. An element $\bar{x} \in M$ is said to be a *fixed point* for f if $f(\bar{x}) = \bar{x}$.

If card $M > 1$, in order to prove the *fixed point theorems*, i.e. to show that $f: M \to M$ has at least one fixed point, we need certain assumptions on M and on f. In this connection one often finds in the literature the expression 'fixed point property', relative to a class of applications and to a class of structures; we will use this expression rarely, but we introduce it nevertheless:

DEFINITION 9.2. Let $\mathscr{S}$ be a class of structures and $\mathscr{F}$ a class of applications of these structures to themselves. The pair $(\mathscr{S}, \mathscr{F})$ is said to have the *fixed point property* if

$$\forall S \in \mathscr{S} \quad \forall f: S \to S, f \in \mathscr{F} \quad \exists \bar{x} \in S: f(\bar{x}) = \bar{x}. \tag{9.1}$$

An example of a pair with the fixed point property is that of the class $\mathscr{S}$ of non-empty complete metric spaces and the class $\mathscr{F}$ of contractions of such spaces into themselves: the Banach theorem for contractions tells us that (9.1) holds.

9.2 FIXED POINT THEOREMS FOR LIPSCHITZ CONTINUOUS APPLICATIONS

9.2.1 SINGLE-VALUED APPLICATIONS: BANACH THEOREM

The Banach theorem for contractions

We will first report the result that we have just cited at the end of the previous section (cf. Banach, 1922, p. 160; see the proof in Part I of this book—theorem 1.3):

THEOREM 9.1 (Banach theorem for contractions). Let (S, d) be a non-empty complete metric space. If $f: S \to S$ is a contraction then there exists one and only one fixed point for f.

This is one of the few results regarding fixed points which ensures uniqueness—fixed point theorems are in fact usually results essentially of existential kind. Note in agreement with the latter statement that the hypothesis of the theorem which is of existence type—a hypothesis heavily used in the proof—is the completeness of the metric space (S, d).

Some corollaries of the Banach theorem

We will now deal with some results related to this theorem, while omitting many others, among which are all the results based on various generalizations of the concept of contraction (see the bibliography quoted at the beginning of this chapter, and in particular Istrăţescu, 1973, p. 147 *et seq.*).

The first result is an elementary corollary of the Banach theorem for contractions:

THEOREM 9.2. Let (S, d) be a complete non-empty metric space. If $f: S \to S$ is such that, for at least one $n \geq 1$, f^n is a contraction then there exists one and only one fixed point for f.

Proof. The uniqueness is obvious. As for the existence, let us note that if $\bar{x}$ is the unique fixed point for f^n (remark that f^n and (S, d) satisfy the conditions of the Banach theorem for contractions) then $f^n(f(\bar{x})) = f(f^n(\bar{x})) = f(\bar{x})$ and hence $f(\bar{x})$ is a fixed point for f^n: since this is unique we have $f(\bar{x}) = \bar{x}$. ∎

Note that Volterra's integral operator satisfies the conditions of the preceding theorem (see, e.g., Hille, 1972, p. 179). For generalizations of the theorem itself, see Roux–Soardi, 1971.

Another interesting result is the following:

THEOREM 9.3. Let (S, d) be a metric space and $f_1 : S \to S$, $f_2 : S \to S$ two applications such that $f_1 \circ f_2 = f_2 \circ f_1$ (i.e., f_1 and f_2 commutate). If f_2 has a unique fixed point $\bar{x}$ then $\bar{x}$ is a fixed point for f_1.

Proof. If $\bar{x} = f_2(\bar{x})$ then $f_1(\bar{x}) = f_1(f_2(\bar{x})) = f_2(f_1(\bar{x}))$ and hence $f_1(\bar{x})$ is a fixed point for f_2: since the latter is unique $\bar{x} = f_1(\bar{x})$. ∎

For other results regarding fixed points common to two or more applications see, among others, Istrățescu, 1973, p. 97 *et seq.*; Smart, 1974, p. 53 *et seq.*

A curious aspect of the two preceding results is that existence is a result of the composition of a hypothesis of existential kind (whose presence is natural) together with a hypothesis of uniqueness kind (which is less expected, but which is heavily utilized in the proof).

Non-expansive applications

Next we will treat briefly non-expansive applications. Here we do not expect to find uniqueness (think of the identity application...) and even existence is difficult: the simple metric structure is usually insufficient to give existence...:

THEOREM 9.4: Let H be a Hilbert space and M a bounded, closed, convex, non-empty subset of H. Every non-expansive application $f : M \to M$ has a fixed point.

For the proof see Browder, 1965a, p. 1274: theorem 1 (in this work, where we can find other results of this type, the author calls contractions those applications which we refer to here as non-expansive) and Smart, 1974, p. 36; cf. also Brézis, 1973, p. 5 for an extension to the case of uniformly convex Banach spaces and for a counter-example which shows that the result is not true for a general Banach space.

We will now present a result regarding *quasi-fixed points* in Banach spaces:

THEOREM 9.5. Let B be a Banach space and M a bounded, convex, closed, non-empty subset of B. If $f : M \to M$ is non-expansive then

$$\forall \varepsilon > 0 \quad \exists x(\varepsilon) \in M : \|f(x(\varepsilon)) - x(\varepsilon)\|_B \leq \varepsilon. \tag{9.2}$$

Proof. Let $R > 0$ be such that $M \subset B(0, R) = \{x \in B : \|x\|_B < R\}$ (M is bounded...) and suppose for convenience that $0 \in M$ (without loss of generality, since the space is linear); further let $r < 1$. The application $\phi = rf$ is a contraction of M in M, since M is a convex set which includes the origin and f is non-expansive. From the Banach theorem for contractions (M is closed...) it follows that ϕ has a unique fixed point: let this be $x(r)$. We have

$$\|f(x(r)) - x(r)\|_B = \|f(x(r)) - \phi(x(r))\|_B = (1-r)\|f(x(r))\|_B \leq (1-r)R, \quad (9.3)$$

and by choosing r such that $(1-r)R < \varepsilon$, for a previously fixed ε, we have proved the theorem. ∎

Let us remark that this theorem does not state that $x(\varepsilon)$ is 'near' a fixed point, but that $x(\varepsilon)$ is a 'quasi-fixed' point... (fixed points may not exist!).

Sequences and fixed points

We will now deal briefly with the problem of the convergence of sequences of fixed points associated with appropriate sequences of applications.

THEOREM 9.6. Let (S, d) be a non-empty metric space and $f_i : S \to S$ applications with at least one fixed point x_i, $i = 1, 2, \ldots$, and let $f_\infty : S \to S$ be a contraction with fixed point x_∞. If $f_i \to f_\infty$ uniformly then $x_i \to x_\infty$.

Proof. Let $\alpha_\infty < 1$ be a Lipschitz constant for f_∞; let $\varepsilon > 0$ be arbitrary and let N be an integer such that

$$\forall i \geq N \quad \forall x \in S \quad d(f_i(x), f_\infty(x)) < \varepsilon(1 - \alpha_\infty) \quad (9.4)$$

(such an N must exist since the sequence f_i converges uniformly to f_∞). Since x_i is a fixed point for f_i and x_∞ is a fixed point for f_∞ we can write, for each i,

$$d(x_i, x_\infty) = d(f_i(x_i), f_\infty(x_\infty)) \quad (9.5)$$

and, from the triangle inequality,

$$d(x_i, x_\infty) \leq d(f_i(x_i), f_\infty(x_i)) + d(f_\infty(x_i), f_\infty(x_\infty)). \quad (9.6)$$

Now, recalling (9.4) (with $x = x_i$) and that f_∞ is a contraction, we have, from (9.6),

$$d(x_i, x_\infty) < \varepsilon(1 - \alpha_\infty) + \alpha_\infty d(x_i, x_\infty) \quad (9.7)$$

and therefore

$$d(x_i, x_\infty) < \varepsilon, \quad (9.8)$$

which proves the theorem. ∎

If the f_i are also contractions and (S, d) is locally compact then the result remains true even when the convergence of the sequence $\{f_i\}$ to f_∞ is only a pointwise convergence (see Nadler, 1968, p. 580). See Nadler, 1968; Istrățescu, 1973, ch. 6 and Brézis, 1973 for other results of this type.

9.2.2 MULTI-VALUED APPLICATIONS

Some nomenclature

We will now deal with multi-valued applications. Definition 9.1 has to be replaced by the following:

DEFINITION 9.3. Let M be a non-empty set and $f: M \to 2^M_\blacksquare$. A point $\bar{x} \in M$ is said to be a fixed point for f if $\bar{x} \in f(\bar{x})$.

For the notation $2^M_\blacksquare$ and others of the same kind see section 20.0. For easier reading of this chapter, we will make precise cross-references wherever the notations, definitions or results of Chapter 20 appear in the text.

Definition 9.2 has an exact analogue here which we will not present.

An extension of the Banach theorem

Let us begin by proving the following theorem (see Nadler, 1969; p. 479).

THEOREM 9.7. Let (S, d) be a complete non-empty metric space. If $f: S \to 2^S_{\blacksquare \text{cl}}$ is a contraction (in the sense of definition 20.9) then it has a fixed point.

Proof. Let us note that if $A, B \in 2^S_{\blacksquare \text{cl}}$ and $a \in A$ then

$$\forall \eta > 0 \quad \exists b \in B: d(a, b) \leq H(A, B) + \eta \tag{9.9}$$

(H is the Hausdorff distance—see definition 20.8; let us remark that if B is compact then (9.9) is true also for $\eta = 0$.) Let $\alpha < 1$ be a Lipschitz constant for f and let p_0 be an arbitrary point of S and p_1 an arbitrary point of $f(p_0)$. From (9.9) it follows that there exists $p_2 \in f(p_1)$ such that

$$d(p_1, p_2) \leq H(f(p_0), f(p_1)) + \alpha; \tag{9.10}$$

further, from (9.9) it also follows that there exists $p_3 \in f(p_2)$ such that

$$d(p_2, p_3) \leq H(f(p_1), f(p_2)) + \alpha^2; \tag{9.11}$$

and so on. Thus, we can construct a sequence p_n of points of S such that, for $n \geq 1$, $p_n \in f(p_{n-1})$ and

$$d(p_n, p_{n+1}) \leq H(f(p_{n-1}), f(p_n)) + \alpha^n \tag{9.12}$$

We can now write

$$d(p_n, p_{n+1}) \leq H(f(p_{n-1}), f(p_n)) + \alpha^n \leq \alpha d(p_{n-1}, p_n) + \alpha^n$$
$$\leq \alpha[H(f(p_{n-2}), f(p_{n-1})) + \alpha^{n-1}] + \alpha^n \leq \ldots$$
$$\leq \alpha^n d(p_0, p_1) + n\alpha^n \tag{9.13}$$

and hence, for every $n, l \geq 1$,

$$d(p_n, p_{n+l}) \leq d(p_n, p_{n+1}) + \ldots + d(p_{n+l-1}, p_{n+l})$$
$$\leq \alpha^n d(p_0, p_1) + n\alpha^n + \ldots + \alpha^{n+l-1} d(p_0, p_1) + (n+l-1)\alpha^{n+l-1}$$
$$= d(p_0, p_1) \sum_{i=0}^{l-1} \alpha^{n+i} + \sum_{i=0}^{l-1} (n+i)\alpha^{n+i}. \tag{9.14}$$

We have shown then that p_n is a Cauchy sequence: let $\bar{p}$ be its limit in S. Since f is continuous (see theorem 20.5(i)) we can conclude also that $f(p_n)$ converges to $f(\bar{p})$, and since $p_n \in f(p_{n-1})$, for every n, it follows that $\bar{p} \in f(\bar{p})$. ∎

This theorem is a generalization of the Banach theorem for contractions since $x \mapsto \{x\}$ is an isometry from (S, d) to $(2^S_{\blacksquare cl}, H)$ (see the observations preceding theorem 20.4). Further generalizations are possible, both by elaborating the concept of contraction (see, e.g., Nadler, 1969, p. 480 et seq.; Istrățescu, 1973, p. 304 et seq.) and by elaborating the notion of the Hausdorff metric itself (see, e.g., Covitz–Nadler, 1970).

Keeping in mind that if (S, d) is complete then $(2^S_{\blacksquare q}, H)$ is also complete (see the observations following theorem 20.4) and that if $f: S \to 2^S_{\blacksquare q}$ is a contraction then so is $f_S: 2^S_{\blacksquare q} \to 2^S_{\blacksquare q}$ (see theorem 20.5(iv)), we can deduce, from the Banach theorem for contractions, that *if (S, d) is a complete non-empty metric space and $f: S \to 2^S_{\blacksquare q}$ is a contraction then f_S has one and only one fixed 'point' $A \subset S$*. One can prove further that the fixed points for f are elements of A.

Sequences and fixed points: multi-valued case

Finally, we state without proof (see Nadler, 1969, p. 484; Istrățescu, 1973, p. 307), the following result regarding sequences of contractions.

THEOREM 9.8. Let (S, d) be a complete non-empty metric space and let $f_i: S \to 2^S_{\blacksquare q}$ be contractions with Lipschitz constants $\alpha_i < 1$ and fixed points x_i. If $f_\infty: S \to 2^S_{\blacksquare q}$ is a contraction and if one of the following conditions is satisfied:

(i) The constants α_i are equal to $\alpha < 1$ and $f_i \to f_\infty$ pointwise,
(ii) $f_i \to f_\infty$ uniformly,
(iii) (S, d) is locally compact and $f_i \to f_\infty$ pointwise,

then the sequence x_i has a subsequence which converges to a fixed point of f_∞.

9.3 FIXED POINT THEOREMS FOR CONTINUOUS APPLICATIONS

9.3.1 THE KNASTER–KURATOWSKI–MAZURKIEWICZ LEMMA AND THE FAN LEMMA

The main aim of this section is to present two important results which, for historical reasons, are called lemmas: the Knaster–Kuratowski–

Mazurkiewicz lemma (KKM for short; see Knaster–Kuratowski–Mazurkiewicz, 1929, p. 134) and the Fan lemma (see Fan, 1961, p. 305). These are two existence results which are expressed through the statement that a given intersection is non-empty: there are other mathematical results of this type and we will give three examples here.

Three results regarding intersections

We will start with a well-known theorem from general topology. Recall that a family $\mathcal{A}$ of non-empty sets has the *finite intersection property* if every finite non-empty subfamily of $\mathcal{A}$ has a non-empty intersection (obviously this does not imply that $\bigcap \mathcal{A} \neq \emptyset$: it is enough to consider the family $]0, \varepsilon[$ ($\varepsilon > 0$) of subsets of $\mathbb{R}$).

THEOREM 9.9. Let E be a compact topological space and $\mathcal{F}$ a family of closed subsets of E. If $\mathcal{F}$ has the finite intersection property then $\bigcap_{F \in \mathcal{F}} F \neq \emptyset$.

The converse is also true, so that compact topological spaces can be characterized as those in which every family of closed sets with the finite intersection property has a non-empty intersection; for the proof of theorem 9.9 and of the converse see, e.g., Kelley, 1962, p. 159.

We will now state again without proof (see e.g., Stoer–Witzgall, 1970, p. 117; Eggleston, 1963, p. 33), a result much used in convex analysis in finite dimensions:

THEOREM 9.10 (Helly's theorem). Let $K_i, i = 1, \ldots, m$, $m \geq n+1$, be m convex sets in $\mathbb{R}^n$. If every $(n+1)$-ple of these has a non-empty intersection then $\bigcap_{i=1}^{m} K_i \neq \emptyset$.

A closely related result, of which we give the easy proof, is the following theorem:

THEOREM 9.11. Let $\mathcal{K}$ be an arbitrary family of compact convex sets in $\mathbb{R}^n$. If any set of $(n+1)$ elements of $\mathcal{K}$ have a non-empty intersection then $\bigcap \mathcal{K} \neq \emptyset$.

Proof. Let $\bar{K} \in \mathcal{K}$ and let us consider the family $\{\bar{K} \cap K : K \in \mathcal{K}\}$ of subsets of $\bar{K}$. Since $\bar{K}$ is a compact space we can say (from theorem 9.9) that every family of closed sets with the finite intersection property has a non-empty intersection. Thus, if $\bigcap_{K \in \mathcal{K}} K = \emptyset$ then there exists a finite sub-family with an empty intersection: let m be the cardinal of this family. Given Helly's theorem, this is impossible, hence the result must be true. ∎

Sperner's lemma

In the proof of the KKM lemma we will use a result of combinatorial type: Sperner's lemma (cf. Sperner, 1928, p. 267; see also Scorza–Dragoni, 1978). We will not prove this result here, but we will introduce some nomenclature which will enable us to understand what it means:

Let us begin by recalling that $m+1$ points $x_0, x_1, \ldots, x_m$ in $\mathbb{R}^n$ are said to be *affinely independent* if the m points $x_1 - x_0, \ldots, x_m - x_0$ are linearly independent ($m > 0$; the singleton $\{x_0\}$ is said to be affinely independent). If $x_0, x_1, \ldots, x_m$ are $m+1$ affinely independent points of $\mathbb{R}^n$ ($0 \leq m \leq n$) the set $S = \text{simp}\{x_0, \ldots, x_m\}$ of all the convex combinations of the points x_i is said to be an *m-simplex* or a *simplex of dimension m*. The points x_i are called *vertices* of S and the sets $\text{simp}\{x_{i_0}, \ldots, x_{i_k}\}$ ($0 \leq k \leq m \leq n$) are called *faces* of S (or rather faces of dimension k of S; any vertex x_i being thus a face of dimension 0).

If x is a point of an m-simplex $S = \text{simp}\{x_0, \ldots, x_m\}$ of $\mathbb{R}^n$ we can uniquely determine $m+1$ real numbers $\lambda_0, \ldots, \lambda_m$ such that $\lambda_i \geq 0$ ($i = 0, \ldots, m$), $\sum_{i=0}^{m} \lambda_i = 1$ and $x = \sum_{i=0}^{m} \lambda_i x_i$: such numbers are called the *baricentric coordinates* of x.

Let now S be an m-simplex of $\mathbb{R}^n$. A family Σ of simplexes contained in S is said to be a *simplex subdivision* of S if the following conditions are satisfied:

(σ_1) S is the union of the m-simplexes of Σ,
(σ_2) the intersection of two simplexes of Σ is a face common to these simplexes,
(σ_3) the faces of the simplexes of Σ are simplexes of Σ.

The faces of dimension 0 of Σ, which are of course the vertices of the simplexes of Σ, are called *nodes* of the simplex subdivision. On the other hand the maximum value of the diameters of the simplexes of dimension 1 of Σ in the usual metric of $\mathbb{R}^n$ is called the *step-length* of the subdivision. It can be proved that given any number $\varepsilon > 0$ there exists a simplex subdivision of S with step-length less than ε.

Again let $S = \text{simp}\{x_0, \ldots, x_m\}$ be an m-simplex of $\mathbb{R}^n$, let Σ be a simplex subdivision of S, and N the set of nodes of Σ. Σ is said to be *labelled* if there is an application $e: N \to \{0, \ldots, m\}$, called the *label* or Σ, which associates with every $z \in N$ a number $e(z) \in \{0, \ldots, m\}$, called the label of z, in such a way that the following condition is satisfied:

(ε) $z \in \text{simp}\{x_{i_0}, \ldots, x_{i_k}\} \Rightarrow e(z) \in \{i_0, \ldots, i_k\}$

(remark in particular that $e(\{x_j\}) = j$). If $T = \text{simp}\{z_0, \ldots, z_j\}$, $j \geq 1$, belongs to Σ we say that $e(T) = \{e(z_0), \ldots, e(z_j)\}$ is the label of T.

Finally, Sperner's lemma states that given any m-simplex $S = \text{simp}\{x_0, \ldots, x_m\}$ of $\mathbb{R}^n$, any simplex subdivision Σ of S, and any label e of Σ, then the number of m-simplexes with the label $\{0, \ldots, m\}$ is odd.

The KKM lemma

We will next deal with the results mentioned in the beginning of this section:

THEOREM 9.12 (KKM lemma). Let $S = \text{simp}\{x_0, \ldots, x_m\}$ be an m-simplex of $\mathbb{R}^m$ and let F_i, $i = 0, \ldots, m$, be $m+1$ closed subsets of $\mathbb{R}^m$ such that

(i) $\forall I \subset \{0, \ldots, m\}$ $\text{simp}\{x_i\}_{i \in I} \subset \bigcup_{i \in I} F_i$.

Then $\bigcap_{i=0}^m F_i \neq \emptyset$.

Proof. As we know, for every $k \geq 1$ there exists a simplex subdivision Σ_k of S such that $\text{diam}(\Sigma_k) \leq 1/k$. We will now proceed to show with the help of the Sperner lemma that for every k there exists an m-simplex of Σ_k which has a vertex in every F_i: the result will then follow by taking the limit with respect to k. Let us first label Σ_k in such a way that 'node $\in F_{e(\text{node})}$', where e is the label of Σ_k. Let then $v^{(k)}$ be a node of Σ_k and $S' = \text{simp}\{x_{i_0}, \ldots, x_{i_s}\}$ the face of S with the minimum dimension that contains $v^{(k)}$. From (i) it follows that $S' \subset \bigcup_{j=0}^s F_{i_j}$, and therefore that $v^{(k)} \in F_{i_j}$ for some i_j: let us call this $\bar{i}_j$. Putting $e(v^{(k)}) = \bar{i}_j$ we obtain a label with the required conditions. From the Sperner lemma it follows then that there exists in Σ_k an m-simplex $\text{simp}\{v_0^{(k)}, \ldots, v_m^{(k)}\}$ with the label $\{0, \ldots, m\}$. We can assume also, without loss of generality, that $e(v_i^{(k)}) = i$ and therefore that $v_i^{(k)} \in F_i$. Let k now tend to infinity. Since S is a compact set, we can extract from the sequence $v_i^{(k)}$ of points of F_i a subsequence which converges to a point $\bar{v}_i \in F_i$ (remember that F_i is closed). Further, $\text{diam}(\Sigma_k) \to 0$ as $k \to +\infty$ and therefore the vertices $v_i(k)$ converge to a single point $\bar{v}_0 = \bar{v}_1 = \ldots = \bar{v}_m \in \bigcap_{i=0}^m F_i$ ∎

The Fan lemma

THEOREM 9.13 (Fan Lemma). Let Y be a Hausdorff topological vector space, let X be a non-empty subset of Y and let $F: X \to 2^Y_{\blacksquare c}$ such that

(i') $\forall \{x_1, \ldots, x_n\} \subset X \text{ conv}\{x_1, \ldots, x_n\} \subset \bigcup_{i=1}^n F(x_i)$,

(ii') $\exists \bar{x} \in X: F(\bar{x})$ is compact.

Then $\bigcap_{x \in X} F(x) \neq \emptyset$.

Proof. Let us remark that the KKM lemma is a particular case of this result: we need only put $Y = \mathbb{R}^m$, $X = \{x_0, \ldots, x_m\}$ and denote by F the application $x_i \mapsto F_i$ (we do not need condition (ii') in the case of the KKM lemma since in that case we essentially work in S). Let us remark also that condition (i') implies that $\forall x \in X$ $x \in F(x)$ (the identity is therefore a selection for F according to definition 20.7) and that $X \subset \bigcup_{x \in X} F(x)$ (condition (i) of theorem 9.12 tells us in an analogous way that $x_i \in F_i$ and $S \subset \bigcup_{i=0}^m F_i \ldots$).

The basic idea of the proof is to make use of (ii') so as to deal with a compact set, make use of theorem 9.9 so as to deal with the finite case, and then to use the KKM lemma. From (ii') and from theorem 9.9, all we need to do to prove the theorem is to show that, for every $\{x_i, \ldots, x_m\} \subset X$, we have

$$\bigcap_{i=1}^{m} (F(\bar{x}) \cap F(x_i)) = F(\bar{x}) \cap \left[\bigcap_{i=1}^{m} F(x_i)\right] \neq \varnothing. \tag{9.15}$$

Now, since the x_i are not in general affinely independent, we cannot directly apply the KKM lemma to prove (9.15): we need then to go about it indirectly. Let us then put $\bar{x} = x_0$ and consider on the one hand the set $\{x_0, \ldots, x_m\}$ and on the other the m-simplex $T = \text{simp}\{e_0, e_1, \ldots, e_m\}$ where e_0 is the origin and $\{e_1, \ldots, e_m\}$ is the canonic base of $\mathbb{R}^m$. Let us also consider the application $\phi: T \to Y$ defined by

$$\phi\left(\sum_{i=0}^{m} \alpha_i e_i\right) = \sum_{i=0}^{m} \alpha_i x_i \quad \left(\alpha_i \geq 0, \sum_{i=0}^{m} \alpha_i = 1\right). \tag{9.16}$$

The application ϕ is in general not injective since the x_i are not necessarily affinely independent and therefore the dimension of the convex set $\text{conv}\{x_0, \ldots, x_m\}$ can be less than m; but ϕ is continuous because the α_i are continuous functions of x_i. Let us now put $G_i = \phi^{-1}(F(x_i))$ and show that $\bigcap_{i=0}^{m} G_i \neq \varnothing$; (9.15) will follow from this (since for $i = 0, \ldots, m$, we have $\phi(G_i) = \phi(\phi^{-1}(F(x_i))) \subset F(x_i)$), and this proves the theorem.

To see that $\bigcap_{i=0}^{m} G_i \neq \varnothing$ we need only note that the G_i satisfy the hypothesis of the KKM lemma: in fact they are closed, since they are inverse images of the closed sets $F(x_i)$ by the continuous function ϕ, and, further, if $I \subset \{0, \ldots, m\}$ then $\text{simp}\{e_i\}_{i \in I} \subset \bigcup_{i \in I} G_i$ (condition (i)) since if $z = \sum_{i \in I} \lambda_i e_i \in \text{simp}\{e_i\}_I$ then $z \in \phi^{-1}(\phi(z)) = \phi^{-1}(\sum_{i \in I} \lambda_i x_i)$ and from (i') it follows that $\phi^{-1}(\sum_{i \in I} \lambda_i x_i) \subset \phi^{-1}(\bigcup_{i \in I} F(x_i)) = \bigcup_{i \in I} \phi^{-1}(F(x_i)) = \bigcup_{i \in I} G_i$. ∎

As we have already seen, the Fan lemma is a generalization of the KKM lemma, a further generalization is the following:

THEOREM 9.14. Let Y be a Hausdorff topological vector space, let X be a non-empty subset of Y and let $F: X \to 2^Y$ (let us remark that $F(x)$ need not be closed!) such that

(i'') $\forall \{x_1, \ldots, x_n\} \subset X \text{ conv}\{x_1, \ldots, x_n\} \subset \bigcup_{i=1}^{n} F(x_i)$,

(ii'') $\exists \bar{x} \in X : F(\bar{x})$ is compact,

(iii'') if L is a subspace of Y of finite dimension then $\forall x \in X \; F(x) \cap L$ is closed,

(iv'') if D is a convex subset of Y then

$$\left(\overline{\bigcap_{x \in X \cap D} F(x)}\right) \cap D = \left(\bigcap_{x \in X \cap D} F(x)\right) \cap D.$$

Then $\bigcap_{x \in X} F(x) \neq \varnothing$.

This result is in effect a generalization of the Fan lemma since if $\forall x \in X$ $F(x)$ is closed then (iii″) and (iv″) are automatically satisfied. For the proof, which uses the KKM lemma but not the Fan lemma, see Brézis–Nirenberg–Stampacchia, 1972, p. 295.

9.3.2 Singled-valued applications: the theorems of Brouwer, Schauder and Tychonov

The Brouwer theorem

The most famous among the fixed point theorems is undoubtedly that of Brouwer (see Brouwer, 1912, p. 115), several proofs of which are known. For example, Istrăţescu, 1973, p. 196 and Dunford–Schwartz, 1958, p. 468 gives a proof using techniques of classical analysis, Cronin, 1964, p. 52 and Miranda, 1949, p. 136 gives a proof using the theory of the topological degree (with this technique one works with the function $g(x) = x - f(x)$, and we observe that in order to prove the Brouwer theorem in the one-dimensional case one applies to this function the Bolzano theorem on the zeros of continuous functions); Boothby, 1971 gives a proof using techniques related to the theory of differential forms. In the latter paper, together with Brouwer's theorem, there is a proof of a result which, like it, has a simple statement but an elaborate proof: *a continuous vector field, tangent to a sphere (= boundary of a ball) of even dimension, takes the value zero in at least one point*; the relationship between these results is not surprising since in algebraic topology they occur together naturally (see e.g. Bourgin, 1963, p. 129 *et seq.*; see also Milnor, 1978; Franklin, 1980; Rogers, 1980, Kannai, 1981).

Here we use the KKM lemma, hence the Sperner lemma: thus we use combinatorial methods. The method followed here seems to us to be the quickest one for proving the theorem; however, this is true only if we already know something about combinatorial methods.

We will now prove Brouwer's theorem, a result which will be followed by one of its extensions which is Tychonov's theorem (see Tychonoff, 1935, p. 770) and then by Schauder's theorem (see Schauder, 1930, p. 173), a result which is 'intermediate' between the two previous results.

THEOREM 9.15 (Brouwer's theorem). Let K be a non-empty convex compact set in $\mathbb{R}^n$. If $f: K \to K$ is a continuous function then f has at least one fixed point, i.e. $\exists \bar{x} \in K : f(\bar{x}) = \bar{x}$.

Proof. This result follows from the 'addition' of the following three lemmas:

LEMMA 1. The fixed point property is preserved by retraction, i.e.: if every continuous function $f: K \to K$ has a fixed point and $\tilde{K}$ is a retraction of K then every continuous function $\phi: \tilde{K} \to \tilde{K}$ has a fixed point.

Proof. Let us remember that a set X is said to be a *retraction* of the set $Y \supset X$ if there exists a continuous application $r: Y \to X$ such that $r(x) = x$ $\forall x \in X$ (i.e.: $r|_X \equiv$ identity) (let us remark that there is a whole retraction theory, which is interesting in that it 'relates' general topology and combinational topology: see e.g., Hu, 1965). Let $r: K \to \tilde{K}$ be a retraction and let us consider the application $i_{\tilde{K}K} \circ \phi \circ r: K \to K$ (see the diagram) which (by assumption) has a fixed point: let $\bar{x}$ be such a point. We have then that

$$\begin{array}{ccc} K & \xrightarrow{r} & \tilde{K} \\ {\scriptstyle i_{\tilde{K}K} \circ \phi \circ r} \downarrow & & \downarrow {\scriptstyle \phi} \\ K & \xleftarrow{i_{\tilde{K}K}} & \tilde{K} \end{array}$$

$\bar{x} = (i_{\tilde{K}K} \circ \phi \circ r)(\bar{x})$, and since $i_{\tilde{K}K}$ is the set inclusion and $r|_{\tilde{K}}$ the identity, we can say that $\phi(\bar{x}) = \bar{x}$. ∎

A result analogous to lemma 1 is the following, which, given its importance, we will prove: *the fixed point property is invariant with respect to homeomorphisms, i.e.: if every continuous function $f: K \to K$ has a fixed point and $\tilde{K}$ is homeomorphic to K then every continuous function $\phi: \tilde{K} \to \tilde{K}$ has a fixed point.* In effect, let $h = K \to \tilde{K}$ be a homeomorphism and consider the application $h^{-1} \circ \phi \circ h: K \to K$ (see the diagram), which by hypothesis has a

$$\begin{array}{ccc} K & \xrightarrow{h} & \tilde{K} \\ {\scriptstyle h^{-1} \circ \phi \circ h} \downarrow & & \downarrow {\scriptstyle \phi} \\ K & \xleftarrow{h^{-1}} & \tilde{K} \end{array}$$

fixed point: let $\bar{x}$ be such a point. Then $\bar{x} = (h^{-1} \circ \phi \circ h)(\bar{x})$ and therefore $h(\bar{x}) = (\phi \circ h)(\bar{x})$, from which it follows that $h(\bar{x})$ is a fixed point for ϕ, and this proves the result.

LEMMA 2. *Let S be an m-simplex of $\mathbb{R}^m$. If $K \subset S$ is a closed convex set then K is a retraction of S.*

Proof. It is enough to consider $r: S \to K$ defined, for every $x \in S$, by $r(x) = P_K(x)$, where P_K is the usual projection operator on K. ∎

LEMMA 3 (Brouwer's theorem for simplexes). *Let $S = \text{simp}\{x_0, \ldots, x_m\}$ be an m-simplex on $\mathbb{R}^m$. If $f: S \to S$ is continuous then it has at least one fixed point.*

Proof. This lemma in fact historically precedes Brouwer's work, and is due to Bohl (see Bohl, 1904). The main idea of the proof is very simple: we find that f has at least one fixed point in S by observing how f 'moves' the various points of S. To do this we represent the points of S in terms of their baricentric coordinates. If $x \in S$ then we can write $x = \sum_{i=0}^{m} \lambda_i x_i$, with $\lambda_i \geq 0$ and $\sum_{i=0}^{m} \lambda_i = 1$: the $\lambda_i = \lambda_i(x)$ are the baricentric coordinates of x, which are, as can be easily seen, continuous functions of x. Let us now put $F_i = \{x \in S : \lambda_i(f(x)) \leq \lambda_i(x)\}$, $i = 0, \ldots, m$, and show that $\bigcap_{i=0}^{m} F_i \neq \emptyset$, from which we can see that there exists a point $\bar{x} \in S$ whose baricentric coordinates are not changed by f and which is therefore fixed. In fact if $\bar{x} \in \bigcap_{i=0}^{m} F_i$ then $\lambda_i(f(\bar{x})) \leq \lambda_i(\bar{x})$, $i = 0, \ldots, m$, and since $\sum_{i=0}^{m} \lambda_i(f(\bar{x})) = \sum_{i=0}^{m} \lambda_i(\bar{x}) = 1$ we have $\lambda_i(f(\bar{x})) = \lambda_i(\bar{x})$, $i = 0, \ldots, m$. To show that $\bigcap_{i=0}^{m} F_i \neq \emptyset$ we use the KKM lemma. It follows immediately that the F_i are closed since λ_i and f are continuous functions; to see that $\text{simp} \{x_i\}_{i \in I} \subset \bigcup_{i \in I} F_i$ whatever $I \subset \{0, \ldots, m\}$ we need only recall that if $y \in \text{simp} \{x_i\}_{i \in I}$ then $\sum_{i \in I} \lambda_i(y) = 1$ and that if, by contradiction, $y \notin \bigcup_{i \in I} F_i$ then $\lambda_i(f(y)) > \lambda_i(y)$, $i = 0, \ldots, m$, and therefore $1 = \sum_{i \in I} \lambda_i(f(y)) > \sum_{i \in I} \lambda_i(y) = 1$. ∎

If any one of the assumptions of the theorem is removed, the theorem may no longer be true (which does not mean, however, that they are necessary...). Let us demonstrate this by means of the three counter-examples which follow: (1) if K is *not convex*: we can take e.g. $K = \{x \in \mathbb{R}^2 : \frac{1}{2} \leq \|x\| \leq 1\}$ and f = rotation; (2) if K is *not compact*: we can take e.g., $K = \mathbb{R}$ and $f(x) = x + 1$ (K is not bounded) or $K =]-1, 1[$ and $f(x) = \frac{1}{2}(x+1)$ (K is not closed); (3) if f is *not continuous*: we can take e.g. $K = [-1, 1]$ and $f(x) = x/2$ if $x \neq 0$ and $f(0) = 1$. Under further (quite restrictive) assumptions on f, one can ensure that there is a unique fixed point; in this connection see Kellogg, 1976.

If we do not wish to use the concept of retraction in the proof of the theorem (and therefore lemmas 1 and 2), we could substitute lemma 1 with the result cited above (before lemma 2) and lemma 2 with the result which states that *two compact convex sets of $\mathbb{R}^m$ with the same dimension are homeomorphic* (see, e.g., Stoer–Witzgall, 1970, p. 124; Warga, 1972; p. 144).

It is worth mentioning also, with respect to the concept of retraction, that Brouwer's theorem is equivalent to the fact that $\partial B(0, 1)$ is not a retraction of $B(0, 1) = \{x \in \mathbb{R}^m : \|x\| \leq 1\}$. In fact, were $\partial B(0, 1)$ a retraction of $B(0, 1)$ it would follow (from lemma 1 and Brouwer's theorem) that $\partial B(0, 1)$ has the fixed point property, which is not true since the application $x \mapsto -x$ from $\partial B(0, 1)$ to $\partial B(0, 1)$ (*antipodal application*) does not have a fixed point. Conversely, if there exists $f : B(0, 1) \to B(0, 1)$ continuous and without fixed point then the application h which with every $x \in B(0, 1)$ associates $h(x) \in \partial B(0, 1)$ determined by the intersection of $\overrightarrow{xf(x)}$ with $\partial B(0, 1)$ would be a retraction of $B(0, 1)$ in $\partial B(0, 1)$ (let us remark that, since $x \neq f(x)$, the line with origin in x and passing through $f(x)$ is always well defined). Further, let us remark that Brouwer's theorem is equivalent to the fact that $B(0,1)$ is not homeomorphic in $\partial B(0,1)$. Finally Brouwer's theorem is also equivalent to the KKM lemma!

Tychonov's theorem

We will next consider a very important generalization of Brouwer's theorem:

THEOREM 9.16 (Tychonov's theorem). Let E be a locally convex Hausdorff real topological vector space and let K be a non-empty compact convex set in E. If $f: K \to K$ is a continuous function then f has at least one fixed point.

In order to prove this we will use the following lemma (due to Fan; see Fan, 1961, p. 309, lemma 4):

LEMMA. Let E be a Hausdorff topological vector space, let X be a compact convex set in E, and let $A \subset X \times X$ be such that

(j) A is closed,
(jj) $\forall x \in X \ (x, x) \in A$ (i.e., $\Delta_X \subset A$),
(jjj) $\forall y \in X \ \{x \in X : (x, y) \notin A\}$ is convex (possibly empty).

Then $\exists y_0 \in X \ X \times \{y_0\} \subset A$.

Proof. Let us first observe that (jjj) tells us that the intersection of the 'horizontal segments' with $X \times X \setminus A$ is convex, or these 'horizontal segments' are contained in A (the set A of fig. 9.1 satisfies (jjj), whereas that of fig. 9.2 does not). The result itself states in fact that there exists at least one 'horizontal segment' contained in A.

Let us put, for $x \in X$, $F(x) = \{y : (x, y) \in A\}$—the set $F(x)$ is the intersection with A of the 'vertical segment' passing through x. If we can show that $\bigcap_{x \in X} F(x) \neq \varnothing$ then we have proved the lemma: in fact $y_0 \in \bigcap_{x \in X} F(x)$ satisfies the thesis since $\forall x \in X : (x, y_0) \in A$, i.e. $X \times \{y_0\} \subset A$. To show that

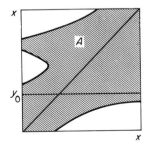

Figure 9.1

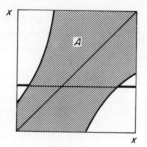

Figure 9.2

$\bigcap_{x \in X} F(x) \neq \emptyset$ we use Fan's lemma. Thus we have to check that the hypotheses of Fan's lemma are satisfied. From (jj) we have that $F(x)$ is non-empty. Besides $F(x)$ is closed, and even compact, since $F(x) = A \cap (\{x\} \times X)$ and A is compact (since it is a closed subset of a compact set, $X \times X$, in a Hausdorff space, $E \times E$) and $\{x\}$ is closed (since E is separate and so therefore is T_1). We will now show that $\forall \{x_1, \ldots, x_m\} \subset X$ conv $\{x_1, \ldots, x_m\} \subset \bigcup_{i=1}^m F(x_i)$. By *reductio ad adsurdum*: let us suppose that $\exists \{x_1, \ldots, x_m\} \subset X \ \exists \bar{x} \in \text{conv}\{x_1, \ldots, x_m\} : \bar{x} \notin \bigcup_{i=1}^m F(x_i)$. Then $\bar{x} \notin F(x_i)$, $i = 1, \ldots, m$, and thus, from the definition of $F(x_i)$, it follows that $(x_i, \bar{x}) \notin A$, $i = 1, \ldots, m$; from this together with (jjj) we have that the $(x_i, \bar{x})$ are in a convex set; this convex set also contains $(\bar{x}, \bar{x})$ (since $\bar{x} \in \text{conv}\{x_1, \ldots, x_m\}$): the point $(\bar{x}, \bar{x})$ does not therefore belong to A, and this contradicts (jj). ∎

Proof of Tychonov's theorem. Let $\{p_\lambda\}_{\lambda \in \Lambda}$ be a separate family of continuous seminorms which defines the locally convex topology of E and let us put, for $\lambda \in \Lambda$,

$$E_\lambda = \{y \in K : p_\lambda(y - f(y)) = 0\}. \tag{9.17}$$

To prove the theorem, all we need to do is to show that $\bigcap_{\lambda \in \Lambda} E_\lambda \neq \emptyset$ (see the proof of Brouwer's theorem for simplexes). In fact, if $\bar{y} \in \bigcap_{\lambda \in \Lambda} E_\lambda$ then $\forall \lambda \in \Lambda \ p_\lambda(\bar{y} - f(\bar{y})) = 0$ and therefore (since the family p_λ is separate) $\bar{y} - f(\bar{y}) = 0$: $\bar{y}$ is therefore a fixed point for f. From the continuity of the seminorms p_λ, it follows that the E_λ are closed subsets of the compact set K, and hence in order to prove that $\bigcap_{\lambda \in \Lambda} E_\lambda \neq \emptyset$ we need only show that $\{E_\lambda\}_{\lambda \in \Lambda}$ has the finite intersection property. Let then $\{\lambda_1, \ldots, \lambda_n\} \subset \Lambda$ and let us show that $\bigcap_{i=1}^m E_{\lambda_i} \neq \emptyset$, i.e. that $\bigcap_{i=1}^m \{y \in K : p_{\lambda_i}(y - f(y)) = 0\} \neq \emptyset$, or, in other words, that $\{y \in K : \sum_{i=1}^m p_{\lambda_i}(y - f(y)) = 0\} \neq \emptyset$. Let us then put now

$$A = \left\{(x, y) \in K \times K : \sum_{i=1}^m p_{\lambda_i}(x - f(y)) \geq \sum_{i=1}^m p_{\lambda_i}(y - f(y))\right\} \tag{9.18}$$

and let us show that A satisfies the assumptions of the lemma, which will prove the result since if $\exists y_0 \ \forall x \in K \ (x, y_0) \in A$, i.e. if $\sum_{i=1}^m p_{\lambda_i}(x - f(y_0)) \geq \sum_{i=1}^m p_{\lambda_i}(y_0 - f(y_0))$ for every x in K, then (putting $x = f(y_0)$) $\sum_{i=1}^m p_{\lambda_i}(f(y_0) -$

$f(y_0)) = 0 \geq \sum_{i=1}^{m} p_{\lambda_i}(y_0 - f(y_0))$ and hence $y_0 \in \bigcap_{i=1}^{m} E_{\lambda_i}$. It is obvious that $(x, x) \in A$ and that, from the continuity of f and of the p_λ, A is closed: (j) and (jj) are therefore satisfied. As for (jjj), let us fix $y \in K$ and show that the set $\{x \in K : (x, y) \notin A\} = \{x \in K : \sum_{i=1}^{m} p_{\lambda_i}(x - f(y)) < \sum_{i=1}^{m} p_{\lambda_i}(y - f(y))\}$ is convex, i.e., that

$$\sum_{i=1}^{m} p_{\lambda_i}(x_k - f(y)) < \sum_{i=1}^{m} p_{\lambda_i}(y - f(y)), \qquad k = 1, 2, \tag{9.19}$$

imply that

$$\sum_{i=1}^{m} p_{\lambda_i}(\theta x_1 + (1 - \theta)x_2 - f(y)) < \sum_{i=1}^{m} p_{\lambda_i}(y - f(y)) \tag{9.20}$$

for every $\theta \in [0, 1]$. This follows immediately:

$$\sum_{i=1}^{m} p_{\lambda_i}(\theta x_1 + (1 - \theta)x_2 - f(y)) = \sum_{i=1}^{m} p_{\lambda_i}(\theta x_1 + (1 - \theta)x_2 - \theta f(y) - (1 - \theta)f(y))$$

$$\leq \sum_{i=1}^{m} \theta p_{\lambda_i}(x_1 - f(y)) + \sum_{i=1}^{m} (1 - \theta)p_{\lambda_i}(x_2 - f(y))$$

$$< \text{(from (9.19))} \sum_{i=1}^{m} \theta p_{\lambda_i}(y - f(y))$$

$$+ \sum_{i=1}^{m} (1 - \theta)p_{\lambda_i}(y - f(y))$$

$$= \sum_{i=1}^{m} p_{\lambda_i}(y - f(y)). \tag{9.21}$$

We have then proved Tychonov's theorem. ∎

For other proofs of this result see Tychonov, 1935; Istrăţescu, 1973, p. 166; Dunford–Schwartz, 1958, p. 456.

Schauder's theorem

An immediate corollary of Tychonov's theorem is the well-known

THEOREM 9.17 (Schauder's theorem). Let B be a normed space and let K be a non-empty convex compact set of B. If $f : K \to K$ is a continuous function then f has at least one fixed point.

For direct proofs of this result see, e.g., Schauder, 1930; Caccioppoli, 1931; Istrăţescu, 1973, p. 134 *et seq.*; see also Birkhoff–Kellogg, 1922; Alexander, 1922, and Caccioppoli, 1930. There are many variations of this result, besides many different formulations and extensions (see, among others, Browder, 1959a and Caccioppoli, 1932); one of the variations, also due to Schauder, is the following:

THEOREM 9.18. Let B be a Banach space and C a non-empty closed convex set of B. If $f: C \to C$ is continuous and $\overline{f(C)}$ is compact then f has at least one fixed point.

Proof. Let $Q = \overline{\text{conv}}\,(\overline{f(C)})$. Q is a subset of C since the latter is a closed convex set and $f(C) \subset C$; further, Mazur's theorem ensures that Q is compact.

Mazur's theorem states in fact that *if B is a Banach space and $A \subset B$ is compact then $\overline{\text{conv}}\,(A)$ is compact*: see Mazur, 1930; Warga, 1972, p. 164 and, for the finite-dimensional case, Stoer–Witzgall, 1970, p. 92; in Warga, 1972, p. 165 there is a counter-example which shows that in the infinite dimensional case one cannot substitute $\overline{\text{conv}}$ with conv. The assumption that the space B is complete (which has not been imposed in theorem 9.17) is necessary in order to ensure the applicability of this result.

From Schauder's theorem it follows then that $f|_Q$ has a fixed point: such a point is, *a fortiori*, a fixed point of f. ∎

The importance of this result lies essentially in the 'scarcity' of strongly compact subsets in Banach spaces ... ; in this context it is important to recall another variation of Schauder's theorem given by theorem 9.4, in which $\overline{f(C)}$ need not even be compact (but where strong restrictions are imposed on f and B!).

Further, although Banach spaces are rich with weakly compact sets, it is not generally convenient to use here the weak topology since the continuity of the functions then becomes an ugly condition. To see that Banach spaces have 'enough' weakly compact sets, one need only recall that the closed unit sphere of a reflexive Banach space is weakly compact; in fact the Alaoglu–Bourbaki theorem (see Alaoglu, 1940; Bourbaki, 1955, p. 114; Horváth, 1966; p. 230) states that *the necessary and sufficient condition for a normed space B to be reflexive is that $\{x \in B : \|x\|_B \leq 1\}$ be weakly compact*. It is relevant to recall here that Kakutani has constructed a homeomorphism of the closed unit sphere of a Hilbert space in itself which does not have a fixed point (see, e.g., Smart, 1974, p. 15; Cronin, 1964, p. 125).

To end this section, we state without proof (see e.g. Brézis, 1968, p. 171) the following variation of Tychonov's theorem (cf. theorem 9.18):

THEOREM 9.19. Let E be a locally convex Hausdorff topological vector space and let C be a closed convex set of E. If $f: C \to C$ is continuous and $\overline{f(C)}$ is compact then f has at least one fixed point.

9.3.3 MULTI-VALUED APPLICATIONS

We will now present the results which extend the theorems of Brouwer, Schauder, and Tychonov to multi-valued applications. These results are due, respectively, to Kakutani (see 1941, p. 457), Bohnenblust and Karlin (see 1950, p. 159), and Glicksberg (see 1952, p. 171).

Kakutani's theorem

THEOREM 9.20 (Kakutani's theorem). Let K be a non-empty convex compact set of $\mathbb{R}^n$. If $f: K \to 2^K_{\blacksquare ck}$ (or $f: K \to 2^K_{\blacksquare qk}$) is a multi-valued u.s.c. application (according to definition 20.11) then f has at least one fixed point, i.e.: $\exists \bar{x} \in K : \bar{x} \in f(\bar{x})$.

Proof. We will divide the proof into three parts as for that of Brouwer's theorem:

LEMMA 1'. The fixed point property in the multi-valued case is conserved by retraction, i.e.: if every multi-valued u.s.c. application $f: K \to 2^K_{\blacksquare ck}$ has a fixed point and $\tilde{K}$ is a retraction of K then every multi-valued u.s.c. application $\phi: \tilde{K} \to 2^{\tilde{K}}_{\blacksquare ck}$ has a fixed point.

Proof. Let $r: K \to \tilde{K}$ be a retraction and consider the multi-valued application $i_{2_{\blacksquare ck}\tilde{K} 2_{\blacksquare ck}K} \circ \phi \circ r : K \to 2^K_{\blacksquare ck}$ (see the diagram), which is obviously u.s.c. and

$$\begin{array}{ccc} K & \xrightarrow{r} & \tilde{K} \\ {\scriptstyle i_{2_{\blacksquare ck}\tilde{K} 2_{\blacksquare ck}K} \circ \phi \circ r}\downarrow & & \downarrow\phi \\ 2^K_{\blacksquare ck} & \xleftarrow{i_{2_{\blacksquare ck}\tilde{K} 2_{\blacksquare ck}K}} & 2^{\tilde{K}}_{\blacksquare ck} \end{array}$$

therefore, by hypothesis, has a fixed point: let $\bar{x}$ be this point. Since $i_{2_{\blacksquare ck}\tilde{K} 2_{\blacksquare ck}K}$ is the set inclusion we can say that $\bar{x}$ is a point of $\tilde{K}$ and that $\bar{x} \in (\phi \circ r)(\bar{x})$: it follows from this and from $r(\bar{x}) = \bar{x}$ that $\bar{x} \in \phi(\bar{x})$. ∎

LEMMA 2'. The same as lemma 2 of Brouwer's theorem.

LEMMA 3' (Kakutani's theorem for simplexes). Let $S = \text{simp}\{x_0, \ldots, x_m\}$ be an m-simplex in $\mathbb{R}^m$. If $f: S \to 2^S_{\blacksquare ck}$ is u.s.c. then it has at least one fixed point.

Proof. The proof is considerably different from that of Brouwer's theorem for simplexes. Here, as in the proof of the KKM lemma, we use the fact, already mentioned, that for every $l \geq 1$ there is a simplex subdivision Σ_l of S such that diam $(\Sigma_l) \leq 1/l$. Now, let us assume that we have a sequence of such simplex subdivisions of S and let us define, for every $l \geq 1$, an application $f^{(l)}: S \to S$ by putting at the vertices $v^{(l)}$ of Σ_l '$f^{(l)}v^{(l)}) =$ a point of $f(v^{(l)})$' and then making a linear extension within each simplex of Σ_l. The $f^{(l)}$ so defined are continuous and hence, by Brouwer's theorem, it follows that for every $l \geq 1$ there exists $\bar{x}^{(l)} \in S$ such that $f^{(l)}(\bar{x}^{(l)}) = \bar{x}^{(l)}$. Assume that $\bar{x}^{(l)} \in \text{simp}\{v_{j_0}^{(l)}, \ldots, v_{j_m}^{(l)}\}$ and put

$$\bar{x}^{(l)} = \sum_{i=0}^{m} \alpha_{j_i}^{(l)} v_{j_i}^{(l)}, \quad \text{with} \quad \alpha_{j_i}^{(l)} \geq 0 \quad \text{and} \quad \sum_{i=0}^{m} \alpha_{j_i}^{(l)} = 1; \quad (9.22)$$

from the linearity (or rather, affinity) of f in each simplex of Σ_l we have

$$f^{(l)}(\bar{x}^{(l)}) = \sum_{i=0}^{m} \alpha_{j_i}^{(l)} f^{(l)}(v_{j_i}^{(l)}). \quad (9.23)$$

Possibly by extracting subsequences (which we will denote in the same way as the sequences ...) we can write

$$\lim_{l \to \infty} v_{j_i}^{(l)} = v_{j_i}, \quad \lim_{l \to \infty} \bar{x}^{(l)} = \bar{x}, \quad \lim_{l \to \infty} \alpha_{j_i}^{(l)} = \alpha_{j_i}$$

$$\left(\text{with} \quad \alpha_{j_i} \geq 0 \quad \text{and} \quad \sum_{i=0}^{m} \alpha_{j_i} = 1\right). \quad (9.24)$$

From (9.24) and from $\lim_{l \to \infty} \text{diam}(\Sigma_l) = 0$ we have again that $v_{j_i} = \bar{x}$ ($i = 0, \ldots, m$) and therefore that $\lim_{l \to \infty} v_{j_i}^{(l)} = \bar{x}$. Further, putting $\lim_{l \to \infty} f^{(l)}(v_{j_i}^{(l)}) = t_{j_i}$, we have

$$\bar{x} = \lim_{l \to \infty} \bar{x}^{(l)} = \lim_{l \to \infty} f^{(l)}(\bar{x}^{(l)})$$

$$= \lim_{l \to \infty} \sum_{i=0}^{m} \alpha_{j_i}^{(l)} f^{(l)}(v_{j_i}^{(l)}) = \sum_{i=0}^{m} \alpha_{j_i} t_{j_i}. \quad (9.25)$$

We will now show that $\bar{x}$ is a fixed point for f, by showing that $t_{j_i} \in f(\bar{x})$ and then by using (9.25) and the fact that $f(\bar{x})$ is convex. Now, from the definition of $f^{(l)}$, we have $(v_{j_i}^{(l)}, f^{(l)}(v_{j_i}^{(l)})) \in G_{SS}(f)$ and thus, since f is u.s.c. (see also theorems 20.14 and 20.15), we have $(\bar{x}, t_{j_i}) \in G_{SS}(f)$. ∎

Generalizations of the theorems of Schauder and Tychonov

We will now deal with the more general fixed point theorem which we will prove (see also Glicksberg, 1952; Istrățescu, 1973, p. 200; Berge, 1959, p. 259).

THEOREM 9.21. Let E be a locally convex Hausdorff topological vector space and let K be a non-empty convex compact set in E. If $f: K \to 2^K_{\blacksquare ck}$ is u.s.c. then f has at least one fixed point.

Proof. Let V be a closed convex symmetric neighbourhood of the origin. The family of open sets $\{x + \mathring{V} : x \in K\}$ covers K, and since this set is compact we can extract from it a finite subcover, i.e.: there exist $x_1, \ldots, x_m \in K$ such that $K \subset \bigcup_{i=1}^{m}(x_i + \mathring{V})$, and *a fortiori* $K \subset \bigcup_{i=1}^{m}(x_i + V)$ (a set $\{x_1, \ldots, x_m\}$ satisfying these conditions is said to be *V-dense* in K).

Let us now put

$$\chi_V = \text{conv}\{x_1, \ldots, x_m\}; \tag{9.26}$$

χ_V is a convex compact subset of K, and further it is of finite dimension, which implies among other things that the topology of E relativized to χ_V is Euclidean. Let us now put, for every $x \in K$,

$$f_V(x) = (f(x) + V) \cap \chi_V \tag{9.27}$$

and show that this expression defines an application $f_V : \chi_V \to 2^{\chi_V}_{\blacksquare ck}$ u.s.c. This is obtained from the following four points:

(1) $f_V(x) \neq \emptyset$ for every $x \in \chi_V$, since $x_i \in \chi_V$, $i = 1, \ldots, m$, and for some i we have $x_i \in f(x) + V$ (in fact, if $t \in f(x)$ then $t \in K$ and hence for some i we have $t \in x_i + V$, i.e. $t = x_i + y$ with $y \in V$, and hence $x_i = t - y$, so that $x_i \in t - V$ and therefore $x_i \in (f(x) - V)$: since $V = -V$ we have then that $x_i \in (f(x) + V)$).
(2) $f_V(x)$ is convex for every $x \in \chi_V$, since $f(x)$, V, and χ_V are convex (obviously, if A and B are convex so is $A + B$).
(3) $f_V(x)$ is closed for every $x \in \chi_V$, since χ_V and V are closed and $f(x)$ is compact (since it is a closed subset of a compact set, K, in a separate space, E; let us remark that if A is closed and B is compact then $A + B$ is closed: see the proof, for example, in Berge, 1959, p. 246; again if A and B are compact, so is $A + B$).
(4) f_V is u.s.c., since we can write (see definitions 20.5 and 20.6) $f_V(x) = [(v \bullet f) \cap c](x)$, where $c(x) = \chi_V$ and $v(x) = x + V$ are obviously u.s.c.

Let us now put, with V being a convex closed symmetric neighbourhood of the origin,

$$f_V^b = \{x \in K : x \in (f(x) + V) \cap K\} \tag{9.28}$$

and show that $\bigcap_V f_V^b \neq \emptyset$. This follows immediately from the following three results:

(1′) $f_V^b \neq \emptyset$ for any neighbourhood V, since $\{x : f_V(x) = x\} \subset f_V^b$ and (from Kakutani's theorem—remembering that the topology of χ_V is Euclidean ...) f_V has at least one fixed point.
(2′) f_V^b is closed for every neighbourhood V, since putting $v(x) = x + V$, $k(x) = K$, $u(x) = \{x\}$ and $h(x) = [(v \bullet f) \cap k \cap u](x)$ (which is u.s.c. and such that $h(x) = \emptyset$ or $h(x) = \{x\}$) we have $f_V^b = \{x \in K : x \in (f(x) + V) \cap K\} = \{x : x \in (f(x) + V) \cap K \cap \{x\}\} = \{x : x \in h(x)\} = \{x : h(x) \neq \emptyset\} = E \setminus h^{-1+}(\emptyset)$, and, since f is u.s.c., $h^{-1+}(\emptyset)$ is open (see theorem 20.8(iv$^+$)).
(3′) the family $\{f_V^b\}_V$ has the finite intersection property, since if $V_1, \ldots, V_l$ are closed convex neighbourhoods of the origin then, putting $V^* = V_1 \cap \ldots \cap V_l$ (which is a closed convex neighbourhood of the origin),

we have

$$f^b_{V_1} \cap \ldots \cap f^b_{V_l} = \{x : x \in (f(x) + V_1) \cap K\} \cap \ldots \cap \{x : x \in (f(x) + V_l) \cap K\}$$
$$= \{x : x \in [(f(x) + V_1) \cap \ldots \cap (f(x) + V_l)] \cap K\}$$
$$= \{x : x \in [f(x) + (V_1 \cap \ldots \cap V_l)] \cap K\} = f^b_{V^*} \neq \varnothing.$$

Now let $\bar{x} \in \bigcap_V f^b_V$ and we show that $\bar{x} \in f(\bar{x})$, which proves the theorem. By *reductio ad absurdum*: let us suppose that $\bar{x} \notin f(\bar{x})$. Then there exists a closed convex symmetric neighbourhood of $\bar{x}$, $I(\bar{x})$ such that $I(\bar{x}) \subset E \setminus f(\bar{x})$: let us put $I_0 = x - I(\bar{x})$ (which is a closed convex symmetric neighbourhood of the origin) and let us show that $\bar{x} \notin f^b_{I_0}$. In fact, if $\bar{x} \in f^b_{I_0}$ then in particular $\bar{x} \in f(\bar{x}) + I_0$, i.e. $\bar{x} = y + \bar{x} - t$ with $y \in f(\bar{x})$ and $t \in I(\bar{x})$, from which it follows that $y = t \in f(\bar{x}) \cap I(\bar{x})$, which is impossible since $I(\bar{x}) \subset E \setminus f(\bar{x})$. ∎

An immediate corollary of theorem 9.21 is

THEOREM 9.22. Let B be a normed space and let K be a non-empty compact convex set in B. If $f : K \to 2^K_{\blacksquare ck}$ is u.s.c. then f has at least one fixed point.

For a direct proof of this result see, e.g., Bohnenblust–Karlin, 1950; Istrăţescu, 1973, p. 161; an important variant of this is the following

THEOREM 9.23. Let B be a Banach space and let C be a closed convex set in B. If $f : C \to 2^C_{\blacksquare ck}$ is u.s.c. and $\overline{f_B(C)} \triangleq \overline{\bigcup_{x \in C} f(x)}$ is compact then f has at least one fixed point.

Proof. Identical to the proof of theorem 9.18. ∎

Let us state yet another result (see Fan, 1961, p. 308):

THEOREM 9.24. Let F be a locally convex Hausdorff topological vector space, let E be a Hausdorff topological vector space and let K be a non-empty compact convex set in E; further let $f : K \to 2^F_{\blacksquare qk}$ be a continuous multi-valued application and $g : K \to F$ a continuous application such that

(i) $\forall x \in K \ f(x) \cap g(x) \neq \varnothing$,
(ii) if $C \subset F$ is closed then $g^{-1}(C)$ is convex (possibly empty).

Then there exists $\bar{x} \in K$ such that $g(\bar{x}) \in f(\bar{x})$.

Let us remark that theorem 9.24, like theorem 9.21, is a generalization of Tychonov's theorem; in fact, if we assume that $E = F$, $g =$ identity, and f single-valued, theorem 9.24 becomes Tychonov's theorem (if $f : K \to 2^F_{\blacksquare qk}$ is single-valued then we can put $f : K \to F$, which—using (i)—becomes $f : K \to K$). Further, theorems 9.21 and 9.24 are not more general one than the other, unless $E = F$, in which case the latter is a particular case of the former.

For other results regarding fixed points for applications defined on subsets of topological vector spaces see, among others, Browder, 1968.

The logical connection between the theorems

OBSERVATION 9.1. In the diagram below we have indicated, with arrows, the logical order followed in the presentation of the main results of section 9.3.

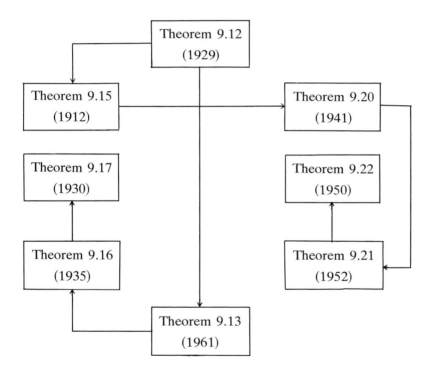

This order is different from the chronological order in which the theorems were discovered, as can be seen from the dates shown in the diagram (which are the respective dates of publication of the original papers—let us remark, however, that Brouwer's theorem, in particular, had already been presented by Brouwer in papers previous to Brouwer, 1912).

In the same column (or: line) of the diagram the results are arranged so that each one is more general than the ones 'above' (or: in 'the left'). A glance at the diagram shows that we could have avoided proving Brouwer's theorem directly by considering it to a corollary of Schauder's theorem, or, alternatively, we could have avoided the treatment of Fan's lemma by considering Tychonov's theorem as a corollary of theorem 9.21.

9.4 FIXED POINT THEOREMS FOR MONOTONE APPLICATIONS

In this section we will deal essentially with two fixed point theorems for monotone applications: the Knaster–Birkhoff theorem (see Birkhoff, 1961, pp. 41 and 54), which requires a lattice structure (which must be complete, an existential hypothesis which is fundamental to the proof), and a result due to Kolodner (see Kolodner, 1968) and used successfully by Tartar (see Tartar, 1974a; 1974b, p. 98) in the study of quasivariational inequalities, which requires only an ordered structure (which must be completely inductive, an existential hypothesis which replaces the previous one)—it must be mentioned, however, that neither of the results is more general than the other. An interesting characteristic of both results is that although they do not ensure the uniqueness of the fixed point, they enable us to formulate rules for 'choosing' fixed points, since they tell us that the sets of such points have maxima and minima.

The Knaster–Birkhoff theorem

Let us deal with the first result:

THEOREM 9.25 (the Knaster-Birkhoff theorem). Let $\langle L, \leq \rangle$ be a complete lattice. If $f: L \to L$ is a monotone non-decreasing application then f has at least one fixed point, i.e.: $\exists a \in L \ \ a = f(a)$. The set Φ of fixed points has a maximum and a minimum.

Proof. Put $S = \{x \in L : x \leq f(x)\}$. This set is non-empty since $\min L \in S$ (remark that the lattice completeness of L ensures the existence of $\min L$: see section 19.2). Now put $a = \sup S$ (which again exists since L is complete). We have, from the definition of lower upper bound, $x \leq a \ \forall x \in S$ and hence, from the monotonicity of f, $f(x) \leq f(a) \ \forall x \in S$; from this and the definition of S we obtain $x \leq f(x) \leq f(a) \ \forall x \in S$ and, therefore, once again from the definition of lower upper bound, $a \leq f(a)$. From this it follows, again from the monotonicity of f, that $f(a) \leq f(f(a))$ and therefore that $f(a) \in S$, and finally that $f(a) \leq a$. Thus we have shown that $a = f(a)$. Thus, putting $\Phi = \{x \in L : f(x) = x\}$, we have $\Phi \neq \emptyset$ (since $a \in \Phi$) and further it is evident that $a = \max \Phi$. The same reasoning applied to the system $\langle L, \geq \rangle$ gives the proof that there exists $b = \min \Phi$. ∎

Let us remark that even when f is strictly monotone we cannot ensure that the fixed point is unique: to see this one has merely to consider the application identity....

The result of Kolodner and Tartar

We will now deal with the second result mentioned at the beginning of the section:

223

THEOREM 9.26. Let $\langle E, \leq \rangle$ be an ordered set. If $f: E \to E$ is monotone non-decreasing and

(i) $\exists a, b \in E; a \leq b: a \leq f(a) \leq f(b) \leq b$ and $[a, b]$ is completely inductive

then f has at least one fixed point, more precisely: $\exists c \in [a, b]: c = f(c)$. The set Φ of the fixed points lying between a and b has a maximum and a minimum.

Proof. Let us first observe that to ensure the existence of a fixed point it is sufficient that $[a, b]$ be either completely (s)-inductive or completely (i)-inductive: in the former case we can ensure that Φ has a minimum and in the second that Φ has a maximum. Here we will tacitly consider the former case: the proof for the second is analogous (and, further, it can be obtained from the other by duality considering the algebraic system $\langle L, \geq \rangle$).

Putting $X = \{u \in [a, b]: u \leq f(u)\}$, $Y = \{u \in [a, b]: f(u) \leq u\}$, $Z = \{u \in X : u \leq y \ \forall y \in Y\}$ (see the scheme shown in fig. 9.3), we will show that: (1) X, Y, and Z are non-empty; (2) X is stable for f (we said that $A \subset B$ is *stable* for $f: B \to B$ if $f(A) \subset A$); (3) Z is stable for f; (4) X with the induced order is completely (s)-inductive; (5) Z with the induced order is completely (s)-inductive; (6) Z has a maximal element, u^*; (7) u^* is a fixed point of f; (8) u^* is the least fixed point for f lying between a and b.

Proof of (1): We need only see that $a \in X$, $b \in Y$ and $a \in Z$. *Proof of* (2): If $x \in X$ then, from the definition of X, $a \leq x \leq b$ and $x \leq f(x)$; from this and the monotonicity of f we have $f(a) \leq f(x) \leq f(b)$ and $f(x) \leq f(f(x))$ and thus, since by hypothesis $a \leq f(a) \leq f(b) \leq b$, we can write $a \leq f(x) \leq b$ and $f(x) \leq f(f(x))$, i.e. $f(x) \in X$. *Proof of* (3): If $z \in Z$ then, from the definition of Z, $z \in X$ and $z \leq y \ \forall y \in Y$; from the stability of X we have $f(z) \in X$ and from the monotonicity of f we have $f(z) \leq f(y) \ \forall y \in Y$, or again, since if $y \in Y$ then $f(y) \leq y$, $f(z) \leq y \ \forall y \in Y$—we have shown then that $f(z) \in Z$. *Proof of* (4): Let $\mathscr{C}$ be a chain in X and $c = \sup \mathscr{C} \in [a, b]$ (note that $\sup \mathscr{C}$ must exist since $[a, b]$ is completely (s)-inductive); if $u \in \mathscr{C}$ then $u \in X$ and hence

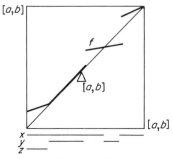

Figure 9.3

$u \leq f(u)$ $\forall u \in \mathscr{C}$, on the other hand $u \leq c$ $\forall u \in \mathscr{C}$ and thus, from the monotonicity of f, $f(u) \leq f(c)$ $\forall u \in \mathscr{C}$: thus $u \leq f(c)$ $\forall u \in \mathscr{C}$; from this and from the definition of the lower upper bound we have $c \leq f(c)$ and thus $c \in X$. *Proof of* (5): Let $\mathscr{C}'$ be a chain in Z and $c' = \sup \mathscr{C}' \in [a, b]$; $c' \in X$ since $\mathscr{C}'$ is also a chain in X, which is completely (s)-inductive; further $\forall u \in \mathscr{C}'$ $u \leq y$ $\forall y \in Y$ and thus, from the definition of lower upper bound, $c' \leq y$ $\forall y \in Y$: thus we have $c' \in Z$. *Proof of* (6): This is an immediate consequence of the Zorn lemma (note that u^* is not necessarily the lower upper bound of Z, which need not exist since Z need not be a chain). *Proof of* (7): From the definition of Z, $u^* \in X$ and thus $u^* \leq f(u^*)$; further, from the stability of Z, $f(u^*) \in Z$: thus $f(u^*) = u^*$ since u^* is maximal. *Proof of* (8): We can see immediately that $\Phi = X \cap Y$ and (since $u^* \in Z$ precedes all the elements of Y and *a fortiori* all the elements of $X \cap Y$) that $u^* = \min \Phi$. ∎

Here, as in the case of the Knaster–Birkhoff theorem, the strict monotonicity of f is not sufficient to ensure that the fixed point is unique; we can, however, ensure uniqueness by imposing assumptions of 'vectorial type' on f, assumptions which, further, are of practical interest since in some important applications of theorem 9.26 the set E has not only an ordered structure but also has a vectorial structure. So, we have the following

THEOREM 9.27. *In the same conditions as theorem* 9.26, *if further E is a vectorial space, f is concave, $f(a) \neq a$, and, denoting by Φ the set of fixed points of f, there exists $\varepsilon > 0$ such that*

$$\min \Phi > a + \varepsilon(b - a), \tag{9.29}$$

then the fixed point of f is unique (that is, Φ has only one element).

Proof. Let $u^* = \min \Phi$ and $u_* = \max \Phi$. Again put

$$\bar{u} = u^* - \frac{\varepsilon}{2}(u_* - u^*). \tag{9.30}$$

Thus we have $a \leq \bar{u} \leq u^*$. From (9.30) we have

$$u^* = \frac{\bar{u} + \frac{\varepsilon}{2} u_*}{1 + \frac{\varepsilon}{2}} \tag{9.31}$$

and from this and the concavity of f we have

$$f(u^*) \geq \frac{f(\bar{u}) + \frac{\varepsilon}{2} f(u_*)}{1 + \frac{\varepsilon}{2}} \tag{9.32}$$

or, since $f(u^*) = u^*$ and $f(u_*) = u_*$,

$$u^* \geq \frac{f(\bar{u}) + \frac{\varepsilon}{2} u_*}{1 + \frac{\varepsilon}{2}}. \tag{9.33}$$

From (9.33) and (9.30) it follows immediately that $f(\bar{u}) \leq \bar{u}$, and given that $a \leq f(a)$ we conclude, by applying theorem 9.26, that f has a fixed point in $[a, \bar{u}]$: since u^* is minimal we must have $u^* = \bar{u}$ and thus, from (9.30), $u^* = u_*$. ∎

If f is convex and $f(b) \neq b$ holds the analogous result, with analogous proof.

To end, we mention a result regarding fixed points for applications defined on ordered sets which does not require assumptions of monotonicity:

THEOREM 9.28. If $\langle E, \leq \rangle$ is a completely (s)-inductive ordered set and $f : E \to E$ is such that $\forall x \in E \; f(x) \geq x$ then $\exists c \in E : c = f(c)$.

For the proof see, e.g., Barsotti, 1968, p. 21.

10
Some Results on the Existence of Solutions of Variational Inequalities

In this chapter we will once again take up the abstract theory of variational inequalities which we abandoned in Chapter 3. We did this not only in order to present some concrete examples, which we have done in the last few chapters of the first part, but also because we lacked adequate structures for continuing with that theory, a lack which the previous chapter has filled in.

The role of Fan's lemma

In dealing with problem 7.37, we mentioned the need for results of existence more general than those which we had then proved. We will now present a very general abstract result, the proof of which follows from Fan's lemma.

To give an idea of the role of results such as Fan's lemma in the study of variational inequalities, we will now give a sketch of a third proof of the Lions–Stampacchia theorem, analogous to that of theorem 2.2. Let us consider, for every $v_0 \in K$, the 'level' set $N(v_0) = \{v \in K : a(v, v - v_0) \leq L(v - v_0)\}$. It is obvious that u satisfies equation (3.28) iff $u \in N(v_0) \ \forall v_0 \in K$; the proof of the existence of a solution is thus reduced to the proof that $\bigcap_{v_0 \in K} N(v_0) \neq \varnothing \ldots$.

10.1 GENERAL RESULTS OF EXISTENCE

The main result

In this section we will deal principally with the following result (see also Browder, 1965b; Hartman–Stampacchia, 1966; Mosco, 1967a):

THEOREM 10.1. If E, $\phi : E \to]-\infty, +\infty]$ and $f : E \times E \to \mathbb{R}$ are such that:

(h1) E is a real Hausdorff topological vector space,

(h2) ϕ is convex, proper and l.s.c.,
(h3) $\forall v \in E \quad \phi(v) < +\infty \Rightarrow f(v, v) \leq 0$,
(h4) $\forall v \in E \quad f(v, \cdot)$ is concave,
(h5) $\forall v \in E \quad f(v, \cdot)$ is u.s.c.,
(h6) $\forall w \in E \quad f(\cdot, w)$ is l.s.c. on the segments,
(h7) $\forall v, w \in E \quad f(v, w) + f(w, v) \geq 0$,
(h8) $\exists B \subset E$, B is a compact set: $\exists v_0 \in B : \forall v \in E \backslash B \quad \phi(v) + f(v, v_0) > \phi(v_0)$,

then the

'PROBLEM 10.1. Find

$$u \in E; \phi(u) + f(u, v) \leq \phi(v) \quad \forall v \in E', \tag{10.1}$$

has a solution. Further the set of solutions of problem 10.1 is a closed convex set which is identical to that of the solutions of the problem:

'PROBLEM 10.2. Find

$$u \in E; \phi(u) - f(v, u) \leq \phi(v) \quad \forall v \in E'. \tag{10.2}$$

Before proving this we need to make a few comments. In the first place, this theorem, despite the numerous hypotheses, takes into account several important situations.

Here we will only mention that problem 10.1 has, as particular cases, among others, the following:

The problem of minimizing the function ϕ in E: we need only put $f \equiv 0$.

The problem of finding the saddle points of a function $g : \mathbb{R}^2 \to \mathbb{R}$: it suffices to put $E = \mathbb{R}^2$, $\phi \equiv 0$ and $f(v, w) = g(v_1, w_2) - g(w_1, v_2)$. In effect, if $u \equiv (u_1, u_2)$ is a solution of problem 10.1 then $g(u_1, w_2) \leq g(w_1, u_2) \; \forall w \equiv (w_1, w_2) \in \mathbb{R}^2$ and so, putting $w_1 = u_1$, we have $g(u_1, w_2) \leq g(u_1, u_2)$ (which means that u is a maximum for g along $u_1 \times \mathbb{R}$) and, putting $w_2 = u_2$, we have $g(u_1, u_2) \leq g(w_1, u_2)$ (which means that u is minimum for g along $\mathbb{R} \times u_2$): u is therefore a saddle-point for g.

And, most important for us, the problems associated with variational inequalities: put, e.g., $\phi = j_K$, where j_K is the indicator function of a convex set K in E, and $f(v, w) = a(v, v - w)$, where $a : E \times E \to \mathbb{R}$ is a bilinear form.

The hypothesis has been selected essentially in order to obtain, as particular cases of theorem 10.1, the Lions–Stampacchia theorem and other analogous results. The connection between the hypotheses of the Lions–Stampacchia theorem and those of theorem 10.1 are not, however, obvious, and require some complicated abstractions and generalizations. Again, in connection with the hypothesis, we mention that for reasons which will become clear later, (h7) and (h8) are known respectively as the *monotonicity hypothesis* and the *coercivity hypothesis*. Further, (h6), which means that for every u, v, w in E the function $t \mapsto f(tu + (1-t)v, w)$ is lower semicontinuous from $\mathbb{R}$ to $\mathbb{R}$, can be called the *hypothesis of lower hemisemicontinuity*.

Finally, the last part of theorem 10.1 is a generalization of Minty's lemma which we proved in Chapter 7.

Minty's lemma

Before proving the theorem, we will prove two lemmas which deal with the 'level sets'
$$\sigma(w) = \{v \in E : \phi(v) + f(v, w) \leq \phi(w)\} \tag{10.3}$$
and
$$\sigma^*(w) = \{v \in E : \phi(v) - f(w, v) \leq \phi(w)\}, \tag{10.4}$$
relative to the inequalities (10.1) and (10.2), respectively.

LEMMA 1. If E, $\phi : E \to]-\infty, +\infty]$ and $f : E \times E \to \mathbb{R}$ satisfy hypotheses (h1), (h2), (h3), (h4), and (h8) then
$$\bigcap_{w \in E} \overline{\sigma(w)} \neq \varnothing. \tag{10.5}$$

Proof. This follows from Fan's lemma. We will check then that the assumptions of Fan's lemma are satisfied. It follows immediately that the sets $\overline{\sigma(w)}$ are closed and non-empty (remark that $w \in \sigma(w)$ since if $\phi(w) = +\infty$ then $\sigma(w) = E$ and if $\phi(w) < +\infty$ then (from (h3)) $f(w, w) \leq 0$ and therefore $\phi(w) + f(w, w) \leq \phi(w)$). To check that (ii') is satisfied, we see that (h8) ensures that there exists a compact set $B \subset E$ and a point $v_0 \in B$ such that $\sigma(v_0) \subset B$, and that from this and (h1) we have that $\overline{\sigma(v_0)} \subset B$ is compact. Finally we will check that (i') is satisfied by showing that
$$\forall w_1, \ldots, w_n \in E \text{ conv}\{w_j\} \subset \bigcup_{j=1}^{n} \sigma(w_j) \subset \bigcup_{j=1}^{n} \overline{\sigma(w_j)}, \tag{10.6}$$
i.e., by showing that if $w = \sum_{j=1}^{n} \lambda_j w_j$ (with $\lambda_j \geq 0$ and $\sum_{j=1}^{n} \lambda_j = 1$) then $w \in \sigma(w_j)$ for at least one j. By *reductio ad absurdum*: let us suppose that, for every j, $w \notin \sigma(w_j)$, i.e., let us suppose that
$$\phi(w) + f(w, w_j) > \phi(w_j), \quad \text{for } j = 1, \ldots, n. \tag{10.7}$$
From (10.7) it follows that $\phi(w_j) < +\infty$, for $j = 1, \ldots, n$, and thus (from (h2)) we have
$$\phi(w) = \phi\left(\sum_{j=1}^{n} \lambda_j w_j\right) \leq \sum_{j=1}^{n} \lambda_j \phi(w_j) < +\infty. \tag{10.8}$$
Again from (10.7) we have (since we can find j such that $\lambda_j = 0$)
$$\lambda_j \phi(w) + \lambda_j f(w, w_j) \geq \lambda_j \phi(w_j), \quad \text{for } j = 1, \ldots, n, \tag{10.9}$$
and hence (since there exists at least one j such that $\lambda_j > 0$)
$$\sum_{j=1}^{n} \lambda_j \phi(w) + \sum_{j=1}^{n} \lambda_j f(w, w_j) > \sum_{j=1}^{n} \lambda_j \phi(w_j). \tag{10.10}$$

From this, and given that ϕ is convex (from (h2): $\phi(\sum_{j=1}^n \lambda_j w_j) \le \sum_{j=1}^n \lambda_j \phi(w_j)$) and from the concavity of $f(w, \cdot)$ (from (h4): $\sum_{j=1}^n \lambda_j f(w, w_j) \le f(w, \sum_{j=1}^n \lambda_j w_j)$) we have

$$\phi(w) + f(w, w) > \phi(w), \tag{10.11}$$

which together with (10.8) gives us the result that $f(w, w) > 0$, which contradicts (h3). ∎

LEMMA 2 (Minty's lemma). If E, $\phi: E \to]-\infty, +\infty]$ and $f: E \times E \to \mathbb{R}$ satisfy the hypotheses (h1), (h2), (h3), (h4), (h6), and (h7) then

$$\bigcap_{w \in E} \sigma(w) = \bigcap_{w \in E} \sigma^*(w). \tag{10.12}$$

Further these sets are convex (possibly empty).

Proof. Let us first note that

$$\sigma(w) \subset \sigma^*(w) \quad \forall w \in E \tag{10.13}$$

since if $v \in \sigma(w)$ then

$$\phi(v) + f(v, w) \le \phi(w) \tag{10.14}$$

and thus, subtracting $f(v, w) + f(w, v)$ from both sides of this inequality and taking into account (h7),

$$\phi(v) - f(w, v) \le \phi(w), \tag{10.15}$$

an inequality which tells us that $v \in \sigma^*(w)$. From (10.13) we have $\bigcap_{w \in E} \sigma(w) \subset \bigcap_{w \in E} \sigma^*(w)$; we will next prove that, conversely, $\bigcap_{w \in E} \sigma^*(w) \subset \bigcap_{w \in E} \sigma(w)$ (noting, however, that in general $\sigma^*(w) \not\subset \sigma(w)$). Let then $v \in \bigcap_{w \in E} \sigma^*(w)$, and therefore $v \in \sigma^*(w) \forall w \in E$, and let us fix $\bar{w}$ in E such that $\phi(\bar{w}) < +\infty$. Further, let us put, for $t \in]0, 1[$,

$$w(t) = t\bar{w} + (1-t)v, \tag{10.16}$$

and note that $v \in \sigma^*(w(t))$, which means that

$$\phi(v) - f(t\bar{w} + (1-t)v, v) \le \phi(t\bar{w} + (1-t)v), t \in]0, 1[, \tag{10.17}$$

and that

$$\phi(w(t)) < +\infty, \quad t \in]0, 1[, \tag{10.18}$$

as follows from (h2), from $\phi(\bar{w}) < +\infty$, and from $\phi(v) < +\infty$ (which must be true since otherwise we would have $\phi \equiv +\infty$, which would be in conflict with (h2)). From (h2) and from (10.17) we have

$$\phi(v) - f(t\bar{w} + (1-t)v, v) \le t\phi(\bar{w}) + (1-t)\phi(v), \quad t \in]0, 1[, \tag{10.19}$$

and adding $tf(v, \bar{w})$ to both sides of this inequality we obtain

$$t[\phi(v) - \phi(\bar{w}) + f(v, \bar{w})] \le tf(v, \bar{w}) + f(t\bar{w} + (1-t)v, v), \quad t \in]0, 1[. \tag{10.20}$$

From this, since (h3) and (h7) imply that $\forall v \in E \ \phi(v) < +\infty \Rightarrow f(v, v) = 0$, we have, taking into account also (10.18),

$$t[\phi(v) - \phi(\bar{w}) + f(v, \bar{w})] \leq tf(v, \bar{w}) + f(t\bar{w} + (1-t)v, v)$$

$$- \frac{1}{1-t} f(t\bar{w} + (1-t)v, t\bar{w} + (1-t)v), \quad t \in]0, 1[, \quad (10.21)$$

or, making use now of (h4),

$$t[\phi(v) - \phi(\bar{w}) + f(v, \bar{w})] \leq tf(v, \bar{w}) + f(t\bar{w} + (1-t)v, v)$$

$$- \frac{1}{1-t} [tf(t\bar{w} + (1-t)v, \bar{w}) + (1-t)f(t\bar{w} + (1-t)v, v)]$$

$$= t\left[f(v, \bar{w}) - \frac{1}{1-t} f(t\bar{w} + (1-t)v, \bar{w})\right], \quad t \in]0, 1[. \quad (10.22)$$

Now, dividing both sides by t and letting t tend to 0, we have, taking into account (h6),

$$\phi(v) - \phi(\bar{w}) + f(v, \bar{w}) \leq f(v, \bar{w}) - f(v, \bar{w}), \quad (10.23)$$

i.e.

$$\phi(v) + f(v, \bar{w}) \leq \phi(\bar{w}), \quad (10.24)$$

and hence $v \in \sigma(\bar{w})$. Since $\bar{w}$ is arbitrary in E (let us remark that the only restriction imposed on $\bar{w}$ is $\phi(\bar{w}) < +\infty$, and that if $\phi(\bar{w}) = +\infty$ then $\sigma^*(\bar{w}) = \sigma(w) = E\ldots$) we have $v \in \bigcap_{w \in E} \sigma(w)$. We have thus proved (10.12). The proof of the last part of the lemma follows immediately. ∎

Proof of theorem 10.1

We now go on to the:

Proof of theorem 10.1. Saying that problem 10.1 has a solution means that there exists $u \in E$ such that $u \in \sigma(w) \ \forall w \in E$, i.e. that $\bigcap_{w \in E} \sigma(w) \neq \emptyset$, or again (from lemma 2) that $\bigcap_{w \in E} \sigma^*(w) \neq \emptyset$. Given this, to prove the first part of the theorem (that a solution of problem 10.1 exists) we need only to show that, for every $w \in E$, $\sigma^*(w)$ is closed, since this implies (from (10.13)) that $\overline{\sigma(w)} \subset \sigma^*(w)$, and hence that $\bigcap_{w \in E} \overline{\sigma(w)} \subset \bigcap_{w \in E} \sigma^*(w)$, and (from lemma 1) $\bigcap_{w \in E} \sigma^*(w) \neq \emptyset$. Further, at the same time we will prove the rest of the theorem, since the equivalence of problem 10.1 and problem 10.2 follows from lemma 2, as does the fact that the sets of the solutions of both problems (which are $\bigcap \sigma(w)$ and $\bigcap \sigma^*(w)$) are convex, and that these sets are closed then follows from the fact that the intersection of closed sets is closed.

To show that $\sigma^*(w)$ is closed, we take in $\sigma^*(w)$ a filter $\{v_\alpha\}_{\alpha \in A}$ with limit v and show that $v \in \sigma^*(w)$; this follows immediately since, from

$$\phi(v_\alpha) - f(w, v_\alpha) \leq \phi(w) \quad \forall \alpha \in A \quad (10.25)$$

and the fact that ϕ and $-f(w, \cdot)$ are lower semicontinuous (i.e. from (h2) and (h5)), we have, taking the limit, $\phi(v) - f(w, v) \leq \phi(w)$. ∎

Two more results

By making the hypothesis (h7) regarding monotonicity strong, we can also ensure the uniqueness of that solution whose existence is given by theorem 10.1. In other words:

THEOREM 10.2. Under the hypothesis of theorem 10.1, making (h7) stronger by means of

(h9) $\forall v, w \in E,$ $v \neq w,$ $f(v, w) + f(w, v) > 0,$

the solution of problem 10.1 (and hence that of problem 10.2) is unique.

Proof. By *reductio ad absurdum*. Let u and u^* be two solutions of problem 10.1. Hence
$$\phi(u) + f(u, w) \leq \phi(w) \quad \forall w \in E, \quad (10.26)$$
$$\phi(u^*) + f(u^*, w) \leq \phi(w) \quad \forall w \in E. \quad (10.27)$$
Putting $w = u^*$ in (10.26), $w = u$ in (10.27), and by adding the respective sides, we obtain
$$\phi(u) + \phi(u^*) + f(u^*, u) + f(u, u^*) \leq \phi(u) + \phi(u^*), \quad (10.28)$$
from which, since $\phi(u) < +\infty$ and $\phi(u^*) < +\infty$ (since otherwise we would have $\phi \equiv +\infty$), we have
$$f(u^*, u) + f(u, u^*) \leq 0, \quad (10.29)$$
which by (h9) implies $u = u^*$. The hypothesis (h9) is naturally called the *hypothesis of strict monotonicity*. ∎

We will state now the following general result of existence, where no assumption of monotonicity is made:

THEOREM 10.3. If E, $\phi: E \to]-\infty, +\infty]$ and $f: E \times E \to \mathbb{R}$ satisfy the hypotheses (h1), (h2), (h3), (h4), (h8), and

(h10) $\forall w \in E \ f(\cdot, w)$ is l.s.c.,

then problem 10.1 has solution.

Proof. The solutions of problem 10.1 are, as we have seen, the elements of $\bigcap_{w \in E} \sigma(w)$. The result follows therefore from lemma 1 and from the fact that (h2) and (h10)) imply that the $\sigma(w)$ are closed. In fact, taking in $\sigma(w)$ a filter $\{v_\alpha\}_{\alpha \in A}$ with limit v we have, from
$$\phi(v_\alpha) + f(v_\alpha, w) \leq \phi(w) \quad \forall \alpha \in A \quad (10.30)$$

and from the fact that ϕ and $f(\cdot, w)$ are lower semicontinuous (i.e. from (h2) and (h10)), taking the limit, $\phi(v) + f(v, w) \leq \phi(w)$ and hence $v \in \sigma(w)$. ∎

This result might seem at first sight to be more general than theorem 10.1, since it does not require monotonicity and, although (h10) is stronger than (h6), (h5) is completely eliminated. However, while theorem 10.1 has many applications, theorem 10.3 is of little interest. This is because the assumption that $f(w, \cdot)$ be u.s.c. is not a very restrictive one in most applications, and on the other hand the 'difference' between (h6) and (h10) is in effect very large. In (h6) the topology of E does not bother us much since we work in finite-dimensional spaces and we can, for example, take in E the weak topology which is the most suitable in order to satisfy (h8); in (h10), on the other hand, the topology of E is essential, and in particular for the weak topology it is a very restrictive assumption. All this will become clearer in the next section.

10.2 PARTICULAR CASES. I

In this section we will take E to be an arbitrary reflexive Banach space with dual E': (h1) is thus satisfied. In general we will work with the weak topology of E because, being the one which has 'most compact sets', it is also the one in which (h8) is most easily satisfied (in this context let us observe that weakly compact sets are the bounded and weakly closed sets, and vice versa).

Let us take ϕ to be defined by

$$\phi(v) \equiv j_K(v), \tag{10.31}$$

where j_K is the indicator of a non-empty closed convex set in E. Note that j_K is l.s.c. for every topology of E between the weak one and the strong one, and further it is a proper convex function, and hence (h2) is satisfied. Finally, let us take f defined by

$$f(v, w) \equiv {}_{E'}\langle Av, v - w \rangle_E, \tag{10.32}$$

where $A: E \to E'$ is an operator (not necessarily linear continuous) in terms of which we will interpret the hypothesis regarding f in theorem 10.1 (and in theorem 10.2).

The hypothesis (h3) is satisfied without further hypotheses on A:

$$\forall v \in E \quad {}_{E'}\langle Av, v - v \rangle_E = {}_{E'}\langle Av, 0 \rangle_E = 0. \tag{10.33}$$

The hypothesis (h4) is also satisfied without further assumptions on A: if we fix v in E we have

$$ {}_{E'}\langle Av, v - w \rangle_E = {}_{E'}\langle Av, v \rangle_E + {}_{E'}\langle -Av, w \rangle_E, \tag{10.34}$$

and, since ${}_{E'}\langle Av, v \rangle_E$ is constant, $w \mapsto {}_{E'}\langle Av, v - w \rangle_E$ is affine and therefore concave. From this it also follows that (h5) is satisfied without further assumptions on A: since $w \mapsto {}_{E'}\langle Av, v-w\rangle_E$ is affine, it is not only u.s.c. but even continuous. This confirms, as we have stated earlier, that (h5) is not a very restrictive assumption in the applications. Let us remark at this point that, under the same conditions, (h10) is really disastrous: supposing E to be a Hilbert space with its weak topology and A to be the identity operator (and therefore linear!), (h10) means that $v \mapsto (v, v-w)_E = \|v\|_E^2 - (v, w)_E$ is weakly continuous—but we know that there exist sequences $v_n \overset{}{\underset{E}{\rightharpoonup}} 0$ such that $\|v_n\|_E = \text{constant} \neq 0$!

We will next deal with (h6). Fixing w in E, we have to find out which assumptions we need to make on A so that the real function of a real variable

$$t \mapsto {}_{E'}\langle A(tz+(1-t)v, tz+(1-t)v-w\rangle_E \qquad (10.35)$$

will be l.s.c. for every z and v in E. A sufficient condition is that A be hemicontinuous, i.e. (see definition 7.2 and the relative comments) that $t \mapsto A(tz+(1-t)v)$ be weakly continuous for every z and v in E. In fact, under these conditions $t \mapsto {}_{E'}\langle A(tz+(1-t)v), w\rangle_E$ is continuous, and since $t \mapsto {}_{E'}\langle tz+(1-t)v, x\rangle_E$ is continuous for every x in E, (10.35) is not only l.s.c. but also continuous.

As for (h7), we require that

$$\forall v, w \in E \qquad {}_{E'}\langle Av, v-w\rangle_E + {}_{E'}\langle Aw, w-v\rangle_E \geq 0, \qquad (10.36)$$

i.e. that

$$\forall v, w \in E \qquad {}_{E'}\langle Av - Aw, v-w\rangle_E \geq 0, \qquad (10.37)$$

which is the same as the monotonicity of A (see definition 7.2): we must require therefore that A be monotone. Let us remark that in this context (h9) is the same as the strict monotonicity of A:

$$\forall v, w \in E, v \neq w, \qquad {}_{E'}\langle Av - Aw, v-w\rangle_E > 0. \qquad (10.38)$$

Finally, we will deal with (h8), which we can write, from what we have already seen, as

$$\exists B \subset E, \text{ bounded and weakly closed:} \quad \exists v_0 \in B: \quad \forall v \in E \backslash B$$
$$\phi_K(v) + {}_{E'}\langle Av, v-v_0\rangle_E > \phi_K(v_0). \qquad (10.39)$$

From (10.39) it follows immediately that we must have $\phi_K(v_0) = 0$, since otherwise $\phi_K(v_0) = +\infty$ and the strict inequality cannot be true. We conclude thus that we must have $v_0 \in B \cap K$ and that we can write (10.39) in the form

$$\exists B \subset E, \text{ bounded and weakly closed:} \quad \exists v_0 \in B \cap K: \quad \forall v \in E \backslash B$$
$$\phi_K(v) + {}_{E'}\langle Av, v-v_0\rangle_E > 0. \qquad (10.40)$$

If K is bounded, we do not need further assumptions on A because (10.40) is true with $K = B$, since then $\phi_K(v) = +\infty$ and this means that the inequality is true for every v_0 in K. The most interesting case, however, is that in which K is not bounded, and in which it will be sufficient for example to require that A be coercive (see definition 7.2):

$$\exists v_0 \in K: \lim_{\|v\|_E \to +\infty, v \in K} {}_{E'}\langle Av, v - v_0 \rangle_E = +\infty \qquad (10.41)$$

Under these conditions (10.40) is in effect satisfied since (10.41) ensures that there exists a sphere $(B\ldots)$ outside which ${}_{E'}\langle Av, v - v_0 \rangle_E > 0 \ldots$.

To end, we mention that, under the above conditions, (10.1) can be written as

$$u \in E; \quad \phi_K(u) + {}_{E'}\langle Au, u - v \rangle_E \leq \phi_K(v) \qquad \forall v \in E, \qquad (10.42)$$

or, more simply but equivalently, as

$$u \in K; \quad {}_{E'}\langle Au, u - v \rangle_E \leq 0 \qquad \forall v \in K. \qquad (10.43)$$

In summary, we can state the following result (which ensures, among other things, that the problem 7.37 has a solution, which is unique if A is strongly monotonic:

THEOREM 10.4. *If E is a reflexive Banach space, K is a non-empty closed convex set in E, and $A: E \to E'$ is a monotone hemicontinuous and coercive operator (the last hypothesis being unnecessary if K is bounded), the inequality (10.43) has solution, which is unique if in addition A is strictly monotone; further*

$$\{u \in K: \forall v \in K \quad {}_{E'}\langle Au, u - v \rangle_E \leq 0\}$$
$$= \{u \in K: \forall v \in K \quad {}_{E'}\langle Av, u - v \rangle_E \leq 0\}. \qquad (10.44)$$

REMARK 10.1. If K is a convex cone with its vertex at the origin, a linear subspace, or an affine variety, inequality (10.43) can be written, respectively, as

$$u \in K; \quad {}_{E'}\langle Au, u \rangle_E = 0 \quad \text{and} \quad \forall v \in K \quad {}_{E'}\langle Au, v \rangle_E \geq 0, \quad (10.45)$$

$$u \in K; \quad \forall v \in K \quad {}_{E'}\langle Au, v \rangle_E = 0, \qquad (10.46)$$

$$u \in K; \quad \forall v \in K \quad {}_{E'}\langle Au, u - v \rangle_E = 0. \qquad (10.47)$$

We will leave the easy proof to the reader.

10.3 PARTICULAR CASES. II

In this section we will suppose that E is a real Hilbert space. We could take E again as any reflexive real Banach space, but this is only apparently more general: indeed, we will not use explicitly the Hilbert structure of E (besides we denote as a duality the scalar product ...) but in any case we come to the conclusion that if our hypotheses are verified then we can define a scalar product in E. As ϕ we take an application of E in $]-\infty, +\infty]$ convex proper and l.s.c., and we consider once more (10.32) but with $A: E \to E'$ linear; initially we will not require the continuity of A (but we will see that this follows from the other conditions that we will impose).

We want to do here something analogous to what was done in the previous section, i.e.: interpret in terms of A the hypothesis of theorem 10.2 regarding f.

We already know that hypotheses (h3), (h4), and (h5) are satisfied without any further assumption on A. From the linearity of A it follows that (h6) is also automatically satisfied: in fact

$$t \mapsto {}_{E'}\langle A(tz+(1-t)v), w\rangle_E = t{}_{E'}\langle Az, w\rangle_E + (1-t){}_{E'}\langle Av, w\rangle_E \quad (10.48)$$

is linear and hence continuous in t, and as a consequence A is hemicontinuous. Let us now consider (h8), which we can write here as

$$\exists B \subset E, B \text{ bounded and weakly closed:} \quad \exists v_0 \in B: \quad \forall v \in E \setminus B$$
$$\phi(v) + {}_{E'}\langle Av, v-v_0\rangle_E > \phi(v_0). \quad (10.49)$$

For this to be satisfied we require here that A (and by now also ϕ!) satisfy

$$\exists v_0 \in E: \phi(v_0) < +\infty \quad \phi(v) + {}_{E'}\langle Av, v-v_0\rangle_E \to +\infty \quad \text{when } \|v\|_E \to +\infty,$$
$$(10.50)$$

which is an adapted form of (10.41). It will be useful to remove ϕ from (10.50) and impose a condition which concerns A only. For this purpose let us put $v = \lambda \bar{v}$, with $\|\bar{v}\|_E = 1$ and $\lambda \in \mathbb{R}$, and rewrite (10.50) in the form

$$\exists v_0 \in E, \phi(v_0) < +\infty: \quad \phi(\lambda\bar{v}) + \lambda^2 {}_{E'}\langle A\bar{v}, \bar{v}\rangle_E - \lambda {}_{E'}\langle A\bar{v}, v_0\rangle_E \to +\infty$$
$$\text{when } \lambda \to +\infty. \quad (10.51)$$

Since $\phi(\lambda\bar{v})$ does not tend to $-\infty$ more rapidly than a linear function (being convex), (10.51) is satisfied as long as

$$\exists v_0 \in E, \phi(v_0) < +\infty: \quad \lambda^2 {}_{E'}\langle A\bar{v}, \bar{v}\rangle_E - \lambda {}_{E'}\langle A\bar{v}, v_0\rangle_E \to +\infty$$
$$\text{when } \lambda \to +\infty, \quad (10.52)$$

which in turn is satisfied as long as

$$\exists v_0 \in E, \phi(v_0) < +\infty: \quad \exists \beta > 0: \quad \forall \bar{v}, \|\bar{v}\|_E = 1 \quad {}_{E'}\langle A\bar{v}, \bar{v}\rangle_E \geq \beta. \quad (10.53)$$

Now the first part of (10.53) tells us only that ϕ is proper, which we already know from (h2) and which we therefore no longer require. We have then that for (10.49) to be satisfied it is sufficient that

$$\exists \beta > 0: \quad \forall \bar{v}, \|\bar{v}\|_E = 1 \quad {}_{E'}\langle A\bar{v}, \bar{v}\rangle_E \geq \beta, \quad (10.54)$$

or, equivalently, given that A is linear, that

$$\exists \beta > 0: \quad \forall v \in E \quad {}_{E'}\langle Av, v\rangle_E \geq \beta \|v\|_E^2. \quad (10.55)$$

A linear operator A which satisfies (10.55) is naturally said to be *coercive*: besides this is the most common way of imposing coercivity on linear operators. Curiously enough, we do not need other assumptions on A in order to satisfy the conditions of theorem 10.1 (or even of theorem 10.2). In fact, if A is linear and coercive, it is strongly monotone (hence (h9) holds and not merely (h7)) since

$$\forall v, w \in E \quad {}_{E'}\langle Av - Aw, v - w\rangle_E = {}_{E'}\langle A(v-w), v-w\rangle_E \geq \beta \, \|v-w\|_E^2 \geq 0, \tag{10.56}$$

being zero only if $v = w$.

Again, since A is linear and coercive, A is in effect a continuous operator, i.e. $A \in \mathcal{L}(E, E')$, and further the following is a scalar product in E:

$$(u, v)_E = \tfrac{1}{2}[{}_{E'}\langle Au, v\rangle_E + {}_{E'}\langle Av, u\rangle_E]. \tag{10.57}$$

Coercivity is, as one can see, an assumption which gives rise to several consequences.

Let us now take ϕ defined by

$$\phi(v) \equiv j(v) + j_K(v), \tag{10.58}$$

where j_K is the indicator function of a non-empty closed convex set $K \subset E$ and $j = E \to {]-\infty, +\infty]}$ is proper convex (on K!) and l.s.c. (we do not have here a real particularization on ϕ).

Under these conditions we can write (10.1) in the form

$$u \in K: \quad \forall v \in K \quad j(u) + {}_{E'}\langle Au, u-v\rangle_E \leq j(v) \tag{10.59}$$

and state

THEOREM 10.5. *If E is a real Hilbert space, $K \subset E$ is a non-empty closed convex set, $j: E \to {]-\infty, +\infty]}$ is convex proper (on K) l.s.c. and $A: E \to E'$ is a continuous and coercive linear operator than there exists one and only one solution of the inequality (10.59).*

To end this section, we state some important corollaries of this theorem, which in turn is a corollary of theorem 10.1.

Let us put then ${}_{E'}\langle Au, u-v\rangle_E = a(u, u-v)$, with $a = E \times E \to \mathbb{R}$ being a continuous and coercive bilinear form, and denote by l a generic element of E'. If $j(v) = {}_{E'}\langle -l, v\rangle_E$ then (10.59) can be written as

$$u \in K; \quad \forall v \in K \quad a(u, u-v) \leq {}_{E'}\langle l, u-v\rangle_E, \tag{10.60}$$

and theorem 10.5 is nothing but the *Lions–Stampacchia theorem*.

Under the same conditions, if further $K = E$, then (10.59) can be written as

$$u \in E; \quad \forall v \in E \quad a(u, v) = {}_{E'}\langle l, v\rangle_E, \tag{10.61}$$

and theorem 10.5 is nothing but the *Lax–Milgram lemma*.

Finally, if $a(u, v) = (u, v)_E$, $K = E$ and $j(v) = {}_{E'}\langle -l, v\rangle_E$, then (10.59) can be written as

$$u \in E; \quad \forall v \in E \quad (u, v)_E = {}_{E'}\langle l, v\rangle_E, \tag{10.62}$$

and theorem 10.5 is nothing but the *Riesz theorem*.

11
Quasivariational Inequalities

11.1 INTRODUCTION

Nash points

To present our first example of a quasivariational problem, we turn to an important question in n-person game theory—the determination of the Nash points for an n-ple of functionals subject to convex constraints. We will consider here a problem already studied by Bensoussan, 1974a, but left partially unsolved in that paper.

For clarity, let us consider here a simplified form of the problem (we take $n = 2$; the difficulties, however, remain essentially the same, and the extension to the general case, which we will discuss briefly in remark 11.1 at the end of this section, is almost automatic).

Let us suppose that we are given

(b1) H_1 and H_2, real Hilbert spaces,
(b2) $a_1: H_1 \times H_1 \to \mathbb{R}$ and $a_2: H_2 \times H_2 \to \mathbb{R}$, continuous bilinear forms, symmetric and non-negative along the diagonal,
(b3) $b_1: H_2 \times H_1 \to \mathbb{R}$ and $b_2: H_2 \times H_1 \to \mathbb{R}$, continuous bilinear forms,
(b4) $L_1: H_1 \to \mathbb{R}$ and $L_2: H_2 \to \mathbb{R}$, continuous linear functionals,

and, in the following problem 11.1:

(b5$_1$) $K_1 \subset H_1$ and $K_2 \subset H_2$, non-empty closed convex sets,

or, in the following problem 11.2:

(b5$_2$) $K \subset H_1 \times H_2$, a non-empty closed convex set.

Let us put $H = H_1 \times H_2$ and, denoting by $v \equiv (v_1, v_2)$ the generic element of H, let us define the functionals $J_1: H_1 \to \mathbb{R}$ and $J_2: H_2 \to \mathbb{R}$ by means of the formulae

$$\left.\begin{array}{l} J_1(v) = a_1(v_1, v_1) + 2b_1(v_2, v_1) - 2L_1(v_1) \\ J_2(v) = a_2(v_2, v_2) + 2b_2(v_2, v_1) - 2L_2(v_2). \end{array}\right\} \quad (11.1)$$

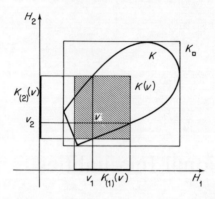

Figure 11.1

Further let us put $K^\square = K_1 \times K_2$, $K_\square = P_{H_1}(K) \times P_{H_2}(K)$, $K_{(1)}(v) = \{z_1 \in H_1 : (z_1, v_2) \in K\}$, $K_{(2)}(v) = \{z_2 \in H_2 : (v_1, z_2) \in K\}$ and $K(v) = K_{(1)}(v) \times K_{(2)}(v)$. Note that if $v \in K$ then $K_{(1)}(v) \subset P_{H_1}(K)$ and $K_2(v) \subset P_{H_2}(K)$ are non-empty closed convex sets and therefore $K(v)$ is a non-empty closed convex set of $K_\square$ (see fig. 11.1). We have thus defined an application $K \ni v \mapsto K(v) \in 2^{K_\square}_{\text{ck}}$ (this symbol denotes the set of non-empty closed convex parts of $K_\square$; cf. 20.0).

Let us now consider the two following problems:

PROBLEM 11.1. Find $u \equiv (u_1, u_2) \in K^\square$ such that

$$\left.\begin{array}{ll} J_1(u) \leq J_1(v_1, u_2) & \forall v_1 \in K_1 \\ J_2(u) \leq J_2(u_1, v_2) & \forall v_2 \in K_2. \end{array}\right\} \qquad (11.2)$$

PROBLEM 11.2. Find $u \equiv (u_1, u_2) \in K$ such that

$$\left.\begin{array}{ll} J_1(u) \leq J_1(v_1, u_2) & \forall v_1 \in K_{(1)}(u) \\ J_2(u) \leq J_2(u_1, v_2) & \forall v_2 \in K_{(2)}(u). \end{array}\right\} \qquad (11.3)$$

We will call the solutions of problem 11.1 the *Nash points* for the pair of functionals (J_1, J_2) under the constraints K_1 and K_2, and we will call the solutions of problem 11.2 the *Nash points* for the same pair of functionals under the constraint K. It is important to note here that while in problem 11.1 we have two independent constraints, the convex sets K_1 and K_2, in problem 11.2 we have an infinite number of interdependent constraints, the convex sets $K(v)$ determined by K. This has a natural consequence on the relative difficulties of the two problems....

We will now try to find necessary and sufficient conditions for the existence of solutions of the problems 11.1 and 11.2. Let us first deal with problem 11.1.

Let us consider then the functionals $G_1: H_1 \to \mathbb{R}$ and $G_2: H_2 \to \mathbb{R}$ defined by

$$\left.\begin{array}{l} G_1(v_1) = J_1(v_1, u_2) \\ G_2(v_2) = J_2(u_1, v_2), \end{array}\right\} \quad (11.4)$$

where $u \equiv (u_1, u_2) \in H$ is fixed. From (11.1) we have

$$\left.\begin{array}{l} G_1(v_1) = a_1(v_1, v_1) + 2[b_1(u_2, v_1) - L_1(v_1)] + c_1 \\ G_2(v_2) = a_2(v_2, v_2) + 2[b_2(v_2, u_1) - L_2(v_2)] + c_2, \end{array}\right\} \quad (11.5)$$

where c_1 is a constant independent of v_1 and c_2 a constant independent of v_2. It follows immediately that if $u \equiv (u_1, u_2)$ is a Nash point for the pair (J_1, J_2) under the constraints K_1 and K_2 then u_1 achieves the minimum of G_1 on K_1 and u_2 achieves the minimum of G_2 on K_2 and, conversely, if u_1 realizes the minimum of G_1 on K_1 and u_2 realizes the minimum of G_2 on K_2 then $u \equiv (u_1, u_2)$ is a Nash point for (J_1, J_2) under the constraints K_1 and K_2. From the hypothesis that a_1 and a_2 are non-negative on the diagonal it follows that the functionals G_1 and G_2 are convex, and therefore (see theorem 3.3) a necessary and sufficient condition that u_1 minimizes G_1 on K_1 and u_2 minimizes G_2 on K_2 is that

$$\left.\begin{array}{l} {}_{H_1'}\langle \nabla G_1(u_1), u_1 - v_1\rangle_{H_1} \leq 0 \quad \forall v_1 \in K_1 \\ {}_{H_2'}\langle \nabla G_2(u_2), u_2 - v_2\rangle_{H_2} \leq 0 \quad \forall v_2 \in K_2, \end{array}\right\} \quad (11.6)$$

which can be written (given that a_1 and a_2 are symmetric and taking into account the example given after definition 3.2) as

$$\left.\begin{array}{l} a_1(u_1, u_1 - v_1) + b_1(u_2, u_1 - v_1) - L_1(u_1 - v_1) \leq 0 \quad \forall v_1 \in K_1 \\ a_2(u_2, u_2 - v_2) + b_2(u_2 - v_2, u_1) - L_2(u_2 - v_2) \leq 0 \quad \forall v_2 \in K_2. \end{array}\right\} \quad (11.7)$$

At this point, defining $a: H \times H \to \mathbb{R}$ and $L: H \to \mathbb{R}$ by the formulae

$$a(w, v) = a_1(w_1, v_1) + a_2(w_2, v_2) + b_2(w_1, v_2) + b_1(v_1, w_2) \quad (11.8)$$

and

$$L(w) = L_1(w_1) + L_2(w_2), \quad (11.9)$$

we can state the following result:

THEOREM 11.1. $u \in K^\square$ is a solution of problem 11.1 iff

$$a(u, u - v) \leq L(u - v) \quad \forall v \in K^\square. \quad (11.10)$$

For the problem 11.2, following exactly the same reasoning, we have the following result:

THEOREM 11.2. $u \in K$ is a solution of problem 11.2 iff

$$a(u, u - v) \leq L(u - v) \quad \forall v \in K(u). \quad (11.11)$$

It is obvious that a is a continuous bilinear form and L is a continuous linear functional: (11.10) is therefore a variational inequality which we know how to study; in particular we can state the following:

THEOREM 11.3. *If a is coercive then there is one and only one solution of problem 11.1.*

On the other hand, (11.11) is in general no longer a variational inequality: it is a variational inequality only when $\forall u \in K$ $K(u) = Q$, with Q being a non-empty closed convex set of H (we are then in the conditions of problem 11.1 since then Q must necessarily be such that $Q = Q_1 \times Q_2$ with Q_1 being a non-empty closed convex set of H_1 and Q_2 being a non-empty closed convex set of H_2). In fact, (11.11) is a new type of entity: we will call it, along with Bensoussan–Goursat–Lions (1973), a *quasivariational inequality*.

Let us remark that the variational inequality (11.10) is a particular case of the quasivariational inequality (11.11), just as the variational equation

$$a(u, v) = L(v) \qquad \forall v \in H, \tag{11.12}$$

which corresponds to the problem of finding the Nash points for the pair (J_1, J_2) without constraints, is a particular case of the variational inequality (11.10). However, it is important to observe that, even when a is symmetric, the problem associated with (11.11) is not in general a problem of calculus of variations.

How does one study the quasivariational inequality (11.11) so as to state, for problem 11.2, a result analogous to theorem 11.3?

The present situation presents a close analogy to that found in the treatment of the Lions–Stampacchia theorem (Chapter 3): we now know how to solve a problem and we want to solve another which is not too wide a generalization of it. In the (first) proof of the Lions–Stampacchia theorem we have reduced the question of the existence of a solution of a variational inequality associated with a non-symmetric bilinear form to the previously solved problem for a symmetric form and to the existence of a fixed point. Here we wish to follow essentially the same technique: we reconduce the study of the quasivariational inequality to the study of a family of variational inequalities and to the finding of a fixed point for an appropriate transformation.

With the quasivariational inequality (11.11) we can associate in a natural way a family (in general infinite) of variational inequalities: if u is fixed in K we will call *variational section* of the quasivariational inequality (11.11) along u the variational inequality

$$a(w, w - v) \leq L(w - v) \qquad \forall v \in K(u). \tag{11.13}$$

Under the hypothesis (which is standard in the variational case, and which we will make here too) of the coerciveness of the form a we can say that (11.13) has one and only one solution, and therefore that the application

$\Sigma: K \to K_\square$ introduced by

$$\Sigma(u) = w, \quad \text{with} \quad w \in K(u) \subset K_\square \text{ solution of (11.13)}, \qquad (11.14)$$

is well defined.

We will call the application Σ the *variational selection* associated with the quasivariational inequality (11.11). Denoting by

$$\sigma_u(v) = \{w \in K(u): a(w, w-v) \le L(w-v)\} \qquad (11.15)$$

the level sets of the variational section (11.13), we can characterize the variational selection Σ by means of the formula

$$\{\Sigma(u)\} = \bigcap_{v \in K(u)} \sigma_u(v). \qquad (11.16)$$

It follows immediately that $u \in K \cap K(u)$ is a solution of (11.11) iff u is a fixed point for Σ. Unfortunately, since Σ is not an application from a set into itself, it is not easy to define the conditions under which fixed points for Σ exist.... We will get around this problem by considering, instead of Σ, the application $\Sigma': K_\square \to K_\square$ defined by

$$\Sigma'(v) = \Sigma(P_K(v)) \qquad (11.17)$$

and noting that the fixed points of Σ' coincide with those of Σ (in effect, if $u = \Sigma'(u)$ then $u \in K(u)$ and therefore $\Sigma'(u) = \Sigma(u)$, i.e. $u = \Sigma(u)$; conversely, if $u = \Sigma(u)$ then $u \in K(u)$ and the refore $P_K(u) = u$ and $\Sigma(u) = \Sigma'(u)$, i.e. $u = \Sigma'(u)$).

We now need a hypothesis which will ensure that Σ' has at least one fixed point. At this point the following question arises: what type of fixed point theorem do we intend using? We do not expect a Lipschitz continuous or a monotonic situation, and thus the theorems of sections 9.2 and 9.4 are useless: so, we have nothing other than Schauder's theorem. Thus $K_\square$ must be a compact set and Σ' must be a continuous application.

We will then assume that K is a compact set, from which it follows that $K_\square$ is also a compact set.

For Σ' to be continuous, we must impose conditions in order that $u \mapsto P_K(u)$ and Σ be continuous. Working with the strong topology (which is a fairly strong limitation since it means that the hypothesis on K is that of stong compactness), we have that $u \mapsto P_K(u)$ is continuous (since the projection operators are non-expansive and therefore strongly continuous); we will show now that Σ is continuous without any further assumption. For this purpose we will show that if $u_n \to u$ then $\Sigma(u_n) \to \Sigma(u)$. In order to do this let us put $w_n = \Sigma(u_n)$ and $w = \lim w_{n_k}$, where w_{n_k} is a convergent subsequence of w_n (remember that $K_\square$ is compact...); if we prove that $w = \Sigma(u)$ then we have completed the proof because the uniqueness of the problem associated with the variational section of (11.11) in u allows us to conclude that w is in effect the limit of w_n. We have $w_{n_k} \in K(u_{n_k})$ and $a(w_{n_k}, w_{n_k} - v) \le L(w_{n_k} - v) \, \forall v \in K(u_n)$. Now, since $K \ni u \mapsto K(u) \in 2^{K_\square}_{\blacksquare \text{ck}}$ is continuous if $2^{K_\square}_{\blacksquare \text{ck}}$

is provided with the Hausdorff metric (which we can do since, at this point, we have fixed a metric topology in H relative to which $K_\square$ is a bounded closed set, and therefore we are in effect working in $2^{K_{ckl}}$), we can write (see remark 20.2) that if $v \in K(u)$ then there exists $v_{n_k} \to v$ with $v_{n_k} \in K(u_{n_k})$. Let then v be an arbitrary element of $K(u)$ and v_{n_k} a sequence satisfying the above conditions: we can write $a(w_{n_k}, w_{n_k} - v_{n_k}) \leq L(w_{n_k} - v_{n_k})$ and, taking the limit (since a and L are linear and continuous), $a(w, w-v) \leq L(w-v)$, which implies that $w = \Sigma(u)$ since v is arbitrary. So, finally, we can state the following:

THEOREM 11.4. *If a is coercive and K is strongly compact then there exists at least one solution of problem 11.2.*

At this point, taking an abstract form of the initial problem, we can consider the following

PROBLEM 11.3. *Let H be a real Hilbert space, $a: H \times H \to \mathbb{R}$ a continuous bilinear form, $L \in H'$, $C \subset H$ a non-empty compact set and $Q: C \to 2^C_{ck}$; find*

$$u \in Q(u); a(u, u-v) \leq L(u-v) \quad \forall v \in Q(u). \quad (11.18)$$

Later we will consider generalizations of this problem, and of the result which solves it, namely the following

THEOREM 11.5. *If a is a coercive and Q is continuous then problem 11.3 has at least one solution.*

The proof of this theorem is practically the same as that of theorem 11.5, or even simpler because the variational selection Σ is already an application of a (in general weakly!) compact set onto itself and Q is continuous (by hypothesis); the fixed point theorem which we need to use is, once again, that of Schauder. Later we will use more elaborate fixed point theorems, and more precisely theorems 9.21 and 9.26, but the basic idea is exactly the same, i.e.: to consider a variational selection and then to look for its fixed points.

Bibliographical note

Bensoussan–Lions, 1973a, considered the first problem in which a quasivariational inequality is explicitly involved; this was a problem of evolution associated with control theory. Bensoussan–Goursat–Lions, 1973, considered the corresponding stationary problem; there the authors introduced the term 'quasivariational inequality' and proved the first existence result (the corresponding uniqueness result has been proved by Laetsch,

1975). For other papers on quasivariational inequalities and control theory see the list at the end of this note (and for general questions about control theory see, e.g., Lions, 1971, and Bensoussan, 1974b).

In Lions, 1976a the reader can find the homogenization of quasivariational inequalities. *Homogenization* is a technique for the study of mathematical models of non-homogeneous materials. These kinds of techniques, which have been developed recently, are beyond our scope. Here we will merely refer to a few papers which deal with these ideas: De Giorgi, 1977; De Giorgi–Franzoni, 1979 (G-*convergence* and Γ-*convergence*); Lions, 1976a; 1977a; 1977b; Bensoussan–Lions–Papanicolaou, 1978 and the notes for the C.R.Ac. Sci. Paris by these authors (*method of multiple scales*); Tartar, 1974c (*energy method*). We also quote Ene–Sanchez-Palencia, 1975, a paper of physical nature where homogenization is used to describe the boundary conditions in phenomena regarding filtration across porous materials, a subject which we deal with in Chapter 12 (see also Sanchez–Palencia, 1980).

In fact, another research subject which has led to numerous studies involving quasivariational inequalities is that dealing with the free-boundary problems associated with filtration across porous materials; we will deal with these in detail in the next chapter, to which we refer the reader to the bibliography.

For other papers in which quasivariational inequalities are studied see Anderson–Friedman, 1977; 1978; Bensoussan, 1974c; 1975; 1976; Bensoussan–Brézis–Friedman, 1977; Bensoussan–Friedman, 1978; Bensoussan–Lions, 1973b; 1973c; 1973d; 1974a; 1974b; 1974c; 1974d; 1974e; 1974f; 1975a; 1975b; 1975c; 1975d; 1975e; 1977; 1978a; 1978b; Biroli, 1976; 1977a; 1977b; 1977c; 1979; 1980; Biroli–Marchi–Norando, 1981; Boccardo, 1979; Boccardo–Dolcetta, 1978; Boccardo–Dolcetta–Matzeu, 1979; Breton–Leguay, 1975; Brezzi, 1979; Caffarelli–Friedman, 1978d; 1979; Charrier–Troianiello, 1978; Charrier–Vivaldi, 1976; 1977; Cortey–Dumont, 1978; 1979; 1980; Dolcetta–Mosco, 1980; Dolcetta–Vivaldi, 1977; 1978; Frehse–Mosco, 1979a; 1982; Friedman, 1975; 1976a; 1976c; 1978; 1980b; Friedman–Kinderlehrer, 1976; Garroni, 1982; Garroni–Hanouzet–Joly, 1982; Garroni–Troianiello, 1979; Garroni–Vivaldi, 1979; Goursat–Maarek, 1976; Goursat–Maurin, 1975; Goursat–Quadrat, 1976; Hanouzet–Joly, 1975b; 1977; 1978; 1979b; 1979c; Joly, 1974; Joly–Mosco, 1974; 1975; 1979; Joly–Mosco–Troianiello, 1974; 1977; Laborde, 1976; Lions, 1973a; 1973b; 1974; 1975a; 1975b; 1975c; 1975d; 1976b; 1976c; 1978; Loinger, 1980; Matzeu–Vivaldi, 1979; 1982; Maurin, 1976; Menaldi, 1977a; 1977b; Miellou, 1976; Mignot, 1975a; 1975b; 1976; Mignot–Puel, 1975a; 1975b; 1976; 1977; Mosco, 1974; 1976; 1977; 1978a; 1978b; 1980; Mossino, 1976; 1978; Mossino–Temam, 1977; Mossino–Zolesio, 1977; Nakoulima, 1977a; 1977b; Robin, 1975; 1976a; 1976b; Shima, 1976; Tartar, 1974a; 1974b.

REMARK 11.1. As we have already mentioned, problems 11.1 and 11.2 constitute a simplified form of those considered in Bensoussan 1974a. We will now deal briefly with the general case. We have n Hilbert spaces H_i ($i = 1, \ldots, n$; $n > 1$), n^2 symmetric continuous bilinear forms $a_{ij} : H_i \times H_j \to \mathbb{R}$ ($i, j = 1, \ldots, n$) such that $a_{ii}(v_i, v_i) \geq 0$ $\forall v_i \in H_i$, $(n^3 - n^2)/2$ continuous bilinear forms $b_{ijk} : H_j \times H_k \to \mathbb{R}$ ($i, j, k = 1, \ldots, n; j > k$), n^2 continuous linear functionals $L_{ij} : H_i \to \mathbb{R}$ ($i, j = 1, \ldots, n$) and, depending on the problem, either n non-empty closed convex sets $K_i \subset H_i$ ($i = 1, \ldots, n$) or one non-empty closed convex set $K \subset \prod_{i=1}^n H_i$. Let us put $H = \prod_{i=1}^n H_i$ and denote by $v \equiv (v_1, \ldots, v_n)$ the generic element of H; let us define the n functionals $J_i : H \to \mathbb{R}$ ($i = 1, \ldots, n$) by means of the formulae $J_i(v) = \sum_{j=1}^n a_{ij}(v_j, v_j) + 2 \sum_{j>k} b_{ijk}(v_j, v_k) - 2 \sum_{j=1}^n L_{ij}(v_j)$ (cf. (11.1)). Further let us put $K^\square = \prod_{i=1}^n K_i$, $K_\square = \prod_{i=1}^n P_{H_i}(K)$ and, introducing the notations $(v \mid i) = (v_1, \ldots, v_{i-1}, v_{i+1}, \ldots, v_n)$ and $(v \mid i \mid z_i) = (v_1, \ldots, v_{i-1}, z_i, v_{i+1}, \ldots, v_n)$, let us put $K(v \mid i) = \{z_i \in H_i : (v \mid i \mid z_i) \in K\} = P_{H_i}[K \cap (\{(v \mid i)\} \times H_i)]$ and $K(v) = \prod_{i=1}^n K(v \mid i)$. Problems 11.1 and 11.2 now take respectively the forms: 'find $u \in K^\square$ such that $J_i(u) \leq J_i(u \mid i \mid v_i)$ $\forall v_i \in K_i$ ($i = 1, \ldots, n$)' and 'find $u \in K_\square$ such that $J_i(u) \leq J_i(u \mid i \mid v_i)$ $\forall v_i \in K(u \mid i)$ ($i = 1, \ldots, n$)', and the results which we proved for those problems can now be stated in terms of the form $a(w, v) = \sum_{i=1}^n a_{ii}(w_i, v_i) + \sum_{j<i} b_{iij}(w_j, v_i) + \sum_{j>i} b_{iji}(v_i, w_j)$ and of the functional $L(w) = \sum_{i=1}^n L_{ii}(w_i)$. In the text we have used $n = 2$ and we have put $a_{11} = a_1$, $a_{22} = a_2$, $a_{12} = a_{21} = 0$, $L_{11} = L_1$, $L_{22} = L_2$, $L_{12} = L_{21} = 0$, $b_{121} = b_1$, $b_{221} = b_2$ and $K(v \mid 1) = K_{(1)}(v)$, $K(v \mid 2) = K_{(2)}(v)$.

In the following sections, we will study quasivariational inequalities by two techniques entirely different from each other; a difference which is due to the difference in the abstract framework in which the problems are placed. We will to a large extent use Tartar, 1974a, in section 11.2 and Joly–Mosco, 1979, in section 11.3.

11.2 TECHNIQUES OF MONOTONICITY

The data

Let us assume that we have given

(d1)   a real Hilbert pseudo-lattice H (or $\langle H, \wedge, \vee \rangle$, or even $\langle H, \leq \rangle$; see section 19.5),

(d2)   a real Hilbert space V such that $V \subseteq^{ds} H \subseteq V'$,

(d3)   a continuous bilinear form $a : V \times V \to \mathbb{R}$ which is H-coercive (in other words such that $\exists \alpha > 0 : \exists \lambda \in \mathbb{R}$: $\forall u \in V$ $a(u, u) + \lambda \|u\|_H^2 \geq \alpha \|u\|_V^2$),

(d4)   two points $a_1, a_2 \in V$ such that $a_1 \leq a_2$,

(d5)   an application $\phi : [a_1, a_2] \times V \to]-\infty, +\infty]$ such that, for every $u \in [a_1, a_2]$, $\phi(u, \cdot) : V \to]-\infty, +\infty]$ is a proper convex functional l.s.c.

Let us also assume that a, V, and ϕ are compatible with the pseudo-lattice structure of H in the sense that the following three conditions are satisfied:

$$a([u]^+, [u]^-) \leq 0 \quad \forall u \in V, \tag{11.19}$$

$$\exists c : \forall v \in V \quad [v]^+ \in V \quad \text{and} \quad \|[v]^+\|_V \leq c \|v\|_V, \tag{11.20}$$

$$u_1 \leq u_2 \Rightarrow \phi(u_1, \cdot) \leq \phi(u_2, \cdot), \tag{11.21}$$

where $\leq$ is the order introduced as in definition 19.19.

The abstract problem and its 'concentizations'

With these data, let us now consider the following

PROBLEM 11.4. Find

$$u \in V;\ a(u, u-v) + \phi(u, u) \leq \phi(u, v) \quad \forall v \in V. \tag{11.22}$$

It is of interest to note that we are dealing with a problem associated with a form which is not necessarily coercive, but only H-coercive: this observation is important because the techniques that we will develop here can be used in the study of the existence of solutions of certain variational problems associated with non-coercive forms (problems which correspond to $\phi(u, v) \equiv \psi(v) \ldots$)

Putting, for $u \in [a_1, a_2]$,

$$K(u) = \{v \in V : \phi(u, v) < +\infty\}, \tag{11.23}$$

we can easily see that (11.21) implies that

$$u_1 \leq u_2, v_1 \in K(u_1) \quad \text{and} \quad v_2 \in K(u_2) \Rightarrow v_1 \wedge v_2 \in K(u_1)$$
$$\text{and} \quad v_1 \vee v_2 \in K(u_2). \tag{11.24}$$

In fact, if $v_1 \in K(u_1)$ and $v_2 \in K(u_2)$ then $\phi(u_1, v_1) < +\infty$ and $\phi(u_2, v_2) < +\infty$ and hence $\phi(u_1, v_1 \vee v_2) < +\infty$ and $\phi(u_2, v_1 \wedge v_2) < +\infty$, since, given that the right-hand side of condition (ii) of definition 19.19 is finite, the left-hand side must be finite.

Conversely, given $K : [a_1, a_2] \to 2^V_{\text{ck}}$ such that (11.24) is satisfied, it follows immediately that the family of proper convex functionals l.s.c.

$$\phi(u, \cdot) = j_{K(u)} \tag{11.25}$$

satisfies (11.21); again, if $f \in V'$, the family defined by

$$\phi(u, v) = j_{K(u)} - {}_{V'}\langle f, v\rangle_V = \begin{cases} -{}_{V'}\langle f, v\rangle_V & \text{if } v \in K(u) \\ +\infty & \text{otherwise} \end{cases} \tag{11.26}$$

also satisfies (11.21).

With this, let us consider the following

PROBLEM 11.5. Given H and V satisfying the conditions of problem 11.4, $f \in V'$ and $K:[a_1, a_2] \to 2^V_{\blacksquare ck}$ satisfying (11.24), find

$$u \in K(u); \forall v \in K(u) \quad a(u, u-v) \leq {}_{V'}\langle f, u-v \rangle_V. \quad (11.27)$$

From what we have just seen, we can say that this problem is a particular case of problem 11.4, the relationship between them being the same as that between theorem 3.5 and the usual problem for variational inequalities, such as that considered in theorem 3.1. Problem 11.5 is, like problem 11.3, a problem associated with a linear quasivariational inequality: however, condition (11.24) would suggest that we use techniques of the 'monotonicity' type rather than of the 'compactness' type as we deal in section 11.1; in other words, it suggests that we use the fixed point theorem 9.26 in place of Schauder's theorem. We have written problem 11.5 so as to make the quasivariational inequality explicit, since its presence in (11.22) is 'conceded'; the problem we are interested in studying is nevertheless the more general problem 11.4, which we will solve by first expressing it as a family (infinite, if K is not a constant) of variational problems by considering the variational selection Σ—or rather, for reasons which will later become clear, Σ_λ—and then proving the existence of fixed points for this.

A theorem of existence

Let then $\lambda \geq 0$ under the conditions of (d3) and let us consider the variational selection Σ_λ associated with the quasivariational problem '*find*

$$u \in K(u); \quad \forall v \in K(u) \quad a(u, u-v) + \lambda(u, u-v)_H$$
$$+ (\phi(u, u) - \lambda(u, u)_H) \leq (\phi(u, v) - \lambda(u, v)_H), \quad (11.28)$$

a problem which has variational sections corresponding to a coercive form, the form $a_\lambda(u, v) = a(u, v) + \lambda(u, v)_H$, but which is equivalent to problem 11.4 which does not have them. We should also note that the two problems are equivalent also from a conceptual point of view, since if $\phi(\cdot, \cdot)$ satisfies (11.21) then the same is true for $\phi(\cdot, \cdot) - \lambda(\cdot, \cdot)_H$. In fact, if $\lambda \geq 0$ and $w_1 \leq w_2$ then $-\lambda(w_1, \cdot)_H \leq -\lambda(w_2, \cdot)_H$, i.e., $\forall u, v \in [a_1, a_2]$ $-\lambda(w_1, u \land v)_H - \lambda(w_1, u \lor v)_H \leq -\lambda(w_1, u)_H - \lambda(w_2, v)_H$; the above statement follows then from the property 2) which we stated immediately following definition 19.19. We can then say that

$$\Sigma_\lambda(u) = \{w \in V : a(w, w-v) + \lambda(w, w-v)_H + \phi(u, w) - \lambda(u, w)_H$$
$$\leq \phi(u, v) - \lambda(u, v)_H \quad \forall v \in V\} \quad (11.29)$$

defines an application $\Sigma_\lambda : [a_1, a_2] \to V$ (we will consider the set $\Sigma_\lambda(u)$ as being identical to its unique element ...). It follows immediately that u is a solution of the problem 11.4 if $u = \Sigma_\lambda(u)$, and this is independent of the value of λ (as long as the conditions on λ are satisfied)—we will now prove that Σ_λ has at least one fixed point, which we will do by using theorem 9.26.

Let us first observe that, in agreement with theorem 19.12, the interval $[a_1, a_2]$, being a bounded set of H, is completely inductive and that, since V is complete, $[a_1, a_2] \cap V$ is also completely inductive. To satisfy the assumptions of theorem 9.26 we now need that Σ_λ (and hence in particular $\Sigma_\lambda|_{[a_1,a_2] \cap V}$) be increasing and that $[a_1, a_2] \cap V$ be stable for it. Σ_k is increasing since, from (11.21), if $u_1 \leq u_2$ then $\phi(u_1, \cdot) - \lambda(u_1, \cdot)_H \leq \phi(u_2, \cdot) - \lambda(u_2, \cdot)_H$ and therefore, from the comparison theorem 19.20, $\Sigma_\lambda(u_1) \leq \Sigma_\lambda(u_2)$. The stability of $[a_1, a_2] \cap V$ relative to $\Sigma_\lambda|_{[a_1,a_2] \cap V}$ does not follow, however, from the data and we must introduce it as an assumption; in other words, let us suppose that

$$a_1 \leq \Sigma_\lambda(a_1) \quad \text{and} \quad \Sigma_\lambda(a_2) \leq a_2 \tag{11.30}$$

and let us introduce in this connection the following nomenclature: if $w \in [a_1, a_2] \cap V$ is such that $w \leq \Sigma_\lambda(w)$ [or: $w \geq \Sigma_\lambda(w)$], then w is said to be a λ-*subsolution* [or: *supersolution*] of problem 11.4. We note here for the sake of curiosity that, in terms of the data of problem 11.5, for (11.30) to be satisfied it is sufficient that

$$a(a_1, \phi) \leq {}_{V'}\langle f, \phi \rangle_V \quad \forall \phi \geq 0 \quad \text{and} \quad \exists w_1 \in K(u_1): \quad a_1 \leq w_1 \tag{11.31}$$

and that

$$a(a_2, \phi) \leq {}_{V'}\langle f, \phi \rangle_V \quad \forall \phi \leq 0 \quad \text{and} \quad \exists w_2 \in K(u_2): \quad a_2 \geq w_2 \tag{11.32}$$

(i.e., if a_1 satisfies (11.31) and a_2 satisfies (11.32) then a_1 and a_2 are, respectively, a λ-subsolution and a λ-supersolution of problem 11.5).

We can now state the following result of existence:

THEOREM 11.6. *If a_1 is a λ-subsolution and a_2 is a λ-super solution of problem 11.4 then the set of solutions of this problem in the interval $[a_1, a_2]$ is non-empty and has a smallest element u_m and a largest element u_M.*

A theorem of uniqueness

The solution is not in general unique (see example 11.2 below); we will, however, state the following result which we will not prove:

THEOREM 11.7. *Under the conditions of the previous theorem, if*

$$u_m \geq a_1 + \varepsilon(a_2 - a_1), \quad \text{with } \varepsilon > 0, \tag{11.33}$$

$$\phi(u, v) < +\infty \quad \text{and} \quad w \leq v \Rightarrow \phi(u, w) < +\infty, \tag{11.34}$$

$$\forall z \in V, z \geq 0 \quad (u, v) \mapsto \phi(u, v) - \phi(u, v - z) \text{ is convex,} \tag{11.35}$$

or if

$$u_M \leq a_2 - \varepsilon(a_2 - a_1), \quad \text{with } \varepsilon > 0, \quad (11.36)$$

$$\phi(u, v) < +\infty \quad \text{and} \quad w \geq v \Rightarrow \phi(u, w) < +\infty, \quad (11.37)$$

$$\forall z \in V, z \geq 0 \quad (u, v) \mapsto \phi(u, v) - \phi(u, v + z) \text{ is convex}, \quad (11.38)$$

the problem 11.4 has a unique solution in $[a_1, a_2]$ (i.e. $u_m = u_M$).

Let us note that (11.34) and (11.35) imply that Σ_λ is concave and (11.37) and (11.38) that Σ_λ is convex (this is the difficult part of the proof...): the result follows then from the uniqueness theorem 9.27. Further, note that, in terms of the data of problem 11.5, conditions (11.34) and (11.35) can be written, respectively, as

$$v \in K(u) \quad \text{and} \quad w \leq v \Rightarrow w \in K(u) \quad (11.39)$$

and

$$v_1 \in K(u_1) \quad \text{and} \quad v_2 \in K(u_2) \Rightarrow [(1-\theta)v_1 + \theta v_2] \in K((1-\theta)v_1 + \theta v_2),$$

$$\theta \in [0, 1]; \quad (11.40)$$

conditions (11.37) and (11.38) have an analogous 'translation'.

A theorem of comparison

Finally, before analysing some examples, let us present a result of comparison of the solutions of problem 11.4 with those of the following problem:

PROBLEM 11.6. Problem 11.4 with ϕ' in place of ϕ such that $\forall u \in [a_1, a_2]$ $\phi(u, \cdot) \leq \phi'(u, \cdot)$.

The result, which follows from the comparison theorem 19.20 together with the fact (which follows from the same theorem) that if $\lambda_1 \leq \lambda_2$ and w is a λ_2-subsolution [or: λ_2-supersolution] then w is a λ_1-subsolution [or: λ_1-supersolution], is as follows:

THEOREM 11.8. If a_1 is a λ-subsolution of problem 11.4 and a_2 is a λ'-supersolution of problem 11.6 then $u_m \leq u'_m$ and $u_M \leq u'_M$, with u'_m and u'_M having the obvious meanings.

Examples

The rest of the section is devoted to the consideration of two examples.

EXAMPLE 11.1. We will take up here the problem dealt with by Bensoussan–Goursat–Lions, 1973, and prove the existence and the uniqueness of the solution using the techniques which we have just developed. We are given a bounded open set $\Omega \subset \mathbb{R}^n$ with the boundary Γ of class $C^{0,1}$, an elliptic differential operator

$$L\square \equiv -\sum_{i,j=1}^{n} \frac{\partial}{\partial x_i}\left[a_{ij}\frac{\partial}{\partial x_j}\square\right] + \sum_{i=1}^{n} b_i \frac{\partial}{\partial x_i}\square + d\square \quad (11.41)$$

with coefficients in $L^\infty(\Omega)$, an operator $M: L^\infty(\Omega) \to L^\infty(\Omega)$ defined by

$$(M(\phi))(x) = 1 + \inf_{x+\xi\in\Omega, \xi\geq 0} \phi(x+\xi), \quad (11.42)$$

and a function $f \in L^\infty(\Omega)$ such that $f \geq 0$ a.e. in Ω. With these data we will consider the following problem, where the conditions are to be interpreted in a purely formal way:

PROBLEM 11.7. Find $u \in H^1(\Omega)$ such that

$$Lu \leq f \quad \text{in } \Omega, \quad (11.43)$$
$$u \leq M(u) \quad \text{in } \Omega, \quad (11.44)$$
$$(Lu - f)(u - M(u)) = 0 \quad \text{in } \Omega, \quad (11.45)$$
$$\frac{\partial u}{\partial \nu_a} \leq 0, \quad (u - M(u))\frac{\partial u}{\partial \nu_a} = 0 \quad \text{on } \Gamma. \quad (11.46)$$

In particular, $\partial u/\partial \nu_a$ must be interpreted formally as the conormal derivative associated with the bilinear form

$$a(u,v) = \int_\Omega \left[\sum_{i,j=1}^{n} a_{ij} \frac{\partial u}{\partial x_i}\frac{\partial v}{\partial x_j} + \sum_{i=1}^{n} b_i \frac{\partial u}{\partial x_i} v + duv\right] dx, \quad (11.47)$$

which we will suppose as being coercive on $H^1(\Omega)$ and which is associated in a standard manner with the operator L. Problem 11.7 recalls the problems with an internal obstacle for variational inequalities, but it is not a variational problem: in fact problem 11.7 is (formally) equivalent to

PROBLEM 11.8. Find $u \in H^1(\Omega)$ such that

$$u \leq M(u); a(u, u-v) \leq {}_{L^2(\Omega)}\langle f, u-v\rangle_{L^2(\Omega)} \quad \forall v \in H^1(\Omega), v \leq M(u). \quad (11.48)$$

This is a quasivariational problem (with an internal obstacle...), as one can see by the fact that, putting

$$K(u) = \{v \in H^1(\Omega) : v \leq M(u)\}, \quad (11.49)$$

the (11.48) can be written

$$u \in K(u); a(u, u-v) \leq {}_{L^2(\Omega)}\langle f, u-v\rangle_{L^2(\Omega)} \quad \forall v \in K(u) \quad (11.50)$$

which is evidently a quasivariational inequality. There is a proof in

Hanouzet–Joly, 1975b, of the fact that if $a_{ij} = \delta_{ij}$ $(i, j = 1, \ldots, n)$, $b_i =$ constant $(i = 1, \ldots, n)$, $d > 0$ and $\Omega =]0, 1[^n$ then problem 11.8 has at least one solution in $W^{2,p}(\Omega)$ for every $p \in]1, +\infty[$ which, further, verifies the relationship $\partial u/\partial \nu_a = 0$, which is more precise than (11.46).

We wish to prove next the existence of a solution of problem 11.8, showing that we are in a position to apply theorem 11.6. We can assume $H = L^2(\Omega)$, which is a Hilbert pseudo-lattice (in fact a Hilbert lattice). We have $V = H^1(\Omega)$ which, in agreement with the results of section 5.5, satisfies the hypothesis. Form (11.47) also satisfies the hypothesis. On the other hand it is evident that the sets $K(u)$ defined in (11.49) are non-empty closed convex sets which satisfy the condition (11.24). Further, if $M : L^2(\Omega) \to L^1_{\text{loc}}(\Omega)$ is an arbitrary increasing operator then (11.49) defines an application $K : L^2(\Omega) \to 2^{H^1(\Omega)}_{\text{ck}}$ which again satisfies (11.24) (we are not in a position to ensure, however, that $K(u) \neq \varnothing$—this must be checked separately every time). With respect to this generalization, let us note further that if $M_1 : L^2(\Omega) \to L^1_{\text{loc}}(\Omega)$ and $M_2 : L^2(\Omega) \to L^1_{\text{loc}}(\Omega)$ are increasing then $K(u) = \{v \in H^1(\Omega) : M_1(u) \leq v \leq M_2(u)\}$ defines an application $K : L^2(\Omega) \to 2^{H^1(\Omega)}_{\text{ck}}$ which satisfies (11.24). All these convex sets (for concrete examples of which we refer the reader to the literature cited) are defined by means of local conditions; as a curiosity we present an application ϕ which determines a family of functionals under the conditions of (11.21) which is defined by means of a global condition: $\phi(u, v) = \int_\Omega \psi(u, v) \, dx$, where $\psi(s, t)$ is convex in t and has derivatives with the appropriate signs (see Tartar, 1974a).

To prove the existence of solution for problem 11.8 we need therefore only show that there exists an interval $[a_1, a_2]$ which is stable for Σ_0 (let us put $\lambda = 0$ since a is coercive on $H^1(\Omega)$ and not only $L^2(\Omega)$-coercive . . .). Let us take a_1 as the solution $u = 0$ of the Neumann problem

$$u \in H^1(\Omega); a(u, v) \leq_{L^2(\Omega)} \langle 0, v \rangle_{L^2(\Omega)} \quad \forall v \in H^1(\Omega), \tag{11.51}$$

and a_2 as the solution $u = \psi$ of the Neumann problem

$$u \in H^1(\Omega); a(u, v) \leq_{L^2(\Omega)} \langle f, v \rangle_{L^2(\Omega)} \quad \forall v \in H^1(\Omega). \tag{11.52}$$

It is evident that $0 \leq \psi$ since $0 \leq f$ and we are under the conditions of the comparison theorem 19.10; we now have to show that $0 \leq \Sigma_0(0)$ and $\psi \geq \Sigma_0(\psi)$, which is evident both directly and by using the sufficient conditions (11.31) and (11.32) which are satisfied, respectively, with $w_1 = 1$ and $w_2 = 0$. We have thus shown that there exists a solution of problem 11.8.

Let us now analyse the question regarding the uniqueness of the solution (in the interval $[0, \psi]$!—theorem 11.7 concerns 'local' uniqueness . . .). Let us first note that the operator M defined in (11.42) is concave, and that, for a general concave operator $M : L^2(\Omega) \to L^1_{\text{loc}}(\Omega)$, (11.49) defines an application $K : L^2(\Omega) \to 2^{H^1(\Omega)}_{\text{ck}}$ which satisfies conditions (11.39) and (11.40) (the concavity of M is immediate: if $\theta \in [0, 1]$, $M(\theta u_1 + (1 - \theta) u_2) = 1 + \inf (\theta u_1 + (1 - \theta) u_2) \geq 1 + \theta \inf u_1 + (1 - \theta) \inf u_2 = \theta + (1 - \theta) + \theta \inf u_1 + (1 - \theta) \inf u_2 = \theta M(u_1) + (1 - \theta) M(u_2))$. On the other hand, condition (11.33) is also satisfied since, as can be shown directly, 0 is not a solution of the problem. We can therefore ensure the uniqueness of the solution of problem 11.8 in the interval $[0, \psi]$.

Finally, we present an application of the comparison theorem 11.8: substituting, in the statement of problem 11.8, f and M by $f' \geq f$ and $M': L^2(\Omega) \to L^1_{\text{loc}}(\Omega)$ defined by

$$(M'(\phi))(x) = 2 + \inf_{x+\xi \in \Omega, \xi \geq 0} \phi(x+\xi), \qquad (11.53)$$

respectively, we obtain a problem which has a solution u' which is greater than or equal to the solution u of problem 11.8.

EXAMPLE 11.2. This example is in effect a set of three counter-examples put together which show that, in general, the solution of problem 11.5 is not unique. We are given $V = H = \mathbb{R}$, $a(u, v) = uv$, $f = 1$ and: (1) $K_1(u) = \{v : v \leq 2u - u^2\}$, $u \in [0, 1]$; (2) $K_2(u) = \{v : 2u - 2 \leq v \leq 2\}$, $u \in [0, 2]$; (3) $K_3(u) = \{v : v \leq 2([u]^+)^2\}$, $u \in [-1, 1]$.

Let us now consider the three problems ($i = 1, 2, 3$):

PROBLEM 11.9.i. Find

$$u \in K_i(u); \, u(u-v) \leq u - v \qquad \forall v \in K_i(u). \qquad (11.54.i)$$

It is evident that we are under the conditions of theorem 11.6. We are not, however, under those of theorem 11.7: in problem 11.9.1, (11.39) and (11.40) are satisfied but (11.33) is not; in problem 11.9.2, (11.33) and (11.40) are satisfied but not (11.39); and, in problem 11.9.3, (11.33) and (11.39) are satisfied but not (11.40). On the other hand, it can be easily seen that 0 and 1 are solutions of problem 11.9.1 in $[0, 1]$; 1 and 2 are solutions

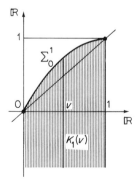

Figure 11.2.1

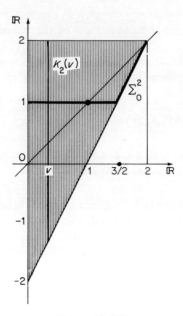

Figure 11.2.2

of problem 11.9.2 in $[0, 2]$; and 0, 1/2, and 1 are solutions of problem 11.9.3 in $[-1, 1]$. Figures 11.2.1–11.2.3 show the variational selections Σ_0^i for these problems ($i = 1, 2, 3$). These examples show, besides, that if one omits one of the three assumptions of theorem 11.7 the theorem is no longer true.

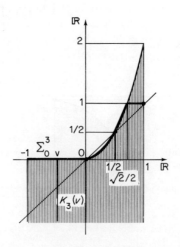

Figure 11.2.3

11.3 TECHNIQUES OF COMPACTNESS

The general problem

Let us suppose that we have

(d1′) a real vector space E,
(d2′) a non-empty convex set $K \subset E$,
(d3′) an application $f: K \times K \to \mathbb{R}$ such that, for every $u \in K$, $f(u, \cdot): K \to \mathbb{R}$ is concave and $f(u, u) \leq 0$,
(d4′) a non-empty set $C \subset K$,
(d5′) an application $\phi: C \times K \to]-\infty, +\infty]$ such that, for every $u \in C$, $\phi(u, \cdot): K \to]-\infty, +\infty]$ is a proper convex functional.

Let us then consider the following

PROBLEM 11.10. Find

$$u \in C; \; \phi(u, u) + f(u, w) \leq \phi(u, w) \quad \forall w \in K. \tag{11.55}$$

Particular cases

The following examples of particular cases of this problem will show how general it is; there is in fact a long and varied series of problems that can be reduced to an inequality of the same type as (11.55). We will leave to the reader the simple proof of the fact that the data of these examples satisfy $(d1'), \ldots, (d5')$.

EXAMPLE 11.3. *Data*: A real vector space E, a convex set $K = C \neq \emptyset$ of E, $\phi \equiv 0$ and $f(u, v) = \langle Au, u - v \rangle$ with $A: E \to E^*$. The problem 11.10 is then the following 'variational' problem: find

$$u \in K; \; \langle Au, u - w \rangle \leq 0 \quad \forall w \in K \tag{11.56}$$

(This is not strictly a variational problem since we have made no assumptions of topological type and one usually associates the term variational with problems formulated on topological spaces. This remark is also true for examples 11.3 *bis*, 11.6, and 11.6 *bis*.)

EXAMPLE 11.3 *bis*. *Data*: E, K, C, and f as in the previous example and $\phi(u, v) \equiv \psi(v)$ with $\psi: K \to]-\infty, +\infty]$ a proper convex functional. The problem 11.10 is once again a 'variational' problem, though more general than the previous one: find

$$u \in K; \; \psi(u) + \langle Au, u - w \rangle \leq \psi(w) \quad \forall w \in K. \tag{11.57}$$

EXAMPLE 11.4. *Data*: A real vector space E, a convex set $K = C \neq \emptyset$ of E, $f \equiv 0$ and $\phi(u, v) = j_{Q(u)}(v)$ with $Q: K \to 2^K_{\blacksquare k}$. The expression (11.55) becomes

$$u \in K; \; j_{Q(u)}(u) \leq j_{Q(u)}(w) \quad \forall w \in K \tag{11.58}$$

and thus (since $Q(u) \neq \varnothing$ and hence for some $w \in K$ we have $w \in Q(u)$) the problem 11.10 is the problem 'find $u \in K$ such that $u \in Q(u)$': this is a fixed point problem for a multi-valued application.

EXAMPLE 11.4 bis. *Data*: E, K, C, and f as in the previous example and $\phi(u, v) = j_{Q(v)}(u)$ with $Q: K \to 2^K$ quasisurjective and such that if $u \in Q(v_1) \cap Q(v_2)$ and $\lambda \in [0, 1]$ then $u \in Q(\lambda v_1 + (1-\lambda)v_2)$. Expression (11.55) now becomes

$$u \in K; \ j_{Q(u)}(u) \leq j_{Q(w)}(u) \qquad \forall w \in K \qquad (11.59)$$

and thus (since Q is quasisurjective and hence, for at least one $w \in K$, $u \in Q(w)$) problem 11.10 is once again the problem 'find $u \in K$ such that $u \in Q(u)$'.

EXAMPLE 11.5. *Data*: $E = \prod_{i=1}^{n} E_i$ with E_i ($i = 1, \ldots, n$) real vector spaces, $K = C = \prod_{i=1}^{n} K_i$ with $K_i \subset E_i$ ($i = 1, \ldots, n$) non-empty convex sets, $f \equiv 0$ and $\phi(u, v) = \sum_{i=1}^{n} \phi_i(u |i| v_i)$ with $\phi_i = K \to \mathbb{R}$ ($i = 1, \ldots, n$) convex in the ith variable. Problem 11.10 is then that of finding the Nash points for the n-ple of the functionals $\phi_1, \ldots, \phi_n$ under the independent constraints $K_1, \ldots, K_n$: find

$$u \in K; \ \phi(u, u) \leq \phi(u, w) \qquad \forall w \in K. \qquad (11.60)$$

EXAMPLE 11.5 bis. *Data*: $E, K,$ and f as in previous example, C a non-empty convex subset of K and $\phi(u, v) = \sum_{i=1}^{n} \phi_i(u |i| v_i) + \sum_{i=1}^{n} j_{C(u|i)}(v)$ with $\phi_i : K \to \mathbb{R}$ ($i = 1, \ldots, n$) again as in the previous example. Problem 11.10 is now the problem of finding the Nash points for the n-ple of the functionals $\phi_1, \ldots, \phi_n$ under the interdependent contraints determined by C: find

$$u \in C; \ \phi(u, u) \leq \phi(u, w) \qquad \forall w \in C(u). \qquad (11.61)$$

EXAMPLE 11.6. *Data*: A real vector space E, a non-empty convex subset K of E, C a non-empty subset of K, $f(u, v) = \langle Au, u - v \rangle$ with $A : E \to E^*$ and $\phi(u, v) = j_{Q(u)}(v)$ with $Q : C \to 2^K_{\blacksquare k}$. Problem 11.10 is now the following 'quasivariational' problem: find

$$u \in C; u \in Q(u) \quad \text{and} \quad \langle Au, u - w \rangle \leq 0 \qquad \forall w \in Q(u). \qquad (11.62)$$

EXAMPLE 11.6 bis. *Data*: $E, K, C,$ and f as in the previous example and $\phi(u, v) = j_{Q(u)}(v) + \psi(v)$ with $Q : C \to 2^K_{\blacksquare k}$ and $\psi : K \to]-\infty, +\infty]$ a proper convex functional. Problem 11.10 is once again a 'quasivariational' problem (which is a generalization of the previous problem in the same way as example 11.3 bis is a generalization of example 11.3): find

$$u \in C; u \in Q(u) \quad \text{and} \quad \psi(u) + \langle Au, u - w \rangle \leq \psi(w) \qquad \forall w \in Q(u) \qquad (11.63)$$

The fundamental theorem

We will now show that, under appropriate hypotheses, problem 11.10 has at least one solution. These hypotheses are, on the one hand, the conditions which we have already imposed on the data of the problem and, on the other, hypotheses which in one way or another are connected with a topology in E, and to which we have not referred so as not to make the description of the examples unnecessarily tedious. An exception to this rule is the hypothesis of monotonicity of f in theorem 11.9 which, though it does not involve any topological concept, we have not yet imposed: this hypothesis is not, however, indispensable in order to prove an existence result, as we will see in theorem 11.10. The hypothesis of theorem 11.9 as well as those of theorem 11.10 have the final aim of ensuring that the application $\Sigma: C \to 2^K$ defined by

$$\Sigma(u) = \{v \in K : \phi(u, v) + f(v, w) \leq \phi(u, w) \quad \forall w \in K\} \qquad (11.64)$$

has at least one fixed point in C; Σ is, in fact, the natural generalization of the concept of variational selection and we intend following here the method indicated in section 11.1 and applied successfully also in section 11.2....

THEOREM 11.9. If E, $\emptyset \neq K \subset E$, $\emptyset \neq C \subset K$, $\phi: C \times K \to]-\infty, +\infty]$ and $f: K \times K \to \mathbb{R}$ are such that

(H1) E is a real locally convex Hausdorff topological vector space and K is a closed convex set,
(H2) $\forall u \in C$ $\phi(u, \cdot): K \to]-\infty, +\infty]$ is a proper convex functional l.s.c.,
(H3) $\forall v \in K$ $f(v, v) \leq 0$,
(H4) $\forall v \in K$ $f(v, \cdot): K \to \mathbb{R}$ is concave,
(H5) $\forall v \in K$ $f(v, \cdot)$ is u.s.c.,
(H6) $\forall w \in K$ $\forall L \subset E$, L vectorial subspace of finite dimension, $f(\cdot, w)|_{K \cap L}$ is l.s.c.,
(H7) $\forall v, w \in K$ $f(v, w) + f(w, v) \geq 0$,
(H8) $\forall u \in C$ $\exists B_u \subset E$ compact: $\exists w_u \in K \cap B_u$: $\forall v \in K \setminus B_u$ $\phi(u, v) + f(v, w_u) > \phi(u, w_u)$,
(H9) $\exists F \subsetneq E$, F locally convex Hausdorff topological vector space: $\exists \emptyset \neq S \subset F \cap C$ compact convex set in F: $\forall u \in S$ $\{v \in K: \forall w \in K$ $\phi(u, v) + f(v, w) \leq \phi(u, w)\} \subset S$,
(H10) $\forall$ filter $S \times S \ni (u_\alpha, v_\alpha) \to (u, v)$ in $F \times F$
       $[\forall w \in K$ $\phi(u_\alpha, v_\alpha) + f(v_\alpha, w) \leq \phi(u_\alpha, w)] \Rightarrow$
       $[\phi(u, v) \leq \liminf_\alpha \phi(u_\alpha, v_\alpha)]$,
(H11) $\forall$ filter $S \times S \ni (u_\alpha, v_\alpha) \to (u, v)$ in $F \times F$
       $[\forall w \in K$ $\phi(u_\alpha, v_\alpha) + f(v_\alpha, w) \leq \phi(u_\alpha, w)] \Rightarrow$
       $[\forall w \in K$ $\exists w_\alpha \in K: \liminf_\alpha (\phi(u_\alpha, w_\alpha) - f(v_\alpha, w_\alpha) + f(v_\alpha, w)) \leq \phi(u, w)]$,

then there exists at least one solution of problem 11.10.

Some comments

Before going on to the proof of this imposing result (imposing not only in its generality but also, unfortunately, in the number of hypotheses made...), we will make some comments and state theorem 11.10, which we will not prove.

The hypotheses (H1),..., (H8) are essentially those of theorem 10.1: we cannot, however, consider theorem 10.1 to be a corollary of theorem 11.9 because in the proof of the latter we make extensive use of the former, which is therefore used as a lemma. We will also use theorem 9.21 as a lemma; this theorem, in its weakened form (hypothesis of continuity, here ensured essentially by (H10) and (H11), rather than upper semicontinuity: theorem 9.24 with $E = F$...), is also a particular case of theorem 11.9. Theorem 11.9 is, as we will see, effectively a 'sum' of theorem 10.1 and theorem 9.21.

Let us remark that the hypothesis (H9), which implies the existence of a compact convex set S such that $\Sigma(S) \subset 2^S$, is the analogue of the stability hypothesis on the existence theorem 11.6; if $K = C$, a sufficient condition for it to be satisfied is that the coercivity hypothesis (H8) should be uniformly true in u, i.e., that

(H8') $\exists B \subset E$ compact: $\exists w_0 \in B \cap K$: $\forall u \in K$ $\forall v \in K \setminus B$

$\phi(u, v) + f(v, w_0) > \phi(u, w_0)$.

In fact, under these conditions it is sufficient to take $F = E$ and $S = K \cap \overline{\mathrm{conv}}(B)$ in order that (H9) be true. Let us remark that in fact S is non-empty since $w_0 \in S$; S is a compact convex set since C is a closed convex set, conv (B) is a compact convex set, and E is a Hausdorff space; S is stable for E since, on the one hand, $\Sigma(u) \subset K$ $\forall u \in K$ and, on the other, $\Sigma(u) \subset B$ $\forall u \in K$ (since if $z \in \Sigma(u)$ and $z \notin B$ then, by the definition of Σ, $\phi(u, z) + f(z, w) \leq \phi(u, w)$ $\forall w \in K$ and, from (H8'), $\phi(u, z) + f(z, w_0) > \phi(u, w_0)$, which is contradictory).

We will now state the often quoted

THEOREM 11.10. *If* E, $\varnothing \neq K \subset E$, $\varnothing \neq C \subset K$, $\phi: C \times K \to]-\infty, +\infty]$ *and* $f: K \times K \to \mathbb{R}$ *satisfy the hypotheses* (H1), (H2), (H3), (H4), (H6), (H8), (H9), (H10),

(H12) $\forall$ filter $S \times S \ni (u_\alpha, v_\alpha) \to (u, v)$ in $F \times F$
$[\phi(u_\alpha, v_\alpha) + f(v_\alpha, w) \leq \phi(u_\alpha, w)$ $\forall w \in K] \Rightarrow$
$[\forall w \in K$ $\exists w_\alpha \in K$: $\limsup_\alpha (\phi(u_\alpha, w_\alpha) - f(v_\alpha, w_\alpha)) \leq \phi(u, w) - f(v, w)]$,

(H13) $\forall u \in S$ $\{v \in K : \phi(u, v) + f(v, w) \leq \phi(u, w)$ $\forall w \in K\}$ *is convex*,

(H14) $\forall u \in C$ $\forall \chi \subset K$ *convex* $\forall$ filter $K \ni v_\alpha \to v \in \chi$ *in* E
$[\phi(u, v_\alpha) + f(v_\alpha, w) \leq \phi(u, w)$ $\forall w \in \chi] \Rightarrow$
$[\phi(u, v) + f(v, w) \leq \phi(u, w)$ $\forall w \in \chi]$,

then there exists at least one solution of problem 11.10.

As we have already stated, we will not give the proof of this result, which is a generalization of theorem 10.3 in the same sense that theorem 11.9 is a generalization of theorem 10.1; for the proof see Joly–Mosco, 1979, §2.3 (in this paper theorem 11.9 has been presented as a consequence of theorem 11.10 ...). The applications of theorem 11.10, like those of theorem 10.3, are strongly limited by a hypothesis of l.s.c.: (H12) is in effect a hypothesis of l.s.c. for f in the pair of variables (besides being a hypothesis of u.s.c. for $\phi(\cdot, w)$) ...; in this context, let us remark that (H10) is a hypothesis of l.s.c. for ϕ in the pair of variables).

Finally we will deal with the:

Proof of theorem 11.9. From (H1), ..., (H8) and from theorem 10.1 it follows that, for every $u \in C$, $\Sigma(u)$ is a non-empty closed convex subset of K, i.e. $\Sigma: C \to 2^K_{\blacksquare ck}$. From (H9) we can then say that $\Sigma: S \to 2^S_{\blacksquare ck}$, where S is a non-empty compact convex set. To fulfill the conditions of theorem 9.21, and thus to prove the theorem, we need only show that $\Sigma|_S$ is u.s.c.

Let us then consider, along with problem 11.10, the problem

PROBLEM 11.11. Find

$$u \in C; \quad \phi(u, u) - f(w, u) \leq \phi(u, w) \quad \forall w \in K. \tag{11.65}$$

Along with the application $\Sigma: C \to 2^K$ let us consider the application $\Sigma^*: C \to 2^K$ associated in a natural way with problem 11.11:

$$\Sigma^*(u) = \{v \in K: \phi(u, v) - f(w, v) \leq \phi(u, w) \quad \forall w \in K\}. \tag{11.66}$$

We will now show that $\Sigma^*|_S$ is closed and thus, since S is compact and from theorem 20.15, that $\Sigma^*|_S$ is u.s.c.: from this it follows that $\Sigma|_S$ is u.s.c. since, from theorem 10.1, $\Sigma \equiv \Sigma^*$. In other words, from theorem 10.1 it follows that problems 11.10 and 11.11 are equivalent: this can be interpreted as an (obvious) generalization of Minty's lemma (problems 11.10 and 11.11 play roles here which are analogous to those of problems 10.1 and 10.2 in theorem 10.1 and in its lemma 2).

Let then $(u_\alpha, v_\alpha)_{\alpha \in A} \in S \times S$ be a filter convergent to (u, v) in $F \times F$: we wish to show that if $\forall \alpha \in A \; v_\alpha \in \Sigma^*(u_\alpha)$ then $v \in \Sigma^*(u)$. If $v_\alpha \in \Sigma^*(u_\alpha)$ then, since $\Sigma^*(u_\alpha) = \Sigma(u_\alpha)$,

$$\phi(u_\alpha, v_\alpha) + f(v_\alpha, w) \leq \phi(u_\alpha, w) \quad \forall w \in K, \tag{11.67}$$

and we are therefore under the conditions of (H10) which ensures that

$$\phi(u, v) \leq \liminf_\alpha \phi(u_\alpha, v_\alpha); \tag{11.68}$$

we are also under the conditions of (H11), and therefore

$$\forall w \in K \; \exists w_\alpha \in K: \quad \liminf_\alpha [\phi(u_\alpha, w_\alpha) - f(v_\alpha, w_\alpha) + f(v_\alpha, w)] \leq \phi(u, w). \tag{11.69}$$

From (11.69) and (H7) it follows that

$$\forall w \in K \quad \exists w_\alpha \in K: \quad \liminf_\alpha [\phi(u_\alpha, w_\alpha) - f(v_\alpha, w_\alpha) + f(v_\alpha, w)$$
$$- f(v_\alpha, w) + f(w, v_\alpha)] \leq \phi(u, w), \quad (11.70)$$

or, in other words,

$$\forall w \in K \quad \exists w_\alpha \in K: \quad \liminf_\alpha [\phi(u_\alpha, w_\alpha) - f(v_\alpha, w_\alpha) + f(w, v_\alpha)] \leq \phi(u, w); \quad (11.71)$$

from this, since (H5) enables us to write $\liminf_\alpha (-f(w, v_\alpha)) = -\limsup_\alpha f(w, v_\alpha) = -f(w, v)$, we also have

$$\forall w \in K \quad \exists w_\alpha \in K: \quad \liminf_\alpha [\phi(u_\alpha, w_\alpha) - f(v_\alpha, w_\alpha)] \leq \phi(u, v) + f(w, v). \quad (11.72)$$

From (11.67) it follows that

$$\forall \alpha \in A \quad \phi(u_\alpha, v_\alpha) + f(v_\alpha, w_\alpha) \leq \phi(u_\alpha, w_\alpha), \quad (11.73)$$

or, in other words,

$$\forall \alpha \in A \quad \phi(u_\alpha, v_\alpha) \leq \phi(u_\alpha, w_\alpha) - f(v_\alpha, w_\alpha), \quad (11.74)$$

and from this, taking the (lower) limit in α, we have (applying (11.68) and (11.72))

$$\phi(u, v) \leq \liminf_\alpha \phi(u_\alpha, v_\alpha) \leq \liminf_\alpha [\phi(u_\alpha, w_\alpha) - f(v_\alpha, w_\alpha)]$$
$$\leq \phi(u, v) + f(w, v), \quad (11.75)$$

which, since w is arbitrary, means that $v \in \Sigma^*(u)$. ∎

Nash points revisited: ∗-continuity

The situation considered in example 11.6 is of particular interest to us and we will therefore analyse it in greater detail: our intention is to present a result which is a direct generalization of theorem 10.4. Let then E be a reflexive real Banach space with a dual E', K a non-empty convex closed set of E, C a non-empty subset of K, and let us put

$$f(u, v) = {}_{E'}\langle Au, u - v \rangle_E, \quad (11.76)$$

with $A: E \to E'$, and

$$\phi(u, v) = j_{Q(u)}(v), \quad (11.77)$$

with $Q: C \to 2^K_{\blacksquare ck}$. As we have already seen in example 11.6, problem 11.10 then becomes

PROBLEM 11.12. Find

$$u \in C; u \in Q(u) \quad \text{and} \quad {}_{E'}\langle Au, u - v \rangle_E \leq 0 \quad \forall v \in Q(u). \quad (11.78)$$

We want to know what conditions have to be imposed on A and Q so that hypotheses (H1)...(H11) are satisfied and the existence of a solution for problem 11.12 ensured. From what we have seen in Chapter 10 (in connection with theorem 10.4...) we conclude that for (H1)...(H8) to be satisfied it is sufficient that A be a monotone, hemicontinuous, and (if K is not bounded) coercive operator. Leaving aside (H9), which must be checked every time, we still have to see how (H10) and (H11) can be satisfied; for this purpose let us introduce the following

DEFINITION 11.1. Let B be a real reflexive Banach space with dual B', X a non-empty convex closed subset of B, Z a non-empty subset of X, T an operator from B into B' and R a multi-valued application from Z into $2^X_{\blacksquare ck}$. Again, let $G \hookrightarrow B$ be a reflexive Banach space and $U \subset G \cap Z$ a non-empty closed convex subset of G. The pair (T, R) is said to be $*$-*continuous* in S if

$\forall$ seq. $U \times U \ni (u_k, v_k) \to (u, v)$ in $G \times G$

$[v_k \in R(u_k)$ and $\forall w \in R(u_k) \langle Tv_k, v_k - w \rangle \leq 0] \Rightarrow$

$[v \in R(u)$ and $\forall w \in R(u) \exists w_k \in R(u_k): \liminf_k \langle Tv_k, w_k - w \rangle \leq 0]$. (11.79)

It is obvious that if (A, Q) is $*$-continuous then (H10) and (H11) are satisfied, so that we can state the following

THEOREM 11.11. Let E be a reflexive real Banach space with dual E', K a non-empty closed convex subset of E, C a non-empty subset of K, $A: E \to E'$ a hemicontinuous and coercive monotonic operator and $Q: C \to 2^K_{\blacksquare ck}$ a multi-valued application. If there exists a reflexive Banach space $F \hookrightarrow E$ and a non-empty compact convex set $S \subset F \cap C$ such that the pair (A, Q) is $*$-continuous in S and $\{v \in Q(u) : \forall w \in Q(u)_E \langle Av, v - w \rangle_E \leq 0\} \subset S$ as long as $u \in S$, then there exists at least one solution of problem 11.12.

Theorem 10.4 is the particular case of this theorem corresponding to $C = K$ and $\forall u \in K\, Q(u) = K$. Let us remark also that even when A is strictly monotone we cannot ensure the uniqueness of the solution.

To end, before going on to analyse an example of the applications of the techniques presented in this section, let us note that *if $A \in \mathcal{L}(E, E')$ then a sufficient condition for the pair (A, Q) to be $*$-continuous on S is that Q be continuous for the Vietoris finite topology in $2^K_{\blacksquare ck}$*. This is an important point since it includes the case of linear quasivariational inequalities such as, for example, that which was involved in problem 11.3 and the one involved in the following problem.

A problem in membrane theory

EXAMPLE 11.7. Let $\Omega \subset \mathbb{R}^n$ be a bounded open set with boundary Γ of class $C^{0,1}$, $h \in H^{1/2}(\Gamma)$, $f \in L^2(\Omega)$, $\lambda > 0$ real and L the elliptic differential operator with coefficients in $L^\infty(\Omega)$ defined by

$$L\square \equiv -\sum_{i,j=1}^{n}\frac{\partial}{\partial x_i}\left(a_{ij}\frac{\partial}{\partial x_j}\square\right)+d\square. \tag{11.80}$$

The following problem, whose conditions are to be interpreted in a purely formal manner, models the flow of a fluid across a semipermeable membrane:

PROBLEM 11.13. Find $u \in H^1(\Omega)$ such that

$$Lu = f \quad \text{in } \Omega, \tag{11.81}$$

$$\gamma_0 u \geqslant h - \lambda \frac{\partial u}{\partial \nu_a} \quad \text{on } \Gamma, \tag{11.82}$$

$$\frac{\partial u}{\partial \nu_a} \geqslant 0 \quad \text{on } \Gamma, \tag{11.83}$$

$$\left[\gamma_0 u - \left(h - \lambda \frac{\partial u}{\partial \nu_a}\right)\right] \frac{\partial u}{\partial \nu_a} = 0 \quad \text{on } \Gamma. \tag{11.84}$$

The derivative $\partial u/\partial \nu_a$ must be interpreted formally as a conormal derivative associated with the continuous bilinear form

$$a(u,v) = \int_\Omega \left[\sum_{i,j=1}^n a_{ij}\frac{\partial u}{\partial x_j}\frac{\partial v}{\partial x_i} + duv\right]dx, \tag{11.85}$$

which we assume to be coercive on $H^1(\Omega)$. The problem 11.13 recalls the problems with obstacle on the boundary for the variational inequalities, but is not a variational problem. Putting, for $u \in H_L^1(\Omega)$,

$$Q(u) = \left\{v \in H^1(\Omega) : \gamma_0 v \geqslant h - \lambda {}_{H^{-1/2}(\Gamma)}\left\langle\frac{\partial u}{\partial \nu_a}, \phi\right\rangle_{H^{1/2}(\Gamma)} \quad \forall \phi \in H^{1/2}(\Gamma)\right\}, \tag{11.86}$$

the formal equivalence between problem 11.13 and the following quasivariational problem (with obstacle on the boundary ...) can be shown:

PROBLEM 11.14. Find

$$u \in H_L^1(\Omega); u \in Q(u) \quad \text{and} \quad a(u, u-v) \leqslant {}_{L^2(\Omega)}\langle f, u-v\rangle_{L^2(\Omega)} \quad \forall v \in Q(u). \tag{11.87}$$

To prove the existence of a solution for this problem we will use theorem 11.11, which, with reference to the operator, we are obviously in a position to apply. Let us then put $E = H^1(\Omega)$, $K = H_L^1(\Omega)$ and $F = H_L^1(\Omega)$; further let us put

$$Z = \left\{v \in H_L^1(\Omega) : Lv = f \quad \text{and} \quad \frac{\partial u}{\partial \nu_a} \geqslant 0\right\}. \tag{11.88}$$

(11.86) defines a multi-valued application $Q: H_L^1(\Omega) \to 2_{ck}^{H^1(\Omega)}$ which restricted to Z has non-empty values. In effect, if $u \in Z$ then the solution of the Dirichlet problem '$Lu = f$ in Ω; $u = h$ on Γ' is an element of $Q(u)$. Again, $Z \neq \emptyset$ since the solution of the Neumann problem '$Lu = f$ in Ω; $\partial u/\partial \nu_a = 0$ on Γ' is an element of Z. Q is also continuous for the Vietoris finite topology since the trace operators are continuous: the $*$-continuity is therefore satisfied. Finally, the stability condition is satisfied with $S = \overline{\text{conv}}\,(\Sigma(Z))$. We can therefore say that problem 11.14 (and consequently also problem 11.13) has a solution.

12
Free-Boundary Problems

12.1 INTRODUCTION

Filtration through porous media

In this chapter we will apply the techniques introduced previously to the study of some *free-boundary problems* for second order elliptic linear differential operators. These are, roughly, differential problems where not only is the solution of an appropriate differential equation unknown but even the domain in which it must be satisfied is unknown. However, a part of the boundary of this domain is known, and on this we impose an appropriate boundary condition (for example a Dirichlet condition), the rest of the boundary being unknown, the *free-boundary*, on which we impose a pair of independent conditions (for example a Dirichlet condition and a Neumann condition). It is important to note that the co-existence of two boundary conditions on the free-boundary, which in 'fixed boundary problems' would lead more often than not to ill-posed problems, is allowed here (it is in fact indispensable) since in a certain sense the free-boundary will 'adapt' to the double condition imposed on it.

In Part I of this book we have already encountered several free-boundary problems: besides those dealt with in Chapter 8 (which we explicitly referred to as 'free-boundary problems') several of those studied in Chapter 7 were, more or less implicitly, problems of this kind (see e.g., the comments regarding examples 7.9 and 7.10).

Further, the fact that variational inequalities translate in general to free-boundary problems has already been pointed out by Lewy and Stampacchia (cf. Lewy–Stampacchia, 1969, part IV).

The problems considered here arise in the study of phenomena regarding filtration of liquids across porous materials; these are of great practical interest, but no rigorous mathematical methods existed for dealing with them until the 1970s. In 1971, Baiocchi (cf. Baiocchi, 1971; 1972) introduced a new method (generalized in 1974—see Baiocchi, 1974a; 1975b) which consists essentially of transforming the free-boundary problem into

263

one on a fixed domain (a variational problem in the 1971 formulation, and a quasivariational problem in that of 1974). This method, which is mathematically rigorous, was then applied to several filtration phenomena by a group of workers at the Laboratorio di Analisi Numerica of the Italian Research National Council (Pavia) and of the Istituto di Matematica of the University of Pavia (see the Bibliographical note below).

Bibliographical note

For a survey of the methods available before the 1970s for the study of problems of filtration through porous materials see Baiocchi, 1974e; Baiocchi–Comincioli–Magenes–Pozzi, 1976; see also Monachov, 1969; Cryer, 1976a; 1976b, 1977.

The work done at Pavia is documented by the following list of papers: Baiocchi, 1971; 1972; 1973a; 1973b; 1973c; 1974a; 1974b; 1974c; 1974d; 1974e; 1975a; 1975b; 1975c; 1976a; 1976b; 1977; 1978; 1979; 1980a; 1980b; 1981a; 1981b; Baiocchi–Brezzi–Comincioli, 1978; Baiocchi–Comincioli–Guerri–Volpi, 1973; Baiocchi–Comincioli–Magenes–Pozzi, 1973; 1976; Baiocchi–Comincioli–Maione, 1975; 1977; Baiocchi–Evans–Frank–Friedman, 1980; Baiocchi–Friedman, 1977; Baiocchi–Magenes, 1974; 1975; Baiocchi–Maione, 1972; Boieri–Gastaldi, 1980a; 1980b; 1981; Brezzi–Caffarelli, 1982; Brezzi–Sacchi, 1976; Comincioli, 1972; 1974a; 1974b; 1975; 1976; 1978; 1979; 1980; Comincioli–Guerri, 1976; Comincioli–Guerri–Volpi, 1971; Comincioli–Torelli, 1979; Gastaldi, 1979; Gilardi, 1976; 1977; 1979; 1980a; 1980b; 1982, Alt–Gilardi, 1981; 1982; Caffarelli–Gilardi, 1980; Magenes, 1972; 1973; 1977; Pietra, 1982; Pozzi, 1974a; 1974b; Quarteroni–Visintin, 1980; Torelli, 1974; 1975a; 1975b; 1977a; 1977b; 1977c; 1977d; Friedman–Torelli, 1977; Visintin, 1979; 1980a; 1980b.

Let us quote also the work of other workers in this field: Alt, 1977a; 1977b; 1979; 1980; Amirat–Atik, 1978; Begis–Glowinski, 1975; Benci, 1973; 1974; Brézis–Kinderlehrer–Stampacchia, 1978; Bruch, 1979; 1980a; 1980b; Bruch–Caffrey, 1979a; 1979b; Bruch–Sayle–Sloss, 1978; Bruch–Sloss, 1978; Caffarelli, 1976b; Caffarelli–Friedman, 1978a; 1978b; 1980; Caffarelli–Rivière, 1977c; Carbone–Valli, 1976; 1977; 1978a; 1978b; Carrillo–Chipot, 1981; 1982; Chipot, 1981, Cryer, 1980; Crank–Ozis, 1980, Friedman, 1976d; 1980a; Friedman–Jensen, 1975; 1976; 1978; Glashoff–Roleff, 1977a; 1977b; Jensen, 1977; 1980b; 1982; Kawarada, 1979; Kikuchi, 1977a; 1977b; Kikuchi–Oden, 1979; Morice, 1975; Popa, 1980; Shimborski, 1975; Stampacchia, 1974; Szeptycki, 1982.

The method has also been successfully applied to free-boundary problems arising in the study of physical problems of a different nature; see Bourgat–Duvaut, 1975; 1977; Brézis–Duvaut, 1973; Brézis–Stampacchia, 1973a; 1976; Tomarelli, 1978; Elliott, 1980; Duvaut, 1973; 1974; 1976; Friedman–Kinderlehrer, 1975; Lions, 1976b; Frémond, 1975; Magenes,

1982; see also the proceedings Magenes, 1980; Fasano–Primicerio, 1982 and their references.

Let us remark at this point that not all the free-boundary problems which are involved in applications can, however, be transformed in this way into variational inequalities. In this connection see Chapter 13, where we have developed an idea introduced in Baiocchi, 1976b for the study of 'all' free-boundary problems which can be transformed in this way into variational inequalities. Again, in this context see Baiocchi, 1978, appendix 1 and Gastaldi, 1979 (for parabolic problems) and, for the inverse approach, i.e. the free-boundary problems which are solved by variational inequalities, see Brézis–Stampacchia, 1973b.

Here, with the aim of presenting the method, we will only study free-boundary problems associated with the phenomenon of filtration of water across an earthen dam; further, we will not consider all the problems of this type which have been analysed by this method, problems to which we will briefly refer in this chapter. Let us mention some other filtration problems that have been treated with this method and we will not refer to them again: in Torelli, 1974 is studied the stationary motion of filtration of water through the walls and the bottom of a channel; in Torelli, 1977b is studied the non-stationary motion of filtration of a compressible liquid in a vertical channel filled with porous material (see also Friedman–Jensen, 1975); Baiocchi–Comincioli–Magenes–Pozzi, 1973 and Baiocchi–Comincioli–Guerri–Volpi, 1973 (for the numerical aspects) and Baiocchi–Comincioli–Maione, 1977 studied the motion of filtration of several immiscible liquids with different densities, a problem which arises from the determination of the position of water-bearing faults opening into the sea.

Finally, we mention that a different formulation of the problem treated here (and which does not explicitly use hypothesis (12.13) which follows) has been given in Alt, 1977a; the problem has been studied by first using 'locally' the transformation presented in Chapter 8, and later a 'balayage' method. This formulation allows us to make much more general assumptions regarding the geometry of the dam and enables us to obtain results on the regularity of the free boundary and on uniqueness (see Alt, 1977b and Caffarelli–Rivière, 1977c).

12.2 THE PHYSICAL PROBLEM

The fluid and the porous media

To fix our ideas, let us consider the following *problem of filtration*. On an impermeable horizontal base, two water basins at different levels are separated by an earthen dam; water filters from the higher to the lower basin by gravity—we are interested in *physical quantities* (such as the distributions of velocity and pressure, carrying capacity, etc.) and *geometrical quantities*

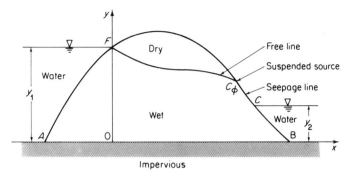

Figure 12.1

(stream lines, the portion of the dam which is wet, etc.) associated with the flow (see fig. 12.1).

In this chapter, water and liquid will be used synonymously, as will porous material and earth. For the study of the properties of porous materials and the process of filtration in porous materials see, e.g., Aravin–Numerov, 1965, Cedergren, 1967; Harr, 1962; Polubarinova–Kochina, 1962; Bear, 1972; 1979.

Let us suppose that the material of which the dam is constructed is incompressible, homogeneous, and isotropic; that the liquid is homogeneous and isotropic; and that the flow is incompressible, non-rotational, stationary, and bidimensional; let us suppose further that the material does not give rise to capillarity phenomena and that there is no evaporation. The homogeneity and isotropy of the dam are to be taken naturally in the macroscopic sense: the porous material consists of particles of solid material of different dimensions put together in an arbitrary fashion and leaving between them pores through which the water flows. On the other hand, the phenomenon is, obviously, as any other physical phenomenon, a three-dimensional one; in effect what we really assume is that the dam is undefined along the axis perpendicular to the plane of the picture and with a constant section; further we assume that the flow has the same characteristics in every section. For the study of effectively three-dimensional phenomena see Stampacchia, 1974 (dam with vertical walls) and Gilardi, 1977 (dam with walls not necessarily vertical—the concavity of the surface of the dam is not required, which we will assume later to be the case). In the following we will use, naturally, two-dimensional rather than three-dimensional nomenclature: thus we will talk of the wet surface of the dam rather than the wet volume, of the phreatic line or free line rather than phreatic surface or free surface, etc. For the case in which the material gives rise to capillarity phenomena see Comincioli–Guerri, 1976 and Comincioli, 1976; for other free boundary problems connected with capillarity see Finn, 1974. For the case in which evaporation is taken into account see Pozzi, 1974a; 1974b; Friedman, 1976d; Jensen, 1977.

Notations

In what follows, in order to simplify the exposition, we will use the notations 'condensed' in fig. 12.2 and in the notation diagram 12.1 (naturally, the figures of this diagram should be considered superimposed on each other). Later we will provide topological details—for the present we need only mention that the sets D, D_1, D_2, D_3, and Ω are taken to be open (from the point of view of physics, it is quite unimportant whether we consider them to be open or closed, but from the mathematical point of view this specification is essential).

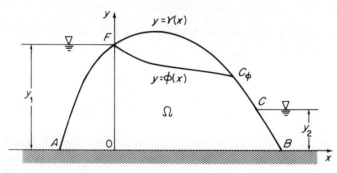

Figure 12.2

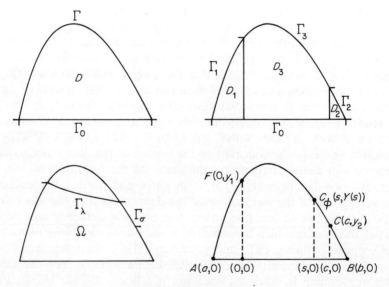

Notation diagram 12.1

Free line, seepage line, and suspended source

Experience shows us that in general the dam is not all wet (however, if the dam has a triangular section and the angle opposite the base is not greater than $\pi/2$ and the highest level is equal to the height of the dam then it is all wet: this can be seen to be a consequence of the so-called conditions— experimental— of input and output: see Harr, 1962, p. 21). The wet part of the dam is bounded above by a line, the *free line*, within the porous material (the line between F and C_ϕ in figs 12.1 and 12.2).

Along this line there is contact between the water and the air contained in the pores of the dry portion of the dam, so that, if we give (as is usual in hydraulics) the value zero to the atmospheric pressure, we can say that in the dry portion of the dam we have zero pressure, and that the pressure is positive in the wet portion (strictly positive, since water is heavy . . .). On the other hand the free line cannot 'touch' the impermeable base, since then there would be no flow—the dam is therefore wet in a neighbourhood of the base. Again experience tells us that, if the material is homogeneous and there is no evaporation, the free line is decreasing and its extremity C_ϕ does not coincide with C: C_ϕ is the so-called *suspended source*, and the part of the wall of the dam determined by C_ϕ and C is the so-called *seepage line* or *emergency line*. Along the seepage line water comes out but cannot enter— this is in fact a line where water leaves the porous material and enters a zone where there is neither water nor porous material, there being direct contact with the atmosphere.

Let us summarize this discussion in the following points:

> *The pressure is zero in the dry part of the dam, positive in the wet part, and certainly positive near the base;* (12.1)

> *The water can go out across the line of seepage but cannot enter.* (12.2)

Darcy's law and its consequences

The flow obeys *Darcy's law*, which states that

$$\mathbf{V} = -k \operatorname{grad}\left(y + \frac{p}{\gamma}\right) \quad \text{in } \Omega, \tag{12.3}$$

where $\mathbf{V} = \mathbf{V}(x, y)$ is the *velocity* of the liquid in the wet portion of the dam, $p = p(x, y)$ is the *pressure*, γ is the *specific weight* of the fluid ($\gamma = \rho g$, where ρ is the *specific mass* and g is the (modulus of the) acceleration due to gravity), and k is the *coefficient of permeability* of the material with respect to the fluid.

Darcy's law is a law of linear behaviour found experimentally (but for which theoretical justifications are also available—see in particular Bear, 1972) and has imprecise limits of validity (which depend on the so-called

Reynolds number). This law implicitly involves the viscosity of the fluid (see below) which can therefore be considered to be non-viscous; further it can be interpreted as a formulation of the Navier–Stokes equation. The law of Darcy is a law of macroscopic behaviour, and it cannot be otherwise because the laws of the general hydrodynamics are themselves of a macroscopic kind and it is impossible (being in the last analysis useless) to analyse the motion of every particle of fluid across the pores of the material. The constant k depends on both the porous material and on the fluid: $k = \tilde{k}\gamma/\eta$, where η is the *viscosity* of the fluid, γ its specific weight, and $\tilde{k}$ is a property of the material known as *physical permeability*. In the general case $\tilde{k}$ (and hence k) is a second order symmetric tensor whose components depend on x and y. The isotropy of the material tells us that $\tilde{k}$ is scalar, and its homogeneity that $\tilde{k}$ is constant. For the study of the motion of filtration in dams made of non-homogeneous materials see Baiocchi–Comincioli–Magenes–Pozzi, 1973 (k constant for horizontal and vertical layers), Benci, 1973; 1974 (k of the form $k(x, y) = \bar{k}(x) \cdot \bar{\bar{k}}(y)$), Baiocchi–Friedman, 1977 (extension of this case), Caffarelli–Friedman, 1978a; 1978b and Baiocchi, 1978 (where a much more general formulation of the problem is given).

Under our conditions we can, without loss of generality, assume $k = 1$, which we will of course do.

Then, putting

$$u = y + \frac{p}{\gamma}, \qquad (12.4)$$

Darcy's law becomes

$$\mathbf{V} = -\text{grad } u \quad \text{in } \Omega, \qquad (12.5)$$

which tells us that u is a *potential of the velocity*. In what follows, where we speak of *equipotential lines* we refer to the lines of level of $u = u(x, y)$. Let us remark that the existence of a kinetic potential is perfectly in agreement with the non-rotationality of the flow. The quantity u is the so-called *piezometric height*, which in the case of fluids at rest is constant (the fundamental principle of hydrostatics; let us recall also that in the case of non-viscous homogeneous fluids in non-rotational, incompressible and stationary (free) motion, the *total height* $u + \mathbf{V}^2/2g$ is constant, where $\mathbf{V}^2/2g$ is the *kinetic height* (Bernoulli's theorem)).

Let us recall that in the general case the equation of continuity, which is a differential formulation of the principle of the conservation of mass, is $D\rho/Dt + \rho \text{ div } \mathbf{V} = 0$ (where D/Dt is the operator of material differentiation or operator of differentiation of Euler). Remark also that $\mathbf{V}(x, y)$ does not represent the velocity of a given particle of the fluid during its flow but the velocity of the different particles at (x, y)—we are in fact adopting the Eulerian point of view in describing the motion. From the incompressibility of the flow it follows that in our case the *equation of continuity* becomes

$$\text{div } \mathbf{V} = 0, \qquad (12.6)$$

which, together with (12.5), allows us to write

$$\Delta u = 0 \quad \text{in } \Omega, \tag{12.7}$$

which tells us that the piezometric height is harmonic (within the dam, of course). From (12.7) it follows that there exists a function $v = v(x, y)$ which is also harmonic, the harmonic conjugate of u, defined (except for an additive constant) by the Cauchy–Riemann equations

$$\partial u/\partial x - \partial v/\partial y = \partial u/\partial y + \partial v/\partial x = 0. \tag{12.8}$$

We will call this function the *stream function* (the complex harmonic function $\Phi = u + iv$ is often called the *complex potential*). The lines of level of v, which are also the lines of force of the vector field **V**, are called *stream lines* (and since the flow is non-rotational the streams' lines coincide with the *trajectories* of the particles of the fluid). The stream lines are orthogonal to the equipotential lines, and drawing these lines for $v = c_1, v = c_2, \ldots, v = c_n$ and $u = d_1, u = d_2, \ldots, u = d_m$ gives us a lattice which is known as the *flow net*; later we will provide some examples of these.

In the permanent regime the stream lines are fixed in space, and we can define as a *stream tube* a surface bounded by two stream lines (we are dealing with two dimensions...)—the liquid flows along the stream tubes as if they had solid walls. An interesting particular case of the stream tube is the whole of the wet part of the dam, since the impermeable base Γ_0 and the free line Γ_λ are stream lines. If the flow of the liquid, besides being permanent, is incompressible, the flux of the field **V** across any section of a stream tube is constant: this constant is called the *carrying capacity* or the *discharge* of the tube.

The *carrying capacity of the dam* is

$$q_d = \int_L \mathbf{V} \cdot \mathbf{n}_L \, dl, \tag{12.9}$$

where L is any line which joins a point on Γ_0 to a point on Γ_λ, and $\mathbf{n}_L$ is the versor of the normal to L. Here we will only give the orientation for the case in which L is a vertical segment: we will take $\mathbf{n}_L \equiv \mathbf{e}_1$ and $dl \equiv dy$; now using (12.5) and (12.8) we can write

$$q(x) = -\int_0^{\phi(x)} \partial u/\partial x \, dy = -\int_0^{\phi(x)} \partial v/\partial y \, dy;$$

$$q(x) = q_d = v|_{\Gamma_0} - v|_{\Gamma_\lambda} \quad \text{if } 0 < x < s. \tag{12.10}$$

Since v is determined except for an additive constant, we can choose this constant so that the stream line Γ_λ corresponds to $v = 0$: the carrying capacity of the dam is then the value (generally unknown) which v has at the base. In some cases the carrying capacity of the dam is known: we will see an example of this when we deal with a dam with vertical walls, in which case $q_d = (y_1^2 - y_2^2)/2c$ (let us note that in this case $a = 0$ and $s = b = c$). It is interesting to note that this is the value obtained for the carrying capacity of the dam shown in fig. 12.1 starting from the so-called Dupuit assumptions. The latter are two simplifying (and contradictory!) assumptions which imply among other things that the free line is a parabola, the Dupuit parabola, which joins F and C (and therefore the seepage line does not exist). For the Dupuit 'solution' see Harr, 1962, p. 50—Harr also gives the approximate solutions of Schaffernak–van Iterson (p. 51), Casagrande (p. 52), and Pavlovsky (p. 55).

We will work principally with the unknown p rather than the unknowns u or v; we will therefore need the carrying capacity in terms of p, which follows immediately from (12.4) and (12.10):

$$q_d = -\frac{1}{\gamma} \int_0^{\phi(x)} \partial p/\partial x \, dy, \qquad 0 < x < s. \tag{12.11}$$

Also, since (cf. (12.1)) $p = 0$ in the dry zone of the dam, and assuming a minimum of regularity for p (in fact it is sufficient that $\partial p/\partial x \in L^1(D)$), we will in general associate with the function p another function, q, defined by

$$q(x) = -\frac{1}{\gamma} \int_0^{Y(x)} \partial p/\partial x \, dy, \tag{12.12}$$

which is preferable to the former in that it uses the known function Y instead of the unknown ϕ.

Let us then put

$$x \mapsto \frac{1}{\gamma} \int_0^{Y(x)} \partial p/\partial x \, dy \text{ is non-decreasing in } [0, c]. \tag{12.13}$$

Let us remark that from (12.4) and (12.7) it follows that the pressure is harmonic in Ω:

$$\Delta p = 0 \quad \text{in } \Omega; \tag{12.14}$$

the pressure is also harmonic (being zero) inside $D \setminus \Omega$, but obviously it is not so in the whole of D.

The conditions at the boundary

We will next deal with the determination of the boundary conditions to associate with equation (12.7), with equations (12.8), and (the one which interests us most) with the equation (12.14). We have four types of 'boundary' which will be treated separately.

(1) *The impermeable base, Γ_0*

The water cannot cross the impermeable base and therefore the normal component of the velocity is zero along Γ_0, i.e. Γ_0 is a stream line. Thus we can put alternatively

$$v = \text{constant on } \Gamma_0, \tag{12.15}$$

or, since $\mathbf{V} \cdot \mathbf{n} = -\partial u/\partial n = \partial u/\partial y$,

$$\partial u/\partial y = 0 \quad \text{on } \Gamma_0. \tag{12.16}$$

(let us recall that the base is horizontal and $\mathbf{n}$ is, as usual, the external normal; this restriction on the base is not essential—for the study of some problems with an 'inclined' base see Baiocchi–Comincioli–Magenes–Pozzi, 1973 and Comincioli, 1975). Let us remark that the constant in (12.15) is not known; in agreement with the comments regarding (12.10) we can introduce the new unknown q_d, the carrying capacity of the dam, and put

$$v = q_d \quad \text{on } \Gamma_0. \tag{12.17}$$

Given (12.4), we can write (12.16) in terms of p:

$$\partial p/\partial y = -\gamma \quad \text{on } \Gamma_0. \tag{12.18}$$

(2) *The boundaries of the basins, Γ_1 and Γ_2*

Let us consider Γ_1. The piezometric height u has on Γ_1 the same value as it has in the adjacent basin. Now, neglecting the velocity of the water in the basin with respect to that within the dam, we can say that the water in the basin is at rest and thus the piezometric height is constant in it: Γ_1 is then an equipotential line, i.e.,

$$u = \text{constant} \quad \text{on } \Gamma_1. \tag{12.19}$$

This constant can, however, be calculated; we need only note that in F there is contact with the atmosphere and thus $p = 0$ in F: from this together with (12.4) we get $u(F) = y_1$ and therefore, from (12.19),

$$u = y_1 \quad \text{on } \Gamma_1. \tag{12.20}$$

In an analogous manner, we find that Γ_2 is an equipotential line and that

$$u = y_2 \quad \text{on } \Gamma_2. \tag{12.21}$$

Using (12.4) we can rewrite (12.20) and (12.21) in terms of p:

$$p = \gamma(y_1 - y) \quad \text{on } \Gamma_1, \tag{12.22}$$

$$p = \gamma(y_2 - y) \quad \text{on } \Gamma_2. \tag{12.23}$$

The method which is explained here has been applied also to the case in which the wall corresponding to the highest level has been partially impermeabilized: see section 8.3; also see Baiocchi–Comincioli–Magenes–Pozzi, 1973, Magenes, 1973; Baiocchi, 1973a; Baiocchi–Magenes, 1975 and Gilardi, 1976.

(3) *The seepage line, Γ_σ*

Since along this line there is contact with the atmosphere

$$p = 0 \quad \text{on } \Gamma_\sigma, \tag{12.24}$$

or, from (12.4),

$$u = y \quad \text{on } \Gamma_\sigma. \qquad (12.25)$$

The seepage line is neither a stream line nor an equipotential. The analytical version of (12.2) can be written in the form

$$\partial u/\partial n \leq 0 \quad \text{on } \Gamma_\sigma, \qquad (12.26)$$

or, equivalently, in the form

$$\partial(p/\gamma + y)/\partial n \leq 0 \quad \text{on } \Gamma_\sigma. \qquad (12.27)$$

(4) *The free line*, Γ_λ

The free boundary is 'the' stream line in contact with the atmosphere and hence

$$v = \text{constant on } \Gamma_\lambda, \qquad (12.28)$$

$$p = 0 \quad \text{on } \Gamma_\lambda. \qquad (12.29)$$

The constant in (12.28) is not known, but if we introduce the unknown q_d and take into consideration (12.17) then we must put

$$v = 0 \quad \text{on } \Gamma_\lambda. \qquad (12.30)$$

In terms of u, (12.28) and (12.29) become (using, respectively, (12.5) and (12.4))

$$\partial u/\partial n = 0 \quad \text{on } \Gamma_\lambda, \qquad (12.31)$$

$$u = y \quad \text{on } \Gamma_\lambda. \qquad (12.32)$$

Further, (12.31) can be written in terms of p in the form

$$\partial(p/\gamma + y)/\partial n = 0 \quad \text{on } \Gamma_\lambda. \qquad (12.33)$$

In summary, on Γ_λ we can impose one of the following pairs of independent conditions: {(12.28), (12.32)}, {(12.31), (12.32)}, {(12.29), (12.33)}.

To end this analysis we will show that, under our conditions, one must have

$$x \mapsto \phi(x) \text{ is strictly decreasing.} \qquad (12.34)$$

For this, we need only see that, denoting by τ the versor of the tangent to Γ_λ oriented in the 'direction of flow' and using (12.31),

$$\mathbf{V} = -\text{grad } u = -\frac{\partial u}{\partial s}\tau \qquad (12.35)$$

and thus

$$-\partial u/\partial s > 0 \quad \text{on } \Gamma_\lambda, \qquad (12.36)$$

from which it follows that u is strictly decreasing in Γ_λ and thus, using (12.32), we have (12.34).

The mathematical problem

We will now give a preliminary formulation of the free-boundary problem in which we are interested, and which is that of finding Ω (or, equivalently, ϕ) and p (or, equivalently, u, or even (u, v)). There are several ways of posing the problem: here we will take as unknowns p and Ω and work with the equation (12.14). In the literature there are other formulations: that which takes u and Ω as unknowns (and one works with equation (12.7) and the conditions (12.16), (12.20), (12.21), (12.25), (12.31), (12.32)), that which takes u, v, Ω, ϕ and q_d as unknowns (and one works with equation (12.8) and the conditions (12.17), (12.20), (12.21), (12.25), (12.30), (12.32)), etc. But for further specifications regarding regularity, the problem in which we are interested is then the following:

PROBLEM 12.1. Find p and Ω such that

Ω is a non-empty open set contained in D, (12.37)

$$\Delta p = 0 \quad \text{in } \Omega, \quad (12.38)$$

$$p = \gamma(y_1 - y) \quad \text{on } \Gamma_1, \quad (12.39)$$

$$p = \gamma(y_2 - y) \quad \text{on } \Gamma_2, \quad (12.40)$$

$$p = 0 \quad \text{on } \Gamma_\sigma, \quad (12.41)$$

$$p = 0 \quad \text{on } \Gamma_\lambda, \quad (12.42)$$

$$\partial(p/\gamma + y)/\partial n = 0 \quad \text{on } \Gamma_\lambda, \quad (12.43)$$

$$\partial p/\partial y = -\gamma \quad \text{on } \Gamma_0. \quad (12.44)$$

In the first place, we observe that problem 12.1 is effectively a free boundary problem: Γ_λ is unknown and on it we impose the two independent conditions (12.42) and (12.43).

The missing conditions

In the next section we will pose the problem precisely; for the present we will make some comments on the 'missing conditions' of the previous formulation, i.e., those conditions (such as (12.1), (12.13), (12.27), and (12.34)) which the analysis of the physical phenomenon has given us and which we have not nevertheless imposed. In principle, if the formulation is exhaustive these 'missing conditions' should derive from those present. Let us see then, in a position between the rigorous and the heuristic one, what actually happens.

Let us first analyse (12.1). Let us assume $p \in C^0(\bar{\Omega})$, which is perfectly acceptable; then, from (12.38), $p \in C^0(\bar{\Omega}) \cap C^\infty(\Omega)$ and we are in a position to apply the maximum (or the minimum...) principle. Let $m \in \bar{\Omega}$ be a point of minimum for p. From the maximum principle we have $m \in \partial\Omega$, since if $m \in \Omega$ then p is constant and cannot satisfy the boundary conditions; from the definition of minimum we have $m \notin \Gamma_0$, since in Γ_0 the derivative with respect to y is a derivative with respect to the inward normal; from the analysis of (12.39), (12.40), (12.41), and (12.42) we have then that $m \in \Gamma_\lambda \cup \Gamma_\sigma$ and $p(m) = 0$. Denoting also by p the trivial extension of p to $\bar{D}$, extension which we can make with continuity since p is zero on Γ_λ, we can write

$$p \geq 0 \quad \text{in } \bar{D}, \qquad p > 0 \quad \text{in } \Omega, \qquad p > 0 \quad \text{on } \Gamma_0, \qquad (12.45)$$

which is essentially (12.1).

The (12.45) allows us to define Ω by means of

$$\Omega = \{(x, y) \in D : p(x, y) > 0\}. \tag{12.46}$$

Further it is obvious that for Ω to effectively represent the wet part of the dam it must have a minimum amount of regularity; in this sense a (very weak) regularity property for Ω is that it be a *subgraph*:

$$(\bar{x}, \bar{y}) \in \Omega \quad \text{and} \quad 0 < y < \bar{y} \Rightarrow (\bar{x}, y) \in \Omega. \tag{12.47}$$

This property need not be true if the material is not homogeneous, but in our case it is a natural one: we will specify it by putting

$$\bar{\phi}(x) = \sup\{y : (x, y) \in \Omega\}, \qquad 0 \leq x \leq c, \tag{12.48}$$

and

$$\Omega = D_1 \cup D_2 \cup \{(x, y) : 0 \leq x \leq c \quad \text{and} \quad 0 < y < \bar{\phi}(x)\} \tag{12.49}$$

(let us remark that the 'graph' of $\bar{\phi}$ is made up of Γ_λ and Γ_σ, and not only of Γ_λ like that of ϕ).

The non-physical solutions

The question regarding conditions (12.13), (12.27), and (12.34) is more delicate: in fact they cannot be deduced from the formulation which we have presented. To face this question let us first note that the problem, as we have posed it, has spurious solutions, i.e. solutions which do not have any physical meaning. One of these solutions is the one which corresponds to the 'completely wet dam'; more precisely, the pair made up of $\Omega \equiv D$ and of the solution of the following mixed Dirichlet–Neumann problem is a solution of problem 12.1:

PROBLEM 12.2. Find p such that

$$\Delta p = 0 \quad \text{in } D, \quad (12.50)$$
$$p = \gamma(y_1 - y) \quad \text{on } \Gamma_1, \quad (12.51)$$
$$p = \gamma(y_2 - y) \quad \text{on } \Gamma_2, \quad (12.52)$$
$$p = 0 \quad \text{on } \Gamma_3, \quad (12.53)$$
$$\partial p/\partial y = -\gamma \quad \text{on } \Gamma_0. \quad (12.54)$$

We observe that this problem is not a free boundary problem, but a 'fixed boundary' problem which we know how to reformulate variationally: *find*

$p \in K; \quad \forall v \in K$

$$\int_D \text{grad } p \text{ grad } (p - v) \, dx \, dy \leq \int_a^b (p(x, 0) - v(x, 0)) \, dx, \quad (12.55)$$

where (note the regularity specification which was absent in problem 12.2)

$K = \{v \in H^1(D): \quad v = \gamma(y_1 - y) \quad \text{on } \Gamma_1,$
$\qquad v = \gamma(y_2 - y) \quad \text{on } \Gamma_2, \quad v = 0 \quad \text{on } \Gamma_3\}. \quad (12.56)$

Problem (12.55), and hence problem 12.2 too, has one and only one solution which, from the maximum principle, is strictly positive in D; this is also a solution of problem 12.1 since, if $\Omega \equiv D$, conditions (12.41), (12.42), and (12.44) become the single condition (12.53).

How can one eliminate this kind of solution? We cannot require that $\Omega \neq D$ since in some cases $\Omega \equiv D$ corresponds to the true solution (e.g. for the 'triangular dam' already considered). The most natural suggestion is to impose

$$x \mapsto \phi(x) \text{ is strictly decreasing}, \quad (12.34)$$

but this condition, besides requiring that ϕ be 'regular', does not eliminate the 'completely wet dam': e.g., if $y_1 = \max_{[a,b]} Y(x)$ and $Y(x)$ is concave (and the 'physical' solution is not then $\Omega \equiv D \ldots$). In place of (12.34) we impose

$$x \mapsto \frac{1}{\gamma} \int_0^{Y(x)} \partial p/\partial x \, dy \text{ is not decreasing in } [0, c] \quad (12.13)$$

or, equivalently,

$$\partial(p/\gamma + y)/\partial n \leq 0 \quad \text{on } \Gamma_\lambda \cup \Gamma_\sigma. \quad (12.27)$$

Before proving that, at least if the boundary $Y(x)$ is 'regular' (which among other things excludes the case of the triangular dam), (12.27) eliminates the spurious solutions, we will prove the equivalence (12.13) $\Leftrightarrow$ (12.27). This results from the following calculations (which are formal, since no regularity is imposed on $\bar{\phi}$):

$$\frac{d}{dx}\frac{1}{\gamma}\int_0^{Y(x)}\frac{\partial p}{\partial x}dy = \frac{1}{\gamma}\frac{d}{dx}\int_0^{\bar{\phi}(x)}\frac{\partial p}{\partial x}dy$$

$$= \frac{1}{\gamma}\int_0^{\bar{\phi}(x)}\frac{\partial^2 p}{\partial x^2}dy + \frac{1}{\gamma}\frac{\partial p}{\partial x}(x,\bar{\phi}(x))\bar{\phi}'(x)$$
(using (12.50))

$$= -\frac{1}{\gamma}\left[\frac{\partial p}{\partial y}(x,\bar{\phi}(x)) - \frac{\partial p}{\partial y}(x,0)\right] + \frac{1}{\gamma}\frac{\partial p}{\partial x}(x,\bar{\phi}(x))\bar{\phi}'(x)$$

$$= -\frac{\partial(p/\gamma + y)}{\partial y}(x,\bar{\phi}(x)) + \bar{\phi}'(x)\frac{\partial(p/\gamma + y)}{\partial x}(x,\bar{\phi}(x))$$

$$= -(1+\bar{\phi}'^2(x))\left[-\bar{\phi}'(1+\bar{\phi}'^2(x))^{-1/2}\frac{\partial(p/\gamma+y)}{\partial x}(x,\bar{\phi}(x))\right.$$

$$\left.+ (1+\bar{\phi}'^2(x))^{-1/2}\frac{\partial(p/\gamma+y)}{\partial y}(x,\bar{\phi}(x))\right]$$

$$= -(1+\bar{\phi}'^2(x))^{1/2}\frac{\partial(p/\gamma+y)}{\partial n}(x,\bar{\phi}(x)). \tag{12.57}$$

We now have to prove that 'problem 12.2 and (12.27)' does not have a solution; it is, however, more convenient to work with u rather than p, and we will therefore consider the equivalent problem '$\Delta u = 0$ in D, $\partial u/\partial y = 0$ on Γ_0, $u = y_1$ on Γ_1, $u = y_2$ on Γ_2, $u = y$ on Γ_3 and $\partial u/\partial n \leq 0$ on Γ_3'. Let $\bar{u}$ be the solution of this problem and M a point of maximum for $\bar{u}$. From $\Delta \bar{u} = 0$ in D and the maximum principle it follows that $M \notin D$, and from $\partial u/\partial y = 0$ on Γ_0 and from the maximum principle of Hopf we have $M \notin \Gamma_0$: and thus $M \in \Gamma$; now $\max_\Gamma \bar{u} = \max_{[a,b]} Y(x)$, and thus one such point M is $\bar{M} \equiv (\bar{x}, Y(\bar{x}))$, where $\bar{x}$ is a point of maximum for $Y(x)$: from the maximum principle of Hopf we have then that $\partial u/\partial n(\bar{M}) > 0$, which is contradictory with $\partial u/\partial n \leq 0$ on Γ_3.

Note. In what follows, for simplicity of exposition, we put $\gamma = 1$; this does not lead to any loss of generality: we are merely changing our unknown from p to p/γ. There is not even incompatibility between choosing $\gamma = 1$ and $k = 1$, given the type of relationship between k and γ: everything comes out with an appropriate (though undoubtedly not very rational...) choice of the units of measure.

12.3 THE MATHEMATICAL PROBLEM

In this section we will formulate problem 12.1 rigorously and prove a result regarding the existence of solution. We will also present a result regarding the uniqueness and the regularity of the solution in the case of a dam with a particular geometry.

12.3.1 Rigorous formulation of the free-boundary problem

The geometry of the problem

Let us first define rigorously the geometry of the problem (cf. figure 12.2 and the Notation Diagram 12.1).

DEFINITION 12.1. Let a, b, c, y_1, and y_2 be five real numbers such that $a<0<c<b$ and $0<y_2<y_1$; let $Y \in C^3([a, b])$ be a concave function such that $Y(a) = Y(b) = 0$, $Y(0) = y_1$, $Y(c) = y_2$ and $Y'(0) \geq 0$. Let us refer by the term *dam* to the set $D = \{(x, y) \in \mathbb{R}^2 : a < x < b \text{ and } 0 < y < Y(x)\}$; further let us put $D_1 = \{(x, y) \in D : a < x < 0\}$, $D_2 = \{(x, y) \in D : c < x < b\}$, $D_3 = \{(x, y) \in D : 0 < x < c\}$, $\Gamma_0 = \{(x, 0) : a < x < b\}$, $\Gamma = \{(x, Y(x)) : a < x < b\}$, $\Gamma_1 = \{(x, Y(x)) : a < x < 0\}$, $\Gamma_2 = \{(x, Y(x)) : c < x < b\}$, and $\Gamma_3 = \{(x, Y(x)) : 0 < x < c\}$.

Integral formulation of the dam problem

Let us next consider the following problem, which is a rigorous reformulation of problem 12.1 in integral form:

PROBLEM 12.3. Find (p, Ω) such that

$$p \in C^0(\bar{D}) \cap H^1(D), \tag{12.58}$$

$$x \mapsto \int_0^{Y(x)} \frac{\partial p}{\partial x} \, dy \text{ is non-decreasing in } [0, c], \tag{12.59}$$

$$p \geq 0 \quad \text{in } \bar{D}, \tag{12.60}$$

$$p > 0 \quad \text{on } \Gamma_0, \tag{12.61}$$

$$p = y_1 - y \quad \text{on } \Gamma_1, \tag{12.62}$$

$$p = y_2 - y \quad \text{on } \Gamma_2, \tag{12.63}$$

$$p = 0 \quad \text{on } \Gamma_3, \tag{12.64}$$

and, putting

$$\Omega = \{(x, y) \in D : p(x, y) > 0\} \tag{12.65}$$

and

$$C_\Gamma = \{\psi \in C^\infty(\bar{D}) : \psi = 0 \text{ in a neighbourhood of } \Gamma\}, \tag{12.66}$$

then

$$\forall \psi \in C_\Gamma \int_\Omega \text{grad}\,(p+y)\,\text{grad}\,\psi \, dx\, dy = 0. \tag{12.67}$$

We will prove that problem 12.3 is effectively a reformulation of problem 12.1 (or rather, that problem 12.3 is a 'completion' of problem 12.1, the inverted commas being due to the fact that we have eliminated any explicit reference to the free boundary...). From (12.58), (12.61), and (12.65) we obtain (12.37). Formulae (12.62) and (12.63), which are to be taken in the sense of continuous functions, are the same as (12.39) and (12.40), respectively. In (12.64), which is also to be taken in the sense of the continuous functions, is included (12.41) and, from the continuity of p and from the definition of Ω, also (12.42) (the points of Γ_λ are points of accumulation of zeros of $p\ldots$). We will now analyse the information contained in (12.67). We have, for every $\psi \in C_\Gamma$,

$$0 = \int_\Omega \operatorname{grad}(p+y) \operatorname{grad} \psi \, dx \, dy = \int_D \operatorname{grad}(p+y) \operatorname{grad} \psi \, dx \, dy$$

$$- \int_{D\setminus\Omega} \operatorname{grad}(p+y) \operatorname{grad} \psi \, dx \, dy$$

$$= \int_D \operatorname{grad}(p+y) \operatorname{grad} \psi \, dx \, dy - \int_{D\setminus\Omega} \frac{\partial \psi}{\partial y} \, dx \, dy$$

$$= \int_D \left[\operatorname{grad}(p+y) \operatorname{grad} \psi - \chi_{D\setminus\Omega} \frac{\partial \psi}{\partial y} \right] dx \, dy. \tag{12.68}$$

From this, now choosing ψ in $C_0^\infty(D)$, we have

$$-\Delta(p+y) + \frac{\partial \chi_{D\setminus\Omega}}{\partial y} = 0 \quad \text{in } \mathcal{D}'(D), \tag{12.69}$$

or

$$-\Delta p - \frac{\partial \chi_\Omega}{\partial y} = 0 \quad \text{in } \mathcal{D}'(D), \tag{12.70}$$

which includes (12.38). Again from (12.67), and from (12.70) (or better from (12.38)), we obtain, by applying a Green's formula,

$$0 = \int_\Omega \operatorname{grad}(p+y) \operatorname{grad} \psi \, dx \, dy$$

$$= \int_\Omega [-\Delta(p+y)]\psi \, dx \, dy + \int_{\partial\Omega} \frac{\partial(p+y)}{\partial n} \psi \, d\sigma$$

$$= \int_{\partial\Omega\setminus\Gamma} \frac{\partial(p+y)}{\partial n} \psi \, d\sigma, \tag{12.71}$$

and therefore

$$\frac{\partial(p+y)}{\partial n} = 0 \quad \text{on } \partial\Omega\setminus\Gamma = \Gamma_0 \cup \Gamma_\lambda \tag{12.72}$$

in the sense of the traces of the functions of $H_\Delta^1(\Omega) \equiv \{u \in H^1(\Omega) : \Delta u \in L^2(\Omega)\}$, which includes both (12.43) and (12.44).

Differential formulation of the dam problem

The integral formulation 12.3 is convenient in order to see that problem 12.1 and problem 12.3 are the 'same' problem, but for further developments the following equivalent differential formulation is more convenient:

PROBLEM 12.4. Find p and Ω such that are satisfied (12.58), (12.59), (12.60), (12.61), (12.62), (12.63), (12.64),

$$\frac{\partial p}{\partial y} = -1 \quad \text{on } \Gamma_0 \tag{12.73}$$

and, defining Ω by means of (12.65),

$$-\Delta p = \frac{\partial \chi_\Omega}{\partial y} \quad \text{in } \mathscr{D}'(D). \tag{12.74}$$

We have already seen that 'problem 12.3 $\Rightarrow$ problem 12.4'. To see that 'problem 12.4 $\Rightarrow$ problem 12.3' we show that (12.73) (which must be interpreted in the sense of the traces of functions of $H^1_\Delta(\Omega)$) and (12.74) imply, through (12.61) and (12.65), (12.67) (and hence, in particular, that the information contained in (12.72) regarding Γ_λ is already present in (12.70)). On the one hand we are now in a position to apply Green's formula (18.31) (since div $(\text{grad } p + \chi_\Omega \text{ grad } y) = \Delta p + \partial \chi / \partial y = 0 \in L^2(D)$ and grad $p + \chi_\Omega$ grad $y \in \{L^2(D)\}^2$), and on the other (12.61) and (12.65) ensure that $\chi_\Omega \equiv 1$ in a neighbourhood of Γ_0: hence we have, for $\psi \in C_\Gamma$,

$$\int_\Omega \text{grad } (p+y) \text{ grad } \psi \, dx \, dy = \int_D \text{grad } (p+y) \text{ grad } \psi \, dx \, dy$$

$$- \int_{D \setminus \Omega} \text{grad } (p+y) \text{ grad } \psi \, dx \, dy$$

$$= \int_D [\text{grad } (p+y) - \chi_{D \setminus \Omega} \text{ grad } y] \text{ grad } \psi \, dx \, dy$$

$$= \int_D [\text{grad } p + \text{grad } y$$

$$- (1 - \chi_\Omega) \text{ grad } y] \text{ grad } \psi \, dx \, dy$$

$$= \int_D [\text{grad } p + \chi_\Omega \text{ grad } y] \text{ grad } \psi \, dx \, dy$$

$$= \int_{\Gamma_0} [\text{grad } p + \chi_\Omega \text{ grad } y] \cdot \mathbf{n} \psi \, d\gamma$$

$$= \int_{\Gamma_0} [\text{grad } p + \text{grad } y] \cdot \mathbf{n} \psi \, d\gamma$$

$$= \int_{\Gamma_0} \left[-\frac{\partial p}{\partial y} - 1 \right] \psi \, d\gamma = 0. \tag{12.75}$$

If one leaves out the intermediate terms, (12.75) is nothing but (12.67). We must recall that in Chapter 7 we have already had to deal with free boundary problems, even though they were not explicitly formulated as such. Thus, the obstacle problem indicated as problem 7.29 is a true and proper free boundary problem, the boundary being that which separates the contact and detachment sets. Now, while in the obstacle problem we have the pair of conditions $-\Delta u \leq 0$ and $u \leq \psi$, in the dam problem we have the condition $-\Delta p = \partial \chi_\Omega/\partial y \leq 0$ (let us remark that, as y increases, χ_Ω goes from 1 to 0) but we also have $p \geq 0$ (rather than $p \leq \psi$): the conditions are not therefore 'well paired' and we cannot then hope to obtain Ω as the detachment set of a variational inequality in the variable p. Another, slightly less heuristic, way of arriving at the same conclusion is based on the fact that were p to solve a variational inequality it should be 'more regular' (e.g. $p \in C^1(\bar{D}))$; but then the pair of conditions $p|_{D\setminus\Omega} \equiv 0$, $\partial(p+y)/\partial n_{\text{int}} = 0$ would imply that Γ_λ is a vertical segment..... This leads to the 'necessity' of transforming the problem, which we will do in the next section.

12.3.2 Transformation of the free-boundary problem

The integral transformation

Let us now introduce the functional transformation which is the basis of the method for solving the free boundary problems which we present here.

Let us remark that the transformation $p \mapsto U$ defined below in (12.76) is closely connected to the transformation $p \mapsto z$ defined in Part I, formula (8.52) (and precisely is $U = -z$); we prefer to work in terms of U here, rather than z, both so as to maintain the notations of Baiocchi 1975b, which we will often refer to, and because the treatment which we develop here is independent of that given in Part I.

DEFINITION 12.2. Let us put, for $p \in C^0(\bar{D}) \cap H^1(D)$ and $(x, y) \in \bar{D}$,

$$U(x, y) = \int_0^y p(x, t) \, dt. \qquad (12.76)$$

This definition of U is clearly equivalent to '$\partial U/\partial y = p$ in D and $U = 0$ on Γ_0', and thus if we know U we can reconstruct all the physical and geometrical quantities which we have considered so far. The first question we will deal with is in effect that of expressing the conditions of problem 12.4 in terms of U, i.e., that of answering the question '*what problem does U solve?*'

The problem for U

By means of a regularity result for the solutions of problem 12.3 (established independently of whether these solutions exist or not) it can be proved that

$$U \in W^{2,r}(D), \qquad \forall r < +\infty. \qquad (12.77)$$

Later we will give some clues for this proof; for further details see Baiocchi, 1975b; p. 595.

We will now try to find the differential equation which U must satisfy. From (12.76) and (12.74) (with notations for the derivative which are different from those so far used in this chapter but which are, however, standard), we have

$$D_y(\Delta U + \chi_\Omega) = \Delta D_y U + D_y \chi_\Omega = \Delta p + D_y \chi_\Omega = 0 \quad \text{in } \mathscr{D}'(D), \tag{12.78}$$

which shows that $\Delta U + \chi_\Omega$ is a distribution independent of the variable y (i.e., of the form $f(x) \otimes 1$). Further, from (12.61) and (12.65), it follows that there exists an open set θ in $\mathbb{R}^2$ such that $\Gamma_0 \subset \theta$ and $\theta \cap D \subset \Omega$: let us put $\theta' = \theta \cap D$ and $\theta'' = \theta' \cup \Gamma_0$. In θ' we have $\chi_\Omega \equiv 1$ and hence, from (12.74), $\Delta p = 0$ in θ': hence we have $p \in C^\infty(\theta')$ (in fact it is $p \in C^\omega(\theta')\ldots$) and, from this together with (12.73), $p \in C^\infty(\theta'')$. We can then say that $U \in C^\infty(\theta'')$ and thus that in θ'' we can calculate the derivatives of U in the classical sense; again, since $\chi_\Omega \equiv 1$ in θ', we have:

$$(\Delta U + \chi_\Omega)(x, y) = D_x^2 \int_0^y p(x, t)\, dt + D_y^2 \int_0^y p(x, t)\, dt$$

$$= \int_0^y p_{xx}(x, t)\, dt + p_y(x, y) + 1, \tag{12.79}$$

and therefore, from (12.73),

$$(\Delta U + \chi_\Omega)(x, 0) = 0. \tag{12.80}$$

$\Delta U + \chi_\Omega$ is therefore a distribution independent of y which coincides, in θ', with a regular function which is zero on Γ_0: by extension we obtain

$$\Delta U + \chi_\Omega = 0 \quad \text{in } \mathscr{D}'(D); \tag{12.81}$$

or rather

$$-\Delta U = \chi_\Omega \quad \text{a.e. in } D. \tag{12.82}$$

The boundary conditions for U

The previous discussion allows us to write:

$$\Omega \text{ contains a neighbourhood of } \Gamma_0. \tag{12.83}$$

From (12.76) and (12.60) we have

$$U_y \geq 0 \quad \text{in } \bar{D}. \tag{12.84}$$

From (12.76) and from (12.62), (12.63), and (12.64) we have, respectively, the following oblique derivative conditions on the boundary:

$$U_y = y_1 - y \quad \text{on } \Gamma_1, \tag{12.85}$$

$$U_y = y_2 - y \quad \text{on } \Gamma_2, \tag{12.86}$$

$$U_y = 0 \quad \text{on } \Gamma_3. \tag{12.87}$$

We add to the previous conditions the Dirichlet condition

$$U = 0 \quad \text{on } \Gamma_0, \tag{12.88}$$

which follows from (12.76).

We will now show that the information contained in (12.59) can be put in terms of U by means of

$$x \mapsto U(x, Y(x)) \text{ is convex in } [0, c]. \tag{12.89}$$

In fact, from (12.77) and from Sobolev's immersion theorem (see Part I) it follows that $U \in C^1(\bar{D})$ and thus, given that Y is of class C^3, the trace of U on Γ is of class C^1. Thus the following calculations make sense with $x \in \,]0, c[$:

$$\frac{d}{dx} U(x, Y(x)) = U_x(x, Y(x)) + Y'(x) U_y(x, Y(x))$$

$$= U_x(x, Y(x)) = \left[D_x \int_0^y p(x, t) \, dt \right]_{y = Y(x)}$$

$$= \int_0^{Y(x)} p_x(x, t) \, dt \tag{12.90}$$

(let us remark that from $U_y = p$ and from (12.60) we have $U_y(x, Y(x)) = 0$ in $]0, c[$). From this together with (12.59) we conclude that

$$\frac{d^2}{dx^2} U(x, Y(x)) \geq 0 \quad \text{in }]0, c[, \tag{12.91}$$

which is to be taken in the sense of distributions and which can be interpreted as an analytic version of (12.89).

We want now to describe Ω in terms of U. Since p is continuous, from (12.60) and (12.65) we have

$$\Omega = \left\{ (x, y) \in D : \int_y^{Y(x)} p(x, t) \, dt > 0 \right\} \tag{12.92}$$

and hence, from the definition of U,

$$\Omega = \{(x, y) \in D : U(x, y) < U(x, Y(x))\}. \tag{12.93}$$

Further, we can write this more precisely since $D_1 \cup D_2 \subset \Omega$:

$$\Omega = D_1 \cup D_2 \cup \{(x, y) \in D : 0 \leq x \leq c \quad \text{and} \quad U(x, y) < U(x, Y(x))\} \tag{12.94}$$

(to see that $D_1 \subset \Omega$ [resp.: $D_2 \subset \Omega$] we recall that (12.84) and (12.93) tell us that Ω is a subgraph, and since p is continuous, from (12.65) and (12.62) [resp.: (12.63)] it follows that there exists a neighbourhood of Γ_1 [resp.: Γ_2] in $\bar{D}$ whose interior points are in Ω).

Let us remark again, even though this does not interest us directly in the formulation of the problem in terms of U, that the carrying capacity of the dam is given by

$$q_d = -U_x(0, y_1) \tag{12.95}$$

and that the level lines, in Ω, of U_x and of $U_x + y$ are, respectively, the stream lines and the equipotentials.

In summary, we can say that U solves the following boundary value problem of mixed Dirichlet-oblique derivative type:

PROBLEM 12.5. Find U such that

$$U \in W^{2,r}(D) \quad \forall r < +\infty, \tag{12.96}$$
$$x \mapsto U(x, Y(x)) \text{ is convex in } [0, c], \tag{12.97}$$
$$U_y \geq 0 \quad \text{in } \bar{D}, \tag{12.98}$$
$$U = 0 \quad \text{on } \Gamma_0, \tag{12.99}$$
$$U_y = y_1 - y \quad \text{on } \Gamma_1, \tag{12.100}$$
$$U_y = y_2 - y \quad \text{on } \Gamma_2, \tag{12.101}$$
$$U_y = 0 \quad \text{on } \Gamma_3, \tag{12.102}$$

and, putting

$$\Omega = D_1 \cup D_2 \cup \{(x, y) \in D : 0 \leq x \leq c \text{ and } U(x, y) < U(x, Y(x))\}, \tag{12.103}$$

it follows that

$$\Omega \text{ contains a neighbourhood of } \Gamma_0, \tag{12.104}$$
$$-\Delta U = \chi_\Omega \quad \text{a.e. in } D. \tag{12.105}$$

Conditions (12.98) to (12.102) are to be taken, because of (12.96), in the sense of functions of class C^1; (12.96) again ensures that the set Ω defined in (12.103) is an open set.

Let us again introduce the following concept:

DEFINITION 12.3. A solution of the problem 12.5 is said to be *regular* if, putting

$$\bar{\phi}(x) = \sup\{y : (x, y) \in \Omega\}, \quad 0 \leq x \leq c, \tag{12.106}$$

the function $[0, c] \ni x \mapsto \bar{\phi}(x)$ is continuous and strictly decreasing and there exists $s \in]0, c[$ such that $\bar{\phi}(x) < Y(x)$ in $]0, s[$ and $\bar{\phi}(x) \equiv Y(x)$ in $[s, c]$.

Let us remark that s is interpreted as the abscissa of the suspended source, C_ϕ, that $\{(x, \bar{\phi}(x)) : x \in]0, s[\}$ is interpreted as the free line, Γ_λ, and that $\{(x, \bar{\phi}(x)) : s < x < c\}$ is interpreted as the seepage line, Γ_σ. The restriction of $\bar{\phi}$ to $[0, s]$ will be denoted simply by ϕ.

The work program

We now need a guide, i.e. a line of research, to study problem 12.5. We will find this guide by transforming the differential equation (12.105) into an appropriate multi-valued differential equation, transformation which will also help us to find out the true nature of the problem.

Let then $M: L^1(D) \to L^1(D)$ be the operator defined by

$$(M(u))(x, y) = \begin{cases} u(x, y) + 1 & \text{if } (x, y) \in D_1 \cup D_2 \\ \text{conv}\,[u(x, Y(x))]^+ & \text{if } (x, y) \in D_3 \end{cases} \quad (12.107)$$

[conv: $L^1(]0, c[) \to L^1(]0, c[)$ is the operator defined by $(\text{conv}\,f)(x) = \sup\{\alpha x + \beta : \alpha, \beta \in \mathbb{R},\ \alpha x + \beta \leq f(x)\text{ a.e. in }]0, c[\}$ (i.e., conv f is the upper envelope of the affine functions which are 'below' f); if $f \geq 0$ then $0 \leq \text{conv}\,f \leq f$; conv is a continuous operator, e.g., from $H^1(]0, c[)$ to itself] and let $\beta: \mathbb{R} \to 2^\mathbb{R}$ be a maximal monotone operator such that

$$\beta(t) = \{1\} \text{ if } t > 0,\ \beta(0) \ni 0. \quad (12.108)$$

We will now prove the following

THEOREM 12.1. If $\bar{U}$ is a solution of problem 12.5 then

$$-\Delta \bar{U} \in \beta(M(\bar{U}) - \bar{U}) \quad \text{a.e. in } D. \quad (12.109)$$

Proof. Let $\bar{U}$ be a solution of problem 12.5. From (12.98) and (12.99) we have

$$\bar{U}(x, Y(x)) \geq \bar{U}(x, y) \geq 0 \quad \text{in } \bar{D}; \quad (12.110)$$

from (12.103) we have

$$\bar{U}(x, Y(x)) > \bar{U}(x, y) \quad \text{in } \Omega \quad (12.111)$$

and

$$\bar{U}(x, Y(x)) = \bar{U}(x, y) \quad \text{in } D \setminus \Omega; \quad (12.112)$$

from (12.10) and (12.97) we have

$$\text{conv}\,[\bar{U}(x, Y(x))]^+ \equiv \bar{U}(x, Y(x)). \quad (12.113)$$

Now, from (12.107), (12.111), and (12.113),

$$M(\bar{U}) > \bar{U} \quad \text{a.e. in } \Omega \quad (12.114)$$

and, from (12.107), (12.112), and (12.113),

$$M(\bar{U}) = \bar{U} \quad \text{in } D \setminus \Omega. \quad (12.115)$$

From (12.114), (12.115), (12.105), and (12.108) we have finally (12.109). ∎

The 'smallest possible' operator β is the operator

$$\beta = h \triangleq \begin{cases} h(t) = \{1\} & \text{if } t > 0 \\ h(t) =]-\infty, 1] & \text{if } t = 0 \\ h(t) = \varnothing & \text{if } t < 0, \end{cases} \quad (12.116)$$

for which equation (12.109) can also be written

$$M(\bar{U}) \geq \bar{U}, \quad -\Delta\bar{U} \leq 1, \quad (\bar{U} - M(\bar{U}))(-\Delta\bar{U} - 1) = 0 \quad \text{a.e. in } D, \quad (12.117)$$

a set of expressions which recalls the interpretation of a quasivariational inequality as a boundary value problem (problem 11.7 should be kept in mind particularly) and shows what the 'true nature' of problem 12.5 is. The 'largest possible' operator β is the monotonic maximal operator associated with the Heaviside function, i.e. the operator

$$\beta = H \triangleq \begin{cases} H(t) = \{1\} & \text{if } t > 0 \\ H(t) = [0, 1] & \text{if } t = 0 \\ H(t) = \{0\} & \text{if } t < 0, \end{cases} \quad (12.118)$$

for which equation (12.109) can also be written

$$-\Delta\bar{U} \in \chi_{D_1} + \chi_{D_2} + \chi_{D_3} H(\text{conv}\,[\bar{U}(x, Y(x))]^+ - \bar{U}(x, y)) \quad \text{a.e. in } D, \quad (12.119)$$

which is the multi-valued differential equation which we will in fact study.

The advantage of the choice $\beta = H$ with respect to $\beta = h$ (and any other choice of β which satisfies (12.108)) lies in the fact that (12.119) contains the information $0 \leq -\Delta\bar{U}$, while (12.117) does not. It is important to observe that, as we have just shown, if $\bar{U}$ is a solution of problem 12.5 then $\bar{U}$ is solution of '*problem* 12.5 *with* (12.119) *instead of* (12.105)', but that if $\bar{U}$ is a solution of this latter problem then for $\bar{U}$ to be also a solution of the problem 12.5 we need to know that $\partial\Omega$ has zero measure (in this connection see the beginning of section 12.3.6).

What does our guide now 'suggest'? It 'suggests' that we divide the proof of the theorem of the existence of a solution of problem 12.5 in three parts. In the first part (section 12.3.3) we will assume the right-hand side of (12.119) to be completely known; i.e., putting the right-hand side $F = \chi_{D_1} + \chi_{D_2} + \chi_{D_3} H(\text{conv}\,[U(x, Y(x))]^+ - U(x, y))$, we will study a linear boundary value problem of the mixed Dirichlet-oblique derivative type associated with the differential equation $-\Delta U = F$.

In the second step (section 12.3.4) we will assume that only $f = \text{conv}\,[U(x, Y(x))]^+$ is known and study a boundary problem of the mixed Dirichlet-nonlinear oblique derivative type associated with the multi-valued differential equation $-\Delta U \in \chi_{D_1} + \chi_{D_2} + \chi_{D_3} H(f - U)$, using naturally the information obtained from the first step. Finally, in the third step (section

12.3.5), we will study the problem 12.5, making use of the results obtained in the previous steps (namely: we will show that the solutions of problem 12.5 are the fixed points of a transformation which is appropriately associated with the problem studied in the second step).

12.3.3 STUDY OF A LINEAR MIXED PROBLEM

The problem

This section is devoted to the study of the following problem:

PROBLEM 12.6. Given $g \in (H_{00}^{1/2}(\Gamma))'$ and $F \in L^2(D)$ find $u \in H^1(D)$ such that

$$-\Delta u = F \quad \text{in } \mathscr{D}'(D), \tag{12.120}$$

$$\gamma_0 u = 0 \quad \text{on } \Gamma_0, \tag{12.121}$$

$$\gamma_a u = g \quad \text{on } \Gamma. \tag{12.122}$$

We must first specify in which sense (12.121) and (12.122) are to be considered (and what (12.122) is...). The problem raised by (12.121) is simple: it must be considered in the sense of (the restrictions of) functions of $H^{1/2}(\partial D)$, i.e. (cf. the comments at the end of Chapter 17) in the sense of functions of $H_{00}^{1/2}(\Gamma_0)$. The meaning to be given to (12.122) is more complicated. We want (naturally...) to study problem 12.6 by variational techniques, or rather: we want problem 12.6 to be an interpretation of an appropriate variational problem. We must choose then an appropriate decomposition for Δ and a bilinear form associated with it; further, this decomposition must not be the standard one since on Γ we have (in problem 12.5, and therefore in this one if we want the two to be related...) a condition on the oblique derivative, and more precisely a condition on u_y. Thus (12.122) must be interpreted in the sense of traces obtained from the Green's formula associated with the bilinear form, i.e.: γ_a is the conormal derivative associated with the bilinear form we choose, form a. So, we must determine at first what that form is, and then prove a formula of Green.

The bilinear form

The argument which will lead us to the bilinear form has to be heuristic. We want the decomposition of Δ to be such that the conormal derivative reproduces an oblique derivative on Γ: from what we have seen in Chapter 7 we find that the appropriate decomposition is (7.21), to which corresponds, e.g., the form (7.122). We must therefore find t such that

$$\frac{\partial u}{\partial n} - t|_\Gamma \frac{\partial u}{\partial \tau} = \frac{\partial u}{\partial y} k(x, y), \tag{12.123}$$

where k is a coefficient which will have to be properly counter-balanced, and whose presence is necessary because we are sure in which direction the differentiation is to be carried out but not the value of the derivative. We have, on Γ,

$$\mathbf{n} \equiv (-Y'(x)/\sqrt{1+Y'(x)^2},\ 1/\sqrt{1+Y'(x)^2}) \tag{12.124}$$

and

$$\boldsymbol{\tau} \equiv (1/\sqrt{1+Y'(x)^2},\ Y'(x)/\sqrt{1+Y'(x)^2}), \tag{12.125}$$

and thus

$$\frac{\partial u}{\partial n} = -\frac{\partial u}{\partial x}\frac{Y'(x)}{\sqrt{1+Y'(x)^2}} + \frac{\partial u}{\partial y}\frac{1}{\sqrt{1+Y'(x)^2}}, \tag{12.126}$$

and

$$\frac{\partial u}{\partial \tau} = \frac{\partial u}{\partial x}\frac{1}{\sqrt{1+Y'(x)^2}} + \frac{\partial u}{\partial y}\frac{Y'(x)}{\sqrt{1+Y'(x)^2}}. \tag{12.127}$$

Putting (12.126) and (12.127) in (12.123), and putting the coefficient of $\partial u/\partial x$ equal to zero, we have

$$t = -Y'(x) \tag{12.128}$$

and

$$k(x, y) = 1 + Y'(x)^2. \tag{12.129}$$

Substituting (12.128) in (7.122) we find then that the bilinear form which we need is

$$a(u, v) = \int_D [(u_x - Y'u_y)v_x + (u_y + Y'u_x)v_y - Y''u_y v]\, dx\, dy. \tag{12.130}$$

We will end our heuristic argument here.

Formula (12.130) defines a continuous bilinear form from $H^1(D) \times H^1(D)$ in $\mathbb{R}$, which is not in general coercive. This non-coercivity is not, however, a real problem since, from (12.121), we see that what interests us is the restriction of a to $V_0 \times V_0$, where

$$V_0 = \{v \in H^1(D):\ \gamma_0 v = 0 \quad \text{on } \Gamma_0\}, \tag{12.131}$$

and a is continuous and coercive in $V_0 \times V_0$. The continuity of a in $V_0 \times V_0$ follows in fact from the continuity of a in $H^1(D) \times H^1(D)$ and from the fact that V_0 is a closed subspace of $H^1(D)$. To see that a is coercive on V_0 we note, on the one hand, that

$$\|v\|_{V_0}^2 = \int_D (v_x^2 + v_y^2)\, dx\, dy \tag{12.132}$$

defines a norm on V_0 equivalent to the one induced by $H^1(D)$ and, on the other, that $C_{\Gamma_0} = \{\psi \in C^\infty(\bar D):\ \psi|_{\Gamma_0} = 0\}$ is dense in V_0 and that, for $\psi \in C_{\Gamma_0}$, we can write successively (using in particular the concavity of Y)

$$a(\psi, \psi) = \int_D (\psi_x^2 - Y'\psi_y\psi_x + \psi_y^2 + Y'\psi_x\psi_y - Y''\psi_y\psi) \, dx \, dy$$

$$= \int_D [(\psi_x^2 + \psi_y^2) - Y''\psi_y\psi] \, dx \, dy$$

$$= \|\psi\|_{V_0}^2 - \int_D Y''\psi_y\psi \, dx \, dy = \|\psi\|_{V_0}^2 - \frac{1}{2}\int_D D_y(Y''\psi^2) \, dx \, dy$$

$$= \|\psi\|_{V_0}^2 - \frac{1}{2}\int_a^b Y''(x)\psi^2(x, Y(x)) \, dx \geq \|\psi\|_{V_0}^2. \tag{12.133}$$

Green's formula

We will now go on to deal with the formula of Green which enables us, among other things, to make sense of (12.122). We will prove then the following theorem, which is a particular case of theorem 18.9:

THEOREM 12.2. *If $u \in H_\Delta^1(D)$ and $v \in V_0$ then the following formula of Green holds:*

$$a(u, v) = -\int_D \Delta u \, v \, dx \, dy + {}_{(H_{00}^{1/2}(\Gamma))'}\langle \gamma_a u|_\Gamma, \gamma_0 u|_\Gamma \rangle_{H_{00}^{1/2}(\Gamma)}. \tag{12.134}$$

Proof. Let us first note that the spaces $H_{00}^{1/2}(\Gamma)$ and $H_{00}^{1/2}(]a, b[)$ are isomorphic since Γ is parametrized in terms of $x \in {]a, b[}$. The same is true of the dual spaces. Under these conditions $\gamma_a u|_\Gamma$ and $\gamma_0 u|_\Gamma$ can be written in terms of x. For $\gamma_0 u|_\Gamma$ this is easy: $\gamma_0 u|_\Gamma = u(x, Y(x))$, with u regular. For $\gamma_a u|_\Gamma$ we will find from our calculations that, again with u regular,

$$\gamma_a u = (1 + Y'(x)^2) u_y(x, Y(x)) \tag{12.135}$$

(note the presence of the factor $1 + Y'(x)^2$!). As we know, C_{Γ_0} is dense in V_0 and $C^\infty(\bar{D})$ is dense in $H_\Delta^1(D)$, so that we need only to consider regular functions and then to conclude, using density, that $u \mapsto \gamma_a u$ (with γ_a as defined in (12.135)) extends in a linear and continuous application from $H_\Delta^1(D)$ to $H_{00}^{1/2}(]a, b[)'$ (note in this context that $v \mapsto \gamma_0 v|_\Gamma$ is linear and continuous from V to $H_{00}^{1/2}(]a, b[) \ldots$). It is in this generalized sense that (12.122) must be interpreted—note that, from $F \in L^2(D)$, $u \in H^1(D)$ and (12.120), we have $u \in H_\Delta^1(D)$.

We must therefore prove that, for $u \in C^\infty(\bar{D})$ and $v \in C_{\Gamma_0}$,

$$a(u, v) = -\int_D \Delta u \, v \, dx \, dy + \int_a^b (1 + Y'(x)^2) u_y(x, Y(x)) v(x, Y(x)) \, dx. \tag{12.136}$$

This formula follows from the following calculations:

$$a(u, v) \triangleq \int_D \{(u_x - Y'u_y)v_x + (u_y + Y'u_x)v_y - Y''u_y v\} \, dx \, dy$$

$$= \int_D \{D_x(v(u_x - Y'u_y)) + D_y(v(u_y + Y'u_x))$$
$$- vu_{xx} + vY'u_{yx} - vu_{yy} - vY'u_{xy}\} \, dx \, dy$$

$$= -\int_D v \Delta u \, dx \, dy + \int_D D_x(v(u_x - Y'u_y)) \, dx \, dy$$
$$+ \int_D D_y(v(u_y + Y'u_x)) \, dx \, dy$$

$$= -\int_D v \Delta u \, dx \, dy + \int_{\partial D} v(u_x - Y'u_y) \, dy - \int_{\partial D} v(u_y + Y'u_x) \, dx$$

$$= -\int_D v \Delta u \, dx \, dy - \int_a^b v(u_x - Y'u_y) Y' \, dx - \int_a^b v(u_y + Y'u_x) \, dx$$

$$= -\int_D v \Delta u \, dx \, dy + \int_a^b v(1 + Y'^2) u_y \, dx \qquad (12.137)$$

Let us remark that, with respect to the step from the fourth to the fifth term, $dx \wedge dy = -dy \wedge dx$.... Let us remark further that the integration on ∂D is done in the direct sense and the functions u_x, u_y and v under $\int_a^b$ are calculated at $(x, Y(x))$—we have not written this out fully so as not to complicate the expressions. ∎

A variational problem

We will now study problem 12.6. To do this, let us consider the following variational problem:

PROBLEM 12.7. Given $F \in L^2(D)$ and $g \in (H_{00}^{1/2}(\Gamma))'$ find

$$u \in V_0; \forall v \in V_0 \quad a(u, v) = \int_D Fv \, dx \, dy + {}_{(H_{00}^{1/2}(\Gamma))'}\!\langle g, \gamma_0 v|_\Gamma \rangle_{H_{00}^{1/2}(\Gamma)}.$$
(12.138)

Problem 12.6 and problem 12.7 are equivalent (in other words, problem 12.6 is the interpretation of problem 12.7 as a boundary value problem). In fact, if u is a solution of problem 12.6 then, from $u \in H^1(D)$ and (12.121), we have $u \in V_0$ and, from (12.120) and (12.122), we have (12.138), by applying Green's formula (12.134). Conversely, if u is a solution of problem 12.7 then $u \in V_0$, and therefore $u \in H^1(D)$ and (12.121) is true; further, taking $v \in \mathcal{D}(D)$ in (12.138) we get (12.120), and from here and again by (12.138) we have (12.122), by applying Green's formula (12.134).

The Lax–Milgram lemma, which we are in a position to apply

since a is a coercive continuous bilinear form and $v \mapsto \int_D Fv \, dx \, dy +{}_{(H_{00}^{1/2}(\Gamma))'}\langle g, \gamma_0 v|_\Gamma \rangle_{H_{00}^{1/2}(\Gamma)}$ is a continuous linear functional in V_0, ensures that problem 12.7 has one and only one solution: thus problem 12.6 also has one and only one solution.

The data g and F

The data g and F have been taken to be very general, but we are interested in fairly particular cases. Thus, from now on, g will be the function defined in the following way (the factor $(1+ Y'(x)^2)$ compensates the one present in $\gamma_a u \ldots$):

$$g(x) = \begin{cases} (1+Y'(x)^2)(y_1 - Y(x)) & \text{if } a < x < 0 \\ 0 & \text{if } 0 \leq x \leq c \\ (1+Y'(x)^2)(y_2 - Y(x)) & \text{if } c < x < b \end{cases} \quad (12.139)$$

(let us recall what we said at the beginning of the proof of theorem 12.2...), and, from now on, the duality present in (12.138) will be written more simply as the integral $\int_a^b g(x)v(x, Y(x)) \, dx$.

As for F, since it 'is' the right-hand side of (12.119), we impose the conditions

$$F \in L^\infty(D), \quad (12.140)$$

$$F \equiv 1 \quad \text{in } D_1 \cup D_2, \quad (12.141)$$

$$F \geq 0 \quad \text{a.e. in } D. \quad (12.142)$$

A regularity result

One can prove by fairly delicate techniques (see Baiocchi, 1975b, p. 595) the following regularity result for the solution of problem 12.6: *if g is given by (12.139) and F satisfies (12.140) and (12.141) then*

$$u \in W^{2,r}(D) \quad \forall r < +\infty. \quad (12.143)$$

While the regularity of u 'far away from the corners of D' follows from the usual regularity results for the solutions of boundary value problems for the Laplace operator, the hypothesis regarding g and F act here as *conditions of compatibility* between these data 'near the corners of D'. The result (12.143) can, for example, be obtained by adapting to this case the reasoning of Grisvard, 1975 (for further details see Baiocchi, 1975b).

Using (12.143) we can now reinterpret the boundary conditions of problem 12.6 (with g given by (12.139)) in the sense of functions of class C^1:

$$u|_{\Gamma_0} = 0, \quad (12.144)$$

$$u_y|_{\Gamma_1} = y_1 - Y(x), \quad (12.145)$$

$$u_y|_{\Gamma_2} = y_2 - Y(x), \quad (12.146)$$

$$u_y|_{\Gamma_3} = 0. \quad (12.147)$$

We state again without proof (which is based on the fact that problem 12.6 is an index problem: see Baiocchi, 1975b, p. 596) that *if g is given by* (12.139), F_1, $F_2 \in L^r(D)(r \in \,]2, +\infty[)$ *and* $F_1 = F_2 = 1$ *in* $D_1 \cup D_2$, *then there exists c_r independent of F_1, F_2 such that*

$$\|u_1 - u_2\|_{W^{2,r}(D)} \leq c_r \|F_1 - F_2\|_{L^r(D)}; \tag{12.148}$$

inequality (12.148) translates a particular continuous dependence of the solution u with respect to F which we will use later.

We will now prove the following additional properties of the solution u of problem 12.6: *if g is given by* (12.139) *and F satisfies* (12.140), (12.141), *and* (12.142) *then*

$$u(x, y) \geq 0 \quad \text{in } \bar{D}, \tag{12.149}$$

$$u_y(x, y) \geq 0 \quad \text{on } \partial D, \tag{12.150}$$

$$u(x, Y(x)) > 0 \quad \text{in }]a, b[\tag{12.151}$$

and

$$u_y(x, 0) > 0 \quad \text{in }]a, b[. \tag{12.152}$$

In the proof we repeatedly use the maximum (or rather, the minimum) principle. Let M be a point of minimum of u in $\bar{D}$. From (12.120), (12.121) and the (classical) maximum principle, we have $M \notin D$ (unless u be constant in $\bar{D}$, which (12.145) and (12.146) exclude). Further $M \notin \Gamma$ since, from the maximum principle of Hopf, every outward oriented derivative in M has to be strictly negative while, from (12.145), (12.146), and (12.147), the derivative u_y (which is outward oriented on Γ) is non-negative. Thus we conclude that $M \in \Gamma_0$ and therefore, from (12.144), that $u(M) = 0$: this proves (12.149). At the same time we have proved (12.151). From (12.145), (12.146), and (12.147) we obtain (12.150) on Γ. Let now P be a point of Γ_0. Since P is a point of minimum for u we have, from the maximum principle of Hopf, that every outward derivative in P must be strictly negative: the u_y (which on Γ_0 is an inward derivative) is therefore strictly positive; by continuity it follows that (12.150) is true in $(a, 0)$ and $(b, 0)$. We have at the same time proved (12.152).

12.3.4 Study of a nonlinear mixed problem

In this section we will study the following problem:

PROBLEM 12.8. *Given* $f \in L^1(]0, c[)$ *find* u *such that* (12.143), (12.144), (12.145), (12.146), (12.147),

$$0 \leq -\Delta u \leq 1 \quad \text{a.e. in } D, \tag{12.153}$$

$$-\Delta u = 1 \quad \text{in } D_1 \cup D_2, \tag{12.154}$$

are satisfied.

In order to show that this problem has a solution (we will come back later to the question of uniqueness) we will show that it can be taken as an interpretation, in terms of boundary value problems, of the following variational problem (g being as defined in (12.139)):

PROBLEM 12.9. Given $f \in L^1(]0, c[)$ find u such that

$u \in V_0$; $\forall v \in V_0$

$$a(u, u-v) + \int_{D_3} [f(x) - u(x, y)]^+ \, dx \, dy \leq \int_{D_3} [f(x) - v(x, y)]^+ \, dx \, dy$$

$$+ \int_a^b g(x)(u(x, Y(x)) - v(x,$$

$$+ \int_{D_1 \cup D_2} (u(x, y) - v(x, y)) \, dx \, dy. \quad (12.155)$$

This problem is closely connected with problem 12.7: the first term in the continuous linear functional (on V_0)

$$L(v) = \int_a^b g(x) v(x, Y(x)) \, dx + \int_{D_1 \cup D_2} v(x, y) \, dx \, dy, \quad (12.156)$$

which serves in giving the boundary conditions, has already been considered in the variational equation (12.138); let us note in this connection that the second term of (12.156) ensures that the datum on $D_1 \cup D_2$ is 1 (cf. (12.141)...). Let us note finally that the presence of the continuous convex functional (on V_0)

$$j_f(v) = \int_{D_3} [f(x) - v(x, y)]^+ \, dx \, dy \quad (12.157)$$

renders the problem nonlinear.

From the results obtained in Chapter 3, problem 12.9 has one and only one solution. We will now show that the solution of problem 12.9 also solves problem 12.8. Putting $v = u - \psi$ in (12.155), with $\psi \in \mathcal{D}(D)$, we have

$$a(u, \psi) + \int_{D_3} [f - u]^+ \, dx \, dy \leq \int_{D_3} [f - u + \psi]^+ \, dx \, dy$$

$$+ \int_{D_1 \cup D_2} \psi \, dx \, dy \quad \forall \psi \in \mathcal{D}(D), \quad (12.158)$$

i.e., since for $\psi \in \mathcal{D}(D)$ is $a(u, \psi) = {}_{\mathcal{D}'(D)}\langle -\Delta u, \psi \rangle_{\mathcal{D}(D)}$

$${}_{\mathcal{D}'(D)}\langle -\Delta u, \psi \rangle_{\mathcal{D}(D)} \leq \int_{D_3} ([f - u + \psi]^+ - [f - u]^+) \, dx \, dy$$

$$+ \int_{D_1 \cup D_2} \psi \, dx \, dy \quad \forall \psi \in \mathcal{D}(D). \quad (12.159)$$

If we now take $\psi \leq 0$ in (12.159) we have (since $[f-u+\psi]^+ \leq [f-u]^+$)

$$_{\mathscr{D}'(D)}\langle -\Delta u, \psi \rangle_{\mathscr{D}(D)} \leq 0 \qquad \forall \psi \in \mathscr{D}(D), \psi \leq 0, \qquad (12.160)$$

i.e.

$$-\Delta u \geq 0 \quad \text{in } \mathscr{D}'(D). \qquad (12.161)$$

If, on the other hand, we take $\psi \geq 0$ in (12.159), we have (since now $[f-u+\psi]^+ \leq [f-u]^+ + [\psi]^+ = [f-u]^+ + \psi$)

$$_{\mathscr{D}'(D)}\langle -\Delta u, \psi \rangle_{\mathscr{D}(D)} \leq \int_D \psi \, dx \, dy = {}_{\mathscr{D}'(D)}\langle 1, \psi \rangle_{\mathscr{D}(D)} \qquad \forall \psi \in \mathscr{D}(D), \psi \geq 0,$$

$$(12.162)$$

i.e.

$$-\Delta u \leq 1 \quad \text{in } \mathscr{D}'(D). \qquad (12.163)$$

We have proved then that

$$0 \leq -\Delta u \leq 1 \quad \text{in } \mathscr{D}'(D) \qquad (12.164)$$

and therefore, since in particular from this it follows that $\Delta u \in L^\infty(D)$, that (12.153) is true. Now if we take $\psi = \pm \chi$ in (12.159), with $\chi \in \mathscr{D}(D)$ and supp $\chi \subset D_1 \cup D_2$, we have (12.154). Since, as we have already pointed out, we have not changed the information on the boundary in going from problem 12.7 to problem 12.9 and since (from (12.153) and (12.154) it follows that there exists an F such that) the conditions (12.140) and (12.141) are satisfied, we can say, in agreement with what we have shown in the preceding section, that the solution u of problem 12.9 satisfies the conditions (12.143) and (12.147).

Let us now assume that f is continuous ($f \in C^0([0, c])$) and let us put

$$\Omega^+ = \{(x, y) \in D_3 : f(x) > u(x, y)\} \qquad (12.165)$$

and

$$\Omega^- = \{(x, y) \in D_3 : f(x) < u(x, y)\}, \qquad (12.166)$$

where u is the solution of problem 12.9. Since f and u are continuous, it follows that Ω^+ and Ω^- are open sets, and we can therefore consider the spaces $\mathscr{D}'(\Omega^+)$ and $\mathscr{D}'(\Omega^-)$. We now want to show that

$$-\Delta u = 1 \quad \text{in } \mathscr{D}'(\Omega^+) \qquad (12.167)$$

and

$$-\Delta u = 0 \quad \text{in } \mathscr{D}'(\Omega^-). \qquad (12.168)$$

In fact, if $\phi \in \mathscr{D}(D)$ and supp $\phi \subset \Omega^+$ then there exists $\lambda_\phi > 0$ such that $|\lambda| \leq \lambda_\phi \Rightarrow f + \lambda \phi \geq u$ in Ω^+ and therefore, putting $\psi = \lambda \phi$ in (12.159), we can write

$$_{\mathscr{D}'(D)}\langle -\Delta u, \lambda \phi \rangle_{\mathscr{D}(D)} \leq \int_{D_3} (f - u + \lambda \phi - [f-u]^+) \, dx \, dy$$

$$+ \int_{D_1 \cup D_2} \lambda \phi \, dx \, dy = \int_D \lambda \phi \, dx \, dy$$

$$\forall \phi \in \mathscr{D}(D), \text{supp } \phi \subset \Omega^+, \quad \text{for } |\lambda| < \lambda_\phi \qquad (12.169)$$

and hence, since $\lambda \le 0$, (12.167). In an analogous manner we can prove (12.168).

Finally, let us note that if f is continuous problem 12.9 is effectively equivalent to problem 12.8, which can now be written in the form

PROBLEM 12.10. Given $f \in C^0([0, c])$ find u such that (12.143) to (12.147) and

$$-\Delta u \in \chi_{D_1} + \chi_{D_2} + \chi_{D_3} H(f - u) \qquad (12.170)$$

are satisfied.

We will now prove this equivalence, which among other things ensures that the solution of problem 12.10 is unique. We have already dealt with the implication 'problem 12.9 $\Rightarrow$ problem 12.10'. Let now u be a solution of problem 12.10; from (12.143) and (12.144) it follows that $u \in V_0$, and the validity of the variational inequality (12.155) follows from the calculations shown below, where v is an arbitrary element of V_0:

$$\begin{aligned}
a(u, u-v) - L(u-v) &= -\int_D \Delta u(u-v) \, dx \, dy + \int_{D_1 \cup D_2} (u-v) \, dx \, dy \\
&= -\int_{D_3} \Delta u(u-v) \, dx \, dy \\
&= \int_{\Omega^+} (u-v) \, dx \, dy + \int_{D_3 \setminus (\Omega^+ \cup \Omega^-)} (f-v) \, dx \, dy \\
&= \int_{\Omega^+} (u-v) \, dx \, dy + \int_{D_3 \setminus \Omega^-} (f-v) \, dx \, dy \\
&\quad - \int_{\Omega^+} (f-v) \, dx \, dy \\
&= -\int_{\Omega^+} (f-u) \, dx \, dy + \int_{D_3 \setminus \Omega^-} (f-v) \, dx \, dy \\
&\le -\int_{D_3} [f-u]^+ \, dx \, dy + \int_{D_3} [f-v]^+ \, dx \, dy \qquad (12.171)
\end{aligned}$$

(note that in going from the first to the second expression we have used Green's formula which we proved in the previous section).

12.3.5 STUDY OF A QUASIVARIATIONAL PROBLEM. EXISTENCE OF THE SOLUTION OF A FREE-BOUNDARY PROBLEM

A fixed point problem

We will henceforth denote by τ the operator defined in $L^1(]0, c[)$ and with values in V_0 which with every $f \in L^1(]0, c[)$ associates the unique solution of the problem 12.9 ($\tau(f) \triangleq u$), and denote by γ_{03} the operator defined in V_0 and with values in $H_{00}^{1/2}(\Gamma_3)(\approx H_{00}^{1/2}(]0, c[))$ which with every $v \in V_0$ associates the restriction of its zero order trace to $\Gamma_3 (\gamma_{03} v \triangleq v(x, Y(x)), 0 < x < c)$.

Starting with the operators τ, γ_{03}, conv and $[.]^+$, we will next construct the operator $\Phi: V_0 \to V_0$ by means of the formula

$$\Phi(v) \triangleq \tau(\text{conv}[\gamma_{03} v]^+), \qquad (12.172)$$

and consider the fixed point problem

PROBLEM 12.11. Find $u \in V_0$ such that

$$\Phi(U) = U. \qquad (12.173)$$

From our previous knowledge, hopefully the solutions of problem 12.5 coincide with those of problem 12.11. One of our two fundamental aims in this section is in fact to prove the equivalence of the problems 12.5 and 12.11; the other aim is naturally to prove that problem 12.11 has a solution, and so answer the question regarding the existence of a solution of problem 12.5.

A quasivariational inequality

It is evident that the following is an equivalent formulation of problem 12.11:

PROBLEM 12.12. Find

$$U \in V_0; \quad \forall v \in V_0 \quad a(U, U-v) + j(U, U) \leq j(U, v) + L(U-v), \qquad (12.174)$$

where

$$j(u, v) = \int_{D_3} [\text{conv}[\gamma_{03} u]^+ - v]^+ \, dx \, dy. \qquad (12.175)$$

The inequality (12.174) is a quasivariational inequality of the type considered in section 11.2: unfortunately the monotonicity techniques are not applicable here since the functional (12.175) depends on the parameter v which varies in V_0, which is not a complete vector lattice. Let us remark that also the fixed points proved in section 9.3 are not applicable either to the study of problem 12.11, since we do not have at our disposal hypotheses regarding compactness and continuity in respect to the same topology of V_0 (Φ is strongly continuous and V_0 is weakly compact). In the proof of the existence theorem for (another equivalent form of writing) problem 12.5 we will use techniques similar to those developed by Bensoussan and Lions in connection with quasivariational inequalities.

The operator τ

We will make a small digression here in order to study some properties of the operator τ which will be of use to us, among other things, in the proof of the equivalence between problem 12.5 and the previous problems.

LEMMA 1. τ is Hölder continuous and monotone in the sense that, $\forall f_1, f_2 \in L^1(]0, c[)$,

$$\|\tau(f_1) - \tau(f_2)\|_{V_0}^2 \leq \|f_1 - f_2\|_{L^1(]0,c[)} \tag{12.176}$$

and

$$f_1 \leq f_2 \text{ a.e. in } [0, c] \Rightarrow (\tau(f_1))(x, y) \leq (\tau(f_2))(x, y) \text{ in } \bar{D}. \tag{12.177}$$

Proof. We first prove (12.176). Let $u_1 = \tau(f_1)$ and $u_2 = \tau(f_2)$ and put $v = u_2$ in the variational inequality (12.155) corresponding to f_1, $v = u_1$ in the variational inequality (12.155) corresponding to f_2, and let us add term by term:

$$a(u_1 - u_2, u_1 - u_2) \leq \int_{D_3} |f_1(x) - f_2(x)| \, dx \, dy; \tag{12.178}$$

from (12.133) we have then

$$\|u_1 - u_2\|_{V_0}^2 \leq \|f_1 - f_2\|_{L^1(]0,c[)} \tag{12.179}$$

and therefore (12.176).

We will next prove (12.177). Since $C^0([0, c])$ is dense in $L^1(]0, c[)$ we can approximate f_1 and f_2 by sequences $\{f_{1,m}\}$ and $\{f_{2,m}\}$ of continuous functions in $[0, c]$ such that $f_{1,m}(x) \leq f_{2,m}(x) \, \forall x \in [0, c]$, $m \in \mathbb{N}$; from this together with (12.176) it is evident that we need only prove (12.177) for functions f_1, $f_2 \in C^0([0, c])$. Let then $f_1(x) \leq f_2(x)$ be two continuous functions in $[0, c]$ and put $C = \{(x, y): (\tau(f_1))(x, y) > (\tau(f_2))(x, y)\}$ (an open set, from (12.143)): in C, from (12.170), we have $\Delta(\tau(f_1) - \tau(f_2)) \geq 0$ and hence, by the maximum principle of Hopf and from conditions (12.144) to (12.147), $C = \emptyset$. ∎

LEMMA 2. For every $f \in L^1(]0, c[)$ we have

$$(\tau(f))(x, Y(x)) > 0 \quad \text{for } a < x < b, \tag{12.180}$$

$$(\tau(f))(x, y) \geq 0 \quad \text{in } \bar{D} \tag{12.181}$$

and

$$D_y(\tau(f))(x, y) \geq 0 \quad \text{in } \bar{D}. \tag{12.182}$$

Proof. (12.180) and (12.181) follow, respectively, from (12.151) and (12.149). In order to prove (12.182) we can restrict ourselves, as we did in the proof of lemma 1, by assuming that f is continuous. Put $C = \{(x, y): D_y(\tau(f))(x, y) < 0\}$ (which from (12.143), is an open set): in $D_3 \cap C$ the function $y \mapsto f(x) - (\tau(f))(x, y)$ is strictly increasing, and therefore the function $H(f(x) - (\tau(f))(x, y))$ is non-decreasing; from this together with

(12.170) it follows that $\Delta D_y \tau(f) \leq 0$ and hence, since $D_y \tau(f)|_{\partial D} \geq 0$ (from (12.150)), Hopf's maximum principle leads us to the conclusion that $C = \emptyset$. ∎

Two more problems

Our digression ends here, and we will now consider the following problem:

PROBLEM 12.13. Find

$$U \in V_0;\ U \geq 0 \quad \text{and} \quad U = \tau(\text{conv } \gamma_{03} U). \tag{12.183}$$

Problem 12.13 is equivalent to problem 12.11. In fact, if U is a solution of problem 12.13 then U is obviously a solution of problem 12.11 and, conversely, if U is a solution of problem 12.11 then, from (12.181), $U \geq 0$, and therefore U is also a solution of problem 12.13.

Another equivalent formulation of problem 12.11 is the following:

PROBLEM 12.14. Find

$$U \in V_0;\ U \geq 0,\ \gamma_{03} U \text{ is convex and } U = \tau(\gamma_{03} U). \tag{12.184}$$

It follows immediately that if U is a solution of problem 12.14 then U is also a solution of problem 12.11; the converse is less evident: we will prove it by showing that if U is a solution of problem 12.13 (which we know to be equivalent to problem 12.11) then

$$U(x, y) \leq (\text{conv } \gamma_{03} U)(x) \quad \forall (x, y) \in \bar{D}_3, \tag{12.185}$$

from which it follows that $\gamma_{03} U \leq \text{conv } \gamma_{03} U$ and hence that $\gamma_{03} U$ is convex (we remark that if U is a solution of problem 12.13 then $U \geq 0$ and hence $\gamma_{03} U \geq 0$).

To prove (12.185) we use the method of *reductio ad absurdum*: let

$$\alpha = \max_{(x,y) \in \bar{D}_3} \{U(x, y) - (\text{conv } \gamma_{03} U)(x)\} > 0. \tag{12.186}$$

From (12.182) this maximum must be achieved in at least one point of Γ_3 and therefore, since $U(0, Y(0)) = (\text{conv } \gamma_{03} U)(0)$ and $U(c, Y(c)) = \text{conv } (\gamma_{03} U)(c)$ (note that $\gamma_{03} U \in C^1([0, c])$ since $U \in C^1(\bar{D})$ and $Y \in C^3([a, b])$), the set

$$C = \{x \in]0, c[:\ U(x, Y(x)) - (\text{conv } \gamma_{03} U)(x) = \alpha\} \tag{12.187}$$

is non-empty. Further, since U and $\gamma_{03} U$ are continuous, the set C is closed. Let x_0 be a point of C. From (12.187) we have $U(x_0, Y(x_0)) > (\text{conv } \gamma_{03} U)(x_0)$ and therefore, since $U = \tau (\text{conv } \gamma_{03} U)$ as follows from the fact that we have assumed U to be a solution of problem 12.13, from the considerations made regarding problem 12.10 (remember that $\gamma_{03} U \in C^1([0, c]) \ldots$), it follows that $-\Delta U = 0$ in a neighbourhood of $(x_0, Y(x_0))$; further $(\text{conv } \gamma_{03} U)(x) < U(x_0, Y(x_0)) \triangleq (\gamma_{03} U)(x_0)$ implies (remember again that $\gamma_{03} U \in C^1([0, c])$...) that $\gamma_{03} U$ is linear in a neighbourhood of $(x_0, Y(x_0))$. From all this it follows that

$$\delta(x, y) = U(x, y) - (\text{conv } \gamma_{03} U)(x) \qquad (12.188)$$

is harmonic in a neighbourhood of $(x_0, Y(x_0))$; further $\delta(x_0, Y(x_0)) = \alpha \geq \delta(x, y)$, from (12.186) and (12.187), and $\delta_y(x_0, Y(x_0)) = 0$, from (12.197). From the maximum principle it follows then that there exists a neighbourhood of $(x_0, Y(x_0))$ in which $\delta(x, y) \equiv \alpha$, i.e., that C contains a neighbourhood of x_0, which is absurd.

Interpretation of a quasivariational problem as a boundary value problem

We are finally in a position to prove the equivalence, which we conjectured at the beginning of this section, between problems 12.5 and 12.11, or rather between problem 12.5 and the equivalent problems 12.11, 12.12, 12.13, and 12.14; we will formalize this result in terms of problem 12.12:

THEOREM 12.3. *The problems* 12.5 *and* 12.12 *are equivalent, or rather, problem* 12.5 *is the interpretation of the quasivariational problem* 12.12 *as a boundary value problem.*

Proof. It is evident that if U is a solution of problem 12.5 then U is also a solution of problem 12.14, and hence of problem 12.12. Conversely, let U be a solution of problem 12.14 and let us prove that U is also a solution of problem 12.5. In fact (12.96) follows from (12.143); (12.97) is included in (12.184); (12.98) follows from (12.182); (12.99), (12.100), (12.101), and (12.102) are, respectively, (12.144), (12.145), (12.146), and (12.147); defining Ω as in (12.103), (12.104) follows from (12.144) and (12.180); finally, (12.105) follows from the following considerations: on the one hand, $-\Delta U = 1$ in Ω, from (12.170), and, on the other, $\Delta U = 0$ in $D \backslash \bar{\Omega}$ since $\Delta U \leq 0$ there, from (12.170), and, from (12.184) (noting also that, in $D \backslash \Omega$, $U(x, y) = (\gamma_{03} U)(x)$, which follows from the fact that $U \geq \gamma_{03} U$ by the definition of Ω and from $U \leq \gamma_{03} U$ by (12.182)), $\Delta U = \Delta \gamma_{03} U = (\gamma_{03} U)'' \geq 0$ (in $D \backslash \bar{\Omega}$ in the sense of distributions, but also a.e. in $D \backslash \Omega$ since it can be shown that $\gamma_{03} U$ is regular, and more precisely that $\gamma_{03} U \in W_{\text{loc}}^{2,\infty}(]0, c[)$: see Baiocchi, 1975b, p. 603—lemma 3.5). ∎

A result of existence

We will now go on to a theorem on the existence of solution for problem 12.5. We will first introduce the following definition (useful also for its numerical aspects):

DEFINITION 12.4. Let us denote by $\{U_m\}$ the sequence defined by recurrence according to the formula

$$U_{m+1} = \tau(\text{conv } \gamma_{03} U_m) \quad m \geq 0, \quad (12.189)$$

where U_0 is the solution of problem 12.7 with g given by (12.139) and $F = \chi_{D_1} + \chi_{D_2}$. Let us denote by $\{U^m\}$ the sequence defined by recurrence according to the formula

$$U^{m+1} = \tau(\text{conv } \gamma_{03} U^m) \quad m \geq 0, \quad (12.190)$$

where U^0 is the solution of problem 12.7 with g given by (12.139) and $F \equiv 1$ in D.

As we have already mentioned, this definition has important numerical relevance. In fact the method exposed here has suggested a new efficient algorithm for the numerical solution of this type of free boundary problems: see Baiocchi, 1975c; Baiocchi–Comincioli–Magenes–Pozzi, 1976; Comincioli–Guerri–Volpi, 1971; Brezzi–Sacchi, 1976 (the latter paper uses finite elements—most of the others use finite differences; see also Baiocchi, 1977), Baiocchi–Comincioli–Guerri–Volpi, 1973 (where a large number of problems are considered along with a survey of heuristic methods for the solution of free boundary problems), Comincioli, 1972 and Baiocchi–Magenes, 1975 (for the dam with an impermeable sheet), Comincioli, 1974a; 1974b; 1975 (for the case of a dam with an inclined left wall and a vertical right wall), Pozzi, 1974b (problems in the presence of evaporation), Comincioli–Guerri, 1976 and Comincioli, 1976 (problems in the presence of capillarity), Comincioli–Torelli, 1979 (non-stationary problem).

We will need the following auxiliary results:

LEMMA 3. For every $f \in L^1(]0, c[)$ we have

$$U_0(x, y) \leq (\tau(f))(x, y) \leq U^0(x, y) \quad \forall (x, y) \in \bar{D}. \quad (12.191)$$

Further

$$U_0 = \tau(0) \quad (12.192)$$

and

$$U^0 = \tau(M), \quad (12.193)$$

where $M \in \mathbb{R}$ is such that

$$U^0(x, y) \leq M \quad \forall (x, y) \in \bar{D}. \quad (12.194)$$

Proof. From (12.170) follows that

$$-\Delta U_0 \leq -\Delta \tau(f) \leq -\Delta U^0 \quad (12.195)$$

and, since the functions U_0, $\tau(f)$, and U^0 satisfy the boundary conditions (12.144) to (12.147), by applying the maximum principle of Hopf, (12.191) follows. Now let $\varepsilon > 0$ be arbitrary; from (12.170) and from the definition of U_0 we can write (since $H(-\varepsilon - U_0) = \{0\}$, which follows from the fact that, given (12.181), $U_0 \geq 0$)

$$-\Delta U_0 = \chi_{D_1} + \chi_{D_2} \in \chi_{D_1} + \chi_{D_2} + \chi_{D_3} H(-\varepsilon - U_0) \qquad (12.196)$$

and, again from (12.170) and from the definition of U^0 (since $H(M + \varepsilon - U^0) = \{1\}$, which follows from the fact that, given (12.194), $M \geq U^0$),

$$-\Delta U_0 = 1 \in \chi_{D_1} + \chi_{D_2} + \chi_{D_3} H(M + \varepsilon - U^0). \qquad (12.197)$$

From (12.196) we have

$$\forall \varepsilon > 0 \qquad U_0 = \tau(-\varepsilon), \qquad (12.198)$$

and from (12.197) we have

$$\forall \varepsilon > 0 \qquad U^0 = \tau(M + \varepsilon); \qquad (12.199)$$

taking the limit as $\varepsilon \to 0^+$, which we can do by virtue of (12.176), from (12.198) we obtain (12.192) and from (12.199) we have (12.193). ∎

LEMMA 4. We have

$$U_m(x, y) \leq U_{m+1}(x, y) \leq U^0(x, y) \qquad \forall (x, y) \in \bar{D}, \quad m = 0, 1, 2, \ldots \qquad (12.200)$$

and

$$U^m(x, y) \geq U^{m+1}(x, y) \geq U_0(x, y) \qquad \forall (x, y) \in \bar{D}, \quad m = 0, 1, 2, \ldots. \qquad (12.201)$$

Further, if U is any solution of problem 12.5, we have

$$U_m(x, y) \leq U(x, y) \leq U^m(x, y) \qquad \forall (x, y) \in \bar{D}, \quad m = 0, 1, 2, \ldots. \qquad (12.202)$$

Proof. The second inequalities in (12.200) and (12.201) follow from (12.191); the first inequalities can be proved by finite induction, as can those written in (12.202); here we will only prove those corresponding to $\{U^m\}$, since the proof of the others are analogous. If M is a real number in the conditions of (12.194) then $\gamma_{03} U^0 \leq M$ and therefore conv $\gamma_{03} U^0 \leq M$: from this it follows, using (12.177), that $\tau(\text{conv } \gamma_{03} U^0) \leq \tau(M)$ and hence, from the definitions of U^0 and U^1, $U^1 \leq U^0$; now, if $U^m \leq U^{m-1}$ (the induction hypothesis) we have conv $\gamma_{03} U^m \leq \text{conv } \gamma_{03} U^{m-1}$ and thus, again from (12.177) and by definition 12.4, $U^{m+1} = \tau(\text{conv } \gamma_{03} U^m) \leq \tau(\text{conv } \gamma_{03} U^{m-1}) = U^m$. This proves the first inequality in (12.201); we will now prove the second inequality in (12.202). If U is a solution of problem 12.5 then, from (12.184), $U = \tau(\gamma_{03} U)$ and therefore, from (12.191), $U \leq U^0$; now, if $U \leq U^m$ (the induction hypothesis) we have $\gamma_{03} U \leq \gamma_{03} U^m$, and hence conv $\gamma_{03} U = \gamma_{03} U \leq \text{conv } \gamma_{03} U^m$, and $U = \tau(\text{conv } \gamma_{03} U) \leq \tau(\text{conv } \gamma_{03} U^m) = U^{m+1}$. ∎

The preceding result (and more precisely (12.200) and (12.201)) shows that the sequences $\{U_m\}$ and $\{U^m\}$ are convergent, and therefore that it makes sense to introduce the following

DEFINITION 12.5. Let us put

$$U_\infty(x, y) = \lim_{m \to \infty} U_m(x, y) \qquad \forall (x, y) \in \bar{D} \qquad (12.203)$$

and

$$U^\infty(x, y) = \lim_{m \to \infty} U^m(x, y) \qquad \forall (x, y) \in \bar{D}. \qquad (12.204)$$

We can now prove

THEOREM 12.4. Problem 12.5 has at least one solution. More precisely, the functions U_∞ and U^∞ are solutions of problem 12.5, being respectively the minimum solution and maximum solution in the sense that

$$U_\infty(x, y) \le U(x, y) \le U^\infty(x, y) \qquad \forall (x, y) \in \bar{D} \qquad (12.205)$$

for every solution U of the problem.

Proof. The (12.205) follows from (12.202). We will now show that U^∞ is a solution of problem 12.5 (we will in fact show that U^∞ is a solution of problem 12.13...): the proof that U_∞ is also a solution is completely analogous. From $U_0 \le U^m \le U^0$ (see (12.191)) it follows that $0 \le \gamma_{03} U^m \le \gamma_{03} U^0$, and therefore that $0 \le \text{conv } \gamma_{03} U^m \le \gamma_{03} U^0$; from (12.176) it follows then that $\{U^m\}$ is strongly bounded in V_0. From the sequence $\{U^m\}$ we can therefore extract a subsequence weakly convergent to $\bar{U}$ in V_0; from the uniqueness of the weak limit and from (12.204) it follows that $\bar{U} = U^\infty$, and since $\{U^m\}$ is monotonic (see (12.201)) it follows that the whole sequence is in fact weakly convergent:

$$U^m \rightharpoonup U^\infty \qquad \text{in } V_0. \qquad (12.206)$$

From this and from (12.148) it follows that

$$U^m \rightharpoonup U^\infty \qquad \text{in } W^{2,r}(D) \qquad \forall r < +\infty \qquad (12.207)$$

and therefore, since Sobolev inclusions are compact, we can write

$$U^m \to U^\infty \qquad \text{in } C^1(\bar{D}). \qquad (12.208)$$

From (12.208) we have in particular that

$$\gamma_{03} U^m \to \gamma_{03} U^\infty \qquad \text{in } C^1([0, c]), \qquad (12.209)$$

and therefore, since conv: $H^1(]0, c[) \to H^1(]0, c[)$ is continuous,

$$\text{conv } \gamma_{03} U^m \to \text{conv } \gamma_{03} U^\infty \qquad \text{in } H^1(]0, c[). \qquad (12.210)$$

From (12.210) and (12.176) we then have

$$U^\infty = \lim U^{m+1} = \lim \tau(\text{conv } \gamma_{03} U^m)$$
$$= \tau(\lim \text{conv } \gamma_{03} U^m) = \tau(\text{conv } \gamma_{03} U^\infty), \qquad (12.211)$$

which together with $U^\infty \geq 0$ (as follows, e.g., from (12.204)) and with $U^\infty \in V_0$ (as follows from (12.206)) shows that U^∞ is a solution of problem 12.13. ∎

Theorem 12.4 ends the discussion regarding the existence of solution for the free boundary problem which we posed in section 12.3.1; in the following section we will analyse the questions regarding uniqueness and regularity of such solution.

12.3.6 Uniqueness and regularity of the solution of a free-boundary problem

Two conjectures

Figure 12.3 shows a flow-net relative to the solution U^∞ for a 'trapezoidal' dam; we have not shown the corresponding flow-net relative to the solution U_∞ because it is practically coincident with the one shown in figure 12.3 (this flow-net as well as those shown later in figures 12.8 and 12.9 was drawn by a Calcomp 936 plotter connected to a Honeywell H6030 computer with a program written by V. Comincioli).

The coincidence of U^∞ and U_∞, and the pictures of the kind shown in fig. 12.3 suggest the following conjectures:

(C1) problem 12.5 has only one solution;
(C2) the (unique) solution of problem 12.5 is regular in the sense of the definition 12.3.

The validity of these conjectures will be proved in section 12.3.7 for a dam with vertical walls; in the general case we will now see that (C1) is true, whereas with respect to (C2) we present here only a partial result.

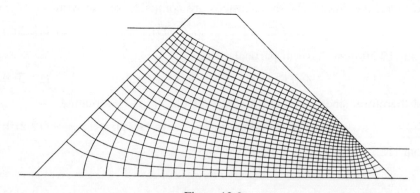

Figure 12.3

The uniqueness of the dam problem

Let us first recall a result regarding 'regularity for the contact set' of the solutions of inequalities with obstacle:

$$\left.\begin{array}{l}\text{let } D^* \text{ be an open set in } \mathbb{R}^n \text{ and let } z \in W^{2,r}(D^*) \forall r < +\infty; \\ \text{further let } z \geq 0 \text{ in } D^* \text{ and, putting } \Omega^* = \{P \in D^* : z(P) > 0\}, \\ \text{let } \Delta z \geq 1 \text{ in } \Omega^*; \text{ then } \mu(D^* \cap \partial \Omega^*) = 0\end{array}\right\} \quad (12.212)$$

(where μ is the Lebesgue measure in $\mathbb{R}^n$). Let us recall that many important results have been obtained in the last few years with respect to these regularity problems; see, e.g., Caffarelli, 1976a; 1976b; 1977; 1978; 1979a; 1979b; Caffarelli–Evans, 1982; Caffarelli–Kinderlehrer, 1980; Caffarelli–Rivière, 1976a; 1976b; 1977a; 1977b; Di Benedetto, 1980; Frehse, 1982; Frehse–Mosco, 1979b; Friedman, 1976b; Friedman–Kinderlehrer, 1975; Kinderlehrer, 1978; 1981; Kinderlehrer–Nirenberg, 1977; Kinderlehrer–Nirenberg–Spruck, 1978b; 1979 and their respective references.

Let then $\{\Omega, p\}$ be a solution of problem 12.4 and let U be the corresponding function defined by (12.76); with $\varepsilon > 0$ arbitrary and fixed, let $D^* = \{(x, y) \in D : \varepsilon < x < c - \varepsilon\}$ and let $z(x, y) = U(x, Y(x)) - U(x, y)$ $\forall (x, y) \in \bar{D}^*$; from (12.96), (12.97), (12.105), and $\gamma_{03} U \in W^{2,\infty}_{\text{loc}}(]0, c[)$ we can apply (12.212) and hence, since ε is arbitrary, we have:

$$\left.\begin{array}{l}\text{if } \{\Omega, p\} \text{ is a solution of problem 12.4 the} \\ \text{set } \partial \Omega \text{ has zero two-dimensional measure.}\end{array}\right\} \quad (12.213)$$

This result (12.213) is a first step towards (C2); we will now show a connection between our formulation and that of Alt, 1977a, which we will use to reinforce (12.213) (but will not prove (C2); cf. (12.217) which follows...) as well as to prove (C1). Following Alt, 1977a, we will refer to a pair (A, f) as a *supersolution* of problem 12.4 if, putting

$$V = \{v \in H^1(D) : v(x, Y(x)) = \gamma(x)\} \quad (12.214)$$

with

$$\gamma(x) = \begin{cases} y_1 - Y(x) & \text{for } a < x < 0 \\ 0 & \text{for } 0 \leq x \leq c \\ y_2 - Y(x) & \text{for } c < x < b, \end{cases} \quad (12.215)$$

we have

$$\left.\begin{array}{l}f \in V; A \text{ is a closed subset of } D; \\ f = 0 \text{ a.e. in } D \setminus A; \forall g \in V - V \text{ one has} \\ \int_A \text{grad}(f + y) \text{ grad } g \, dx \, dy \geq 0.\end{array}\right\} \quad (12.216)$$

Again, let us put for the supersolution (A, f):

$$J(A, f) = \int_A |\text{grad}(f + y)|^2 \, dx \, dy. \quad (12.217)$$

In Alt, 1977a, it has been shown that the functional J has a minimum in the family of supersolutions; and that every minimum point (A, f) is such that, putting $\Omega = \mathring{A}$, $p = f|_\Omega$, the pair $\{p, \Omega\}$ solves problem 12.4 (for the validity of (12.59) see Alt, 1977b, theorem 2.8); vice versa, by virtue of (12.213) every solution $\{\Omega, p\}$ is such that, putting $A = \bar{\Omega}$, $f = \tilde{p}$ (extension to 0 of p in $D\backslash\Omega$) (A, f) is a supersolution and minimizes the functional J given by (12.217) in the family of supersolutions (cf. Alt, 1977a); then, using the results of Alt, 1977b, we have

$$\left. \begin{array}{l} \text{if } \{p, \Omega\} \text{ is a solution of problem 12.4 then} \\ \bar{\Omega} = \bar{D}_1 \cup \bar{D}_2 \cup \{(x, y) \in D_3 : 0 \leq y \leq \phi(x)\} \text{ with } \phi \\ \text{such that the arcs with graphs in } D \text{ are analytic.} \end{array} \right\} \quad (12.218)$$

Further it can be proved (cf. Alt, 1979; Caffarelli–Rivière, 1977c and personal communications) that the minimum of J in the family of supersolutions is unique, and therefore we have

THEOREM 12.5. *Problem 12.4 has one and only one solution.*

In particular we have proved that conjecture (C1) is true. As for (C2), (12.213) and (12.218) are not sufficient to give us a positive answer. In the next section we will show directly (i.e. without using the results of Alt) that conjectures (C1) and (C2) are true for a dam D with vertical walls.

12.3.7 DAM WITH VERTICAL WALLS

The geometry of the problem

We will first describe in a rigorous manner the new geometry of the dam (see also fig. 12.4 and the notation diagram 12.2—it will become clear later why the point E has also been given a name):

DEFINITION 12.6. Let c, d, y_1, and y_2 be four real numbers such that $0 < c$

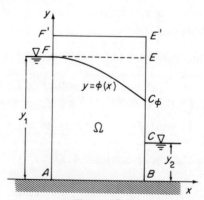

Figure 12.4

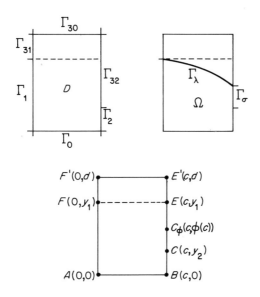

Notation diagram 12.2

and $0 < y_2 < y_1 \leq d$. The rectangle $D =]0, c[\times]0, d[$ will be called a *dam with vertical walls*. Further let us put $\Gamma_0 = \{(x, 0) : 0 < x < c\}$, $\Gamma_1 = \{(0, y) : 0 < y < y_1\}$, $\Gamma_2 = \{(c, y) : 0 < y < y_2\}$, $\Gamma_{31} = \{(0, y) : y_1 < y < d\}$, $\Gamma_{32} = \{(c, y) : y_2 < y < d\}$ and $\Gamma_{30} = \{(x, d) : 0 < x < c\}$.

The new boundary conditions

With this new geometry the boundary conditions (12.100), (12.101) and (12.102) are considerably modified since the derivative with respect to y is no longer oblique, but tangential on the walls and normal at the top of the dam. Thus, integrating (12.100) and (12.101) we get, respectively,

$$U = (y_1^2/2) - (y_1 - y)^2/2 \quad \text{on } \Gamma_1 \quad \text{and} \quad U = (y_2^2/2) - (y_2 - y)^2/2 \quad \text{on } \Gamma_2;$$
(12.102)

must be broken up into three parts (just like $\Gamma_3 \ldots$): on Γ_{30} the condition is still $U_y = 0$ (but it is now a Neumann condition and no longer one of oblique derivative kind); on Γ_{31} and Γ_{32} we get, by integration and using the continuity of $U|_{\partial D}$ and the conditions on Γ_1 and Γ_2, $U = y_1^2/2$ on Γ_{31} and $U = y_2^2/2$ on Γ_{32}. The above considerations show also that the effective height of the dam, d, is not involved in the boundary conditions (which was to be expected since we assume that the porous material is incompressible and ignore capillarity phenomena...): we can then put, without any loss of generality, $d = y_1$ (and therefore $F' \equiv F$, $E' \equiv E$, $\Gamma_{31} = \emptyset$ and $\Gamma_{30} = \{(x, y_1) : 0 < x < c\}$, this latter set being the discontinuous line in figure 12.4).

The problem in the new geometry

Let us now restate problem 12.5 with the necessary modifications:

PROBLEM 12.15. Find U such that

$$U \in W^{2,r}(D) \quad \forall r < +\infty, \tag{12.219}$$

$$x \mapsto U(x, y_1) \text{ is convex in } [0, c], \tag{12.220}$$

$$U_y \geq 0 \quad \text{in } \bar{D}, \tag{12.221}$$

$$U = 0 \quad \text{on } \Gamma_0, \tag{12.222}$$

$$U = \frac{y_1^2}{2} - \frac{(y_1 - y)^2}{2} \quad \text{on } \Gamma_1, \tag{12.223}$$

$$U = \frac{y_2^2}{2} - \frac{(y_2 - y)^2}{2} \quad \text{on } \Gamma_2, \tag{12.224}$$

$$U = \tfrac{1}{2} y_2^2 \quad \text{on } \Gamma_{32}, \tag{12.225}$$

$$U_y = 0 \quad \text{on } \Gamma_{30}, \tag{12.226}$$

and, putting

$$\Omega = \{(x, y) \in D : 0 < x < c \text{ and } U(x, y) < U(x, y_1)\}, \tag{12.227}$$

it must follow that

$$\Omega \text{ contains a neighbourhood of } \Gamma_0, \tag{12.228}$$

$$-\Delta U = \chi_\Omega \quad \text{a.e. in } D. \tag{12.229}$$

As is natural, the concept of regular solution introduced in definition 12.3 has also to be modified; precisely:

DEFINITION 12.7. A solution of problem 12.15 is said to be *regular*, if, putting

$$\left.\begin{array}{l}\phi(x) = \sup\{y : (x, y) \in \Omega\}, \, 0 < x < c; \\ \phi(0) = \lim_{x \to 0^+} \phi(x); \, \phi(c) = \lim_{x \to c^-} \phi(x),\end{array}\right\} \tag{12.230}$$

the function $[0, c] \ni x \mapsto \phi(x)$ is strictly decreasing, continuous and $\phi(0) = y_1$ and $\phi(c) > y_2$.

In this case, as is evident, the suspended source is a point with abscissa c: further the free line is the curve $\Gamma_\lambda = \{(x, \phi(x)) : 0 < x < c\}$ and the seepage line is $\Gamma_\sigma = \{(c, y) : y_1 < y < \phi(c)\}$.

Problem 12.15 is a mixed Dirichlet–Neumann boundary value problem, which is not exactly a particular case of problem 12.5 but which can be interpreted (just as the geometry described in definition 12.6 is a limit case of that described in definition 12.1) as a limit case of this problem.

It can be easily seen that whatever we have done in the previous section can also be done, with considerable simplifications, with regard to problem 12.15. Thus, in particular, we can say that problem 12.15 has a maximal solution, U^∞, and a minimal solution, U_∞. We will now prove that $U_\infty \equiv U^\infty$, i.e. that problem 12.15 has only one solution, and that this unique solution is regular in the sense of definition 12.7.

Remark

Dams with vertical walls were the objective of the first papers in which the method of changing the unknown function described in section 12.3.2 was applied (cf. chapter 8 ...). In those papers, instead of condition (12.220) there was the (equivalent of the) condition '$x \mapsto U(x, y_1)$ is linear in $[0, c]$' (which expresses the constancy of the carrying capacity along the whole dam); with this one can formulate the problem variationally (further at that time the concept of quasivariational inequality had not yet been introduced)—we will come back to this question. A problem 'intermediate' between problems 12.5 and 12.15 is the one in which the dam has a vertical 'right-wall' and a sloping 'left-wall'; the problem can in this case be formulated in terms of a family of variational inequalities depending on a parameter: see Baiocchi–Comincioli–Magenes–Pozzi, 1973 and Comincioli, 1974a; 1975; 1974b; (a numerical comparison between this formulation and the quasivariational one).

Uniqueness of the solution

We will first prove some preliminary results (in what follows we will denote by γ_{030} the zero order trace operator on Γ_{30}).

LEMMA 1. $\gamma_{030} U^0$ is concave.

Proof. Let us put
$$z^0 = y + D_y U^0 \qquad (12.231)$$
and try to find out which problem is solved by z^0. From the definition of U^0 we have
$$\Delta z^0 = 0, \qquad (12.232)$$
and from (12.193) and (12.182) it follows that
$$z^0 \geq 0 \quad \text{in } \bar{D}. \qquad (12.233)$$

Again from the definition of U^0 we see that z^0 satisfies the boundary conditions imposed in problem 12.15, and hence with the differential equation (12.232) we must associate the conditions

$$z^0 = y_1 \quad \text{on } \Gamma_1, \qquad (12.234)$$
$$z^0 = y_2 \quad \text{on } \Gamma_2, \qquad (12.235)$$
$$z^0 = y \quad \text{on } \Gamma_{32}, \qquad (12.236)$$
$$z^0 = y_1 \quad \text{on } \Gamma_{30}, \qquad (12.237)$$
$$z^0_y = 0 \quad \text{on } \Gamma_0. \qquad (12.238)$$

Let us point out that the conditions (12.234), (12.235), (12.236), and (12.237) follow, respectively, from conditions (12.223), (12.224), (12.225), and (12.226); condition (12.238) follows from (12.222), from which we can write $U^0_{xx} = 0$ on Γ_0, and from the definition of U^0, from which we can write $-\Delta U^0 = 1$ in D and consequently $z^0_y = U^0_{yy} + 1 = U^0_{xx}$ on Γ_0.

Now, from the maximum principle of Hopf it follows that $\max_{\bar{D}} z^0 = y_1$ is achieved in $\Gamma_1 \cup \Gamma_{30} \cup \{(0, y_1)\}$, and therefore that $\delta = z^0 - y_1$ is strictly negative in D. From this and from (12.237) (which shows that $\delta = 0$ in Γ_{30}) it follows that we have necessarily

$$\gamma_{030} \delta_y \geq 0, \tag{12.239}$$

and so, by the definition of z^0,

$$\gamma_{030} U^0_{yy} + 1 \geq 0, \tag{12.240}$$

i.e., since $\Delta U^0 = -1$,

$$\gamma_{030} U^0_{yy} - \Delta U^0 \geq 0. \tag{12.241}$$

Finally we have then that $\gamma_{030} U^0_{xx} \leq 0$ and hence the result. ∎

LEMMA 2. $\gamma_{030} U^\infty$ *is linear; further* $U^1 = U^\infty$.

Proof. From lemma 1 we have that

$$\lambda(x) = (\text{conv } \gamma_{030} U^0)(x) \tag{12.242}$$

is linear in $[0, c]$. By definition we have $U^1 = \tau(\lambda(x))$, and thus if we can prove that $(\gamma_{030} U^1)(x) \geq \lambda(x)$ then we have finished since under these conditions $(\text{conv } \gamma_{030} U^1)(x) = \lambda(x)$, and therefore $U^1 = U^2 = \ldots = U^\infty$, and since $\gamma_{030} U^\infty$ is convex the first part of the lemma is proved. (We note that the identity of U^1 and U^∞ is of a considerable numerical interest: it tells us in effect that the process of iteration for the calculation of U^∞ stops at the first step.) Let now (by *reductio ad absurdum*) $\emptyset \neq \gamma \subset \Gamma_{30}$ be an interval (remember that we are dealing with continuous functions...) in which $(\gamma_{030} U^1)(x) < \lambda(x)$; the proof of lemma 1 can be immediately adapted to prove that $\gamma_{030} U^1$ is concave in γ, which is absurd since $\gamma_{030} U^1$ is continuous in $[0, c]$ (let us remark that the linearity of λ is used to ensure that $\Delta U^1 = 0$ where $U^1 = \lambda$ (and thus to obtain the analogue of (12.232)), and that the step corresponding to that from (12.240) to (12.241) is possible because in a neighbourhood of γ we have $U^1 < \lambda$ and hence in this neighbourhood $-\Delta U^1 = 1$ (recall problem 12.10)). ∎

Since $\gamma_{030} U^\infty$ is linear, from (12.223) and (12.225) one can explicitly calculate this trace, which is

$$(\gamma_{030} U^\infty)(x) = \frac{y_1^2}{2} + \frac{(y_2^2 - y_1^2)}{2c} x \qquad x \in [0, c] \tag{12.243}$$

(from this and from (12.95) we get the so-called Dupuit formula for the carrying capacity of the dam: $q_d = (y_1^2 - y_2^2)/2c$.
We can now say that $U^\infty|_{\partial D} = g$, where g is the function defined by

$$g(x, y) = \begin{cases} \dfrac{y_1^2}{2} - \dfrac{(y_1-y)^2}{2} & \text{on } \Gamma_1 \\ 0 & \text{on } \Gamma_0 \cup \{(0,0), (c,0)\} \\ \dfrac{y_2^2}{2} - \dfrac{(y_2-y)^2}{2} & \text{on } \Gamma_2 \\ \dfrac{y_2^2}{2} & \text{on } \Gamma_{32} \cup \{(c, y_2)\} \\ \dfrac{y_1^2}{2} - \dfrac{y_1^2 - y_2^2}{2c} x & \text{on } \Gamma_{30} \cup \{(0, y_1), (c, y_1)\} \end{cases} \quad (12.244)$$

LEMMA 3. We have

$$D_x[\gamma_{030}U^\infty - U^\infty] \le 0 \quad \text{on } \bar{D}, \quad (12.245)$$

$$D_y[\gamma_{030}U^\infty - U^\infty] \le 0 \quad \text{on } \bar{D}. \quad (12.246)$$

Proof. The inequality (12.246) is a direct consequence of (12.221). To prove (12.245) let us put

$$h(x, y) = (\gamma_{030}U^\infty)(x) - U^\infty(x, y) \quad (12.247)$$

and

$$\Omega_x = \{(x, y) \in D : D_x h(x, y) > 0\}. \quad (12.248)$$

Ω_x is an open set since $D_x h(x, y)$ is continuous. Further, denoting by Ω^∞ the detachment set corresponding to the solution U^∞ (and by Ω_∞, which we will use later, that corresponding to the solution U_∞), we have $\Omega_x \subset \Omega^\infty$ since if (by *reductio ad absurdum*) $(x_0, y_0) \in \Omega_x \setminus \Omega^\infty$ then $D_x h(x_0, y_0) > 0$ (since $(x, y_0) \in \Omega_x$) and $h(x_0, y_0) = 0$ (since $(x_0, y_0) \notin \Omega^\infty$): from this it follows that $h(x, y_0) < 0$ for $x < x_0$ in a neighbourhood of (x_0, y_0), which is absurd since $h(x, y) \ge 0$ everywhere in $\bar{D}$. Let us now suppose (again by *reductio ad absurdum*) that $\Omega_x \ne \emptyset$. From $\Omega_x \subset \Omega^\infty$, from (12.229) and from the linearity of $\gamma_{030}U^\infty$ it follows that

$$\Delta D_x h(x, y) = D_x \Delta h(x, y) = 0 \quad \text{in } \Omega_x. \quad (12.249)$$

From here it follows that the maximum

$$\mu = \max_{\bar{\Omega}_x} \{D_x h(x, y)\} \quad (12.250)$$

(which exists since $D_x h$ is continuous in $\bar{\Omega}_x$) id achieved at a point $(x_0, y_0) \in \partial \Omega_x$, and we must have $\mu > 0$ from the definition of Ω_x; further $(x_0, y_0) \in \partial \Omega_x \cap \partial D$ since $D_x h(x, y) = 0$ in $\partial \Omega_x \cap D$, from the definition of Ω_x and from the continuity of $D_x h$. We will now prove that $(x_0, y_0) \notin \partial D$ in order to obtain the absurd. $(x_0, y_0) \notin \Gamma_0$ since, from (12.222) and (12.243), $D_x h < 0$ in

Γ_0; $(x_0, y_0) \notin \Gamma_{30}$ since, from the very definition of h, $D_x h = 0$ in Γ_{30}; $(x_0, y_0) \notin \Gamma_{32}$ since $D_x h \leq 0$ in Γ_{32}, as one can see from the fact that $h \geq 0$ in $\bar{D}$ and, from (12.225) and the definition of h, $h = 0$ in Γ_{32}; finally, to see that $(x_0, y_0) \notin \Gamma_1 \cup \Gamma_2$, one need only consider that in $\Gamma_1 \cup \Gamma_2$ $D_{xx} h(x, y) = 0$ (which follows from the definition of h, (12.223) and (12.224) and, since $\Gamma_1 \cup \Gamma_2 \subset \partial \Omega^\infty$, from (12.229)), and that, from the maximum principle of Hopf, if $(x_0, y_0) \in \Gamma_1$ then $D_{xx} h(x_0, y_0) < 0$ and if $(x_0, y_0) \in \Gamma_2$ then $D_{xx} h(x_0, y_0) > 0$. ∎

LEMMA 4. Let us put, for every point $P_0 \equiv (x_0, y_0) \in D$,

$$Q_{P_0}^+ = \{(x, y) \in D : x > x_0, \quad y > y_0\}, \tag{12.251}$$

$$Q_{P_0}^- = \{(x, y) \in D : x < x_0, \quad y < y_0\}. \tag{12.252}$$

Then, for every point $P \in D$,

$$P \in D \setminus \Omega^\infty \Rightarrow Q_P^+ \subset D \setminus \overline{\Omega^\infty}, \tag{12.253}$$

$$P \in D \cap \partial \Omega^\infty \Rightarrow Q_P^- \subset \Omega^\infty. \tag{12.254}$$

Proof. The intrinsically geometric significance of this result is evident in figure 12.5 (the presence of the axes $X = x + y$ and $Y = x - y$ will be explained later). Let us first prove (12.253). Let then $P_0 \equiv (x_0, y_0) \in D \setminus \Omega^\infty$; from the definition of Ω^∞ and from (12.247) we have $h(x_0, y_0) = 0$. Since $h \geq 0$ in D and, from (12.245), h is not increasing with respect to x, $h(x, y_0) = 0$ in $[x_0, c]$; thus, since $h \geq 0$ in D and from the fact, which follows from (12.246), that h does not increase with respect to y, $h = 0$ in $[x_0, c] \times [y_0, y_1] = \overline{Q_{P_0}^+}$. From the definition of Ω^∞ we have then that $\overline{Q_{P_0}^+} \subset \bar{D} \setminus \Omega^\infty$ and thus $Q_{P_0}^+ = (\overline{Q_{P_0}^+})^0 \subset (\bar{D} \setminus \Omega^\infty)^0 \subset (\bar{D})^0 \setminus \overline{\Omega^\infty} = D \setminus \overline{\Omega^\infty}$.

Next we will prove (12.254). Let (by *reductio ad absurdum*) $P_0 \in D \cap \partial \Omega^\infty$ be such that $Q_{P_0}^- \not\subset \Omega^\infty$; then there exists $P_1 \in D \setminus \Omega^\infty$ with $P_1 \in Q_{P_0}^-$, and hence

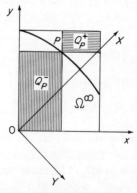

Figure 12.5

$P_0 \in Q_{P_1}^+$: from (12.253) it then follows that $P_0 \in D\backslash\overline{\Omega^\infty}$, which is contradictory to the assumption $P_0 \in D \cap \partial\Omega^\infty$. ∎

LEMMA 5. We have
$$\partial\Omega^\infty \cap \Gamma_{30} = \varnothing. \qquad (12.255)$$

Further $\partial\Omega^\infty \cap D$ contain neither horizontal nor vertical segments.

Proof. We will first prove that $\partial\Omega^\infty \cap D$ does not contain horizontal segments; the proof regarding vertical segments is completely analogous. Let (by *reductio ad absurdum*) γ be such a segment and let y_0 be its ordinate. From the definition of Ω^∞ and from the continuity of h (defined in (12.247)), it follows that $h \equiv 0$ in γ; further we must have $D_y h = 0$ in γ because, from lemma 4, 'above' γ_0, $h \equiv 0$. Now, from the uniqueness of the Cauchy problem associated with the equation $\Delta h = -1$ (see (12.229)) and the initial conditions just obtained, on all of Ω^∞ (which, from lemma 4, is connected) h has the form $h(x, y) = -\frac{1}{2}(y - y_0)^2$ and therefore, from (12.247) and (12.243), U^∞ has the form

$$U^\infty(x, y) = \tfrac{1}{2}(y - y_0)^2 + \frac{y_1^2}{2} - \frac{(y_2^2 - y_1^2)}{2c} x, \qquad (12.256)$$

which is contradictory to (12.244). The same argument shows that $\partial\Omega^\infty$ does not contain a segment of Γ_{30}. We must now ensure that Γ_{30} does not contain an isolated point of $\partial\Omega^\infty$; with this aim we will show that if $(x_0, y_1) \in \partial\Omega^\infty$, with $x_0 \in \,]0, c]$, then the segment $\{(x, y_1) : 0 < x < x_0\}$ is contained in $\partial\Omega^\infty$, which is contradictory to what we have just shown. Now let $x_0 \in \,]0, c]$ be such that $(x_0, y_1) \in \partial\Omega^\infty$. Then $h(x_0, y) > 0$, for $y \in [0, y_1[$, since if $x_0 = c$ then, from lemma 4, $\Omega^\infty = D$, and if, for $x_0 < c$, $h(x_0, y_0) = 0$ were true (by *reductio ad absurdum*) then, from (12.246), we would have $h(x_0, y) = 0$ in $[y_0, y_1]$ and hence $(x_0, y_1) \notin \partial\Omega^\infty$ (which follows from the fact that, from the first part of lemma, the vertical segment $\{(x_0, y) : y_0 < y < y_1\}$ is contained in the interior of $D\backslash\overline{\Omega^\infty}$). From this and from (12.245) it follows that $h(x, y) > 0$ in $R = \,]0, x_0[\,\times\,]0, y_1[$, and therefore $R \subset \Omega^\infty$, and consequently we conclude that $\{(x, y_1) : 0 < x < x_0\} \subset \partial\Omega^\infty$. ∎

We can now prove that conjecture (C1) is true in the case of problem 12.15. Naturally, the proof which follows is independent of the discussion in section 8.2; one should also note the difference between this uniqueness result (relative to a quasivariational problem) and the uniqueness result obtained in theorem 8.3 (relative to a variational problem).

THEOREM 12.6. *Problem 12.15 has only one solution.*

Proof. We will prove this result by showing that $\gamma_{030} U_\infty$ is linear, which implies that $\gamma_{030} U_\infty \equiv \gamma_{030} U^\infty$, and therefore that $U^\infty \equiv U_\infty$, and hence (from

(12.205)) the result. To show that $\gamma_{030}U_\infty$ is linear we will prove that

$$\Omega_\infty \subset \Omega^\infty; \tag{12.257}$$

from this it follows in fact that, in a neighbourhood of Γ_{30}, $(\gamma_{030}U_\infty)'' = \Delta U_\infty = 0$ and hence that $\gamma_{030}U_\infty$ is linear. To prove (12.257) we will show that

$$U_\infty - \gamma_{030}U_\infty \geq U^\infty - \gamma_{030}U^\infty \quad \text{in } D, \tag{12.258}$$

since from this inequality and from the definitions of Ω_∞ and Ω^∞ it follows that wherever $U^\infty = \gamma_{030}U^\infty$, i.e. in $D\setminus\overline{\Omega^\infty}$, we have also $U_\infty = \gamma_{030}U_\infty$, and therefore $D\setminus\overline{\Omega^\infty} \subset D\setminus\overline{\Omega_\infty}$. Let then, by *reductio ad absurdum*,

$$\varnothing \neq C = \{(x, y) \in D : \delta = U_\infty - \gamma_{030}U_\infty - (U^\infty - \gamma_{030}U^\infty) < 0\}. \tag{12.259}$$

Since the functions we are dealing with are continuous, C is an open set, and from $U^\infty \leq \gamma_{030}U^\infty$ in $\bar{D}$ and from the definition of Ω_∞ we have $C \subset \Omega_\infty$; thus, since $\gamma_{030}U^\infty$ is linear, since $\gamma_{030}U_\infty$ is convex, and since $-1 \leq \Delta U^\infty$, it follows that

$$\Delta \delta = \Delta U_\infty - \Delta \gamma_{030}U_\infty - \Delta U^\infty + \Delta \gamma_{030}U^\infty \leq 0. \tag{12.260}$$

On the other hand, $\delta \geq 0$ in $\partial C \cap \Gamma_i$, $i = 1, 0, 2, 32$, since U_∞ and U^∞ satisfy the same boundary conditions and, since $U^\infty \geq U_\infty$, $\gamma_{030}U^\infty \geq \gamma_{030}U_\infty$; from the definition of trace we have $\delta = 0$ on Γ_{30}; and from the definition of C we have $\delta = 0$ on $\partial C \cap D$. Thus we can say that δ solves, on every connected component of C, a problem of the type 'δ subharmonic strictly negative on the interior, δ non-negative on the boundary': from the maximum principle we then have a contradiction. ∎

Equivalent formulations of problem 12.15

We can now say that problem 12.15 can be written as a Dirichlet problem associated with the multi-valued equation

$$-\Delta U \in H(g(x, y_1) - U) \quad \text{in } D \tag{12.261}$$

and with data g (defined in (12.244)) on ∂D. This problem can be written in variational form in the following way:

PROBLEM 12.16. Find

$$U \in K; \ \forall v \in K \int_D \operatorname{grad} U \operatorname{grad}(U - v) \, dx \, dy + \int_D [g(x, y_1) - U]^+ \, dx \, dy$$

$$\leq \int_D [g(x, y_1) - v]^+ \, dx \, dy, \tag{12.262}$$

where

$$K = \{v \in H^1(D) : \gamma_0 v = g \quad \text{on } \partial D\}. \tag{12.263}$$

It is essentially in this form that the 'problem of the dam with vertical walls' was initially treated. We remark also that the problem 12.16, which is with respect to problem 12.12 what problem 12.15 is with respect to problem 12.5, can be written, equivalently, as a minimization problem:

PROBLEM 12.17. Minimize on K the functional

$$J(v) = \tfrac{1}{2} \int_D |\text{grad } v|^2 \, dx \, dy + \int_D [g(x, y_1) - v]^+ \, dx \, dy. \qquad (12.264)$$

Regularity of the free-boundary

Let us now continue our series of lemmata in order to prove, for the case of problem 12.15, the conjecture (C2). Let us denote by U the solution of problem 12.15 and by Ω the corresponding detachment set.

LEMMA 6. *The function ϕ defined in (12.230) is continuous and strictly decreasing in $[0, c]$; in particular the graph of ϕ is Lipschitz continuous relative to the axes $X \equiv x + y$, $Y \equiv x - y$.*

Proof. Equation (12.228) ensures that for every $x \in \,]0, c[$ the set $\{y : (x, y) \in \Omega\}$ is non-empty, and hence that (12.230) effectively defines a function in $]0, c[$ (also since, from lemma 5, $\partial \Omega \cap D$ does not contain vertical segments); further this function is monotone by virtue of lemma 4, and hence the limits written in (12.230) exist. Thus we conclude that (12.230) defines a monotone decreasing function in $[0, c]$. The Lipschitz continuity nature of the graph of ϕ relative to the axes X and Y (see fig. 12.5), and consequently the continuity of ϕ, follow again from lemma 4. By virtue of lemma 5 ϕ is strictly monotone, which ensures that its graph (relative to the axes x and y) does not contain horizontal segments. ■

LEMMA 7. $\phi(0) = y_1$ and $\phi(c) > y_2$.

Proof. From $\Gamma_1 \cup \Gamma_2 \subset \partial \Omega$ (which follows from the fact that U is continuous, from (12.244), and from the definition of Ω) one can immediately see that $\phi(0) = y_1$ and $\phi(c) \geq y_2$. We will give a brief outline of the proof that $\phi(c) > y_2$ using the techniques of one complex variable (for a complete proof see Friedman–Jensen, 1978), the so-called techniques of the hodograph plane. Making the conformal transform of $\Omega \subset \mathbb{R}^2$ into $\Omega^* \subset \mathbb{C}$ by means of

$$(x, y) \in \mathbb{R}^2 \mapsto (p = u_x = U_{yx}, q = v_x = U_{xx}) \in \mathbb{C} \qquad (12.265)$$

(see fig. 12.6) we see that we must have $C_\phi \neq C$, since otherwise Ω^* would be empty, and so would Ω. We have at the same time proved that $U \notin W^{2,\infty}(D)$ since $u_x = U_{yx}$ cannot remain bounded in a neighbourhood of (c, y_2), to which the improper point of the plane (p, q) belongs. ■

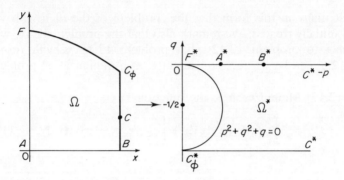

Figure 12.6

Let us remark, with respect to the techniques used in the last part of the previous proof, that, in the hydrodynamics of porous materials, one usually refers to the plane of variables (x, y) as the 'physical plane' and to that of the variables (u_x, v_x) as the 'hodograph plane'. One of the heuristic techniques for solving the problem of filtration in porous materials, the so-called 'method of the hodograph plane', consists of transforming the problem formulated on the physical plane to one formulated on the hodograph plane (see Bear, 1972, § 7.3; Polubarinova–Kochina, 1962; ch. 2).

We can prove

THEOREM 12.7. *The solution of problem 12.15 is regular in the sense of definition 12.7; further the restriction of ϕ to the interval $]0, c[$ is an analytic function.*

Proof. The first part of the theorem follows from a simple sum of lemmata 6 and 7. To prove the second part we will once again use techniques of complex variable; by conformally transforming $\Omega \subset \mathbb{R}^2$ into $\Omega' \subset \mathbb{C}$ by means

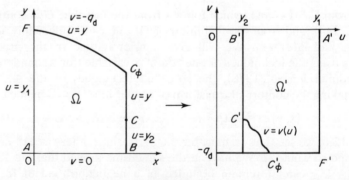

Figure 12.7

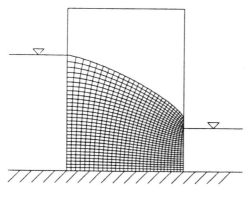

Figure 12.8

of

$$(x, y) \in \mathbb{R}^2 \mapsto (u = U_y + y, \quad v = U_x) \in \mathbb{C} \qquad (12.266)$$

(see fig. 12.7: these functions u and v are, with respect to those denoted with the same symbols in section 12.2, shifted by a quantity $q_d = (y_1^2 - y_2^2)/2c$). Note that the transformation (12.265) is the derivative of this transformation—the argument used in the present proof comes from Caccioppoli (see Miranda, 1969). The inverse conformal transform of (12.266), which can be written

$$(u, v) \in \mathbb{C} \mapsto (x(u, v), y(u, v)) \in \mathbb{R}^2, \qquad (12.267)$$

is such that the harmonic function $y(u, v)$ can be extended analytically in a neighbourhood of $\overline{C_\phi F'}$ (together with its harmonic conjugate $x(u, v)$), because in it, it takes the value of u: hence $u = u(x, y)$ also can be extended analytically in a neighbourhood of $\overline{FC_\phi}$, and hence the theorem is true. ■

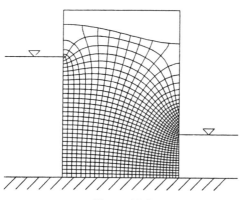

Figure 12.9

Finally we note that the free line, or rather the function ϕ, also has the following properties:

$$\phi \text{ is concave,} \tag{12.268}$$

$$\phi'(0) = 0 \quad \text{and} \quad \phi'(c) = -\infty, \tag{12.269}$$

and so, in particular, the experimental conditions of 'input and output' are satisfied (cf. Harr, 1962; p. 21). For the proof of these properties see Friedman–Jensen, 1978; see also Cryer, 1980.

In fig. 12.8 we have shown a 'flow-net' for a dam with vertical walls—note the monotonocity of the free line and the presence of the suspended source (which was already evident in the flow-net shown in fig. 12.3). In fig. 12.9 we have shown a 'flow-net' for the same dam but taking into account the phenomena of capillarity: near the free boundary one can see a negative pressure zone (the so-called *capillary fringe*).

13
Free-Boundary Problems and Variational Inequalities

In this chapter we will deal briefly with the relationships between free boundary problems and variational inequalities, along the lines suggested in Baiocchi 1976b. In other words, we will analyse the conditions under which a free boundary problem associated with a second order differential operator can be transformed into a variational inequality on a fixed domain.

The operators grad, Grad, div, and Div

Let us first introduce some notations. Let Ξ be an open set of $\mathbb{R}^n$, $n \geq 1$. If $u \in \mathscr{D}'(\Xi)$ we will denote by grad u the vector

$$\text{grad } u \equiv \begin{bmatrix} \frac{\partial u}{\partial x_1} \\ \vdots \\ \frac{\partial u}{\partial x_n} \end{bmatrix} \in \{\mathscr{D}'(\Xi)\}^n, \qquad (13.1)$$

and by Grad u the vector

$$\text{Grad } u \equiv \begin{bmatrix} u \\ \frac{\partial u}{\partial x_1} \\ \vdots \\ \frac{\partial u}{\partial x_n} \end{bmatrix} \in \{\mathscr{D}'(\Xi)\}^{n+1}. \qquad (13.2)$$

If $\mathbf{v} \equiv (v_1, \ldots, v_n) \in \{\mathscr{D}'(\Xi)\}^n$ we denote by div $\mathbf{v}$ the distribution

$$\text{div } \mathbf{v} \equiv \sum_{i=1}^{n} \frac{\partial v_i}{\partial x_i} \in \mathscr{D}'(\Xi), \qquad (13.3)$$

and if $\mathbf{w} \equiv (u, \mathbf{v}) \equiv (u, v_1, \ldots, v_n) \in \{\mathscr{D}'(\Xi)\}^{n+1}$ we denote by Div $\mathbf{w}$ the distribution

$$\text{Div } \mathbf{w} \equiv u - \text{div } \mathbf{v} \in \mathscr{D}'(\Xi). \tag{13.4}$$

Obviously, the operator $-\text{div}: \{\mathscr{D}'(\Xi)\}^n \to \mathscr{D}'(\Xi)$ is the transpose of the operator grad: $\mathscr{D}'(\Xi) \to \{\mathscr{D}'(\Xi)\}^n$:

$$_{\{\mathscr{D}'(\Xi)\}^n}\langle \text{grad } u, \boldsymbol{\phi} \rangle_{\{\mathscr{D}(\Xi)\}^n} = \sum_{i=1}^{n} {}_{\mathscr{D}'(\Xi)}\langle \partial u/\partial x_i, \phi_i \rangle_{\mathscr{D}(\Xi)}$$

$$= \sum_{i=1}^{n} {}_{\mathscr{D}'(\Xi)}\langle u, -\partial \phi_i/\partial x_i \rangle_{\mathscr{D}(\Xi)} = {}_{\mathscr{D}'(\Xi)}\langle u, -\text{div } \boldsymbol{\phi} \rangle_{\mathscr{D}(\Xi)}$$

$$\forall \boldsymbol{\phi} \equiv (\phi_1, \ldots, \phi_n) \in \{\mathscr{D}(\Xi)\}^n, \quad (13.5)$$

and, in an analogous manner, the operator $\text{Div}: \{\mathscr{D}'(\Xi)\}^{n+1} \to \mathscr{D}'(\Xi)$ is the transpose of the operator Grad: $\mathscr{D}'(\Xi) \to \{\mathscr{D}'(\Xi)\}^{n+1}$:

$$_{\{\mathscr{D}'(\Xi)\}^{n+1}}\langle \text{Grad } u, \boldsymbol{\chi} \rangle_{\{\mathscr{D}(\Xi)\}^{n+1}} = {}_{\mathscr{D}'(\Xi)}\langle u, \nu \rangle_{\mathscr{D}(\Xi)} + {}_{\{\mathscr{D}'(\Xi)\}^n}\langle \text{grad } u, \boldsymbol{\phi} \rangle_{\{\mathscr{D}(\Xi)\}^n}$$

$$= {}_{\mathscr{D}'(\Xi)}\langle u, \nu \rangle_{\mathscr{D}(\Xi)} + {}_{\mathscr{D}'(\Xi)}\langle u, -\text{div } \boldsymbol{\phi} \rangle_{\mathscr{D}(\Xi)}$$

$$= {}_{\mathscr{D}'(\Xi)}\langle u, \text{Div } \boldsymbol{\chi} \rangle_{\mathscr{D}(\Xi)} \quad \forall \boldsymbol{\chi} \equiv (\nu, \boldsymbol{\phi}) \in \{\mathscr{D}(\Xi)\}^{n+1}. \tag{13.6}$$

The operator L

We denote by $[\lambda]$ the matrix

$$[\lambda] \equiv \begin{bmatrix} d & \vdots & b_1 & \ldots & b_n \\ \hdashline -c_1 & \vdots & a_{11} & \ldots & a_{1n} \\ \vdots & \vdots & \vdots & & \vdots \\ -c_n & \vdots & a_{n1} & \ldots & a_{nn} \end{bmatrix} \equiv \begin{bmatrix} \delta & \vdots & \beta \\ \hdashline -\gamma & \vdots & \alpha \end{bmatrix}, \tag{13.7}$$

on the elements of which we will make the following assumptions:

$$a_{ij}, b_i, c_i, d \in W^{1,\infty}(\Xi) \quad (i, j = 1, \ldots, n). \tag{13.8}$$

Let us now introduce the differential operator which we will deal with here:

DEFINITION 13.1. Let us denote by L the linear differential operator with coefficients a_{ij}, b_i, c_i, d defined as follows:

$$\text{dom}(L) = \{v \in L^2(\Xi) : [\lambda] \text{ Grad } v \in \{L^2(\Xi)\}^{n+1}\}, \tag{13.9}$$

$$\forall v \in \text{dom}(L) \quad Lv = \text{Div}[\lambda] \text{ Grad } v \quad \text{in } \mathscr{D}'(\Xi). \tag{13.10}$$

The definition makes sense since if $v \in L^2(\Xi)$ then Grad $v \in \{H^{-1}(\Xi)\}^{n+1}$ and therefore, from (13.8), the product $[\lambda]$ Grad v which is involved in the definition of the domain of L is meaningful. Again, in general dom $(L) \supset H^1(\Xi)$, being dom $(L) = H^1(\Xi)$ if the matrix $[\alpha]$ is positive definite, in which case L is an elliptic operator. Formally, (13.10) and (7.41) are the same: it is in fact the search for this coincidence which led us to write the elements of [λ] in the strange way which we have adopted.

The free boundary problem

Let us now consider the following free boundary problem associated with the operator L (for a definition of the space $L^2_{\text{div}}(\Xi)$ see Chapter 18):

PROBLEM 13.1. Given $f \in L^2(\Xi)$, $\psi \in \text{dom}(L)$ and $\pi \in L^2_{\text{div}}(\Xi)$ find $\{\Omega, \tilde{u}\}$ such that

$$\Omega \text{ is a non-empty open subset of } \Xi, \quad (13.11)$$

$$\tilde{u} \in \text{dom}(L), \quad (13.12)$$

$$L\tilde{u}|_\Omega = f|_\Omega, \quad (13.13)$$

$$\text{supp}(\tilde{u} - \psi) \subset \bar{\Omega}, \quad (13.14)$$

$$\left.\frac{\partial u}{\partial \nu_L}\right|_{\partial\Omega \cap \Xi} = \pi \cdot \mathbf{n}_\Omega|_{\partial\Omega \cap \Xi}. \quad (13.15)$$

Formally, in terms of $u = \tilde{u}|_\Omega$ the conditions (13.13), (13.14), (13.15) can be written, respectively, as:

$$Lu = f \quad \text{in } \Omega, \quad (13.16)$$

$$u = \psi \quad \text{on } \partial\Omega \cap \Xi, \quad (13.17)$$

$$\frac{\partial u}{\partial \nu_L} = \pi \cdot \mathbf{n}_\Omega \quad \text{on } \partial\Omega \cap \Xi. \quad (13.18)$$

Writing them in this way, it is obvious that problem 13.1 is effectively a free boundary problem: on the free part of $\partial\Omega$ (i.e. on $\partial\Omega \cap \Xi$) we have imposed two conditions, (13.17) and (13.18). On the other hand, the problem is not complete: to the conditions on the free boundary we must add conditions on the fixed part of $\partial\Omega$ (i.e. on $\partial\Omega \cap \partial\Xi$).

The condition (13.18) is to be taken, naturally, in the sense of the traces of $L^2_{\text{div}}(\Omega)$: formally we have

$$\left.\frac{\partial u}{\partial \nu_L}\right|_{\partial\Omega \cap \Xi} = [-\gamma \mid \alpha] \operatorname{Grad} u \cdot \mathbf{n}_\Omega \quad (13.19)$$

and, from (13.8) and (13.9),

$$\operatorname{div}[-\gamma \mid \alpha] \operatorname{Grad} u = -\operatorname{Div}[\lambda] \operatorname{Grad} u - u[\delta \mid \beta] \operatorname{Grad} u$$

$$= -f - u[\delta \mid \beta] \operatorname{Grad} u \in L^2(\Omega) \quad (13.20)$$

(cf. (18.13)...). We note also that (13.18) implies that

$$[\alpha] \operatorname{grad}(u - \psi) \operatorname{grad}(u - \psi) + \{[-\gamma \mid \alpha] \operatorname{Grad} \psi - \pi\} \operatorname{grad}(u - \psi) = 0$$

$$\text{on } \partial\Omega \cap \Xi, \quad (13.21)$$

where the normal $\mathbf{n}_\Omega$ to $\partial\Omega$ is not explicitly involved: in fact, from (13.17), (13.18), and (13.19) and using the fact that $\text{grad}\,(u-\psi)$ is either zero or parallel to $\mathbf{n}_\Omega$, we can write (formally)

$$\begin{aligned}
0 &= [-\gamma \mid \alpha]\,\text{Grad}\,u\,\text{grad}\,(u-\psi) - \boldsymbol{\pi}\,\text{grad}\,(u-\psi) \\
&= ([-\gamma \mid \alpha]\,\text{Grad}\,u - \boldsymbol{\pi})\,\text{grad}\,(u-\psi) \\
&= ([-\gamma \mid \alpha]\,\text{Grad}\,(u-\psi) + [-\gamma \mid \alpha]\,\text{Grad}\,\psi - \boldsymbol{\pi})\,\text{grad}\,(u-\psi) \\
&= ([-\gamma](u-\psi) + [\alpha]\,\text{grad}\,(u-\psi) + [-\gamma \mid \alpha]\,\text{Grad}\,\psi - \boldsymbol{\pi})\,\text{grad}\,(u-\psi) \\
&= [\alpha]\,\text{grad}\,(u-\psi)\,\text{grad}\,(u-\psi) + \{[-\gamma \mid \alpha]\,\text{Grad}\,\psi - \boldsymbol{\pi}\}\,\text{grad}\,(u-\psi). \quad (13.22)
\end{aligned}$$

Problem 13.1 is a problem with a non-null obstacle —ψ— which evidently has as a particular case the problem with a null obstacle but which, on the other hand, is not more general, since it is effectively equivalent to an appropriate problem with a null obstacle. More precisely, the following problem 13.2 is equivalent to the problem 13.1:

PROBLEM 13.2. Given $\mathbf{l} \equiv (\boldsymbol{\pi}'', \boldsymbol{\pi}') \in \{L^2(\Xi)\}^{n+1}$ find $\{\Omega, p\}$ such that

$$\Omega \text{ is a non-empty open subset of } \Xi, \quad (13.23)$$

$$p \in \text{dom}\,(L), \quad (13.24)$$

$$\text{supp}\,p \subset \bar{\Omega}, \quad (13.25)$$

$$Lp = \text{Div}\,(\mathbf{l}\chi_\Omega) \text{ in } \Xi. \quad (13.26)$$

This equivalence is to be understood in the sense that $\{\Omega, \bar{u}\}$ is a solution of the problem 13.1 iff $\{\Omega, \bar{u}-\psi\}$ is a solution of problem 13.2 with

$$\boldsymbol{\pi}' = \boldsymbol{\pi} - [-\gamma \mid \alpha]\,\text{Grad}\,\psi \quad (13.27)$$

and

$$\boldsymbol{\pi}'' = f + \text{div}\,\boldsymbol{\pi} + [\delta \mid \beta]\,\text{Grad}\,\psi. \quad (13.28)$$

It is interesting to note that (13.21) can be written in terms of p as

$$[\alpha]\,\text{grad}\,p\,\text{grad}\,p - \boldsymbol{\pi}'\,\text{grad}\,p = 0 \quad \text{on } \partial\Omega \cap \Xi, \quad (13.29)$$

which shows that $\partial/\partial\nu_L$ does not depend on the decomposition $\text{Div}\,[\lambda]\,\text{Grad}$ of L but only on the symmetric part of $[\alpha]$, which is independent of the decomposition of the operator (this also shows that it is correct to write $\partial/\partial\nu_L$ instead of the usual $\partial/\partial\nu_a\ldots$).

Transformation of the free boundary problem

Next, we will give conditions under which we can transform the free boundary problem into a problem associated with a variational inequality. Incidentally we use pseudo-differential operators: for the notion and a survey of pseudo-differential operators see Kohn, 1974 (and the bibliography contained in it) and Treves, 1978. Let us suppose then that

(i) *it is possible to find a differential operator Q and an operator (differential if possible, and if not pseudo-differential) R such that*

$$L(Q\square) \equiv \text{Div}(\mathbf{I}R\square), \tag{13.30}$$

(ii) *it is possible to solve, with appropriate regularity, the problem* 'find $\{\Omega, w\}$ such that

$$\text{supp } w \subset \bar{\Omega}, \tag{13.31}$$

$$Rw = \chi_\Omega'. \tag{13.32}$$

It follows then that

$$p = Qw \tag{13.33}$$

is a solution of the problem 13.2 and that

$$\tilde{u} = Qw + \psi \tag{13.34}$$

is a solution of the problem 13.1.

Hence the problem '(13.31), (13.32)' can be written in the form 'find $\{\Omega, w\}$ such that

$$\Omega = \{x \in \Xi : w(x) < 0\}, \tag{13.35}$$

$$w \le 0 \text{ in } \Xi, \quad Rw \le 1 \text{ in } \Xi, \quad w(1 - Rw) = 0 \quad \text{in } \Xi', \tag{13.36}$$

i.e., as a problem associated with a variational inequality (an incomplete problem: among other things the conditions of the fixed part of the boundary are missing). Let us remark that the condition $w \le 0$ in Ξ is not a restriction if R satisfies a maximum principle.

In the case of the free boundary problem studied in Chapter 12 we have $\Xi = D$, $L = -\Delta$, $f = 0$, $\psi = y$ and $\pi = 0$. Further, problem 12.4 is essentially problem 13.2 with $\mathbf{I}$ such that Div $(\mathbf{I}) = \partial/\partial y$: in this case $R = L$ and $Q = \partial/\partial y$ satisfy (13.30) (and (13.33) is essentially the inverse of (12.76) ...).

To end, let us note that (13.21) shows that free boundary problems such as those considered in Daniljuk, 1971, i.e. with (13.15) substituted by a condition of the type

$$[\alpha] \text{ grad }(u - \psi) \text{ grad }(u - \psi) = 1 \quad \text{in } \partial\Omega \cap \Xi, \tag{13.37}$$

do not enter into the formulation given here.

For yet other types of free-boundary problems see Alt–Caffarelli–Friedman, 1982a; 1982b; 1982c; Amerio, 1978; Amerio–Prouse, 1975; Berestycki–Brézis, 1976; Brézis–Sibony, 1971; Cimatti, 1977; McAllister–Rohde, 1975; Mossino, 1976; Temam, 1976a; 1976b (see also the previously mentioned bibliography Cryer, 1977 and the bibliography 'Free boundary problems' compiled from *Zentralblatt für Mathematik/ Mathematics Abstracts* by Fachinformationszentrum Energie Physik Mathematik GmbH (Karlsruhe, 1981)).

PART III

Technical Tools

14
Seminorms

In various places (e.g. Chapter 4) we have come across vector spaces whose topology was neither of the Hilbert nor of the Banach type.

One of the ways of introducing topologies on vector spaces is by way of seminorms; here we will present briefly some of the concepts without entering into details (we will not, for example, discuss duality); for further information see, for example, Aubin, 1977; 1979a; Bourbaki, 1966; Cotlar–Cignoli, 1974; Edwards, 1965; Grothendieck, 1958; Horváth, 1966; Köthe, 1970; 1979; Kelley–Namioka, 1963; Treves, 1967a; Yosida, 1971.

The notion of a topological vector space

On a single set we can define different structures (algebraic, topological, etc.). These structures can be considered independently of each other, but the most interesting case is that in which they all contribute in creating a more complex structure of which they are parts—we say then that the structures are 'compatible'. Here we will limit ourselves to explain this concept of compatibility in the case where the structures are a topological one and a linear one. Keeping in mind the case of the space $\mathbb{C}^n$, we will introduce the following definition:

DEFINITION 14.1. Let E be a vector space over $\mathbb{C}$, provided with a topology. E is said to be a *topological vector space* if the two structures are compatible in the following sense:

(VT1)   the vectorial addition $(u, v) \mapsto u + v$ is continuous from $E \times E$ to E,
(VT2)   the multiplication by a scalar $(\lambda, u) \mapsto \lambda u$ is continuous from $\mathbb{C} \times E$ to E,

where $E \times E$ and $\mathbb{C} \times E$ are provided with the respective product topologies.

In what follows, as in this definition, we will consider vector spaces over $\mathbb{C}$, but we can substitute $\mathbb{C}$ by $\mathbb{R}$ at any point.

Systems of neighbourhoods and open sets

An immediate consequence of definition 14.1 is that if $V \subset E$ is a neighbourhood of a point $a \in E$ then there exists a neighbourhood V_0 of the point 0 such that $V = a + V_0$ (we need only note that (VT1) implies that shifts are homeomorphisms of E in E, or in other words that the applications $u \mapsto u + a$ and $u \mapsto u - a$ are continuous from E to E). This shows that in order to know the system of neighbourhoods of any point in a topological vector space, it is sufficient to know those of the point 0, which in turn suggests that (in principle...) the most appropriate way of introducing the topology itself is that which uses the concept of a neighbourhood as a preliminary notion.

Let us recall then the following definition:

DEFINITION 14.2. A topological space is a pair $(E, \mathcal{U})$ where E is a non-empty set and $\mathcal{U}: E \to 2^E$ is a function which with every point $u \in E$ associates a non empty family $\mathcal{U}(u)$ of subsets of E, which will be called *neighbourhoods* of u, in such a way that the following axioms are satisfied:

(U1) $u \in U$ for every $U \in \mathcal{U}(u)$,
(U2) if $U_1, U_2 \in \mathcal{U}(u)$ then $U_1 \cap U_2 \in \mathcal{U}(u)$,
(U3) if $U_1 \in \mathcal{U}(u)$ and $U_1 \subset U_2$ then $U_2 \in \mathcal{U}(u)$,
(U4) for every $U_1 \in \mathcal{U}(u)$ there exists $U_2 \in \mathcal{U}(u)$ such that $U_1 \in \mathcal{U}(w)$ for every $w \in U_2$.

Starting from $(E, \mathcal{U})$ we then define *open subset* of E any $O \subset E$ such that $\forall x \in O, O \in \mathcal{U}(x)$. Denoting by $\mathcal{O}$ the family of open sets, the pair $(E, \mathcal{O})$ will then satisfy the alternative definition of topological space:

DEFINITION 14.2'. A topological space is a pair $(E, \mathcal{O})$ where $\mathcal{O} \subset 2^E$ is a family of subsets of E, which we will call *open sets* of E, which satisfy the following axioms:

(O1) $E, \emptyset \in \mathcal{O}$
(O2) if $O_1, O_2 \in \mathcal{O}$ then $O_1 \cap O_2 \in \mathcal{O}$,
(O3) if $O_\alpha \in \mathcal{O}$, $\alpha \in A$, then $\bigcup_{\alpha \in A} O_\alpha \in \mathcal{O}$.

Conversely, starting from $(E, \mathcal{O})$ satisfying definition 14.2', we can construct $\mathcal{U}$ by setting $\forall x \in E$ $\mathcal{U}(x) = \{U \subset E : \exists O \in \mathcal{O} \text{ with } x \in O \subset U\}$. It is easy to see that the couple $(E, \mathcal{U})$ so obtained satisfies definition 14.2 and gives rise to the original family of open sets; so that definition 14.2 and definition 14.2' are 'equivalent' to each other.

Let us recall again that given two topologies in E, i.e. two structures $(E, \mathcal{U}_1)$ and $(E, \mathcal{U}_2)$, the former is said to be finer than the latter if $\forall u \in E$ $\mathcal{U}_2(u) \subset \mathcal{U}_1(u)$.

Filters and filter bases

The concept of the neighbourhood of a point and that of an open set belong to general topology; let us now recall two concepts from general set theory:

DEFINITION 14.3. Let E be a non-empty set. A *filter* in E is a family $\mathscr{F} \subset 2^E$ of subsets of E such that

(F1) if $A \in \mathscr{F}$ and $A_1 \subset A_2$ then $A_2 \in \mathscr{F}$,
(F2) if $A_1, A_2 \in \mathscr{F}$ then $A_1 \cap A_2 \in \mathscr{F}$,
(F3) $\emptyset \notin \mathscr{F}$.

If E is a topological space, a particularly important example of a filter in E is that provided by the system $\mathscr{U}(a)$ of neighbourhoods of a point $a \in E$.

DEFINITION 14.4. Let E be a non-empty set. A non-empty family $\mathscr{B} \subset 2^E$ of subsets of E is a *filter base* if it satisfies the following conditions:

(BF1) if $B_1, B_2 \in \mathscr{B}$ then there exists $B \in \mathscr{B}$ such that $B \subset B_1 \cap B_2$,
(BF2) $\emptyset \notin \mathscr{B}$.

The name given to $\mathscr{B}$ derives from the fact that the family $\hat{\mathscr{B}} = \{F \in \mathscr{P}(E): \exists B \in \mathscr{B}, B \subset F\}$ is a filter (the filter generated by the base $\mathscr{B}$).

The concept of a filter is closely connected with that of a sequence; among other things, if $\{u_m\}$ is a sequence in E then the family $U_p = \{u_p, u_{p+1}, \ldots\}$, $p \in \mathbb{N}$ is a filter base in E, the *elementary filter* associated with $\{u_m\}$ (the image, through the application $p \mapsto u_p$ from $\mathbb{N}$ to E, of the *Fréchet filter* in $\mathbb{N}$, which is the one which has as a base the family $F_p = \{p, p+1, \ldots\}$, $p \in \mathbb{N}$).

Fundamental system of the neighbourhoods of a point

We have already seen that in order to describe the topology of a topological vector space we only need to define how the neighbourhoods of the origin 0 are made up; in practice—as we will see in the next theorem—topologies are described through the so-called fundamental systems of the neighbourhoods of the origin. Let us then recall.

DEFINITION 14.5. Let E be a topological space and $a \in E$. We will define as *fundamental system of the neighbourhoods* of a a basis of the filter $\mathscr{U}(a)$ of the neighbourhoods of a.

Thus, a family $\mathscr{T}(a)$ of neighbourhoods of a is a fundamental system of the neighbourhoods of a if $\forall V \in \mathscr{U}(a) \; \exists W \in \mathscr{T}(a) \; W \subset V$. A particularly important example of a fundamental system of the neighbourhoods of a point a is that made up of the family of open neighbourhoods of a.

Fundamental theorem

We will now state without proof the fundamental result which we have already referred to and which is a profound consequence of the compatibility hypotheses (VT1) and (VT2):

THEOREM 14.1. Every topological vector space E has a fundamental system of neighbourhoods of the origin $\mathcal{T}(0)$ such that:
(i) if $V \in \mathcal{T}(0)$ then $\forall u \in E \; \exists \alpha > 0 \; u \in \lambda V \; \forall \lambda \in \mathbb{C}, \; |\lambda| \geq \alpha$,
(ii) if $V \in \mathcal{T}(0)$ then $\lambda V \subset V \; \forall \lambda \in \mathbb{C}, \; |\lambda| \leq 1$,
(iii) if $V \in \mathcal{T}(0) \; \exists U \in \mathcal{T}(0) \; U + U \subset V$.

Conversely, if E is a vector space over $\mathbb{C}$ and $\mathcal{T}(0)$ is a filter base in E which satisfies (i), (ii), and (iii) then there exists a unique topology in E such that E is a topological vector space and $\mathcal{T}(0)$ a fundamental system of the neighbourhoods of the origin.

For a proof, see e.g. Horváth, 1966, p. 81. A set V which satisfies (ii) is said to be *balanced*, and a set V which satisfies (i) is said to be *absorbent*. This latter concept can be generalized as follows: if $U, V \subset E$, we say that V absorbs U if $\exists \alpha > 0 \; U \subset \lambda V \; \forall \lambda \in \mathbb{C}, \; |\lambda| \geq \alpha$ (in (i) we have $U = \{u\}$). The importance of this generalization lies in the fact that it allows us to introduce the concept of boundedness independently of a metric: a subset L of a topological vector space E is said to be *bounded* if it is absorbed by every neighbourhood of the origin.

Hausdorff spaces

Let us recall that a topological space $(E, \mathcal{U})$ is said to be a *separate space*, a *T2 space*, or a *Hausdorff space* if $\forall u, v \in E, \; u \neq v \; \exists U \in \mathcal{U}(u), \; V \in \mathcal{U}(v) \; U \cap V = \emptyset$. In the case of topological vector spaces this condition can be written in the simpler form $\forall u \neq 0 \; \exists V \in \mathcal{U}(0) \; u \notin V$, a simplification which is a consequence of the compatibility of the topological and linear structures of E. Another consequence of this compatibility is that every Hausdorff topological vector space is *regular* (i.e. it satisfies the axioms of separation T_1 and T_3) since in any topological vector space every neighbourhood of the origin contains a closed neighbourhood of the origin.

Convergent filters. Completeness

We have seen that there is a close connection between filters and sequences; as a further demonstration of this fact we introduce:

DEFINITION 14.6. Let $(E, \mathcal{U})$ be a topological space and $\mathcal{F}$ a filter in E. The filter $\mathcal{F}$ is said to *converge* to $a \in E$ and a is called the *limit* of $\mathcal{F}$ if $\forall U \in \mathcal{U}(a) \; \exists F \in \mathcal{F} : F \subset U$.

It can be easily seen that $\mathscr{F}$ converges to $a \in E$ iff $\mathscr{F}$ is finer than the filter $\mathscr{U}(a)$ of all the neighbourhoods of a, and that the limit of a filter on a Hausdorff space, if it exists, is unique (in fact this condition is necessary and sufficient and characterizes Hausdorff spaces). The concept of a filter is essential in non-metrizable topological spaces: in fact while for metrizable structures convergent sequences are sufficient in order to characterize the topology, this is not so for non-metrizable spaces.

Let us now introduce the concept which generalizes that of Cauchy sequences

DEFINITION 14.7. Let E be a topological vector space and A a subset of E. A filter $\mathscr{F}$ in A is said to be a *Cauchy filter* if for every neighbourhood V of the origin there exists $F \in \mathscr{F}$ such that $F - F \subset V$. A is said to be *complete* if every Cauchy filter in A converges to a point in A.

We will state without proof (see Horváth, 1966, p. 131) that if E is a Hausdorff topological vector space then there exists a Hausdorff topological vector space $\hat{E}$ (unique except for isomorphisms) such that E is isomorphic to a subspace dense in $\hat{E}$. $\hat{E}$ is called the *completion* of E.

Locally convex topological vector spaces

The topological vector spaces which are of practical importance in analysis satisfy the conditions given in the following definition:

DEFINITION 14.8. A topological vector space is said to be *locally convex* if it has a fundamental system of the neighbourhoods of the origin consisting of convex sets.

It follows immediately that in a locally convex topological vector space there exists a fundamental system of the neighbourhoods of the origin made up of absorbent, balanced, closed, and convex sets, i.e. of *barrels*. Conversely, a locally convex topological vector space is said to be a *barrelled space* if every barrel is a neighbourhood of the origin. The importance of this class of spaces lies in the fact that they constitute the suitable framework for the Banach–Steinhaus theorem (which in the framework of Banach spaces reads *if a family* $\{f_\lambda\}$, $\lambda \in \Lambda$, *of continuous linear functionals in a Banach space is uniformly bounded, in the sense that* $\forall u \in B$ $\exists M(u)$ $\forall \lambda \in \Lambda$ $|f_\lambda(u)| \leq M(u)$, *then there exists an* M *such that* $|f_\lambda(u)| \leq M \|u\|$ $\forall u \in B$, $\lambda \in \Lambda$)—further a space is barrelled iff an appropriate generalization of this theorem holds in it.

We now state (again without proof) the following result, a corollary of theorem 14.1:

THEOREM 14.2. Let E be a vector space, $\mathscr{A}$ a family of absorbent, balanced, and convex subsets of E and $\mathcal{T}(0)$ the family of the finite intersections of sets of the form λV, with $V \in \mathcal{A}$ and $\lambda > 0$: there exists a unique topology in E for which E is a locally convex topological vector space and for which $\mathcal{T}(0)$ is a fundamental system of neighbourhoods of the origin.

Seminorms

We now introduce a fundamental tool in the theory of locally convex topological spaces:

DEFINITION 14.9. Let E be a vector space over the field $\mathbb{C}$. A *seminorm* in E is defined as a function $p: E \to \mathbb{R}$ such that

(SN1) $p(u) \geq 0 \quad \forall u \in E,$
(SN2) $p(u+v) \leq p(u) + p(v) \quad \forall u, v \in E,$
(SN3) $p(\alpha u) = |\alpha| p(u) \quad \forall \alpha \in \mathbb{C}, u \in E.$

It follows immediately that

$$p(0) = 0 \tag{14.1}$$

since $p(0) = p(0 \cdot u) = 0 \cdot p(u) = 0$. Let us remark that if conversely

(N) $p(u) = 0 \Rightarrow u = 0$

then we get the definition of a *norm*.

Note that the axiom (SN1) is useless in that from (SN2) and (SN3) it follows more generally that

$$p(u-v) \geq |p(u) - p(v)| \quad \forall u, v \in E \tag{14.2}$$

(in fact from (SN2) we have $p(u-v) + p(v) \geq p(u)$ and hence that $p(u-v) \geq p(u) - p(v)$; noting now that from (SN3) we have $p(u-v) = p((-1)(v-u)) = |-1| p(v-u) = p(v-u)$, we have once again from (SN2) also that $p(u-v) \geq p(v) - p(u)$): putting $v = 0$ in (14.2) we obtain (SN1).

On the other hand, axioms (SN1), (SN2), and (SN3) are equivalent to the single axiom

(SN) $p(\alpha u + \beta v) \leq |\alpha| p(u) + |\beta| p(v) \quad \forall \alpha, \beta \in \mathbb{C}, u, v \in E.$

This condition (SN) follows, in fact, from (SN2) and (SN3). Conversely, putting $\alpha = \beta = 1$ in (SN), we get (SN2); putting $\beta = 0$ in (SN) we get $p(\alpha u) \leq |\alpha| p(u)$ and we can also write that $p(\alpha^{-1} \alpha u) \leq |\alpha^{-1}| p(\alpha u)$: from this it follows that $p(\alpha u) = |\alpha| p(u)$, $\alpha \neq 0$; keeping in mind what we have seen above, (SN1) also follows.

Seminorms and neighbourhoods

To see how seminorms can be utilized in describing the topology of a locally convex topological vector space, we will first prove the following result:

THEOREM 14.5. Let E be a vector space over the field $\mathbb{C}$ and p a seminorm in E. The set

$$M = \{u \in E : p(u) \leq 1\} \quad (14.3)$$

contains the origin, and is convex, balanced, and absorbent.

Proof. $0 \in M$ since $p(0) = 0 \leq 1$. The convexity of M follows from the fact that for any u, $v \in M$ and $0 < \alpha < 1$ we have $p(\alpha u + (1-\alpha)v) \leq \alpha p(u) + (1-\alpha)p(v) \leq \alpha + (1-\alpha) = 1$ and thus $\alpha u + (1-\alpha)v \in M$. That M is balanced follows from the fact that if $u \in M$ and $|\alpha| \leq 1$ then $p(\alpha u) = |\alpha| p(u) \leq |\alpha| \leq 1$ and therefore $\alpha u \in M$. Finally, to see that M is absorbent we need only note that if $u \in E$ is such that $p(u) = \alpha \neq 0$ (if $p(u) = 0$ then of course $u \in M$) then $p(\alpha u) = 1$ and therefore $\alpha^{-1} u \in M$. ∎

Now let $\{p_\lambda\}$, $\lambda \in \Lambda$ be any family of seminorms defined in E and consider the family of sets of the form $M_\lambda = \{u \in E : p_\lambda(u) \leq 1\}$. Starting from this family of sets we can obtain, through finite intersections of homothetics of the sets themselves, the family

$$V\lambda_1, \ldots, \lambda_n; \varepsilon_1, \ldots, \varepsilon_m = \{u \in E : p_{\lambda_j}(u) \leq \varepsilon_j, 1 \leq j \leq n\} \quad (14.4)$$

which, in agreement with theorem 14.2, constitutes a fundamental system of neighbourhoods of the origin for a locally convex topology in E.

We have shown then that, starting from a family of seminorms, one can define a topology in a vector space such that the space is a locally convex topological vector space. In the extreme case in which the family of seminorms is made up of all the seminorms defined in the vector space, this topology is the finest locally convex topology that one can introduce in this space; other examples of topologies defined through families of seminorms can be found in Chapter 4—this is in fact the commonest (in addition to being the most convenient) way of introducing a locally convex topology in a vector space.

Minkowski functionals

Conversely, we will now show that a locally convex topology can always be described through an appropriate family of seminorms. With this aim we first introduce the definition:

DEFINITION 14.10. Let E be a vector space over the field $\mathbb{C}$ and M a subset of E. The *Minkowski functional* associated with M is the application $f_M : E \to [0, +\infty]$ defined by

$$\left. \begin{array}{lll} f_M(u) = +\infty & \text{if } \forall \alpha > 0 & u \notin \alpha M \\ f_M(u) = \inf_{\substack{u \in \alpha M \\ \alpha > 0}} \alpha & \text{if } \exists \alpha > 0 & u \in \alpha M \end{array} \right\} \quad (14.5)$$

We will now prove the following important result:

THEOREM 14.4. Let E be a vector space over the field $\mathbb{C}$ and M a balanced, convex, and absorbent subset of E. The Minkowski functional associated with M is a seminorm in E.

Proof. From (14.5) it follows immediately that if M is absorbent then f_M is a real function, as required by the definition 14.9; it also follows immediately from (14.5) that (SN1) is satisfied. Since M is balanced, we have

$$\forall w \in E \quad \forall \varepsilon > 0 \quad \frac{w}{f_M(w)+\varepsilon} \in M, \qquad (14.6)$$

and since M is convex we have (it is evident that the coefficients of the combination are in agreement with the definition of a convex set)

$$\forall u, v \in E \quad \forall \varepsilon > 0$$

$$\frac{f_M(u)+\varepsilon}{f_M(u)+f_M(v)+2\varepsilon} \frac{u}{f_M(u)+\varepsilon} + \frac{f_M(v)+\varepsilon}{f_M(u)+f_M(v)+2\varepsilon} \frac{v}{f_M(v)+\varepsilon} \in M; \qquad (14.7)$$

from this it follows that $u+v \in (f_M(u)+f_M(v)+2\varepsilon)M$ and therefore, from the definition of f_M, since ε is arbitrary, that (SN2) is true. Again from (14.6), since ε is arbitrary, it follows that $\forall u \in E \ \forall \lambda \in \mathbb{C} \ f_M(\lambda u) = |\lambda| f_M(u)$, i.e. that (SN3) is true. ∎

Let us recall at this point that in a locally convex topological vector space there exists a fundamental system of the neighbourhoods of the origin made up of barrels and let us consider the following facts, E being an arbitrary topological vector space: (1) let $A \subset E$ be a balanced, absorbent and convex set: A is a neighbourhood of the origin iff the Minkowski functional f_A is continuous; (2) let $V \subset E$ be a balanced, absorbent, and convex neighbourhood of the origin: $\bar{V} = \{u \in E : f_V(u) \leq 1\}$. We can now say that the family of Minkowski functionals associated with the sets of a fundamental system of neighbourhoods of the origin made up of barrels consists of continuous seminorms (from $(1; \Rightarrow)$) and that the topology defined by this family coincides with the original one since $V = \{u \in E : f_V(u) \leq 1\}$ is a neighbourhood of the origin (from $(1; \Leftarrow)$) and a barrel (from (2)).

Saturated families of seminorms

Let E be a vector space and $\mathcal{S} = \{p_\lambda\}$, $\lambda \in \Lambda$ a family of seminorms in E. $\mathcal{S}$ is said to be a *saturated family* of seminorms if, for any finite subfamily $\{p_i\}$, $i = 1, \ldots, n$, the seminorm defined by $p(u) = \max_i p_i(u)$ belongs to $\mathcal{S}$ (the definition makes sense since p is evidently a seminorm in E). The family of all continuous seminorms in a locally convex topological vector space is saturated and we can therefore say that every locally convex topology can be defined through a saturated family of seminorms.

The axiom of separation

In terms of the concept of a seminorm, Hausdorff spaces are characterized by the following theorem:

THEOREM 14.5. Let E be a locally convex topological vector space whose topology is defined by means of a family $\{p_\lambda\}$, $\lambda \in \Lambda$ of seminorms. E is a Hausdorff space iff this family satisfies the following *axiom of separation*:

(H) $\forall u \in E \setminus \{0\}$ $\exists \lambda \in \Lambda\; p_\lambda(u) \neq 0$.

Proof. If (H) is true then, taking $u \neq 0$ and λ under the conditions of (H) and putting $\alpha = p_\lambda(u)$, the set $\{u : p_\lambda(u) \leq \alpha/2\}$ is a neighbourhood of the origin which does not contain u and E is therefore a Hausdorff space. Conversely, if E is a Hausdorff space and $u \neq 0$ then there exists a neighbourhood V of the origin which does not contain u; since V must necessarily contain a set of the type $V\lambda_1, \ldots, \lambda_n$; $\varepsilon_1, \ldots, \varepsilon_n$, we can conclude that there exists λ_j such that $p_{\lambda_j}(u) \neq 0$ since otherwise $u \in V$. ∎

Metrizability. Fréchet spaces

In a locally convex topological vector space E whose topology is defined by means of a family $\{p_\lambda\}$, $\lambda \in \Lambda$, a set $L \subset E$ is bounded iff $\forall \lambda \in \Lambda\; \exists M_\lambda \in \mathbb{R}$ $\forall u \in L\; p_\lambda(u) \leq M_\lambda$. It can be proved (see, e.g. Horváth, 1966; p. 109) that if E is a locally convex topological vector space in which there is a neighbourhood of the origin which is bounded then the topology of E can be defined by means of a single seminorm. This seminorm is a norm iff E is a Hausdorff space and in that case we say that the space is *normable*. Let us note also that if the topology of a locally convex topological vector space can be defined by menas of a finite family $\{p_i\}$, $i = 1, \ldots, n$, of seminorms then it can be defined by means of a single seminorm, such as $p(u) = \max_i p_i(u)$ or $p(u) = \sum_i p_i(u)$.

More generally, a topological space is said to be *metrizable* if there exists a metric d in E which induces a topology in E which coincides with the one previously considered. It can be shown (see e.g. Horváth, 1966; p. 111) that if a Hausdorff topological space is such that it has a countable fundamental system of the neighbourhoods of the origin then it is metrizable, and it is so through a metric d which is 'shift invariant' in the sense that $\forall u, v, a \in E$ $d(u+a, v+a) = d(u, v)$ (E with such a metric is called a *linear metric space*). For locally convex topological vector spaces this result can be written as follows: if E is a Hausdorff locally convex topological vector space whose topology is defined by means of a sequence of seminorms $\{p_n\}$, $n \in \mathbb{N}$, the formula $d(u, v) = \sum_{n=0}^{\infty} 2^{-n} p_n(u-v)/(1+p_n(u-v))$ defines an invariant metric which induces in E the original topology.

The function $[\cdot] : u \to [u] = d(u, 0)$ is called a *quasi-norm* and satisfies $[u] \geq 0$, $[u] = 0 \Leftrightarrow u = 0$, $[u+v] \leq [u] + [v]$.

Complete metrizable locally convex topological vector spaces are known as *Fréchet spaces*.

The Hahn–Banach theorem

Finally, let us state without proof the following result which is an extension of theorem 1.1:

THEOREM 14.6 (Hahn–Banach theorem). Let E be a vector space over $\mathbb{C}$, p a seminorm in E, and V a linear variety of E. If $f: V \to \mathbb{C}$ is a linear functional such that $\forall v \in V\ |f(v)| \leq p(v)$ then there exists a linear functional $g: E \to \mathbb{C}$ such that $g|_V \equiv f$ and $\forall v \in E\ |g(v)| \leq p(v)$.

For a proof see e.g. Horváth, 1966, p. 176. This result can be further generalized since p need not be a seminorm but need only, in the case of real spaces, be a subadditive non-negative functional, zero at the origin and such that $\forall u \in E\ p(\lambda u) \to 0$ when $\lambda \to 0^+$ (see Kranz, 1972).

15
Regularization and Partition of the Unit

15.1 REGULARIZATION

A widely used technique in the study of problems which involve 'irregular' functions consists of substituting these functions with sequences of 'regular' functions which converge to them in convenient topologies: this is the so-called *technique of regularization*.

The function $\phi_\Omega(x)$

Among the 'regular' functions most used in this connection (especially in questions of a theoretical nature . . .) are the elements of $C_0^\infty(\Omega)$; the following result is therefore an interesting one:

THEOREM 15.1. *If $\Omega \subset \mathbb{R}^n$ is a non-empty open set then there exists $\phi \in C_0^\infty(\Omega)$ such that $\phi \geq 0$ in Ω and*

$$\int_\Omega \phi(x)\,dx = 1. \tag{15.1}$$

Proof. We first show that, for $\Omega = \mathbb{R}^n$, *there exists in $C_0^\infty(\mathbb{R}^n)$ a function which is non-identically zero.* Let then $\phi_{\mathbb{R}^n}:\mathbb{R}^n \to \mathbb{R}$ be the function defined by

$$\phi_{\mathbb{R}^n}(x) = \begin{cases} e^{-1/(1-\|x\|^2)} & \text{if } \|x\| < 1 \\ 0 & \text{if } \|x\| \geq 1, \end{cases} \tag{15.2}$$

where $\|\cdot\|$ is the Euclidean norm of $\mathbb{R}^n$, and let us show that $\phi_{\mathbb{R}^n} \in C_0^\infty(\mathbb{R}^n)$. It follows immediately that supp $\phi_{\mathbb{R}^n} = \{x \in \mathbb{R}^n : \|x\| \leq 1\}$ is a compact set of $\mathbb{R}^n$. We will now show that $\phi_{\mathbb{R}^n} \in C^\infty(\mathbb{R}^n)$: given the symmetry of $\phi_{\mathbb{R}^n}$ we need only consider the one-dimensional case and hence, since it is evident that $\phi_\mathbb{R} \in C^\infty(\mathbb{R}\setminus\{-1, 1\})$ and also that $\phi_\mathbb{R}$ is an even function, it is sufficient that we study the behaviour of $\phi_\mathbb{R}$ in a neighbourhood of the point $x = 1$. Given that the function is identically zero in $[-1, +\infty[$, the problem reduces to proving that

$$\lim_{x \to 1^-} \phi_\mathbb{R}^{(k)}(x) = 0, \quad \text{for every } k \in \mathbb{N}. \tag{15.3}$$

Now, as can be easily seen by finite induction

$$\phi_{\mathbb{R}}^{(k)}(x) = e^{-1/(1-\|x\|^2)} \frac{P_k(x)}{Q_{2k}^{(x)}} \qquad \forall k \in \mathbb{N}, \qquad \text{in }]-1, 1[, \tag{15.4}$$

where P_m and Q_m are polynomials in x of degree m, and hence (15.3) is satisfied. We have thus shown that $\phi_{\mathbb{R}^n} \in C_0^\infty(\mathbb{R}^n)$. Let now $\Omega \subset \mathbb{R}^n$ be a non-empty open set and $x_0 \in \Omega$. Given the structure of the Euclidean topology of $\mathbb{R}^n$ we can be sure that

$$\exists \varepsilon > 0 \quad I_\varepsilon(x_0) = \{x \in \mathbb{R}^n : \|x - x_0\| \leq \varepsilon\} \subset \Omega, \tag{15.5}$$

and therefore that the function $\phi : \Omega \to \mathbb{R}$ defined by

$$\phi_\Omega(x) = \phi_{\mathbb{R}^n}\left(\frac{x - x_0}{\varepsilon}\right), \qquad x \in \Omega, \tag{15.6}$$

is an element of $C_0^\infty(\Omega)$, since it is indefinitely differentiable and supp $\phi_\Omega = I_\varepsilon(x_0)$ is a compact set. We have shown therefore that if Ω is a non-empty open set there exists $\phi_\Omega \in C_0^\infty(\Omega)$ such that $\phi_\Omega \neq 0$; a simple inspection of ϕ_Ω tells us further that $\phi_\Omega(x) \geq 0 \ \forall x \in \Omega$ and hence that

$$0 < \int_\Omega \phi_\Omega(x) \, dx < +\infty \tag{15.7}$$

(let us remark that ϕ_Ω is integrable in Ω since it is continuous and bounded in it, and further that the integral on Ω reduces to an integral on $I_\varepsilon(x_0)$): we can then consider the function $\phi : \Omega \to \mathbb{R}$ defined by

$$\phi(x) = \frac{\phi_\Omega(x)}{\int_\Omega \phi_\Omega(x) \, dx}, \tag{15.8}$$

which, obviously, obeys the conditions stated above. ∎

To the functions $\phi_{\mathbb{R}^n}$, ϕ_Ω, and ϕ considered in this proof are often given the suggestive name of '*bell-shaped*' *functions* due to the form of their graphs for $n = 1$ and $n = 2$—fig. 15.1 shows that of $\phi_{\mathbb{R}}$.

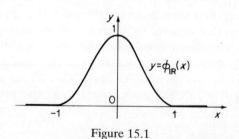

Figure 15.1

Mollifiers. Regularization theorem

Before proving the fundamental result of this section, the basis of the 'regularization technique' mentioned above, we need an important definition (in which by trivial extension of $u:\Omega \to \mathbb{R}$ to $\mathbb{R}^n$ we mean the function $\tilde{u}:\mathbb{R}^n \to \mathbb{R}$ such that $\tilde{u} \equiv u$ in Ω and $\tilde{u} \equiv 0$ in $\mathbb{R}^n \setminus \Omega$):

DEFINITION 15.1. Let $\Omega \subset \mathbb{R}^n$ be an open set, $u \in L^1(\Omega)$ and $\tilde{u}$ the trivial extension of u to $\mathbb{R}^n$. If ε is a positive real number and $\phi \in C_0^\infty(\mathbb{R}^n)$ has mean 1, the function

$$u_\varepsilon(x) = \int_{\mathbb{R}^n} \tilde{u}(x - \varepsilon y)\phi(y)\,dy. \tag{15.9}$$

is said to be the ε-*regularization* or ε-*mollifier* of u with respect to the function ϕ.

Let us remark that $u_\varepsilon(x)$ is a well defined function of $\mathbb{R}^n$ in $\mathbb{R}$ since, for every fixed x, $\tilde{u}(x - \varepsilon y)\phi(y) \in L^1(\mathbb{R}^n)$, because $\tilde{u} \in L^1(\mathbb{R}^n)$ and $\phi \in C_0^\infty(\mathbb{R}^n)$— we are in fact dealing with a *convolution*. We will henceforth denote the trivial extension of u also by u, so that (15.9) can be written

$$u_\varepsilon(x) = \int_{\mathbb{R}^n} u(x - \varepsilon y)\phi(y)\,dy, \tag{15.10}$$

and we will indicate by u_ε both this function as well as its restriction to Ω. Further, we will take as ϕ the function

$$\phi(x) = \frac{\phi_{\mathbb{R}^n}(x)}{\displaystyle\int_{\mathbb{R}^n} \phi_{\mathbb{R}^n}(x)\,dx}, \tag{15.11}$$

which does not conceptually imply a loss of generality of the following result:

THEOREM 15.2. (Regularization theorem): Let $\Omega \subset \mathbb{R}^n$ be an open set, $K \subset\subset \Omega$ a compact set and $u \in L^1(\Omega)$ such that $u|_{\Omega \setminus K} \equiv 0$. The following results follow:

(i) if $\varepsilon < d(K, \mathbb{R}^n \setminus \Omega)$ then $u_\varepsilon \in C_0^\infty(\Omega)$;
(ii) if $u \in C^k(\Omega)$, $k \in \mathbb{N}$, then $u_\varepsilon \to u$, as $\varepsilon \to 0$, in $\mathscr{D}^k(\Omega)$;
(iii) if $u \in L^p(\Omega)$, $p \in [1, +\infty[$, then $u_\varepsilon \to u$, as $\varepsilon \to 0$, in $L^p(\Omega)$.

Proof. To prove (i) we first show that, for fixed ε, for every $\alpha \in \mathbb{N}^n$, there exists and is continuous $D^\alpha u_\varepsilon$ and hence $u_\varepsilon \in C^\infty(\Omega)$. With this aim we transform (15.10) into an expression in which the variable x is not related to the function u (which we cannot differentiate) by means of the change of variable

$$z = x - \varepsilon y. \tag{15.12}$$

Since the Jacobian of this transformation is $(-1)^n \varepsilon^{-n}$ we can write

$$u_\varepsilon(x) = \varepsilon^{-n} \int_{\mathbb{R}^n} u(z)\phi\left(\frac{x-z}{\varepsilon}\right) dz, \qquad (15.13)$$

which satisfies the required conditions. Function $u_\varepsilon(x)$ is obviously continuous. Let us show that its first derivative with respect to x_i (i.e., $D^\alpha u_\varepsilon = \partial u_\varepsilon / \partial x_i$ with $\alpha = (\delta_{i1}, \ldots, \delta_{ii}, \ldots, \delta_{in})$, where δ_{ij} is the Kronecker symbol) exists. Indicating by $\{e_i\}$ the canonical base of $\mathbb{R}^n$ we can say that (for fixed x) this derivative, if it exists, is given by:

$$\frac{\partial u_\varepsilon(x)}{\partial x_i} = \lim_{h \to 0} \frac{u_\varepsilon(x + he_i) - u_\varepsilon(x)}{h}$$

$$= \lim_{h \to 0} \varepsilon^{-n} \int_{\mathbb{R}^n} u(z) \frac{\phi\left(\frac{x + he_i - z}{\varepsilon}\right) - \phi\left(\frac{x-z}{\varepsilon}\right)}{h} dz, \qquad (15.14)$$

where $h \in \mathbb{R}$. Now since $\phi \in C_0^\infty(\mathbb{R}^n)$ we can write

$$\lim_{h \to 0} u(z) \frac{\phi\left(\frac{x+he_i-z}{\varepsilon}\right) - \phi\left(\frac{x-z}{\varepsilon}\right)}{h} = \frac{1}{\varepsilon} \frac{\partial \phi}{\partial x_i}\left(\frac{x-z}{\varepsilon}\right) u(z) \qquad (15.15)$$

and therefore there exists $c > 0$ such that

$$\left| u(z) \frac{\phi\left(\frac{x+he_i-z}{\varepsilon}\right) - \phi\left(\frac{x-z}{\varepsilon}\right)}{h} \right| \leq c |u(z)|; \qquad (15.16)$$

we have then the conditions to apply Lebesgue's dominated convergence theorem (see, e.g., Rudin, 1970, p. 26) and write (15.14) in the form

$$\frac{\partial u_\varepsilon(x)}{\partial x_i} = \frac{1}{\varepsilon^n} \int_{\mathbb{R}^n} u(z) \lim_{h \to 0} \frac{\phi\left(\frac{x+he_i-z}{\varepsilon}\right) - \phi\left(\frac{x-z}{\varepsilon}\right)}{h} dz$$

$$= \frac{1}{\varepsilon^{n+1}} \int_{\mathbb{R}^n} u(z) \frac{\partial \phi}{\partial x_i}\left(\frac{x-z}{\varepsilon}\right) dz, \qquad (15.17)$$

which enables us to deduce the existence and the continuity of the derivative in question. Further, since

$$\frac{\partial \phi}{\partial x_i}\left(\frac{x-z}{\varepsilon}\right) = \psi\left(\frac{x-z}{\varepsilon}\right) \qquad (15.18)$$

with $\psi \in C_0^\infty(\mathbb{R}^n)$, this reasoning can be automatically extended to any other derivative: we have thus shown that $u_\varepsilon \in C^\infty(\mathbb{R}^n)$. Let us now find the support of u_ε. Equation (15.10) states that, for fixed x, in order that $u_\varepsilon(x) = 0$ it is sufficient that for every y such that $\|y\| < 1$ one has $x - \varepsilon_y \in \Omega \setminus K$, i.e., $u_\varepsilon(x)$ is

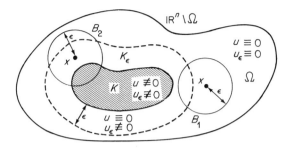

Figure 15.2

zero for values of x which cannot be 'transported' in K by the transformation $x \mapsto x - \varepsilon y$ with $\|y\| < 1$, or even more suggestively, with the fig. 15.2 in mind, we can say that $u_\varepsilon(x)$ is zero for points x for which there is a corresponding sphere of type B_1 and can be non-zero at points for which there is a corresponding sphere of type B_2. Now the points for which there is a sphere of the first type are characterized by the fact that $d(x, K) > \varepsilon$, and hence taking $\varepsilon < d(K, \mathbb{R}^n \setminus \Omega)$ ($d(K, \mathbb{R}^n \setminus \Omega) > 0$ as can be easily seen from the continuity of $d : \mathbb{R}^n \times \mathbb{R}^n \to \mathbb{R}$) we can say that there exists a compact set $K_2 \subset \Omega$ which contains the support of u_ε, i.e.: $u_\varepsilon \in C_0^\infty(\Omega)$.

We will next prove (ii). To say that $u_\varepsilon \to u$ in the topology of $\mathscr{D}^k(\Omega)$ means, according to definition 4.3, that there exists a compact set $K^* \subset \subset \Omega$ such that, for every ε, supp $u_\varepsilon \subset K^*$ and, for every $\alpha \in \mathbb{N}^n$ such that $|\alpha| \leq k$, $D^\alpha u_\varepsilon \to D^\alpha u$ uniformly in K^*. Now such a K^* exists: since we want ε to tend to zero we can assume $\varepsilon \leq \sigma = \frac{1}{2} d(K, \mathbb{R}^n \setminus \Omega)$ and hence take $K^* = K_\sigma = \{x \in \Omega : d(x, K) \leq \sigma\}$.

We then have to prove that, for fixed α such that $|\alpha| \leq k$,

$$\forall \delta > 0 \quad \exists \sigma \geq \bar{\varepsilon} > 0 \quad \forall x \in K^* \quad \varepsilon < \bar{\varepsilon} \Rightarrow |D^\alpha u_\varepsilon(x) - D^\alpha u(x)| \leq \delta. \tag{15.19}$$

We will now examine the difference $|D^\alpha u_\varepsilon(x) - D^\alpha u(x)|$. Given that ϕ has mean 1 we can write

$$|D^\alpha u_\varepsilon(x) - D^\alpha u(x)| = \left| \int_{\mathbb{R}^n} D^\alpha u(x - \varepsilon y) \phi(y) \, dy - D^\alpha u(x) \right|$$

$$= \left| \int_{\mathbb{R}^n} (D^\alpha u(x - \varepsilon y) - D^\alpha u(x)) \phi(y) \, dy \right|$$

$$\leq \int_{\mathbb{R}^n} |D^\alpha u(x - \varepsilon y) - D^\alpha u(x)| \, \phi(y) \, dy. \tag{15.20}$$

The function $D^\alpha u$ is continuous in Ω and therefore (from the Heine–Cantor theorem) it is uniformly continuous in $K^* = K_\sigma$, i.e.,

$$\forall \delta > 0 \quad \exists \sigma \geq \bar{\varepsilon} > 0 \quad \forall x', x'' \in K^* \|x' - x''\| < \bar{\varepsilon} \Rightarrow |D^\alpha u(x') - D^\alpha u(x'')| \leq \delta. \tag{15.21}$$

Now, for fixed x, $x - \varepsilon y$ varies in a sphere with centre x and radius ε and therefore, taking $\varepsilon < \bar{\varepsilon}/2$, (15.21) enables us to write

$$|D^\alpha u(x - \varepsilon y) - D^\alpha u(x)| \leq \delta. \tag{15.22}$$

Now from (15.20) we have

$$|D^\alpha u_\varepsilon(x) - D^\alpha u(x)| \leq \int_{\mathbb{R}^n} \delta\phi(y)\,dy = \delta, \tag{15.23}$$

and we have therefore proved (ii).

Let us next consider (iii). We want to prove that if $u \in L^p(\Omega)$ then

$$\forall \delta > 0 \quad \exists \bar{\varepsilon} > 0 \quad \varepsilon < \bar{\varepsilon} \Rightarrow \|u - u_\varepsilon\|_{L^p(\Omega)} \leq \delta. \tag{15.24}$$

With this aim we will first prove an inequality which is very important even on its own: *if $u \in L^p(\Omega)$ then*

$$\|u_\varepsilon\|_{L^p(\Omega)} \leq c \|u\|_{L^p(\Omega)} \tag{15.25}$$

with c constant independent of ε. In fact, by applying the Schwarz–Hölder inequality to (15.10) (and since the reasoning followed in proving part (i) enable us to reduce the integral on $\mathbb{R}^n$ to an integral on the unit sphere) we get

$$|u_\varepsilon(x)| \leq \left[\int_{\|y\|\leq 1} |u(x-\varepsilon y)|^p\,dy\right]^{1/p} \left[\int_{\|y\|\leq 1} |\phi(y)|^{p^*}\,dy\right]^{1/p^*}, \tag{15.26}$$

where p and p^* are conjugate exponents, and hence

$$\|u_\varepsilon\|^p_{L^p(\Omega)} = \|u_\varepsilon\|^p_{L^p(\mathbb{R}^n)} \leq \|\phi\|^p_{L^{p^*}(\mathbb{R}^n)} \int_{\mathbb{R}^n} \left[\int_{\|y\|\leq 1} |u(x-\varepsilon y)|^p\,dy\right] dx. \tag{15.27}$$

Applying to (15.27) the Fubini theorem and recalling that the Lebesgue integral is invariant with respect to translations, we get

$$\|u_\varepsilon\|^p_{L^p(\Omega)} \leq \|\phi\|^p_{L^{p^*}(\mathbb{R}^n)} \int_{\|y\|\leq 1} \left[\int_{\mathbb{R}^n} |u(x-\varepsilon y)|^p\,dx\right] dy$$

$$= \|\phi\|^p_{L^{p^*}(\mathbb{R}^n)} \|u\|^p_{L^p(\mathbb{R}^n)} \int_{\|y\|\leq 1} 1\,dy = c^p \|u\|^p_{L^p(\mathbb{R}^n)}, \tag{15.28}$$

which, recalling that u is identically zero outside K, is (15.25) with

$$c = \|\phi\|_{L^{p^*}(\mathbb{R}^n)} \cdot (\text{measure}(\{y : \|y\| \leq 1\}))^{1/p}. \tag{15.29}$$

We know that $C_0^0(\Omega)$ *is dense in* $L^p(\Omega)$, $p \in [1, +\infty[$, i.e. every function $u \in L^p(\Omega)$ can be approximated (in the topology of $L^p(\Omega)$) by a sequence of continuous functions with compact support contained in Ω:

$$\forall \delta > 0 \quad \forall u \in L^p(\Omega) \quad \exists v \in C_0^0(\Omega) \quad \|u - v\|_{L^p(\Omega)} \leq \delta. \tag{15.30}$$

(see e.g. Adams, 1975, p. 28; the result is not true for $p = +\infty$: a sequence of continuous functions which converges in $L^\infty(\Omega)$ has as limit a continuous function, while $L^\infty(\Omega) \neq C^0(\Omega)\ldots$).

Let us now take v so that it satisfies (15.30) and let us consider its ε-regularization. We can write

$$\|u - u_\varepsilon\|_{L^p(\Omega)} = \|u - v + v - v_\varepsilon + v_\varepsilon - u_\varepsilon\|_{L^p(\Omega)}$$
$$\leq \|u - v\|_{L^p(\Omega)} + \|v - v_\varepsilon\|_{L^p(\Omega)} + \|v_\varepsilon - u_\varepsilon\|_{L^p(\Omega)} \quad (15.31)$$

or, since the operation of regularization is linear,

$$\|u - u_\varepsilon\|_{L^p(\Omega)} \leq \|u - v\|_{L^p(\Omega)} + \|v - v_\varepsilon\|_{L^p(\Omega)} + \|(u - v)_\varepsilon\|_{L^p(\Omega)}, \quad (15.32)$$

or again, applying (15.25) to the function $u - v \in L^p(\Omega)$ and using (15.30),

$$\|u - u_\varepsilon\|_{L^p(\Omega)} \leq (1 + c)\delta + \|v - v_\varepsilon\|_{L^p(\Omega)}. \quad (15.33)$$

Now, part (ii) of the theorem ensures that there exists $\bar{\varepsilon}$ such that if $\varepsilon < \bar{\varepsilon}$ then

$$\|v - v_\varepsilon\|_{L^p(\Omega)} \leq \text{measure } (\text{supp}(v - v_\varepsilon)) \cdot \max_\Omega |v - v_\varepsilon| \leq \text{measure } (K_{\bar{\varepsilon}})\delta,$$
$$(15.34)$$

which, together with (15.33), proves the required result. ∎

In this proof we have used the fact that $C_0^0(\Omega)$ is dense in $L^p(\Omega)$. As a first application of this theorem we will prove a finer density result:

THEOREM 15.3. $C_0^\infty(\Omega)$ is dense in $L^p(\Omega)$, $p \in [1, +\infty[$.

Proof. We have to show that every function $v \in L^p(\Omega)$ can be approximated, in the norm of $L^p(\Omega)$, by a sequence of functions of $C_0^\infty(\Omega)$. Let then $v \in L^p(\Omega)$ and consider the sequence of compact sets

$$K_m = \{x \in \Omega : d(x, \mathbb{R}^n \setminus \Omega) \geq 1/m \text{ and } \|x\| \leq m\} \quad (15.35)$$

and the sequence of functions

$$v_m = v\chi_{K_m}, \quad (15.36)$$

where χ_{K_m} is the characteristic function of K_m. It is obvious that $v_m \in L^p(\Omega)$, $K_m \subset K_{m+1} \subset\subset \Omega$ and $\Omega = \bigcup_{m=1}^\infty K_m$ (a sequence of compact sets which satisfies such conditions is said to be 'invading'). The continuity of the integral with respect to the domain of integration enables us to conclude that $v_m \to v$ in $L^p(\Omega)$ since

$$\int_\Omega |v(x) - v_m(x)|^p \, dx = \int_{\Omega \setminus K_m} |v(x)|^p \, dx. \quad (15.37)$$

Now, considering for every fixed m the sequence $v_{m,\varepsilon}$ of the ε-regularizations of v_m, we can say, according to theorem 15.2 (which we can apply since v_m is zero outside K_m), that $v_{m,\varepsilon}$ has a subsequence which converges to v in $L^p(\Omega)$. Now, since for every m and every ε, $v_{m,\varepsilon} \in C_0^\infty(\Omega)$, the theorem is proved. ∎

15.2 PARTITION OF THE UNIT

Localization

The solution of a differential problem is often more regular than is required in the formulation of the problem itself. Thus, e.g., the Cauchy problem '$y' = y$ in $]0, 1[$, $y(0) = 1$' is well-posed if one has to find y in $C^1(]0, 1[) \cap C^0([0, 1])$, but one can easily see that the solution belongs to $C^\infty([0, 1])$. However, the study of the regularity of a function is not normally so simple, and, further, the function may be less regular globally than it is on the portion of its domain which is of particular interest for us. For this and other reasons it is often convenient to 'break up' the domain of the function and study its regularity in each part separately; obviously this 'break up' must satisfy conditions such that, e.g., from the regularity of each of the parts one can determine the global regularity.

Partition

The so-called *technique of partition of the unit* gives an answer to this problem and we will deal with it now.

DEFINITION 15.2. Let $K \subset \mathbb{R}^n$ be a compact set and $\{\Omega_i\}_{i=1,\ldots,m}$ an open covering of K (i.e. $\Omega_i \subset \mathbb{R}^n$ ($i = 1, \ldots, m$) is an open set and $K \subset \bigcup_{i=1}^m \Omega_i$; recall that from every open covering of K one can extract a finite covering). A *partition of the unit* of class C^∞ subordinate to the covering $\{\Omega_i\}$ is a family of functions $\{\phi_i\}_{i=1,\ldots,m}$ such that

(i) $\phi_i \in C_0^\infty(\Omega_i)$, $i = 1, \ldots, m$,
(ii) $0 \leq \phi_i \leq 1$, $i = 1, \ldots, m$,
(iii) $\sum_{i=1}^m \phi_i(x) \equiv 1$ in a neighbourhood of K.

The term 'partition of the unit' is obviously due to (iii). More generally one can consider, instead of K, an open set Ω (the covering will then not necessarily be finite) or a manifold—see Schwartz, 1966, p. 22 and De Rham, 1960, p. 4. The definition 15.2 shows us how to study the regularity of a function: one chooses a convenient covering $\{\Omega_i\}$ and one studies on Ω_i the regularity of $u = \sum_i u\varphi_i$ by means of the regularity of the function $u_i = u\phi_i$. We must, however, prove that, given any covering of K by means of a finite number of open sets, there exists a partitioning of the unit subordinate to it. To prove this we will use the following.

343

LEMMA. Let Ω be an open set and $K \subset\subset \Omega$. There exists $\psi \in C_0^\infty(\Omega)$ such that $0 \leq \psi(x) \leq 1$ in Ω and $\psi(x) \equiv 1$ in a neighbourhood of K, i.e.,

$$\exists \rho > 0 \quad \exists \psi \in C_0^\infty(\Omega) \quad 0 \leq \psi(x) \leq 1 \quad \text{and} \quad d(x, K) < \rho \Rightarrow \psi(x) = 1. \tag{15.38}$$

Proof. If $\delta < \tfrac{1}{2} d(K, \mathbb{R}^n \setminus \Omega)$ then the set $K_\delta = \{x \in \Omega : d(x, K) \leq \delta\}$ is a compact set. Let χ_{K_δ} be its characteristic function and let $\chi_{K_{\delta,\varepsilon}}$ be its ε-regularization. We can talk of the ε-regularization of χ_{K_δ} because this function is measurable (it is the characteristic function of a compact set, and compact sets are measurable) and is summable (since $\mu(K_\delta) < +\infty$): $\chi_{K_\delta} \in L^1(\Omega)$. We will prove that, by a convenient choice of ε, we can take $\psi = \chi_{K_{\delta,\varepsilon}}$. It is obvious that $0 \leq \chi_{K_{\delta,\varepsilon}}(x) \leq 1$, because of (15.10) together with

$$1 - \chi_{K_{\delta,\varepsilon}}(x) = \int_{\mathbb{R}^n} \phi(y)\,dy - \int_{\mathbb{R}^n} \chi_{K_\delta}(x - \varepsilon y)\phi(y)\,dy$$
$$= \int_{\mathbb{R}^n} [1 - \chi_{K_\delta}(x - \varepsilon y)]\phi(y)\,dy \geq 0. \tag{15.39}$$

Further, in order that $\chi_{K_{\delta,\varepsilon}} \in C_0^\infty(\Omega)$ we can take $\varepsilon < \delta/2$. We will see now that, with this choice of ε, $\chi_{K_{\delta,\varepsilon}}$ becomes identically equal to 1 in a neighbourhood of K, i.e. at least in $K_{\delta/2} = \{x \in \Omega : d(x, K) \leq \delta/2\}$. We can in fact write, for $x \in K_{\delta/2}$,

$$\chi_{K_{\delta,\varepsilon}}(x) = \int_{\|y\| \leq 1} \chi_{K_\delta}(x - \varepsilon y)\phi(y)\,dy = \int_{\|y\| \leq 1} \phi(y)\,dx = 1, \tag{15.40}$$

since if $\varepsilon < \delta/2$ and $\|y\| \leq 1$ then $x - \varepsilon y \in K_\delta$ and therefore $(x - \varepsilon y) = 1$. Now,

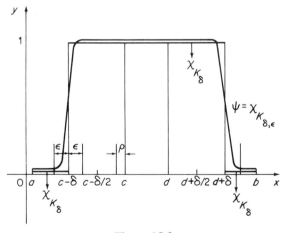

Figure 15.3

taking, e.g., $\rho = (\delta/2) - \varepsilon$ and $\psi = \chi_{K_{\delta,\varepsilon}}$, with $\varepsilon < \delta/2$, we get (15.38). Figure 15.3 shows a scheme which gives one an idea of what ψ is like in the one-dimensional case (in the figure $\Omega =]a, b[$, $K = [c, d]$, $K_\delta = [c - \delta, d + \delta]$ and $K_{\delta/2} = [c - \delta/2, c + \delta/2]$). ∎

THEOREM 15.4 (Theorem of the partition of the unit). *For every compact set $K \subset \mathbb{R}^n$ and for each of its finite open covering $\{\Omega_i\}_{i=1,\ldots,m}$ there exists a partition of the unit of class C^∞ subordinate to it.*

Proof. It is easy to verify the existence of a family $\{K_i\}_{i=1,\ldots,m}$ of compact sets such that $K \subset \bigcup_{i=1}^m K_i$ and $K_i \subset \Omega_i$, $i = 1, \ldots, m$. The lemma ensures the existence of m functions ψ_i such that

(j) $\psi_i \in C_0^\infty(\Omega)$, $i = 1, \ldots, m$,
(jj) $0 \leq \psi_i(x) \leq 1$, $i = 1, \ldots, m$,
(jjj) $\psi_i(x) \equiv 1$ in a neighbourhood of K_i, $i = 1, \ldots, m$.

Let us put then

$$\left. \begin{array}{l} \phi_1 = \psi_1 \\ \phi_i = \psi_i \prod_{j=1}^{i-1} (1 - \psi_j), \quad i = 2, \ldots, m, \end{array} \right\} \quad (15.41)$$

and let us show that the functions ϕ_i defined by this way constitute a partition of the unit subordinate to the covering $\{\Omega_i\}_{i=1,\ldots,m}$. It is evident that the condition (i) of definition 15.2 is satisfied since the ψ_i satisfy (j) and the operations considered in (15.41) neither alter the regularity nor increase the support of the functions. For analogous reasons (ii) is satisfied. As for (iii):

$$\sum_{i=1}^m \phi_i = \phi_1 + \sum_{i=2}^m \left[\psi_i \prod_{j=1}^{i-1} (1 - \psi_j) \right]$$

$$= \phi_1 - \sum_{i=2}^m \left[(-\psi_i) \prod_{j=1}^{i-1} (1 - \psi_j) \right]$$

$$= \phi_1 - \sum_{i=2}^m \left[(-1 + 1 - \psi_i) \prod_{j=1}^{i-1} (1 - \psi_j) \right]$$

$$= \phi_1 + \sum_{i=2}^m \prod_{j=1}^{i-1} (1 - \psi_j) - \sum_{i=2}^m \prod_{j=1}^i (1 - \psi_j)$$

$$= \phi_1 - \prod_{j=1}^m (1 - \psi_j) + (1 - \phi_1)$$

$$= 1 - \prod_{j=1}^m (1 - \psi_j), \quad (15.42)$$

and taking x in $\bigcup_{i=1}^m K_{\rho_i}$, where K_{ρ_i} is the neighbourhood of K_i in which ψ_i is identically equal to 1, we obtain

$$\prod_{j=1}^{m}(1-\psi_j)=0, \qquad (15.43)$$

since, for some value of j, $\psi_j(x)=1$. This proves the theorem. ∎

This proof is constructive and, further, gives us more information than the statement itself. The functions ϕ_i can be chosen so that in the neighbourhood of a given point they are all identically zero except one, which here will then be identically equally to 1, which in certain cases is convenient.

16
On the Regularity of the Open Sets

In the study of functional spaces, and in particular that of Sobolev spaces, we often find results which depend on the regularity of the boundary of the open sets on which the elements of the spaces are defined. Here we will present some characterizations of the open sets most frequently used in functional analysis, approaching the question from a more general point of view and recalling the concept of a manifold with and without boundary.

16.1 MANIFOLDS. OPEN SETS OF CLASS C^k AND $C^{k,\mu}$. CONE AND SEGMENT PROPERTIES

Local charts, atlases, and manifolds

We will first present the fundamental definitions about manifolds:

DEFINITION 16.1. Let M be a Hausdorff topological space. An n-dimensional local chart on M is a triplet (V, U, ϕ) where V is an open set in M, U is an open set of $\mathbb{R}^n$ and $\phi: V \to U$ is a homeomorphism. An n-dimensional atlas on M is a family $\mathscr{A} = \{(V_\alpha, U_\alpha, \phi_\alpha)\}_{\alpha \in A}$, of local charts of dimension n, such that $M = \bigcup_{\alpha \in A} V_\alpha$. Finally, an n-dimensional topological manifold is a pair $(M, \mathscr{A})$ where $\mathscr{A}$ is an n-dimensional atlas on M.

The dimension of a manifold is unique (and is therefore one of its characteristics) since, from a well known theorem due to Brouwer (see, e.g., Aleksandrov, 1957, p. 213; Hurewicz–Wallman, 1948, p. 97), $\mathbb{R}^n$ and $\mathbb{R}^m$ are homeomorphic iff $n = m$. The manifolds defined above are called real manifolds or manifolds modelled on $\mathbb{R}^n$; more generally, we can consider manifolds modelled on any Banach space B (see, e.g., Lang, 1962, p. 16): in particular, taking $B = \mathbb{C}^n$ we get an n-dimensional complex manifold (see, e.g., Sorani, 1969, p. 79). Topological manifolds are also known as manifolds of class C^0 and the space M, the 'support' of the structure of manifold, is said to be locally Euclidean.

If $x \in V_\alpha$ we will call local coordinates of x the real numbers $\phi_{\alpha_1}(x), \ldots, \phi_{\alpha_n}(x)$, where ϕ_{α_i} ($i = 1, \ldots, n$) are the components of ϕ_α. If for a pair $(\alpha, \beta) \in A^2$ of distinct indices $V_\alpha \cap V_\beta \neq \emptyset$ we can represent $x \in V_\alpha \cap V_\beta$ by two systems of local coordinates, and one naturally asks how they are related: it follows immediately that the function $\phi_\beta \circ \phi_\alpha^{-1}$, responsible for the change of local coordinates, is (modulo restrictions) a homeomorphism between $\phi_\alpha(V_\alpha \cap V_\beta)$ and $\phi_\beta(V_\alpha \cap V_\beta)$.... Now $\phi_\beta \circ \phi_\alpha^{-1}$ is a function between two open sets of $\mathbb{R}^n$ and one can therefore talk of its differentiability—this leads to the following definition (where ω is such that $k < +\infty < \omega$, for every $k \in \mathbb{N}$).

DEFINITION 16.2. Let M be a Hausdorff topological space. We will say that an n-dimensional atlas $\mathscr{A} = \{(V_\alpha, U_\alpha, \phi_\alpha)\}_{\alpha \in A}$ on M is an *atlas of class C^k*, $k \in \mathbb{N} \cup \{+\infty, \omega\}$ if the following condition of compatibility between local charts is satisfied: '$\forall \alpha, \beta \in A$ if $V_\alpha \cap V_\beta \neq \emptyset$ then $\phi_\beta \circ \phi_\alpha^{-1}$ is (modulo restrictions) a diffeomorphism of class C^k between $\phi_\alpha(V_\alpha \cap V_\beta)$ and $\phi_\beta(V_\alpha \cap V_\beta)$'. We define as an *$n$-dimensional manifold of class C^k* a pair $(M, \mathscr{A})$ where $\mathscr{A}$ is an n-dimensional atlas of class C^k on M.

This definition is the same as the previous one for $k = 0$ (with some etymological incoherence, homeomorphisms are here referred to as diffeomorphisms of class C^0...).

On a single topological space one can have more than one atlas of class C^k: two atlases of class C^k, $\mathscr{A}$ and $\mathscr{B}$, are said to be *(k)-equivalent* (or to induce on M the same *(k)-differential structure*) if the atlas $\mathscr{A} \cup \mathscr{B}$ is of class C^k. This is clearly an equivalence relationship and therefore the definition 16.2 introduces an element of an equivalence class; in order to avoid a further passage through the quotient one can define manifold of class C^k as the pair $(M, \bar{\mathscr{A}})$ where $\bar{\mathscr{A}}$ is the union of all atlases of class C^k which are (k)-equivalent to a given atlas $\mathscr{A} : \bar{\mathscr{A}}$ is therefore the only maximal extension of $\mathscr{A}$.

Manifolds of class C^k with $1 \leq k \leq +\infty$ are generically called *differentiable manifolds* and those of class C^ω are called *analytic manifolds* (and the atlases of class C^ω are said to be analytic).

Henceforth we will refer both to M and to $(M, \mathscr{A})$ as manifolds, wherever this does not lead to any misunderstanding.

The simplest example of an n-dimensional manifold is that in which M is an open set $\Omega \subset \mathbb{R}^n$ provided with the topology induced by $\mathbb{R}^n$ and the atlas, called for obvious reasons the *canonical atlas*, consists of the single chart $(V = \Omega, U = \Omega, \phi = \text{identity})$. This is an analytic manifold since the problem of the compatibility between local charts does not arise. Further, this manifold is said to be *immersed* since its 'support' is a subset of a Euclidean space; besides, a vast class of differentiable manifolds can be considered to be immersed in a convenient Euclidean space, namely: a result due to Whitney (see, e.g., Auslander–Mackenzie, 1963, p. 116) states that *if $(M, \mathscr{A})$ is an n-dimensional differentiable manifold and M satisfies the second axiom of numerability then M is diffeomorphic to a closed subset of $\mathbb{R}^{2n+1}$*.

Manifolds with boundary

From what we have said above it follows that it is a useless complication to consider an open set Ω as a manifold; this cannot, however, be said, in general, for the boundary Γ of Ω. Now, there is a close connection between Γ and Ω from the topological point of view, and one naturally asks whether, if we consider Ω and (if possible) Γ as manifolds, this connection is maintained in some sense. The answer to this question leads us to the consideration of manifolds with boundary. We will indicate by $\mathring{\mathbb{R}}_+^n$ the semi-space $\{x \in \mathbb{R}^n : x_n > 0\}$ and by $\partial \mathbb{R}_+^n$ the hyperplane $\mathbb{R}^{n-1} \times \{0\}$ ($\mathbb{R}_+^n = \{x \in \mathbb{R}^n : x_n \geq 0\} \neq (\mathbb{R}_+)^n$!).

DEFINITION 16.3. Let M be a Hausdorff topological space. $(M, \mathcal{A})$ is said to be an *n-dimensional manifold of class C^k with boundary*, $k \in \mathbb{N} \cup \{+\infty, \omega\}$, if $\mathcal{A} = \{(V_\alpha, U_\alpha, \phi_\alpha)\}_{\alpha \in A}$ is a family (which we will again refer to as an atlas of class C^k) of triplets (which we will again call local charts), made up of an open set V_α of M, an open set U_α of $\mathbb{R}_+^n$ (with the topology induced by $\mathbb{R}^n$) and a homeomorphism $\phi_\alpha : V_\alpha \to U_\alpha$, such that $\bigcup_{\alpha \in A} V_\alpha = M$ and $\forall \alpha, \beta \in A$ if $V_\alpha \cap V_\beta \neq \emptyset$ then $\phi_\beta \circ \phi_\alpha^{-1}$ is a diffeomorphism of class C^k between $\phi_\alpha(V_\alpha \cap V_\beta)$ and $\phi_\beta(V_\alpha \cap V_\beta)$. If $x \in M$ has a neighbourhood V such that there exists a local chart $(V_\alpha, U_\alpha, \phi_\alpha)$ such that $V \subset V_\alpha$ and $\phi_\alpha(V) \subset \mathring{\mathbb{R}}_+^n$, x is said to be an *internal point* of M and the set $\mathring{M}$ of points internal to M is said to be the *interior* of M. If $x \in M$ has a neighbourhood V' such that there exists a chart $(V_{\alpha'}, U_{\alpha'}, \phi_{\alpha'})$ with $V' \subset V_{\alpha'}$, and $\phi_{\alpha'}(x) \in \partial \mathbb{R}_+^n$, x is said to be a *boundary point* of M and the set ∂M of all boundary points of M is called the *boundary* of M.

Here, as in definition 16.2, the function that we require to be a diffeomorphism is the restriction to $\phi_\alpha(V_\alpha \cap V_\beta)$ of the function $\phi_\beta \circ \phi_\alpha^{-1}$. Further, since $\phi_\alpha(V_\alpha \cap V_\beta)$ and $\phi_\beta(V_\alpha \cap V_\beta)$ are open sets of $\mathbb{R}_+^n$ but not necessarily of $\mathbb{R}^n$, the concept of diffeomorphism must be clarified: *if $\mathcal{O}_1$ and $\mathcal{O}_2$ are open sets of $\mathbb{R}_+^n$, $f : \mathcal{O}_1 \to \mathcal{O}_2$ is said to be a diffeomorphism of class C^k if there exists $F : \mathbb{R}^n \to \mathbb{R}$ such that $f : F|_{\mathcal{O}_1}$ and F is a diffeomorphism of class C^k.*

In the context of the considerations following definition 16.2 we can require that $\mathcal{A}$ be a maximal atlas of class C^k. Further, a manifold in the sense of definition 16.2 is also a manifold with boundary (which is empty) since $\mathbb{R}^n$ and $\mathring{\mathbb{R}}_+^n$ are (analytically) diffeomorphic. Manifolds in the sense of definition 16.2 are also called manifolds without boundary—we will not use this nomenclature. It is evident that $\mathring{M}$ and ∂M are manifolds of class C^k and of dimensions n and $n-1$, respectively; also it is obvious that $\partial M = M \setminus \mathring{M}$.

Open sets and their boundaries

Returning to the question which led to the preceding definition, we can say that *if $\Omega \subset \mathbb{R}^n$ is an open set whose boundary Γ is an $(n-1)$-dimensional* manifold then $\bar{\Omega} = \Omega \cup \Gamma$ is a manifold with boundary $\partial \bar{\Omega} \subset \Gamma$ and, further, that *if the closure $\bar{\Omega}$ of an open set $\Omega \subset \mathbb{R}^n$ is an n-dimensional manifold with boundary then $\partial \bar{\Omega} \subset \Gamma = \bar{\Omega} \setminus \Omega$*. In order to look for conditions on Ω such that $\partial \bar{\Omega} = \Gamma$ we arrive at the following definition:

DEFINITION 16.4. An open set $\Omega \subset \mathbb{R}^n$ is said to have a *boundary Γ of class C^k*, $k \in \mathbb{N} \cup \{+\infty, \omega\}$, or more simply Ω is said to be an *open set of class C^k*, if the following condition is satisfied:

(c) for every $x \in \Gamma$ there exist two open sets $V \subset \mathbb{R}^n$ and $U \subset \mathbb{R}^n$ and a diffeomorphism $\phi : \mathbb{R}^n \to \mathbb{R}^n$ of class C^k such that $x \in V$, $\phi(V \cap \bar{\Omega}) = U \cap \mathbb{R}^n_+$ and $\phi(x) \in \partial \mathbb{R}^n_+$.

It is obvious that if Ω is an open set of class C^k then $\bar{\Omega}$ is a manifold with boundary of class C^k (immersed) and $\partial \bar{\Omega} = \Gamma$. If all the diffeomorphisms considered in (c) are of class $C^{k,\mu}$ (i.e., ϕ and ϕ^{-1} are of class $C^{k,\mu}$) one talks of *open sets of class $C^{k,\mu}$* —in the particular case where $k = 0$ one talks of μ-*Hölder continuous open sets*, and if $\mu = 1$ one talks of *Lipschitz continuous open sets*. The following theorem gives us a characterization of open sets of class C^k which is frequently used in analysis; essentially it reinterprets the identity $\partial \bar{\Omega} = \Gamma$ through the identity $\Omega = \overset{\circ}{\bar{\Omega}}$ which can be taken as saying that Ω *is locally on one side only with respect to its boundary*.

THEOREM 16.1. *An open set $\Omega \subset \mathbb{R}^n$ is of class C^k iff the following conditions are satisfied:*

(c1) $\Omega = \overset{\circ}{\bar{\Omega}}$,
(c2) for every $x \in \bar{\Omega}$ there exist two open sets $V \subset \mathbb{R}^n$ and $U \subset \mathbb{R}^n$ and a diffeomorphism $\phi : \mathbb{R}^n \to \mathbb{R}^n$ of class C^k such that $x \in V$ and $\phi(V \cap \bar{\Omega}) = U \cap \mathbb{R}^n_+$.

Proof. We will first show that the conditions (c1), (c2) are necessary. We have already said that (c) implies (c2), which tells us that $\bar{\Omega}$ is a manifold with boundary of class C^k. To see that (c) implies (c1) we will use the technique of *reductio ad absurdum*. Since the inclusion $\Omega \subset \overset{\circ}{\bar{\Omega}}$ is always true, we will assume that $\overset{\circ}{\bar{\Omega}} \not\subset \Omega$, i.e., we assume that there exists $x \in \overset{\circ}{\bar{\Omega}} \setminus \Omega$. It follows that $x \in \bar{\Omega} \setminus \Omega = \Gamma$; let (V, U, ϕ) be the local chart of which under these conditions (c) ensures us the existence. Now from $x \in \bar{\Omega} \cap V$ and from $x \in \overset{\circ}{\bar{\Omega}}$ it follows that x has an open neighbourhood W such that $W \subset \bar{\Omega} \cap V$; then $\phi(W)$ is an open neighbourhood (in $\mathbb{R}^n$) of $\phi(x)$ such that $\phi(W) \subset \mathbb{R}^n_+$, from which it follows that $\phi(x) \notin \partial \mathbb{R}^n_+$, which is absurd.

We will next show that (c1) and (c2) are sufficient. Let $x \in \partial \Omega$ and let (V, U, ϕ) be the local chart whose existence is ensured by (c2). It is obvious that we only have to prove that $\phi(x) \in \partial \mathbb{R}^n_+$ and, since if $x \in \bar{\Omega} \cap V$ then $\phi(x) \in \mathbb{R}^n_+ \cap U$, we need only prove that $\phi(x) \notin \overset{\circ}{\mathbb{R}}^n_+ \cap U$. If $\phi(x) \in \overset{\circ}{\mathbb{R}}^n_+ \cap U$ then $\phi(x)$ has an open neighbourhood $W \subset U \cap \mathbb{R}^n_+$ and therefore, from $x \in$

$\phi^{-1}(W) \subset \bar{\Omega}$, it follows that $x \in \overset{\circ}{\Omega}$ since ϕ^{-1} is continuous, and, from (c1), that $x \in \Omega$, which is absurd. ∎

In the literature the open sets U of condition (c) are often more precisely defined so as to simplify some of the calculations; of these definitions, which are more elaborate than ours but equivalent to them, we will present that given in Nečas, 1967, p. 14 for a bounded open set (for other definitions of this type, see Prodi, 1956, p. 37, and Lions–Magenes, 1961b, p. 56):

DEFINITION 16.5. Let $\Omega \subset \mathbb{R}^n$ be a bounded open set. Ω is said to be an *open set of class* $C^{k,\mu}$ if there exist two real numbers $b > 0$ and $h > 0$, m systems of coordinate axes $(x_{\alpha_1}, \ldots, x_{\alpha_{n-1}}, x_{\alpha_n}) \equiv (x'_\alpha, x_{\alpha_n})$, $\alpha = 1, \ldots, m$, and m real functions ϕ_α such that, with $\Delta_\alpha = \{x'_\alpha \in \mathbb{R}^{n-1} = |x_{\alpha_i}| < b,\ i = 1, \ldots, n-1\}$, $\phi_\alpha \in C^{k,\mu}(\bar{\Delta}_\alpha)$ and with $\Gamma_\alpha = \{(x'_\alpha, x_{\alpha_n}) \in \mathbb{R}^n : x'_\alpha \in \Delta_\alpha$ and $\phi_\alpha(x'_\alpha) = x_{\alpha_n}\}$, $V_\alpha^+ = \{(x'_\alpha, x_{\alpha_n}) \in \mathbb{R}^n : x'_\alpha \in \Delta_\alpha$ and $\phi_\alpha(x'_\alpha) < x_{\alpha_n} < \phi_\alpha(x'_\alpha) + h\}$ and $V_\alpha^- = \{(x'_\alpha, x_{\alpha_n}) \in \mathbb{R}^n : x'_\alpha \in \Delta_\alpha$ and $\phi_\alpha(x'_\alpha) - h < x_{\alpha_n} < \phi_\alpha(x'_\alpha)\}$ we have $\Gamma = \bigcup_{\alpha=1}^m \Gamma_\alpha$ and, for every α, $V_\alpha^+ \subset \Omega$ and $V_\alpha^- \subset \mathbb{R}^n \setminus \bar{\Omega}$.

This definition states essentially that the boundary Γ of Ω can be locally represented as the graph of a function of class $C^{k,\mu}$ and that, again locally, Ω is the epigraph of this function. Figure 16.1 shows the geometrical meaning of the conditions which we have imposed.

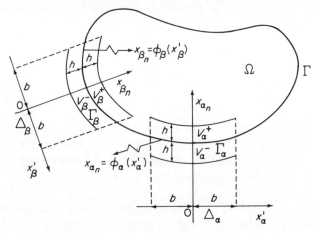

Figure 16.1

The segment and cone properties

We will now present two concepts of regularity of open sets which are quite different from that considered in definition 16.5. We will denote by $C_x(\lambda, \omega, \mathbf{d})$ the interior of a cone of revolution with the vertex at x, height λ, opening ω and the axis pointing towards the versor $\mathbf{d}$; we recall that an open

351

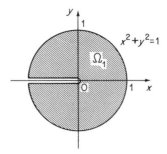

Figure 16.2

covering of a set $C \subset \mathbb{R}^n$ is said to be *locally finite* if every point of C has a neighbourhood which intersects with only a finite number of covering open sets.

DEFINITION 16.6. An open set $\Omega \subset \mathbb{R}^n$ is said to have the *cone property* if

(k) $\quad \exists \lambda > 0 \quad \exists \omega > 0 \quad \forall x \in \Gamma \quad \exists \mathbf{d} \ C_x(\lambda, \omega, \mathbf{d}) \subset \Omega$

(for obvious reasons, Ω is also said to have the *property of internal cone*; Ω is said to have the *property of external cone* if $\mathbb{R}^n \setminus \bar{\Omega}$ satisfies (k)). The set Ω is said to have the *segment property* if

(s) Γ has a locally finite covering $\{\Gamma_\alpha\}_{\alpha \in A}$ and there exists a family $\{y_\alpha\}_{\alpha \in A}$ of points of $\mathbb{R}^n$ such that $\forall x \in \bar{\Omega} \cap \Gamma_\alpha \ \forall t \in]0, 1[\ x + ty_\alpha \in \Omega$.

We will briefly examine the relationship that exists between the different types of regularity of open sets which we have presented.

It is obvious that if Ω is of class C^1 then Ω satisfies (k) and (s), while the converse is false, as can be seen from the open set $\Omega_1 = \{(x, y) \in \mathbb{R}^2 : x^2 + y^2 < 1\} \setminus \{(x, 0) \in \mathbb{R}^2 : x \leq 0\}$ (see fig. 16.2), which satisfies (k) and (s) but not (c1) and therefore is not even of class C^0. Further, the *snowflake curve* or *Koch's curve* (see fig. 16.3 for the first steps in its construction) enables us to construct an open set of class C^0 which satisfies neither (k) nor (s).

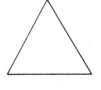  

Figure 16.3

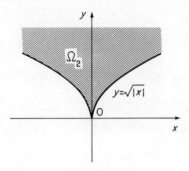

Figure 16.4

The Hölder continuity nature of Ω does not imply that it has the cone property, as can be seen from the open set $\Omega_2 = \{(x, y) \in \mathbb{R}^2 : y > \sqrt{|x|}\}$ (see fig. 16.4); nor does this, even if associated with the class C^0, imply that Ω is Hölder continuous, as can be seen from the open set $\Omega_3 = \{(x, y) \in \mathbb{R}^2 : |x| < 1$ and $y < 1/|\log|x||\}$ (see fig. 16.5).

The relationship between being Lipschitz continuous and having the cone property are, however, more interesting, as can be seen from the following result (see Gagliardo, 1958, p. 104):

THEOREM 16.2. Let $\Omega \subset \mathbb{R}^n$ be a bounded open set. If Ω has a Lipschitz continuous boundary then Ω has the cone property. If Ω has the cone property then Ω can be represented in the form $\Omega = \bigcup_{i=1}^{m} \Omega_i$ with the Ω_i being Lipschitz continuous.

For a further discussion of open sets with the cone property see Campanato, 1964, p. 138; Pucci, 1964, p. 313 and Agmon, 1965, pp. 11 et seq.

Let us note, further, that if Ω has a Lipschitz continuous boundary then there exists, almost everywhere on Γ, the normal (see e.g. Nečas, 1967, p. 88).

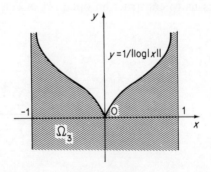

Figure 16.5

16.2 DISTRIBUTIONS ON A MANIFOLD. SPACES $L^p(\Gamma)$

Functions of class C^k on a manifold

The concept of continuous function was originally introduced in connection with real functions of one real variable and later extended, in a very natural manner, to functions between any two topological spaces. The possibility of making this extension shows that the concept of continuity is linked exclusively to the topological structure of $\mathbb{R}$, and not to other structures defined on $\mathbb{R}$ (and perhaps hidden by the flourising richness of the structure of the complete ordered field $\mathbb{R}$).

Instead, the concept of differentiable function, also first introduced for real functions of a real variable, cannot, however, be extended so naturally to functions defined on any topological space. In fact, in order to define the derivative of a function $f: \mathbb{R} \to \mathbb{R}$ one has to use not only the topological structure of $\mathbb{R}$ but also its algebraic structure: this is evident when one thinks of forming difference ratios.

There are essentially two ways of extending (with respect to spaces on which the functions are defined) the concept of derivative. The first consists of constructing a theory of differentiation analogous to that developed for functions defined on $\mathbb{R}^n$ and applicable to functions defined on some topological vector spaces; one thus obtains the concept of the Fréchet derivative, the Gâteaux derivative, etc. (for a survey of the different concepts of derivative see Averbukh–Smolyanov, 1968). The second way, which we will deal with briefly here, uses the concept of a differentiable manifold, and is one of the reasons why such a structure has been introduced.

DEFINITION 16.7. Let $(M, \mathcal{A})$ be an n-dimensional manifold of class C^k, $k \in \mathbb{N} \cup \{+\infty, \omega\}$. $f: M \to \mathbb{R}$ is said to be a *function of class* C^r, $0 \leq r \leq k$, if for every chart $(V_\alpha, U_\alpha, \phi_\alpha)$ in $\mathcal{A} = \{(V_\alpha, U_\alpha, \phi_\alpha)\}_{\alpha \in A}$ the function $f_\alpha = f \circ \phi_\alpha^{-1}$ is of class C^r in U_α.

We note that different structures of differentiable manifold can be defined on M, and a function can be of class C^r relative to one of them but not to another: we will assume that a given structure has been fixed for once and for all through a precise atlas.

We have then related the concept of differentiability on the topological space M to differentiability in the usual sense of $\mathbb{R}^n$ by using the structure of manifold defined on M, i.e., by exploiting the possibility of representing M in terms of local charts. This definition further suggests that we 'represent' the functions $f: M \to \mathbb{R}$ themselves by means of families of functions defined on open sets of $\mathbb{R}^n$. This is possible because, given $f: M \to \mathbb{R}$, the family $\{f_\alpha = f \circ \phi_\alpha^{-1}: U_\alpha \to \mathbb{R}\}_{\alpha \in A}$ is automatically defined and, as one can easily check, satisfies

$$\forall \alpha, \beta \in A \quad \forall x \in \phi_\alpha(V_\alpha \cap V_\beta) \quad f_\alpha(x) = f_\beta \circ (\phi_\beta \circ \phi_\alpha^{-1})(x); \quad (16.1)$$

and, conversely, given a family $\{f_\alpha : U_\alpha \to \mathbb{R}\}_{\alpha \in A}$ which satisfies the compatibility condition (16.1), there exists a unique function $f = M \to \mathbb{R}$ such that $f_\alpha = f \circ \phi_\alpha^{-1}$ in U_α: in effect for every $y \in M$ there exists $\alpha \in A$ such that $y \in V_\alpha$ and putting $f(y) \triangleq f_\alpha \circ \phi_\alpha(y)$ we have, from (16.1), that the function $f = \{f_\alpha\}_{\alpha \in A}$ is well defined.

Distributions on a manifold

At this point it is very natural to give the following definition:

DEFINITION 16.8. Let $(M, \mathscr{A})$ be an n-dimensional manifold of class C^∞ or C^ω. A family $u = \{u_\alpha\}_{\alpha \in A}$ of distributions $u_\alpha \in \mathscr{D}'(U_\alpha)$ is said to be a *distribution* on $(M, \mathscr{A})$ if

$$\forall \alpha, \beta \in A \quad u_\alpha = u_\beta \circ (\phi_\beta \circ \phi_\alpha^{-1}) \quad \text{in } \phi_\alpha(V_\alpha \cap V_\beta). \tag{16.2}$$

We will denote by $\mathscr{D}'(M)$ the space of distributions on $(M, \mathscr{A})$.

If $M = \Omega \subset \mathbb{R}^n$ and $\mathscr{A}$ is the canonical atlas on the open set Ω then the space $\mathscr{D}'(M)$ is identical to the space $\mathscr{D}'(\Omega)$ considered in the first part. The space $\mathscr{D}'(M)$ can be considered to be the dual of a space $\mathscr{D}(M)$ of certain differential forms on M (see, e.g., De Rham, 1960, pp. 43 and 47) or of certain functions defined on M (see, e.g., Choquet–Bruhat, 1973, p. 43). Further, if M is a manifold of class C^k, $k < +\infty$, one can define in an entirely analogous manner the *distributions of order k* on M and the space $\mathscr{D}'^k(M)$. In particular if $k = 0$ one can define the *Radon measures* on M, and therefore the space $L^p(M)$; a serious problem arises here: since continuous functions do not necessarily transform sets of measure zero into sets of measure zero, it is not sure that two families that are equal a.e. define the same function on M—the homeomorphisms ϕ_α must be at least bilipschitz continuous, i.e. the manifold must be of class $C^{0,1}$.

Functions of class L^p on a manifold

The space $L^p(M)$ can be defined more directly, in a manner analogous to definition 16.7. Here we will introduce it in the particular case which is of greatest interest to us—i.e., that in which M is the boundary Γ (or $\partial \bar{\Omega}$, or simply $\partial \Omega$) of a bounded open set $\Omega \subset \mathbb{R}^n$ of class $C^{0,1}$. If Ω is bounded then Γ is a compact set and hence from the family of open sets V of the condition (c) (definition 16.4) we can extract a finite family with the same characteristics. Let $\{V_\alpha\}_{\alpha = 1,\ldots,m}$ be this family and let $\{\phi_\alpha\}_{\alpha = 1,\ldots,m}$ be the associated family of diffeomorphisms. With the family $\{V_\alpha\}_{\alpha = 1,\ldots,m}$ we associate, besides, a partitioning of the unit (see 15.2). Let $\{\xi_\alpha\}_{\alpha = 1,\ldots,m}$ be this partition. Let us put $\Gamma_\alpha = \Gamma \cap V_\alpha$, $\beta_\alpha = \phi_\alpha|_{\Gamma_\alpha}$ and $\theta_\alpha = \xi_\alpha|_{\Gamma_\alpha}$: this gives us a covering of Γ through the set Γ_α relatively open in Γ and a partition of the unit (of the same class as Γ, unless Γ is analytic) subordinate to it. With these notations we introduce

DEFINITION 16.9. Let $f : \Gamma \to \mathbb{R}$ be a function defined on the boundary Γ of a bounded open set of class $C^{0,1}$. We say that $f \in L^p(\Gamma)$, $p \in [1, +\infty]$, if $f_\alpha = f \circ \beta_\alpha^{-1} \in L^p(\beta_\alpha(\Gamma_\alpha))$ for every $\alpha = 1, \ldots, m$. We provide $L^p(\Gamma)$ with the norm

$$\|f\|_{L^p(\Gamma)} = \left(\sum_{\alpha=1}^m \|(\theta_\alpha f) \circ \beta_\alpha^{-1}\|_{L^p(\beta_\alpha(\Gamma_\alpha))}^p \right)^{1/p} \tag{16.3}$$

(with the usual modification for $p = +\infty$).

The functions of $L^p(\Gamma)$ cannot be considered to be restrictions to Γ of functions of $L^p(\mathbb{R}^n)$, since Γ has zero n-dimensional measure. We naturally ask whether one can introduce on Γ a measure, say σ, such that by constructing from σ a theory of integration one obtains once more the concept of a summable pth power function on Γ; the answer is yes—however here we will merely define the measure without developing the theory of integration (we will use the same notations as in definition 16.9):

DEFINITION 16.10. Let $\Omega \subset \mathbb{R}^n$ be a bounded open set with a boundary Γ of class $C^{0,1}$. $T \subset \Gamma$ is said to be σ-*measurable* if, for every $\alpha = 1, \ldots, m$, $\beta_\alpha(T \cap \Gamma_\alpha)$ is a measurable set of $\mathbb{R}^{n-1}$; and the number

$$\sigma(T) = \sum_{\alpha=1}^m \mu[\beta_\alpha(T \cap \Gamma_\alpha)] \tag{16.4}$$

is called the *measure* of T (μ is the ordinary $(n-1)$-dimensional Lebesgue measure).

The numerical value of the measure $\sigma(T)$ of T (also called, for obvious reasons, the *superficial measure* of T) depends in general on the atlas used for describing Γ, but the measurability of T itself does not depend on the atlas—the class of sets of zero measure is also invariant.

We will now present a definition of $L^p(\Gamma)$ based on the definition 16.5, the notations of which we will also adopt:

DEFINITION 16.11. Let $\Omega \subset \mathbb{R}^n$ be a bounded open set of class $C^{0,1}$ and let $f: \Gamma \to \mathbb{R}$ be a function defined a.e. (with respect to Γ) on the boundary Γ of Ω. Under these conditions, for $\alpha = 1, \ldots, m$, $f(x'_\alpha, \phi_\alpha(x'_\alpha))$ is defined almost everywhere in Δ_α and if $f(x'_\alpha, \phi_\alpha(x'_\alpha)) \in L^p(\Delta_\alpha)$, $p \in [1, +\infty]$, we say that $f \in L^p(\Gamma)$. On $L^p(\Gamma)$ we introduce the norm

$$\|f\|_{L^p(\Gamma)} = \left[\sum_{\alpha=1}^m \int_{\Delta_\alpha} |f(x'_\alpha, \phi_\alpha(x'_\alpha))|^p \, dx'_\alpha \right]^{1/p} \tag{16.5}$$

(with the usual modification for $p = +\infty$).

With the norm (16.5) the space $L^p(\Gamma)$ is a Banach space (see Nečas, 1967, p. 82). The numerical value of the norm of $f \in L^p(\Gamma)$ depends on the particular description of Γ, but by changing the family of coordinate axes and of bilipschitz continuous functions which enable us to give this description we obtain equivalent norms (see Nečas, 1967, p. 119). Under the same conditions as of definition 16.11 we can define the surface integral of a function $f \in L^1(\Gamma)$ by means of the relation

$$\int_\Gamma f \, d\sigma = \sum_{\alpha=1}^m \int_{\Delta_\alpha} f(x'_\alpha, \phi_\alpha(x'_\alpha)) \theta_\alpha(x'_\alpha, \phi_\alpha(x'_\alpha)) \left[1 + \sum_{i=1}^{m-1} (\partial \phi_\alpha / \partial x_i)^2 \right]^{1/2} dx'_\alpha \tag{16.6}$$

(where the θ_α are functions of a partitioning of the unit subordinate to Γ_α); the value of the surface integral is independent both of the charts as well as that of the partition of the unit associated with them (see Nečas, 1967, p. 122). Finally, the *surface element*

$$d\sigma_\alpha = \left[1 + \sum_{i=1}^{n-1} (\partial \phi_\alpha / \partial x_i)^2 \right]^{1/2} dx'_\alpha, \tag{16.7}$$

starting from which one can construct a theory of integration on Γ, is well defined since the derivatives of the bilipschitz continuous functions ϕ_α exist and are bounded almost everywhere.

17
The Maximum Principle and its Applications

17.1 INTRODUCTION

In this chapter we will deal, roughly, with results of the following type: *if a function u satisfies the inequality* $L(u) \geq 0$ *in an open set* Ω, *where L is an appropriate differential operator, then u does not take its maximum in* Ω, in which case u is said to satisfy a *maximum principle*. Naturally, one can consider the symmetric concept of *minimum principle*, associated with the inequality $L(u) \leq 0$. Working with the concept of minimum or that of maximum, as we do, is mainly a question of taste: if u satisfies a maximum principle then $-u$ satisfies a minimum principle and the results can therefore automatically be transported from one case to the other one.

The great interest of this type of results lies, on the one hand, in the fact that frequently the maxima and the minima of the solutions of differential problems have a profound significance, and, on the other, in their usefulness in proving the uniqueness of the solutions of some problems. Here we will deal exclusively with second order elliptic operators—for higher order elliptic operators see, e.g., Agmon, 1960 and for parabolic and hyperbolic operators see, e.g., Protter–Weinberger, 1967, chs 3 and 4.

17.2 MAXIMUM PRINCIPLE IN $\mathbb{R}$

Let us first consider the simplest possible case. It is well known that if a function $u: \,]a, b[\to \mathbb{R}$ of class C^2 in $]a, b[$ takes a maximum (local or global) in $c \in \,]a, b[$ then $u'(c) = 0$ and $u''(c) \leq 0$. From this we conclude that if $u \in C^2(]a, b[)$ satisfies the differential inequality

$$u''(x) > 0 \quad \text{in }]a, b[\tag{17.1}$$

then u does not take a maximum in $]a, b[$. We can then say that a function which satisfies (17.1) satisfies a maximum principle. If, on the other hand, the interval $]a, b[$ is bounded and $u \in C^2(]a, b[) \cap C^0([a, b])$ then (from the Weierstrass theorem: *continuous functions transform compact sets into compact sets*) u takes a maximum, μ, in $[a, b]$ and thus $u(x) < \mu$ in $]a, b[$ and

$u(a) = \mu$ or $u(b) = \mu$ (or both ...); further we can say that if $u(a) = \mu$ and the right derivative exists in a then $u'_R(a) < 0$ and, in an analogous manner, if $u(b) = \mu$ and the left derivative exists in b then $u'_L(b) > 0$. These are the types of results we are interested in generalizing in several directions.

As a first generalization, we see that if a function $u \in C^2(]a, b[)$ satisfies the inequality

$$u''(x) + g(x)u'(x) > 0 \quad \text{in }]a, b[, \qquad (17.2)$$

g being a bounded function, then u satisfies a maximum principle. Instead, the fact that $u \in C^2(]a, b[)$ satisfies an inequality such as

$$u''(x) + g(x)u'(x) + h(x)u(x) > 0 \quad \text{in }]a, b[, \qquad (17.3)$$

g and h being bounded functions, does not imply that u does not have a maximum in $]a, b[$: consider, e.g., the case $]a, b[=]0, \pi[$, $g(x) \equiv 0$, $h(x) \equiv 2$ and $u(x) = \sin x$. One faces then the problem of imposing further conditions (on u, g, and h) such that a maximum principle can be associated with (17.3). We will return to this problem later (theorem 17.3) under even more general conditions: we will in fact assume that u satisfies the inequality

$$u''(x) + g(x)u'(x) + h(x)u(x) \geq 0 \quad \text{in }]a, b[\qquad (17.4)$$

instead of the strict inequality (17.3).

The inequalities (17.1), (17.2), and (17.3) are in fact too restrictive and the maximum principles associated with them are of little practical value: we need maximum principles associated with inequalities which are not strict. Such results, however, are more complicated to prove and we will prove only the simplest of them (and refer the reader to the bibliography for the others):

THEOREM 17.1. If $u \in C^2(]a, b[)$ satisfies the inequality

$$u''(x) \geq 0 \quad \text{in }]a, b[\qquad (17.5)$$

and takes a maximum μ in $]a, b[$ then $u(x) \equiv \mu$ in $]a, b[$. In particular, if $-\infty < a < b < +\infty$ and $u \in C^2(]a, b[) \cap C^0([a, b])$ is not constant then it takes its maximum μ only in a or in b; further, if $u(a) = \mu$ (or: $u(b) = \mu$) and u has a right (or: left) derivative in a (or: b) then $u'_R(a) < 0$ (or: $u'_L(b) > 0$).

Proof. The result is obvious if $u'' \equiv 0$. Let us assume then that $u'' \not\equiv 0$ and that $u(c) = \mu$, where $c \in]a, b[$. Under these conditions $u'(c) = 0$ and we can write

$$u'(x) = \int_c^x u''(t)\, dt, \qquad (17.6)$$

which together with (17.5) shows that $u'(x) \leq 0$ in $]a, c[$ and $u'(x) \geq 0$ in

]c, b[, which contradicts the hypothesis that c is a point of maximum. This proves the first part.

To prove the second part we recall that according to the Weierstrass theorem there exists $c \in [a, b]$ such that $u(c) = \mu$: the first part enables us then to conclude that $c \in \{a, b\}$. To study the sign of the lateral derivative at the point of maximum let us assume that this is taken in b: $u(b) = \mu$. It follows immediately that we cannot have $u'_L(b) < 0$; further if $u'_L(b) = 0$ then we can write

$$u'(x) = \int_b^x u''(t)\,dt \qquad (17.7)$$

and thus, from (17.5), we must have $u'(x) \leq 0$ in $]a, b[$, i.e. u is non-increasing: this contradicts the assumption that u takes a maximum in b and shows that we must have $u'_L(b) > 0$. Analogous reasoning shows that if $u(a) = \mu$ then $u'_R(a) < 0$. ∎

A first generalization of this result is

THEOREM 17.2. If $u \in C^2(]a, b[)$ satisfies the inequality

$$u''(x) + g(x)u'(x) \geq 0 \quad \text{in }]a, b[, \qquad (17.8)$$

g being a bounded function, and takes a maximum μ in $]a, b[$ then $u(x) \equiv \mu$ in $]a, b[$. The second part of theorem 17.1 still holds.

The proof of this theorem can be found in, e.g., Protter–Weinberger, 1967, pp. 2 and 4. Also see the same book (pp. 6 and 7) for the proof of a further extension of theorem 17.1 which is the following theorem:

THEOREM 17.3. If $u \in C^2(]a, b[)$ satisfies the inequality (17.4), where g is a bounded function and h is a non-positive bounded function, and assumes a non-negative maximum value $\mu (\geq 0)$ in $c \in]a, b[$ then $u(x) \equiv \mu$ in $[a, b]$ (moreover $\mu = 0$ if $h(x) \not\equiv 0$). In particular, if $-\infty < a < b < +\infty$, $u \in C^2(]a, b[) \cap C^0([a, b])$ is not constant and there exists $d \in]a, b[$ such that $u(d) \geq 0$, then the same conclusions as the second part of theorem 17.1 hold.

This result has two strong restrictions: $h(x) \leq 0$ and $\mu \geq 0$; these are, however, essential, as the following examples show:

(1) $]a, b[=]0, \pi[$, $g(x) \equiv 0$, $h(x) \equiv 1$ and $u(x) = \sin x$ (which takes the maximum $\mu = 1$ in $x = \pi/2$);
(2) $]a, b[=]-1, 1[$, $g(x) \equiv 0$, $h(x) \equiv -1$ and $u(x) = -e^x - e^{-x}$ (which takes the maximum $\mu = -2$ in $x = 0$).

Other extensions of the preceding maximum principles are possible—we will not, however, deal with them (in this context see Protter–Weinberger,

1967, ch. 1) and we prefer to present an example of the possible applications of theorem 17.3:

THEOREM 17.4. Suppose that u_1 and u_2 are two solutions of the following

PROBLEM 17.1. 'Given $f \in \mathcal{F}([a, b])$ (with $-\infty < a < b < \infty$), $\phi_1, \phi_2 \in \mathbb{R}$ and $\theta_1, \theta_2 \in [0, \pi/2]$, find $u \in C^2(]a, b[) \cap C^1([a, b])$ such that

$$u''(x) + g(x)u'(x) + h(x)u(x) = f(x) \quad \text{in }]a, b[\qquad (17.9)$$

$$-u'_R(a) \cos \theta_1 + u(a) \sin \theta_1 = \phi_1 \qquad (17.10)$$

$$u'_L(b) \cos \theta_2 + u(b) \sin \theta_2 = \phi_2, \qquad (17.11)$$

where $h \leq 0$ and g are bounded functions'.

Then $u_1 \equiv u_2$, unless $h \equiv 0$ and $\theta_1 = \theta_2 = 0$, and in this case $u_1 = u_2 + k$, where $k \in \mathbb{R}$.

Proof. Before the proof, let us remark that if $\theta_1 = \theta_2 = 0$ then problem 17.1 is a Neumann problem, and if $\theta_1 = \theta_2 = \pi/2$ it is a Dirichlet problem; in the latter case we can look for u in $C^2(]a, b[) \cap C^0([a, b])$, since this is sufficient in order that the boundary conditions (17.10) and (17.11) be meaningful.

It can be easily seen that the function $\bar{u} = u_1 - u_2$ is a solution of the following

PROBLEM 17.2. 'Given $\theta_1, \theta_2 \in [0, \pi/2]$ find $u \in C^2(]a, b[) \cap C^1([a, b])$ (with $-\infty < a < b < +\infty$) such that

$$u''(x) + g(x)u'(x) + h(x)u(x) = 0 \quad \text{in }]a, b[\qquad (17.12)$$

$$-u'_R(a) \cos \theta_1 + u(a) \sin \theta_1 = 0 \qquad (17.13)$$

$$u'_L(b) \cos \theta_2 + u(b) \sin \theta_2 = 0, \qquad (17.14)$$

where $h \leq 0$ and g are bounded functions'.

Further, if $h(x) \equiv 0$ and $\theta_1 = \theta_2 = 0$ then every constant $k \in \mathbb{R}$ is also a solution of problem 17.2, the converse being true if $k \neq 0$, i.e., if $k \neq 0$ is a solution of problem 17.2 then (from (17.12)) $h \equiv 0$ and (from (17.13) and (17.14), given that $u(a) = u(b) = k$ and $u'_R(a) = u'_L(b) = 0$) $\theta_1 = \theta_2 = 0$: we conclude thus that the only case in which $\bar{u}$ is a non-zero constant is that in which $h \equiv 0$ and $\theta_1 = \theta_2 = 0$. We will now show that, but for this exceptional case, the only solution of problem 17.2 is the function which is identically zero and hence $\bar{u} \equiv 0$. Let us assume then that problem 17.2 has a non-constant solution $\bar{u}$ such that $\bar{u}(c) > 0$ with $c \in]a, b[$. Theorem 17.3 (which we can apply since g and h are bounded, $h \leq 0$ and $\mu = \max_{[a,b]} \bar{u}(x) \geq \bar{u}(c) > 0$) then ensures that the maximum of $\bar{u}$ is taken in $\{a, b\}$; if $\bar{u}(b) = \mu$ the same theorem ensures that $\bar{u}'_L(b) > 0$, which together with $\bar{u}(b) = \mu > 0$ contradicts (17.14); in the same way it can be shown that the maximum cannot be taken in a and hence that there does not exist $c \in]a, b[$ such that

$\bar{u}(c) > 0$. Using the same reasoning for the function $-\bar{u}$ we conclude that there cannot exist $c \in]a, b[$ such that $\bar{u}(c) < 0$ and hence that $\bar{u} \equiv 0$. ∎

17.3 MAXIMUM PRINCIPLE IN $\mathbb{C}$

The theory of functions of one complex variable is full of surprises. Often, apparently weak assumptions lead to very strong results. Thus, the assumption of holomorphy of a function $f: \Omega \to \mathbb{C}$, $\Omega \subset \mathbb{C}$ being an open set, implies the existence of all the derivatives of f and that f is analytic. The same assumption also implies that if $|f(z)|$ has a maximum in $z_0 \in \Omega$ then $f(z)$ is constant in Ω, a result which is known as the *maximum modulus principle* and which we will deal with now. Apparently, this result is not a maximum principle according to the considerations which we have already made in the introduction (section 17.1), since the presence of a differential relationship is not obvious: we will see, however, that it is present, though hidden, and at the same time that there is a profound connection between the results mentioned, a connection which depends naturally on the common hypothesis of holomorphy.

To prove the maximum modulus principle we will use one of the most elegant results of the theory of functions of one complex variable, the Cauchy formula, which we will state without proof (and which can be found in any textbook of complex analysis: e.g. Nevanlinna–Paatero, 1969, p. 134; Courant, 1967, p. 79; Dieudonné, 1968a, p. 211):

THEOREM 17.5 (Cauchy formula). Let $\Omega \subset \mathbb{C}$ be a simply connected open set, $f: \Omega \to \mathbb{C}$ a holomorphic function in Ω and γ a closed path contained in Ω. For every point $z_0 \in \Omega \setminus \gamma$ the following formula holds:

$$f(z_0) \, \text{ind} \, (\gamma, z_0) = \frac{1}{2\pi i} \int_\gamma \frac{f(z)}{z - z_0} \, dz, \tag{17.15}$$

where

$$\text{ind} \, (\gamma, z_0) = \frac{1}{2\pi i} \int_\gamma \frac{dz}{z - z_0}. \tag{17.16}$$

Let us recall that by the term *path* we mean a continuous piecewise differentiable curve, with the orientation induced by one of its parametrizations—a typical example of a closed path is the circumference centred in $a \in \mathbb{C}$ and with radius $r > 0$: $\gamma(a, r) = \{z : z = \gamma(t) = a + re^{it}, t \in [0, 2\pi]\}$, which is said to be positively oriented.

The value ind (γ, z_0), which is called the *index* of the path γ with respect to the point z_0, is an integer (for a curious geometrical interpretation of it see, e.g., Nevanlinna–Paatero, 1969, p. 121; Dieudonné, 1968a, p. 207). The most interesting case is no doubt that in which γ is a closed *Jordan curve* piecewise differentiable and positively oriented: under these conditions ind $(\gamma, z_0) = 1$ if z_0 'is always seen to the left while following the path γ' and ind $(\gamma, z_0) = 0$ otherwise. This is so for the circumference considered above, for which (17.15) can be written, if we assume $r < \text{dist}(a, \mathbb{C}\setminus\Omega)$ and $|z_0 - a| < r$, as

$$f(z_0) = \frac{1}{2\pi i} \int_{\gamma(a, r)} \frac{f(z)}{z - z_0} dz. \tag{17.17}$$

This formula tells us that the values of a holomorphic function in the interior of a circle are determined exclusively by the values it takes on the circumference. Putting $z_0 = a$ in (17.17) we have the so-called *Gauss's mean value formula*

$$f(a) = \frac{1}{2\pi} \int_0^{2\pi} f(a + re^{it}) dt, \tag{17.18}$$

which states essentially that the value of a holomorphic function at the centre of a circle is equal to the arithmetic mean of the values it takes on the circumference. We will use the particular cases (17.17) and (17.18) of (17.15) in proving the already mentioned following result:

THEOREM 17.6 (Maximum modulus principle). *Let $f : \Omega \to \mathbb{C}$ be a holomorphic function in the connected open set $\Omega \subset \mathbb{C}$. If the function $|f(z)| : \Omega \to \mathbb{R}$ takes a maximum μ in $a \in \Omega$ then $f(z)$ is constant in Ω (with modulus μ).*

Proof. Let r be a positive real number such that $r < \rho = \text{dist}(a, \mathbb{C}\setminus\Omega)$. The circumference $\gamma(a, r)$ is contained in Ω and from (17.18) we can write

$$0 = |f(a)| - \mu = \left| \frac{1}{2\pi} \int_0^{2\pi} f(a + re^{it}) dt \right| - \mu \leq \frac{1}{2\pi} \int_0^{2\pi} |f(a + re^{it})| dt - \mu$$

$$\leq \frac{1}{2\pi} \int_0^{2\pi} \mu \, dt - \mu = 0 \tag{17.19}$$

and therefore

$$\frac{1}{2\pi} \int_0^{2\pi} (\mu - |f(a + re^{it})|) dt = 0. \tag{17.20}$$

From (17.20) we conclude that

$$\mu = |f(a + re^{it})|, \tag{17.21}$$

and from this, together with the fact that $r \in \,]0, \rho[$ is arbitrary, we have $|f(z)| = \mu$ in the disc $D = \{z \in \mathbb{C} : |z - a| < \rho\}$. Now, a holomorphic function with a constant modulus is constant: we need only take the derivative in x and y of $|f|^2 = \mu^2$ and use the Cauchy–Riemann system (see (17.23) and (17.24)). So, we conclude therefore that $f(z) = f(a)$ in D, and hence in the open set Ω which can be covered by a family of 'linked' discs. ∎

The preceding result can be easily generalized to functions of n complex variables—see, e.g., Bochner–Martin, 1948, p. 107.

We will now try to discover why the single assumption of holomorphy of f is sufficient to prove the maximum principle. Putting $z = x + iy$ and splitting f in its real part $u(x, y) = \mathcal{R}e(f(z))$ and imaginary part $v(x, y) = \mathcal{I}m(f(z))$ we can write

$$f(z) = u(x, y) + iv(x, y). \tag{17.22}$$

It is well known that if f is holomorphic then the functions u and v satisfy the so-called *Cauchy–Riemann differential equations*

$$\frac{\partial u}{\partial x} = \frac{\partial v}{\partial y} \tag{17.23}$$

$$\frac{\partial u}{\partial y} = -\frac{\partial v}{\partial x}; \tag{17.24}$$

and the converse is also true: if u and v are differentiable and satisfy equations (17.23) and (17.24) then the function f defined by (17.22) is holomorphic and

$$f'(z) = \frac{\partial u}{\partial x} + i\frac{\partial v}{\partial x} = \frac{\partial v}{\partial y} - i\frac{\partial u}{\partial y}. \tag{17.25}$$

Now from equations (17.23) and (17.24) it follows that

$$\Delta u = 0 \tag{17.26}$$

and that

$$\Delta v = 0, \tag{17.27}$$

i.e.: the real part and the imaginary part of a holomorphic function satisfy the Laplace equation, and are therefore harmonic functions. It is natural to ask whether (17.26) and (17.27) are the differential relationships that one could expect to have in association with a function which satisfies a maximum principle.

The answer is yes, and this is even more evident if we introduce the *Cauchy–Riemann differential operators*

$$\partial \equiv \frac{1}{2}\left(\frac{\partial}{\partial x} - i\frac{\partial}{\partial y}\right) \tag{17.28}$$

$$\bar{\partial} \equiv \frac{1}{2}\left(\frac{\partial}{\partial x} + i\frac{\partial}{\partial y}\right) \tag{17.29}$$

and the *complex Laplace operator* $\partial\bar{\partial}$, since it follows that if f is a holomorphic function then

$$\partial\bar{\partial} f = 0, \tag{17.30}$$

i.e.: the holomorphic functions are *complex harmonic functions*. We see then that the maximum modulus principle is a 'natural' consequence of the (very strong, though apparently weak) assumption of holomorphy.

The harmonic functions u and v, the real and imaginary parts, respectively, of the holomorphic function f, are called *harmonic conjugates* (more precisely: v is said to be the harmonic conjugate of u and u is said to be the harmonic conjugate of $-v$). It is interesting to note that if u is an harmonic function in $\Omega \subset \mathbb{R}^2$ then the function

$$v = \int_{(x_0,y_0)}^{(x,y)} \left(-\frac{\partial u}{\partial y} dx + \frac{\partial u}{\partial x} dy \right), \tag{17.31}$$

where (x_0, y_0) is a point fixed in Ω and the integral is taken along a path with the extremities (x_0, y_0) and (x, y) and completely contained in Ω, is a harmonic conjugate of u (the only one, except for the addition of a constant). This shows that every harmonic function $u : \Omega \subset \mathbb{R}^2 \to \mathbb{R}$ can be considered to be the real part of a holomorphic function $f : \Omega \subset \mathbb{C} \to \mathbb{C}$ (with $\Omega \subset \mathbb{R}^2$ and $\Omega \subset \mathbb{C}$ identified through the relationship $(x, y) \leftrightarrow z$ defined by $z = x + iy$) and hence $u \in C^\infty(\Omega)$. This is one of the many examples of results regarding harmonic functions in two real variables which can be obtained in complex analysis. Another example, particularly important for us, is the following one:

THEOREM 17.7. *If $u : \Omega \to \mathbb{R}$ is a harmonic function in the open set $\Omega \subset \mathbb{R}^2$ and has a maximum in Ω then it is constant in Ω.*

Proof. If v is a harmonic conjugate of u then the function $e^f = e^{u+iv}$ is holomorphic in Ω, and therefore if $|e^f| = e^u$ has a maximum in Ω then, from theorem 17.6 and from the monotonicity of the exponential function, it follows that u is constant in Ω. ∎

Let us remark that the proof can be performed, following that of theorem 17.6, starting from the 'real version' of Gauss's mean value formula

$$u(a) = \frac{1}{2\pi} \int_0^{2\pi} u(a + re^{it}) \, dt, \tag{17.32}$$

which can be obtained by equating the real parts of (17.18): this is the usual method of proving the n-dimensional analogue of theorem 17.3, and we will take it up in the following section.

An immediate consequence of theorem 17.7 is that if a harmonic function in Ω has a minimum in it then it is constant in Ω; thus, *if $u \in C^2(\Omega) \cap C^0(\bar{\Omega})$ is a non-constant harmonic function in a bounded set $\Omega \subset \mathbb{R}^2$ then its maximum and its minimum are taken on the boundary Γ of Ω.* It follows from this that if $u, v \in C^2(\Omega) \cap C^0(\bar{\Omega})$ are two harmonic functions in a bounded open set Ω and $u = v$ on Γ then $u = v$ in $\bar{\Omega}$ (this is evident if one considers the function $w = u - v$). Let us remark that the assumption that Ω is bounded is essential for the maximum and minimum to exist (and in that case they are taken on Γ) as can be seen from the following example: in

$\Omega = \{(x, y) \in \mathbb{R}^2 : x \in \,]0, \pi[\}$ the function

$$u(x, y) = e^y \sin x \tag{17.33}$$

is harmonic and positive in Ω but zero on the boundary.

17.4 MAXIMUM PRINCIPLE IN $\mathbb{R}^n$

If a function $u : \Omega \to \mathbb{R}$ of class C^2 in the open set $\Omega \subset \mathbb{R}^n$ has a maximum in $c \in \Omega$ then $\partial u/\partial x_i(c) = 0$ $(i = 1, \ldots, n)$ and $\partial^2 u/\partial x_i^2(c) \leq 0$ $(i = 1, \ldots, n)$. Hence, if $u \in C^2(\Omega)$ satisfies the inequality

$$\Delta u > 0 \quad \text{in } \Omega \tag{17.34}$$

then u does not have a maximum in Ω; the solutions of (17.34) therefore satisfy a maximum principle. In an analogous manner we conclude that the functions which satisfy the inequality

$$\Delta u + \sum_{i=1}^n a_i \frac{\partial u}{\partial x_i} > 0 \quad \text{in } \Omega, \tag{17.35}$$

where $a_i = a_i(x) : \Omega \to \mathbb{R}$ are bounded functions, satisfy a maximum principle.

As in the one-dimensional case, we are more interested in inequalities which are not strict, as are (17.34) and (17.35). Further, in the two-dimensional case we have seen that if in (17.34) we substitute $>$ with $\geq$ we obtain the same result, modulo the constant functions. Theorem 17.7, which enables us to reach this conclusion, can be generalized to the n-dimensional case by making use of the mean value formula in n variables (cf. (17.32))

$$u(a) = \frac{1}{\omega_n r^{n-1}} \int_{S(a, r)} u(x)\, ds, \tag{17.36}$$

where $\omega_n = 2\pi^{n/2}/\Gamma(n/2)$ and $S(a, r)$ is the hypersphere with centre a and radius r defined by the equation $\|x - a\| = r$. We can then state the following

THEOREM 17.8. *Let $\Omega \subset \mathbb{R}^n$ be an open set with boundary Γ. If $u \in C^2(\Omega)$ satisfies the inequality*

$$\Delta u \geq 0 \quad \text{in } \Omega \tag{17.37}$$

and takes a maximum μ in Ω then $u(x) \equiv \mu$ in Ω. In particular, if Ω is bounded and $u \in C^2(\Omega) \cap C^0(\bar{\Omega})$ is not constant then its maximum value μ is taken exclusively in Γ.

The solutions of the inequality (17.37) are known as *subharmonic functions* because if Ω is bounded and $v \in C^2(\Omega) \cap C^0(\bar{\Omega})$ is a harmonic function in Ω which on Γ takes the same values as a solution u of (17.37) (continuous in $\bar{\Omega}$) then $u \leq v$ in Ω (to see that this is so we have to apply theorem 17.8 to the function $w = u - v$). In an analogous manner, the solutions of the inequality $\Delta u \leq 0$ are known as *superharmonic functions*.

Given theorem 17.1, it is natural to ask whether we can add to the preceding statement something concerning the behaviour of u at the points of maximum. This is possible and is very important; we must first, however, introduce some new concepts.

DEFINITION 17.1. Let $\Omega \subset \mathbb{R}^n$ be an open set with boundary Γ. A point $\bar{x} \in \Gamma$ is said to be a *Hopf point* if there exists $x_0 \in \Omega$ and $r > 0$ such that the sphere $B(x_0, r) = \{x \in \mathbb{R}^n : \|x - x_0\| < r\}$ satisfies the following conditions:

(i) $B(x_0, r) \subset \Omega$
(ii) $\bar{x} \in \overline{B(x_0, r)}$

The open set Ω is said to be an *Hopf open set* if all the points of Γ are Hopf points.

DEFINITION 17.2. Let $\Omega \subset \mathbb{R}^n$ be an open set with boundary Γ and $\bar{x} \in \Gamma$ a Hopf point. Further let $x_0 \in \Gamma$ and $r > 0$ be under the conditions of definition 17.1 and l be a point of $B(x_0, r)$. The point l, or the vector $\mathbf{l} = \overrightarrow{l\bar{x}}$, is said to determine an *outward direction* from Ω in $\bar{x}$ if there exist $\bar{t} > 0$ and $\bar{r} > 0$ such that $\{x \in \mathbb{R}^n : x = \bar{x} + t\mathbf{l}, t \in]0, \bar{t}]\} \subset B(\bar{x}, \bar{r}) \cap (\mathbb{R}^n \setminus \bar{\Omega})$. If $\mathbf{l}$ determines an outward direction from Ω in $\bar{x}$ the limit

$$\frac{\partial^- u}{\partial \mathbf{l}}(\bar{x}) = \liminf_{t \to 0^+} \frac{u(\bar{x} - t\mathbf{l}) - u(\bar{x})}{t} \qquad (17.38)$$

is said to be the *lower derivative* with respect to $\mathbf{l}$ in $\bar{x}$.

As is evident, if Γ has an external normal ν in $\bar{x}$ this determines an outward direction from Ω in $\bar{x}$ (put $l = x_0$), and if u is sufficiently regular (17.38) contains, as a particular case, the definition of the derivative of u with respect to the outward normal in $\bar{x}$, $\partial u / \partial \nu(\bar{x})$. In fig. 17.1 the points $\bar{x}_1$ (which satisfy the previous conditions) and $\bar{x}_2$ (which do not) are Hopf points of Γ, while $\bar{x}_3$ is not; referring once more to this figure, all the points of the (open) sphere B_1 determine outward directions from Ω in $\bar{x}_1$, while the points of B_2 which determine outward directions from Ω in $\bar{x}_2$ are only those in the shaded 'cone'.

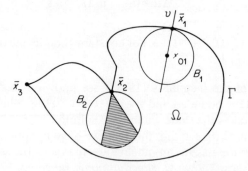

Figure 17.1

We are now in a position to complete the statement of theorem 17.8:

THEOREM 17.8 bis. *In addition to the statement of theorem* 17.8, *if* $\bar{x}$ *is a Hopf point of* Γ *such that* $u(\bar{x}) = \mu$ *then, whatever the outward direction* $\mathbf{l}$ *from* Ω *in* $\bar{x}$ *is, we have* $\partial^- u/\partial \mathbf{l}(\bar{x}) > 0$.

For a proof see, e.g., Hopf, 1952 and Protter–Weinberger, 1967, p. 65—in these references the following more general result has been proved:

THEOREM 17.9 (Hopf maximum principle). *Let* $\Omega \subset \mathbb{R}^n$ *be an open set with boundary* Γ *and* $u \in C^2(\Omega)$ *a solution of the inequality*

$$\sum_{i,j=1}^{n} a_{ij} \frac{\partial^2 u}{\partial x_i \partial x_j} + \sum_{i=1}^{n} b_i \frac{\partial u}{\partial x_i} \geq 0 \quad \text{in } \Omega, \tag{17.39}$$

where the functions $a_{ij} = a_{ij}(x)$ *and* $b_i = b_i(x)$ *are bounded in* Ω *and further the* a_{ij} *are such that the differential operator associated with* (17.39) *is uniformly strongly elliptic in* Ω. *If* u *takes a maximum* μ *in* Ω *then* $u(x) \equiv \mu$ *in* Ω. *In particular, if* Ω *is bounded and* $u \in C^2(\Omega) \cap C^0(\bar{\Omega})$ *is not constant then its maximum* μ *is taken exclusively in* Γ; *further, if* $\bar{x}$ *is a Hopf point of* Γ *and* $u(\bar{x}) = \mu$ *then* $\partial^- u/\partial \mathbf{l}(\bar{x}) > 0$, *whatever the outward direction* $\mathbf{l}$ *from* Ω *in* $\bar{x}$ *is.*

In the same way that theorem 17.3 generalizes theorem 17.2, the following result (see Pucci, 1957; 1958; Protter–Weinberger, 1967, pp. 64 and 67) generalizes theorem 17.9:

THEOREM 17.10. *Let* a_{ij} *and* b_i *be as in theorem* 17.9 *and* c *a non-positive bounded function. If* $u \in C^2(\Omega)$ *is a solution of*

$$\sum_{i,j=1}^{n} a_{ij} \frac{\partial^2 u}{\partial x_i \partial x_j} + \sum_{i=1}^{n} b_i \frac{\partial u}{\partial x_i} + cu \geq 0 \quad \text{in } \Omega \tag{17.40}$$

which takes a maximum value $\mu \geq 0$ *in* Ω *then* $u(x) \equiv \mu$ *in* Ω (*moreover* $\mu = 0$ *if* $c(x) \neq 0$). *In particular, if* Ω *is bounded,* $u \in C^2(\Omega) \cap C^0(\bar{\Omega})$ *is not constant and there exists* $x_0 \in \Omega$ *such that* $u(x_0) \geq 0$, *then the same conclusions as in the second part of theorem* 17.9 *are true.*

As in theorem 17.3, the restrictions $c(x) \leq 0$ and $\mu \geq 0$ are essential, as can be seen from the following two examples: (1) the function $u(x, y) = \sin x \sin y$, which takes a maximum $\mu = 1$ in $(x, y) = (\pi/2, \pi/2)$, satisfies $\Delta u + 2u = 0$ in the square $\Omega = \{(x, y) \in \mathbb{R}^2 : 0 < x < \pi \text{ and } 0 < y < \pi\}$; (2) the function $u(x, y) = -e^{x+y} - e^{-x-y}$, which takes a maximum $\mu = -2$ in $(x, y) = (0, 0)$, satisfies $\Delta u - u = 0$ in $\Omega = \{(x, y) \in \mathbb{R}^2 : x^2 + y^2 < 1\}$.

Further extensions of the maximum principles that we have presented are possible but, as in the one-dimensional case, we will not deal with them (see, e.g., Protter–Weinberger, 1967, ch. 2) but rather consider an example of an application of the results presented:

THEOREM 17.11. Suppose that u_1 and u_2 are two solutions of the following problem:

PROBLEM 17.3. 'Let $\Omega \subset \mathbb{R}^n$ be a bounded open set with boundary Γ, and Γ_D and Γ_N two subsets of Γ such that $\Gamma_N \cap \Gamma_D = \varnothing$ and $\Gamma = \Gamma_D \cup \Gamma_N$, and let us assume that every point of Γ_N is a Hopf point. Given $f \in \mathscr{F}(\bar{\Omega})$, $g_1 \in \mathscr{F}(\Gamma_N)$ and $g_2 \in \mathscr{F}(\Gamma_D)$, find $u \in C^2(\Omega) \cap C^1(\bar{\Omega})$ such that

$$\sum_{i,j=1}^{n} a_{ij} \frac{\partial^2 u}{\partial x_i \partial x_j} + \sum_{i=1}^{n} b_i \frac{\partial u}{\partial x_i} + cu = f \quad \text{in } \Omega \tag{17.41}$$

$$\frac{\partial u}{\partial \nu} = g_1 \quad \text{in } \Gamma_N \tag{17.42}$$

$$u = g_2 \quad \text{in } \Gamma_D, \tag{17.43}$$

where the functions a_{ij}, b_i, and c satisfy the conditions of theorem 17.10'.

Then $u_1 \equiv u_2$, unless $c(x) \equiv 0$ and $\Gamma_D = \varnothing$, in which case $u_1 = u_2 + k$ with $k \in \mathbb{R}$.

For the proof see Protter–Weinberger, 1967, p. 70.

17.5 COMPLEMENTS

In the previous sections we have always assumed that the functions we were dealing with were very regular, which is contrary to our philosophy in most of this book. In this section we will briefly discuss 'irregular' functions again, in order to present an important formulation of the maximum principle applicable to them.

Let then Ω be a bounded connected open set of $\mathbb{R}^n$ and let us consider the differential operator

$$L\square \equiv -\sum_{i,j=1}^{n} \frac{\partial}{\partial x_i}\left(a_{ij}\frac{\partial}{\partial x_j}\square\right) + \sum_{i=1}^{n} b_i \frac{\partial}{\partial x_i}\square + \sum_{i=1}^{n} \frac{\partial}{\partial x_i}(c_i \square) + d\square \tag{17.44}$$

which we will assume to be uniformly strongly elliptic in Ω and, besides, such that (cf. section 7.2) $a_{ij} \in L^\infty(\Omega)$, $b_i \in L^n(\Omega)$, $c_i \in L^n(\Omega)$ and $d \in L^{n/2}(\Omega)$ (this if $n > 2$—the usual modifications should be made for $n=1$ and $n=2$). We know, from Chapter 7, that under these conditions $L \in \mathscr{L}(H^1(\Omega), H^{-1}(\Omega))$ and that we can consider the bilinear form $a(u,v): H^1(\Omega) \times H^1(\Omega) \to \mathbb{R}$ defined by

$$a(u,v) = \int_\Omega \left[\sum_{i,j=1}^{n} a_{ij} \frac{\partial u}{\partial x_j}\frac{\partial v}{\partial x_i} + \sum_{i=1}^{n} b_i v \frac{\partial u}{\partial x_i} - \sum_{i=1}^{n} c_i u \frac{\partial v}{\partial x_i} + duv \right] dx. \tag{17.45}$$

From what we have done in the previous sections it is natural to think of associating a maximum principle with the solutions of the inequality

$$Lu \leq 0 \quad \text{in } \Omega, \tag{17.46}$$

i.e. to the inequality

$$a(u, v) \leq 0 \quad \forall v \in H_0^1(\Omega), v \geq 0 \quad \text{a.e. in } \Omega. \tag{17.47}$$

(note that the principal part of L has the negative sign and we should not therefore consider the inequality $Lu \geq 0$ if we wish to consider a maximum principle).

We have then the following result (where (17.48) is analogous to the hypothesis of non-positivity of c in theorem 17.10):

THEOREM 17.12. Let $u \in H^1(\Omega) \cap L^\infty(\Omega)$ be a solution of (17.47), where the coefficients a_{ij}, b_i, c_i, and d $(i, j = 1, \ldots, n)$ satisfy the assumptions made previously and further

$$e = d + \sum_{i=1}^{n} \frac{\partial c_i}{\partial x_i} \geq 0 \quad \text{in the sense of } \mathcal{D}'(\Omega), \tag{17.48}$$

and let $\mu = \operatorname{ess\,sup}_\Omega u > 0$. If there exists $\bar{x} \in \Omega$ such that $\forall r > 0$ $\operatorname{ess\,sup}_{B(\bar{x},r) \cap \Omega} u = \mu$ (we then say that u takes its maximum in $\bar{x}$) then $u = \mu$ a.e. in Ω (and in this case $e = 0$ in the sense of $\mathcal{D}'(\Omega)$ and $a(u, v) = 0$ $\forall v \in H_0^1(\Omega)$).

For a proof see Stampacchia, 1965, p. 206; see also Nečas, 1967, p. 323 and Chicco, 1967; 1970; 1975 for this and other similar results.

18
On Green's Formulae

The boundary conditions of natural kind, which are implicitly written in variational equations and in variational inequalities can, often, be made explicit by applying an appropriate extension of the so-called *Green's first formula*. Here we will deal with the formula which enables us to make the above conditions explicit in the Neumann problems and other similar problems; the techniques we will use are, however, essentially those we use in constructing formulae suitable for studying many other problems, such as the mixed problems (which we will come back to briefly at the end of this chapter).

The divergence theorem

Let us first recall the proof of Green's first formula (in the classical framework) since, in order to make generalizations, it will be useful to have it as a guide. Let us start with a well known result (by $\mathbf{z} \in E^n$, where E is a function space, we denote that $\mathbf{z}$ is a vector whose components (z_i, such that $\mathbf{z} = \{z_i\}_{i=1,\ldots,n}$) belong to E; and by $(\cdot, \cdot)$ we denote the scalar product in $\mathbb{R}^n$):

THEOREM 18.1 (The divergence theorem or Ostrogradsky–Gauss theorem). Let $\Omega \subset \mathbb{R}^n$ be a bounded connected open set with boundary Γ of class $C^{0,1}$, ν the versor of the external normal to Γ, and $\mathbf{f} \in [C^1(\bar{\Omega})]^n$. Then the following formula holds:

$$\int_\Omega \operatorname{div} \mathbf{f} \, dx = \int_\Gamma (\mathbf{f}|_\Gamma, \nu) \, d\sigma. \qquad (18.1)$$

In the classical framework this theorem is, in general, proved for open sets with boundaries of class C^1 or piecewise C^1: see, e.g., Kellogg, 1929, ch. 4; Fleming, 1965, pp. 264 and 265. In the latter reference the theorem has been proved in the framework of the theory of the integration of differential forms on manifolds, which is the natural framework for discussing the important set of results which consists of the theorems of Ostrogradsky–Gauss, Stokes, and Green (not to be confounded with the formulae which we are considering here); in this context all the results cited are corollaries of a very general and abstract formulation of Stokes' theorem (see Fleming, 1965, p. 273; Spivak, 1965, pp. 124 and 134). The extension to open sets of class $C^{0,1}$ entails, besides making the proof more difficult, a reinterpretation of the integrals present in (18.1), which must now be taken as Lebesgue integrals and not as Riemann integrals, but we will still be essentially in the classical framework (this result can be proved by using techniques analogous to those used in Nečas, 1967, p. 121).

Green's first formula

From formula (18.1) we now will derive the so-called Green's first formula. By 'substituting' in (18.1) $\mathbf{f} = v \cdot \mathbf{w}$, where $v \in C^1(\bar{\Omega})$ and $\mathbf{w} \in [C^1(\bar{\Omega})]^n$, we have

$$\int_\Omega v \operatorname{div} \mathbf{w} \, dx + \int_\Omega (\mathbf{w}, \operatorname{grad} v) \, dx = \int_\Gamma v|_\Gamma (\mathbf{w}|_\Gamma, \nu) \, d\sigma. \tag{18.2}$$

Let us now assume that $\mathbf{w}$ is of the form

$$\mathbf{w} = \operatorname{grad} u, \tag{18.3}$$

with $u \in C^2(\bar{\Omega})$, i.e.: we assume that $\mathbf{w}$ is a conservative field which can be obtained from the potential u. From (18.2) and (18.3) we then have

$$\int_\Omega v \operatorname{div} \operatorname{grad} u \, dx + \int_\Omega (\operatorname{grad} u, \operatorname{grad} v) \, dx = \int_\Gamma v|_\Gamma (\operatorname{grad} u|_\Gamma, \nu) \, d\sigma, \tag{18.4}$$

which we can write in the more compact form

$$\int_\Omega v \Delta u \, dx + \int_\Omega (\operatorname{grad} u, \operatorname{grad} v) \, dx = \int_\Gamma v|_\Gamma \frac{\partial u}{\partial \nu} \, d\sigma, \tag{18.5}$$

using the definition of the Laplacian (in its standard decomposition)

$$\Delta \triangleq \operatorname{div} \operatorname{grad} \tag{18.6}$$

and the definition of the derivative with respect to the normal

$$\frac{\partial u}{\partial \nu} \triangleq (\operatorname{grad} u|_\Gamma, \nu) = \sum_{i=1}^n \frac{\partial u}{\partial x_i}\bigg|_\Gamma \cos(\nu, x_i). \tag{18.7}$$

We will call the formula (18.5) *Green's first formula* or simply *Green's formula*.

We have therefore proved the following

THEOREM 18.2. *Let $\Omega \subset \mathbb{R}^n$ be a bounded connected open set with boundary Γ of class $C^{0,1}$ and ν the versor of the external normal to Γ. For every $v \in C^1(\bar{\Omega})$ and every $u \in C^2(\bar{\Omega})$ the formula (18.5) is true.*

If further $v \in C^2(\bar{\Omega})$ then we can exchange v and u in (18.5) and thus obtain a formula (18.5 bis); subtracting (18.5 bis) from (18.5) we obtain a new important formula: *Green's second formula*

$$\int_\Omega (v\Delta u - u\Delta v)\,dx = \int_\Gamma \left(v|_\Gamma \frac{\partial u}{\partial \nu} - u|_\Gamma \frac{\partial v}{\partial \nu}\right) d\sigma.$$

This formula can also be made much more general, but we will not make such a generalization here (see Lions–Magenes, 1972a, pp. 114 *et seq.*). Let us remark that the classical Green's formulae are in fact three—the last, however, is not as interesting as the first two in the modern formulation of the theory of boundary value problems for partial differential equations.

In the one-dimensional case (18.5) can be written, with $\Omega =]a,b[$, $-\infty < a < b < +\infty$, in the form

$$\int_a^b v(x)u''(x)\,dx = v(b)u_L'(b) - v(a)u_R'(a) - \int_a^b u'(x)v'(x)\,dx, \quad (18.8)$$

which is the well known *formula of integration by parts*. Thus, even with several variables one often finds in the literature the phrase 'integrating by parts' which means 'applying an appropriate formula of Green'.

A Green's formula for the operator L (classic)

The presence of the operator Δ in (18.5) is due to the choice made in (18.3); (18.2) can, however, be used in order to construct formulae in which differential operators much more general than Δ are involved. Thus, we will now construct a formula analogous to (18.5) for the operator

$$L\square \equiv -\sum_{i,j=1}^n \frac{\partial}{\partial x_i}\left(a_{ij}\frac{\partial}{\partial x_j}\square\right) + \sum_{i=1}^n b_i\frac{\partial}{\partial x_i}\square + \sum_{i=1}^n \frac{\partial}{\partial x_i}(c_i\square) + d\square, \quad (18.9)$$

with $a_{ij}, c_i \in C^1(\bar{\Omega})$ and $b_i, d \in C^0(\bar{\Omega})$ $(i,j = 1,\ldots,n)$, with which we associate the bilinear form

$$a(u,v) = \int_\Omega \left[\sum_{i,j=1}^n a_{ij}\frac{\partial u}{\partial x_j}\frac{\partial v}{\partial x_i} + \sum_{i=1}^n b_i\frac{\partial u}{\partial x_i}v - \sum_{i=1}^n c_i u\frac{\partial v}{\partial x_i} + duv\right] dx. \quad (18.10)$$

In place of (18.3) we will now take

$$\mathbf{w} = \left\{-\sum_{j=1}^n a_{ij}\frac{\partial u}{\partial x_j} + c_i u\right\}_{i=1,\ldots,n}, \quad (18.11)$$

with $u \in C^2(\bar{\Omega})$; from this together with (18.2) we have

$$\int_\Omega v\left[-\sum_{i,j=1}^n \frac{\partial}{\partial x_i}\left(a_{ij}\frac{\partial u}{\partial x_j}\right) + \sum_{i=1}^n \frac{\partial}{\partial x_i}(c_i u)\right] dx$$

$$+ \int_\Omega \left[-\sum_{i,j=1}^n a_{ij}\frac{\partial u}{\partial x_j}\frac{\partial v}{\partial x_i} + \sum_{i=1}^n c_i u\frac{\partial v}{\partial x_i}\right] dx$$

$$= \int_\Gamma v|_\Gamma\left[-\sum_{i,j=1}^n a_{ij}|_\Gamma \frac{\partial u}{\partial x_j}\bigg|_\Gamma \cos(\nu,x_i) + \sum_{i=1}^n c_i|_\Gamma u|_\Gamma \cos(\nu,x_i)\right] d\sigma. \quad (18.12)$$

Now, defining the *conormal derivative* of u by the formula

$$\frac{\partial u}{\partial \nu_a} \triangleq (\mathbf{w}|_\Gamma, \nu) = \sum_{i,j=1}^{n} a_{ij}|_\Gamma \left.\frac{\partial u}{\partial x_j}\right|_\Gamma \cos(\nu, x_i) - \sum_{i=1}^{n} c_i|_\Gamma u|_\Gamma \cos(\nu, x_i), \quad (18.13)$$

(which reduces to (18.7) when $a_{ij} = \delta_{ij}$ and $c_i = 0$—we denote by $\partial u/\partial \nu_a$ the conormal derivative of u since it depends on the decomposition of the operator L and hence on the bilinear form, a, associated with the decomposition) it can be easily seen that from (18.9), (18.10), and (18.12) we have

$$a(u, v) = \int_\Omega (Lu)v \, dx + \int_\Gamma v|_\Gamma \frac{\partial u}{\partial \nu_a} d\sigma, \quad (18.14)$$

which is a generalization of (18.5)—once again in the classical set up—and which we will also call *Green's formula*. We have thus proved the following

THEOREM 18.3. Let $\Omega \subset \mathbb{R}^n$ be a bounded connected open set with boundary Γ of class $C^{0,1}$ and let ν be the versor of the external normal to Γ. If a_{ij}, $c_i \in C^1(\bar{\Omega})$ and b_i, $d \in C^0(\bar{\Omega})$ $(i, j = 1, \ldots, n)$ then, with definition (18.13), (18.14) is true, whatever $v \in C^1(\bar{\Omega})$ and $u \in C^2(\bar{\Omega})$.

With this result we end the part of this chapter dealing with the classical framework.

The case of irregular functions

We will now generalize theorem 18.3 by drastically weakening the regularity assumptions on the coefficients a_{ij}, b_i, c_i, d $(i, j = 1, \ldots, n)$ and on the functions u and v. In other words we want to obtain a formula which extends (18.14) to the case in which $v \in H^1(\Omega)$, $u \in H^1_L(\Omega) = \{u \in H^1(\Omega) : Lu \in L^2(\Omega)\}$ and the coefficients of L satisfy the assumptions

(H1) $a_{ij} \in L^\infty(\Omega)$;
(H2) $b_i \in L^\infty(\Omega)$;
(H3) $c_i \in L^n(\Omega)$ if $n > 2$; $c_i \in L^{2+\varepsilon}(\Omega)$ if $n = 2$ (with $\varepsilon > 0$);
       $c_i \in L^2(\Omega)$ if $n = 1$;
(H4) $d \in L^n(\Omega)$ if $n > 2$; $d \in L^{2+\varepsilon}(\Omega)$ if $n = 2$ (with $\varepsilon > 0$);
       $d \in L^2(\Omega)$ if $n = 1$.

Let us remark that, in theorem 7.2 we assumed, instead of (H2), (H4) respectively

(H2)' $b_i \in L^n(\Omega)$ if $n > 2$; $b_i \in L^{2+\varepsilon}(\Omega)$ if $n = 2$ (with $\varepsilon > 0$);
        $b_i \in L^2(\Omega)$ if $n = 1$;
(H4)' $d \in L^{n/2}(\Omega)$ if $n > 2$; $d \in L^{1+\varepsilon}(\Omega)$ if $n = 2$ (with $\varepsilon > 0$);
        $d \in L^1(\Omega)$ if $n = 1$;

we could also here weaken (H2), (H4) in such a form, but this would require a different choice both for the space $H^1_L(\Omega)$ and for $L^2_{\text{div}}(\Omega)$ (which will be defined in the next section); e.g. we could define:

$$H_L^1(\Omega) = \{v \in H^1(\Omega) : Lv \in \Xi^{-1}(\Omega)\};$$
$$L_{\mathrm{div}}^2(\Omega) = \{\mathbf{w} \in (L^2(\Omega))^n : \mathrm{div}\,\mathbf{w} \in \Xi^{-1}(\Omega)\}$$

(see Lions–Magenes, 1972a for the spaces $\Xi^s(\Omega)$ and for such type of modifications). In order to avoid the definition of spaces Ξ we will confine ourselves to hypotheses (H1), (H2), (H3), (H4).

The first way one can think of generalizing theorem 18.3 is, keeping in mind the proof of theorem 18.3 itself, that of extending (18.2) to the case in which $v \in H^1(\Omega)$ and $\mathbf{w} \in [H^1(\Omega)]^n$ and then taking

$$\mathbf{w} = \left\{ -\sum_{j=1}^{n} a_{ij} \frac{\partial u}{\partial x_j} + c_i u \right\}_{i=1,\ldots,n}, \qquad (18.15)$$

with $u \in H_L^1(\Omega)$. Unfortunately this cannot be done; the above extension of (18.2) is perfectly feasible (see Nečas, 1967, p. 121) and we can say that, for every $v \in H^1(\Omega)$ and every $\mathbf{w} \in [H^1(\Omega)]^n$, the formula

$$\int_\Omega v \,\mathrm{div}\,\mathbf{w}\, dx + \int_\Omega (\mathbf{w}, \mathrm{grad}\, v)\, dx = \int_\Gamma \gamma_0 v (\gamma_0 \mathbf{w}, \nu)\, d\sigma \qquad (18.16)$$

is true, but the substitution (18.15) cannot be made because, in general, $\mathbf{w} \notin [H^1(\Omega)]^n$ and therefore we cannot consider $\gamma_0 \mathbf{w} = \{\gamma_0 w_i\}_{i=1,\ldots,n}$. It is important to note that the irregularity of the coefficients a_{ij} and c_i complicates things, but not in a decisive manner: even in the particularly simple case in which $a_{ij} = -\delta_{ij}$ and $c_i = 0$ we cannot give a meaning to $\gamma_0 \mathbf{w}$ (on the other hand, under the previous conditions, $(\gamma_0 \mathbf{w}, \nu)$ can be interpreted as $\gamma_1 u$ if $u \in H^2(\Omega) \ldots$).

We need here a further extension of the concept of trace which would enable us to give a precise meaning to $\gamma_0 \mathbf{w}$ or, still better, to $(\gamma_0 \mathbf{w}, \nu)$. We will now make this extension, but let us immediately say that the method we are using in order to generalize theorem 18.3 is not the one we have just abandoned, though quite similar to it (we will not, however, assume that (18.16) is known).

The space $L_{\mathrm{div}}^2(\Omega)$

Let us remark that if $\mathbf{w}$ can be written in the form (18.15) then $\mathbf{w} \in [L^2(\Omega)]^n$ and $\mathrm{div}\,\mathbf{w} \in L^2(\Omega)$, and one naturally considers the space $L_{\mathrm{div}}^2(\Omega)$ defined as follows:

DEFINITION 18.1. Let $\Omega \subset \mathbb{R}^n$ be an open set. We denote by $L_{\mathrm{div}}^2(\Omega)$ the space $\{\mathbf{w} \in [L^2(\Omega)]^n : \mathrm{div}\,\mathbf{w} = \sum_{i=1}^n (\partial w_i/\partial x_i) \in L^2(\Omega)\}$ provided with the norm of the graph

$$\|\mathbf{w}\|_{L_{\mathrm{div}}^2(\Omega)}^2 = \|\mathbf{w}\|_{[L^2(\Omega)]^n}^2 + \|\mathrm{div}\,\mathbf{w}\|_{L^2(\Omega)}^2 \qquad (18.17)$$

where, naturally, $\|\mathbf{w}\|_{[L^2(\Omega)]^n}^2 = \sum_{i=1}^n \|w_i\|_{L^2(\Omega)}^2$.

We will now give, for the elements of the space $L^2_{\text{div}}(\Omega)$, a 'trace theorem', following the formulation due to Lions and Magenes. For a parallel to the theory developed in Lions–Magenes, 1972a, note that if $v \in H^1(\Omega)$ and $\Delta v \in L^2(\Omega)$ then the problem of the trace of $\partial v/\partial \nu$ coincides with that of the trace of (grad v, ν), and grad $v \in L^2_{\text{div}}(\Omega)$. We will first give two properties of this space. The first is of a structural nature and the second is a density result which will play a fundamental role in what follows.

THEOREM 18.4. $L^2_{\text{div}}(\Omega)$ is a Hilbert space with the scalar product

$$(\mathbf{u}, \mathbf{v})_{L_{\text{div}}^2(\Omega)} = (\mathbf{u}, \mathbf{v})_{[L^2(\Omega)]^n} + (\text{div } \mathbf{u}, \text{div } \mathbf{v})_{L^2(\Omega)}. \tag{18.18}$$

Proof. It can be easily seen that (18.18) is a scalar product, that associated with the norm (18.17), and therefore $L^2_{\text{div}}(\Omega)$ is a pre-Hilbert space. We will now show that $L^2_{\text{div}}(\Omega)$ is a complete space. Let $\mathbf{v}_m$ be a Cauchy sequence in $L^2_{\text{div}}(\Omega)$. It is then a Cauchy sequence in $[L^2(\Omega)]^n$ and, since this is a complete space, it converges in it: let $\mathbf{v}$ be its limit. Further $v_m = \text{div } \mathbf{v}_m$ is also a Cauchy sequence, this time in $L^2(\Omega)$, and therefore converges: let v be its limit. Since the operator div: $[L^2(\Omega)]^n \to \mathscr{D}'(\Omega)$ is continuous it follows that $v = \text{div } \mathbf{v}$ and hence $\mathbf{v} \in L^2_{\text{div}}(\Omega)$. ∎

THEOREM 18.5. If $\Omega \subset \mathbb{R}^n$ is an open set of class $C^{0,1}$ then $[\mathscr{D}(\bar{\Omega})]^n$ is dense in $L^2_{\text{div}}(\Omega)$.

Proof. Suppose, for simplicity, that the vector spaces considered are real: the proof, however, can be modified for the case in which they are complex (the difference lies only in the theorem of representation that one uses). We will prove the theorem by showing that if $F \in (L^2_{\text{div}}(\Omega))'$ is zero on $[\mathscr{D}(\bar{\Omega})]^n$ then F is identically zero. From Riesz's theorem there exists $\mathbf{f} \in L^2_{\text{div}}(\Omega)$ such that

$$_{(L_{\text{div}}^2(\Omega))'}\langle F, \mathbf{w}\rangle_{L_{\text{div}}^2(\Omega)} = (\mathbf{f}, \mathbf{w})_{L_{\text{div}}^2(\Omega)}$$
$$= (\mathbf{f}, \mathbf{w})_{[L^2(\Omega)]^n} + (\text{div } \mathbf{f}, \text{div } \mathbf{w})_{L^2(\Omega)} \quad \forall \mathbf{w} \in L^2_{\text{div}}(\Omega). \tag{18.19}$$

Now, denoting by $\tilde{\mathbf{f}}$ the trivial extension of $\mathbf{f}$ to $\mathbb{R}^n$, we can write

$$0 = (\tilde{\mathbf{f}}, \boldsymbol{\phi})_{[L^2(\mathbb{R}^n)]^n} + (\widetilde{\text{div } \mathbf{f}}, \text{div } \boldsymbol{\phi})_{L^2(\mathbb{R}^n)}$$
$$= \sum_{i=1}^n {}_{\mathscr{D}'(\mathbb{R}^n)}\left\langle \tilde{f}_i - \frac{\partial(\widetilde{\text{div } \mathbf{f}})}{\partial x_i}, \phi_i \right\rangle_{\mathscr{D}(\mathbb{R}^n)} \quad \forall \boldsymbol{\phi} \in [\mathscr{D}(\mathbb{R}^n)]^n, \tag{18.20}$$

and in consequence

$$\tilde{f}_i - \frac{\partial(\widetilde{\text{div } \mathbf{f}})}{\partial x_i} = 0 \quad \text{in } \mathscr{D}'(\mathbb{R}^n), i = 1, \ldots, n, \tag{18.21}$$

which implies, given the regularity of Ω, that div $\mathbf{f} \in H^1_0(\Omega)$. Now let ψ_m be a sequence of elements of $\mathscr{D}(\Omega)$ converging to div $\mathbf{f}$ in $H^1(\Omega)$; we can then write

$$(\operatorname{div} \mathbf{f}, \operatorname{div} \mathbf{w})_{L^2(\Omega)} = \lim_m (\psi_m, \operatorname{div} \mathbf{w})_{L^2(\Omega)}$$

$$= -\lim_m \sum_{i=1}^n \left\langle \frac{\partial \psi_m}{\partial x_i}, w_i \right\rangle_{\mathscr{D}'(\Omega), \mathscr{D}(\Omega)}$$

$$= -(\operatorname{grad} \operatorname{div} \mathbf{f}, \mathbf{w})_{[L^2(\Omega)]^n} \qquad \forall \mathbf{w} \in L^2_{\operatorname{div}}(\Omega). \quad (18.22)$$

It follows, from (18.19) and (18.21), that

$$_{(L^2_{\operatorname{div}}(\Omega))'}\langle F, \mathbf{w}\rangle_{L^2_{\operatorname{div}}(\Omega)} = (\mathbf{f} - \operatorname{grad} \operatorname{div} \mathbf{f}, \mathbf{w})_{[L^2(\Omega)]^n}$$

$$= \sum_{i=1}^n \int_\Omega \left(f_i - \frac{\partial (\operatorname{div} \mathbf{f})}{\partial x_i} \right) w_i \, dx \qquad \forall \mathbf{w} \in L^2_{\operatorname{div}}(\Omega), \quad (18.23)$$

which proves the theorem. ∎

The trace operators $\bar{\gamma}$ and γ_a

Let us now consider the operator

$$\theta : \mathbf{w} \in [\mathscr{D}(\bar{\Omega})]^n \mapsto \theta(\mathbf{w}) \triangleq (\mathbf{w}|_\Gamma, \nu) \in H^{-1/2}(\Gamma) \quad (18.24)$$

and show that it can be extended in a unique way to an operator $\bar{\gamma} \in \mathscr{L}(L^2_{\operatorname{div}}(\Omega), H^{-1/2}(\Gamma))$. The operator θ is well defined (note that $\theta(\mathbf{w}) \in L^2(\Gamma) \subset H^{-1/2}(\Gamma)$) and linear, so that, in order to prove the existence and the uniqueness of the extension $\bar{\gamma}$, we need only show that θ is continuous from $[\mathscr{D}(\bar{\Omega})]^n$, with the topology induced by $L^2_{\operatorname{div}}(\Omega)$, in $H^{-1/2}(\Omega)$, with its natural topology, and then recall that $[\mathscr{D}(\bar{\Omega})]^n$ is dense in $L^2_{\operatorname{div}}(\Omega)$. From (18.2) we have, for every $\mathbf{w}$ fixed in $[\mathscr{D}(\bar{\Omega})]^n$,

$$\left| \int_\Gamma v|_\Gamma \theta(\mathbf{w}) \, d\sigma \right| = \left| \int_\Gamma v|_\Gamma (\mathbf{w}|_\Gamma, \nu) \, d\sigma \right| \le \left| \int_\Omega v \operatorname{div} \mathbf{w} \, dx \right|$$

$$+ \left| \int_\Omega (\mathbf{w}, \operatorname{grad} v) \, dx \right| \le \|v\|_{H^1(\Omega)} \|\operatorname{div} \mathbf{w}\|_{L^2(\Omega)}$$

$$+ \|v\|_{H^1(\Omega)} \sum_{i=1}^n \|w_i\|_{L^2(\Omega)} \le c \|v\|_{H^1(\Omega)} \|\mathbf{w}\|_{L^2_{\operatorname{div}}(\Omega)} \qquad \forall v \in \mathscr{D}(\bar{\Omega}),$$
$$(18.25)$$

where c is an appropriate constant [namely, c is such that $(1/c) \|\mathbf{w}\|_{L^2_{\operatorname{div}}(\Omega)} \le \|\operatorname{div} \mathbf{w}\|_{L^2(\Omega)} + \sum_{i=1}^n \|w_i\|_{L^2(\Omega)} \le c \|\mathbf{w}\|_{L^2_{\operatorname{div}}(\Omega)} \ \forall \mathbf{w} \in L^2_{\operatorname{div}}(\Omega)$] and hence we can say that the linear functional

$$v \in \mathscr{D}(\bar{\Omega}) \mapsto \int_\Gamma v|_\Gamma \cdot \theta(\mathbf{w}) \, d\sigma \in \mathbb{R} \quad (18.26)$$

is continuous on $\mathscr{D}(\bar{\Omega})$ for the topology of $H^1(\Omega)$; then, since $\mathscr{D}(\bar{\Omega})$ is dense in $H^1(\Omega)$, we can extend it in the continuous linear functional

$$v \in H^1(\Omega) \mapsto \int_\Gamma \gamma_0 v \theta(\mathbf{w}) \, d\sigma \in \mathbb{R}, \quad (18.27)$$

and we can therefore consider the continuous linear functional

$$\phi \in H^{1/2}(\Gamma) \mapsto \int_\Gamma \phi \theta(\mathbf{w}) \, d\sigma \in \mathbb{R}, \qquad (18.28)$$

which is obtained from the previous one by noting that it depends on v only through its trace $\gamma_0 v$ and that every function $\phi \in H^{1/2}(\Gamma)$ can be written in the form $\phi = \gamma_0(\mathcal{R}\phi)$, with $\mathcal{R} \in \mathcal{L}(H^{1/2}(\Gamma), H^1(\Omega))$. From all this we conclude that we can write

$$|_{H^{-1/2}(\Gamma)}\langle \theta(\mathbf{w}), \phi \rangle_{H^{1/2}(\Gamma)}| \leq c \, \|\mathcal{R}\|_{\mathcal{L}(H^{1/2}(\Gamma), H^1(\Omega))} \|\phi\|_{H^{1/2}(\Gamma)} \|\mathbf{w}\|_{L^2_{\text{div}}(\Omega)} \qquad (18.29)$$

and hence

$$\|\theta(\mathbf{w})\|_{H^{-1/2}(\Gamma)} \leq c \, \|\mathcal{R}\|_{\mathcal{L}(H^{-1/2}(\Gamma), H^1(\Omega))} \|\mathbf{w}\|_{L^2_{\text{div}}(\Omega)}, \qquad (18.30)$$

an inequality which tells us that θ is continuous and which enables us to consider $\bar\gamma \in \mathcal{L}(L^2_{\text{div}}(\Omega), H^{-1/2}(\Gamma))$.

Let us consider (18.2) once again. Since $\gamma_0 v$ and $\bar\gamma \mathbf{w}$ are, respectively, the extension by continuity of $v|_\Gamma$ and of $(\mathbf{w}|_\Gamma, \nu)$, we can write

$$_{H^{-1/2}(\Gamma)}\langle \bar\gamma \mathbf{w}, \gamma_0 v \rangle_{H^{1/2}(\Gamma)} = \int_\Omega v \, \text{div } \mathbf{w} \, dx + \int_\Omega (\mathbf{w}, \text{grad } v) \, dx, \qquad (18.31)$$

which will replace (18.16). Before going on to the generalization of theorem 18.3, let us define, starting with the operator $\bar\gamma$, an operator $\gamma_a \in \mathcal{L}(H^1_L(\Omega), H^{-1/2}(\Gamma))$ which generalizes the concept of the *conormal derivative*:

DEFINITION 18.2. If $u \in H^1_L(\Omega)$, put $\gamma_a u = \bar\gamma \mathbf{w}$, where $\mathbf{w}$ is the vector obtained from u by applying (18.15).

If u, a_{ij} and c_i are sufficiently regular then $\gamma_a u = \partial u/\partial \nu_a$, and therefore the operator γ_a is in effect a generalization of the operator $\partial/\partial \nu_a$.

The generalized Green's formula

We can now state and prove the following generalization of theorem 18.3:

THEOREM 18.6 (Theorem of the generalized Green's formula). Let $\Omega \subset \mathbb{R}^n$ be a bounded connected open set with boundary Γ of class $C^{0,1}$. If the coefficients a_{ij}, b_i, c_i, d ($i, j = 1, \ldots, n$) of the operator L defined in (18.9) satisfy the conditions (H1), (H2), (H3), (H4) then Green's formula

$$a(u, v) = \int_\Omega (Lu) v \, dx + {}_{H^{-1/2}(\Gamma)}\langle \gamma_a u, \gamma_0 v \rangle_{H^{1/2}(\Gamma)}, \qquad (18.32)$$

is true for every $u \in H^1_L(\Omega)$ and every $v \in H^1(\Omega)$.

Proof. The proof is immediate: one only has to substitute (18.15) in (18.31) and note that the effect of the parts associated with the coefficients b_i and d is the same in both $a(u, v)$ and in $\int_\Omega (Lu) v \, dx$. ∎

Two trace theorems

The operators $\bar{\gamma}$ and γ_a are essentially *trace operators* and it is then natural to ask whether the traces corresponding to them can be 'inverted' with continuity; the answer is yes, and this is the fundamental purpose of the following *trace theorems*:

THEOREM 18.7. Let $\Omega \subset \mathbb{R}^n$ be a bounded connected open set with boundary Γ of class $C^{0,1}$ and external normal ν. Then the following propositions are true:

(I) there exists a unique operator $\bar{\gamma} \in \mathscr{L}(L^2_{\text{div}}(\Omega), H^{-1/2}(\Gamma))$ such that if $\mathbf{w} \in [\mathscr{D}(\bar{\Omega})]^n$ then $\bar{\gamma}(\mathbf{w}) = (\mathbf{w}|_\Gamma, \nu)$;
(II) there exists $\bar{\mathscr{R}} \in \mathscr{L}(H^{-1/2}(\Gamma), L^2_{\text{div}}(\Omega))$ such that $\bar{\gamma}(\bar{\mathscr{R}}) = $ identity; in particular, given any $g \in H^{-1/2}(\Gamma)$ there exists at least one vector $\mathbf{w} = \bar{\mathscr{R}}g \in L^2_{\text{div}}(\Omega)$ such that $\bar{\gamma}(\mathbf{w}) = g$.

THEOREM 18.8. Let $\Omega \subset \mathbb{R}^n$ be a bounded connected open set with boundary Γ of class $C^{0,1}$ and external normal ν; further let L be the differential operator (18.9) where the coefficients a_{ij}, b_i, c_i, d $(i, j = 1, \ldots, n)$ satisfy the conditions (H1), (H2), (H3), (H4). Then the following propositions are true:

(I) there exists a unique operator $\gamma_a \in \mathscr{L}(H^1_L(\Omega), H^{-1/2}(\Gamma))$ such that if $u \in \mathscr{D}(\bar{\Omega})$ and $a_{ij}, c_i \in \mathscr{D}(\bar{\Omega})$ then $\gamma_a(u) = \partial u / \partial \nu_a$;
(II) there exists $\mathscr{R}_a \in \mathscr{L}(H^{-1/2}(\Gamma), H^1_L(\Omega))$ such that $\gamma_a(\mathscr{R}_a) = $ identity; in particular, given any $g \in H^{-1/2}(\Gamma)$ there exists at least one function $u = \mathscr{R}_a g \in H^1_L(\Omega))$ such that $\gamma_a(\mathscr{R}_a g) = g$.

Proofs. Part (I) of both theorems has already been proved. To prove the existence of the 'inverses' $\bar{\mathscr{R}}$ and $\mathscr{R}_a$ let us consider the following problem:

PROBLEM 18.1. Given $g \in H^{-1/2}(\Gamma)$, find $\lambda \in \mathbb{R}$ and $u \in H^1(\Omega)$ such that

$$a(u, v) + \lambda(u, v)_{L^2(\Omega)} = {}_{H^{-1/2}(\Gamma)}\langle g, \gamma_0 v \rangle_{H^{1/2}(\Gamma)} \quad \forall v \in H^1(\Omega). \quad (18.33)$$

With λ large enough this problem is well-posed since, in that case, in the left-hand side of (18.33) we have the value that a continuous bilinear form coercive on $H^1(\Omega)$ takes on u and v, and, on the other hand, in the right-hand side we have the value that the functional $g \circ \gamma_0 \in (H^1(\Omega))'$ takes on v. This problem is not in the framework presented in Section 7.2.2 since we consider the dual of $H^1(\Omega)$, which is not a normal space; the problem 18.1 is, however, in the framework presented in Chapter 3.

It can be easily seen, through the application of (18.32), that the problem

18.1 is equivalent to the following problem:

PROBLEM 18.2. Given $g \in H^{-1/2}(\Gamma)$ and $\lambda \in \mathbb{R}$ find $u \in H^1(\Omega)$ such that

$$Lu + \lambda u = 0 \quad \text{in } \Omega \tag{18.34}$$

$$\gamma_a u = g \quad \text{on } \Gamma. \tag{18.35}$$

The latter problem is also therefore well-posed if λ is large enough, which we will always henceforth assume to be the case. Now let $u = u(g)$ be the solution of problem 18.2; it is obvious that $u(g) \in H_L^1(\Omega)$ and, given that problem 18.2 is well-posed, the application $g \mapsto u(g)$ is continuous from $H^{-1/2}(\Gamma)$ to $H_L^1(\Omega)$. To complete the proof of theorems 18.7 and 18.8 we need only take as $\mathcal{R}_a$ the application $u \mapsto u(g)$ and as $\bar{\mathcal{R}}$ the application $u \mapsto \operatorname{grad} u(g)$. ∎

The space $H_{00}^{1/2}(\Gamma)$ and its dual

Finally, we will present, without proof, the Green's formula suitable for the study of the mixed problems considered in the Chapter 7:

THEOREM 18.9. Let $\Omega \subset \mathbb{R}^n$ be a bounded connected open set with boundary Γ of class $C^{0,1}$, let Γ_0 be a connected open set of Γ with a positive $(n-1)$-dimensional measure and $\Gamma_1 = \Gamma \backslash \bar{\Gamma}_0$. If the coefficients a_{ij}, b_i, c_i, d $(i, j = 1, \ldots, n)$ of the operator L satisfy the conditions (H1), (H2), (H3), (H4) then the Green's formula

$$a(u, v) = \int_\Omega (Lu)v \, dx + {}_{(H_{00}^{1/2}(\Gamma_1))'}\langle \gamma_a u|_{\Gamma_1}, \gamma_0 u|_{\Gamma_1}\rangle_{H_{00}^{1/2}(\Gamma_1)} \tag{18.36}$$

holds, whatever $u \in H_L^1(\Omega)$ and $v \in V_{\Gamma_0} = \{u \in H^1(\Omega): \gamma_0 u = 0 \text{ on } \Gamma_0\}$, where $H_{00}^{1/2}(\Gamma_1) = \{\gamma_0 v|_{\Gamma_1} : v \in V_{\Gamma_0}\}$ must be provided with the topology determined by the norm

$$\|\psi\|_{H_{00}^{1/2}(\Gamma_1)} = \inf \{\|v\|_{H^1(\Omega)} : v \in V_{\Gamma_0} \quad \text{and} \quad \psi = \gamma_0 v|_{\Gamma_1}\}. \tag{18.37}$$

The use of the spaces $H_{00}^{1/2}(\Gamma_1)$ and $(H_{00}^{1/2}(\Gamma_1))'$ is necessary because the operation of restriction to $\Gamma_1 \subset \Gamma$ of the elements of $H^{-1/2}(\Gamma)$ is not valued on $H^{-1/2}(\Gamma_1)$ but in a larger space, namely $(H_{00}^{1/2}(\Gamma_1))'$ (a dual property of the fact that the trivial extension $u \mapsto \tilde{u}$ operates with values in $H^{1/2}(\Gamma)$, *not* starting from $H^{1/2}(\Gamma_1)$ but from $H_{00}^{1/2}(\Gamma_1)$; i.e.: $u \mapsto \tilde{u} \in \mathcal{L}(H_{00}^{1/2}(\Gamma_1), H^{1/2}(\Gamma))$ and $u \in H^{1/2}(\Gamma_1) \not\Rightarrow \tilde{u} \in H^{1/2}(\Gamma)$). For a treatment of these spaces see Lions–Magenes, 1972a; pp. 66 *et seq.*; here we will merely note that $\phi \in H_{00}^{1/2}(\Gamma_1)$ iff the trivial extension of ϕ to Γ belongs to $H^{1/2}(\Gamma)$.

REMARK 18.1. In many applications the space $L_{\text{div}}^2(\Omega)$ must be replaced by:

$$L_{p,\text{div}}^2(\Omega) = \{\mathbf{w} \in [L^2(\Omega)]^n : \operatorname{div} \mathbf{w} \in L^p(\Omega)\},$$

p being such that $H^1(\Omega) \subset L^p(\Omega)$. The results we prove for $L^2_{\text{div}}(\Omega)$ still hold for $L^2_{p,\text{div}}(\Omega)$.

REMARK 18.2. The extension by continuity is not the only way of defining the trace of a function; we will mention here two other possible ways:

(1) By using the concept of convergence on the mean of the values taken on 'parallel manifolds' (see Cimmino, 1937; 1952); see also Lions–Magenes, 1972a, p. 191 for the relationship between this concept and the extension by continuity.
(2) By starting from Łojasiewicz's definition of the value of a distribution at a point (see Łojasiewicz, 1957 and the references given there)—this theory has been used both in the study of boundary value problems (see, e.g., Boutet de Monvel–Geymonat, 1971) and in the study of initial value problems (see, e.g., Ferreira–Oliveira, 1964) and enables one also to construct an interesting theory of integration in the field of distributions (see, e.g., Silva, 1964; Ferreira, 1967).

19
Ordered Structures

19.1 BASIC DEFINITIONS

In this section we will recall some well known definitions. The main purpose of this is to establish the nomenclature that we will use: for a more detailed and systematic treatment see, e.g. Bourbaki, 1963 and the bibliography cited at the beginning of sections 19.2 and 19.3.

DEFINITION 19.1. Let A be a non-empty set. A binary relationship $\leq$ in A is said to be an *order* if

(ω_1) $\forall x \in A$ $x \leq x$ (reflexivity),
(ω_2) $\forall x, y \in A$ $[x \leq y \text{ and } y \leq x] \Rightarrow x = y$ (antisymmetry),
(ω_3) $\forall x, y, z \in A$ $[x \leq y \text{ and } y \leq z] \Rightarrow x \leq z$ (transitivity).

An order which satisfies

(ω_4) $\forall x, y \in A$ $x \leq y$ or $y \leq x$ (trichotomy)

is a *total order*, and an order which satisfies

(ω_5) $\forall X \subset A,$ $X \neq \emptyset$ $\exists a \in X : \forall x \in X$ $a \leq x$

is a *well-ordering*.

Let us recall that a *binary relationship* ρ in a set C is any part ρ of C^2; if $(x, y) \in \rho$ we write $x\rho y$ and say that x is in relation ρ with y. Thus we can say that an order $\leq$ is a part of A^2 which has the properties which we have defined by (ω_1), (ω_2), and (ω_3) ((ω_1), e.g., is the expression of the fact that the diagonal $\Delta_A = \{(x, x) : x \in A\}$ of A^2 is contained in $\leq$; for the other conditions see Bourbaki, 1963; section 1). If $(x, y) \in \leq$, x and y are said (in relation $\leq$, or) to be comparable: again, to make things clearer, x is said to precede y, y to follow x, x is less or equal to y, etc. If $x \leq y$ we can also write $y \geq x$—the relationship $\geq$ is an order in A, as can be easily seen. As usual, we write $x < y$ (or: $x > y$) to denote that $x \leq y$ and $x \neq y$ (or: $x \geq y$ and $x \neq y$). Thus (ω_4) can be written in the form '(ω_4'): $\forall x, y \in A$ $x < y$, $x = y$ or $x > y$', a formulation which justifies the name trichotomy which we have given it. In agreement with the nomenclature introduced later (definition 19.2) we can interpret condition (ω_5) by saying that 'every non-empty subset of A has a minimum': the existence of a minimum (or the first element) of every non-empty subset implies that it is possible to define the successor of every element, which, in a certain sense, justifies the use of the adjective 'well'.

The algebraic system $\langle A, \leq \rangle$ (or, if there is no risk of confusion, A) is called an *ordered set*, a *totally ordered set* or a *well ordered set* depending on whether, respectively, $\leq$ is an order, a total order, or a well-ordering (let us remark that well-ordered sets are totally ordered: we need only put $X = \{x, y\}$ in (ω_5)...). The literature has a large variety of nomenclature which often creates confusion. Thus, some authors refer to what we have called total order as order and what we have called order as partial order. Totally ordered sets are also known as *linearly ordered sets* or chains; besides, we use the term chain in a much wider context here (see definition 19.3).

We have then defined what we mean by an ordered set, a totally ordered set and a well ordered set; we have not, however, explained how to order, totally order or well order a given set. Further, there is a big difference between the concept of order and those of total order or well-ordering. Thus, it is obvious that every non-empty set A can be ordered, i.e., it is possible to construct an algebraic system $\langle A, \leq \rangle$ with $\leq$ order in A: all we need to do is to consider the system $\langle A, \Delta_A \rangle$.

Of greater interest is the fact that, using the axiom of Zermelo, one can show that every non-empty set can be well ordered; further, the statement that *every non-empty set can be well ordered* (the theorem of well ordering) is equivalent to the axiom of Zermelo!

Let us recall that Zermelo's axiom states that *the Cartesian product of a non-empty family of non-empty sets is non-empty* (i.e., in every set of the family we can choose an element: it is this possibility of an arbitrary choice (either successive or simultaneous) which gives the name 'axiom of choice' by which this axiom is often referred to). Since the possibility of 'well ordering' a set is a property which is not very intuitive, many mathematicians have in the past been against the axiom of choice (which in effect implies the possibility of well ordering); however functional analysis would suffer greatly in the absence of this axiom: e.g. the Hahn–Banach theorem and the Tychonov theorem on the product of compact spaces would no longer be true! Now, the consistency of the Zermelo–Fraenkel axiomatic having been proved, often one uses an equivalent property in place of the axiom of choice or of the theorem of well ordering: the Zorn's lemma (see theorem 19.1 which follows). Let us remark that also the Tychonov theorem on the product of compact spaces is equivalent to the axiom of choice (see Kelley, 1950).

The fact that *every non-empty set can be totally ordered* (the theorem of total ordering) can be proved by using the theorem of well ordering and then the fact that well-ordered sets are totally ordered, or by applying to the system $\langle A, \Delta_A \rangle$ a result due to Szpilrajn (see Szpilrajn, 1930) which states that *every ordering can be extended into a total ordering*. For the connection between the possibility of extending orders and the axiom of choice see, e.g., Rubin–Rubin, 1963, pp. 24 and 101.

Let us now introduce the following definition:

DEFINITION 19.2. Let $\langle A, \leq \rangle$ be an ordered set and let $X \subset A$. We will say that

(i) $a \in X$ is a *maximal* (or: *minimal*) element of X if, for every $x \in X$, $a \leq x$ (or: $x \leq a$) $\Rightarrow x = a$;

(ii) $a \in X$ is the *maximum* (or: *minimum*) of X if, for every $x \in X$, $x \leq a$ (or: $a \leq x$);

(iii) $a \in A$ is an *upper bound* (or: *lower bound*) of X if, for every $x \in X$, $x \leq a$ (or: $a \leq x$);

(iv) $a \in A$ is the *supremum* or *least upper bound* (or: *infimum* or *greatest lower bound*) of X if a is the minimum (or: maximum) of the set of upper bounds (or: lower bounds) of X.

It can be easily seen that the maximum (or: minimum) of a set X, if it exists, is unique and it is then the unique maximal (or: minimal) element of X. Further the least upper bound (or: lower bound) of X, if it exists, is also unique and if it belongs to X it coincides with the maximum (or: minimum) of X. We normally denote by max X, min X, sup X, and inf X, respectively, the maximum, minimum, least upper bound, and greatest lower bound of X. If the set of upper bounds (or: lower bounds) of a set X is non-empty, X is said to be *bounded above* (or: *below*); if X is bounded below and above we say that X is a *bounded set*. In mathematics it is important to place every concept and result in its proper context—so, where should we place the concept of boundedness? Here it is linked to order, in metric spaces it is linked to the metric, in linear spaces to absorbing sets, etc. The concept of boundedness, axiomized in the bornological structures (see e.g. Hogbe-Nlend, 1977, p. 18; Waelbroeck, 1971, p. 23), seems to avoid a classification. . . .

Returning to ordered sets, the case $X = \emptyset$ deserves special treatment since the set $\emptyset$ presents a pathological behaviour—naturally it does not have maximal or minimal elements, maximum or minimum; on the other hand, all the elements of A are upper bounds and lower bounds of $\emptyset$, and if min A and max A exist then sup $\emptyset = $ min A and inf $\emptyset = $ max A (this is the only case of a set X such that inf $X \not\leq $ sup X. . .).

DEFINITION 19.3. Let $\langle A, \leq \rangle$ be an ordered set. A non-empty subset $L \subset A$ is said to be a *chain* of A if the algebraic system $\langle L, \leq \rangle$ is a totally ordered set. The algebraic system $\langle A, \leq \rangle$ is said to be *(s)-inductive* (or: *(i)-inductive*) if every chain of A is bounded above (or: below) and $\langle A, \leq \rangle$ is said to be *completely (s)-inductive* (or: *completely (i)-inductive*) if every chain of A has a least upper bound (or: greatest lower bound); again, $\langle A, \leq \rangle$ is said to be *inductive* (or: *completely inductive*) if it is (s)-inductive and (i)-inductive (or: completely (s)-inductive and completely (i)-inductive).

Of course, in $\langle L, \leq \rangle$ we denote again by $\leq$ the order induced on L by the one defined in A. The order so induced is $\leq \cap L^2$, but to simplify notations this set is still denoted by $\leq$.... In principle we denote by $\leq$ all the relationships of order which we will deal with: the use of two different symbols to denote as many relationships of order will however be necessary when different relationships exist on the same set.

We will now present a very important theorem:

THEOREM 19.1 (Zorn's lemma). An (s)-inductive (or: (i)-inductive) set has at least one maximal (or: minimal) element.

To end this section let us recall that if $\langle A, \leq \rangle$ is an ordered set and a, $b \in A$, with $a \leq b$, the set $[a, b] = \{x \in A : a \leq x \leq b\}$ (or: $]a, b[= \{x \in A : a < x < b\}$) is called the closed (or: open) *interval* with origin at a and extremity at b; in an analogous manner one can define the intervals $]a, b]$ and $[a, b[$. These intervals are bounded (all the non-empty of these have greatest lower bound a and least upper bound b): in an analogous manner one can define unbounded intervals. Let us note that, contrary to what happens with intervals of $\mathbb{R}$, an open interval $]a, b[$ can be empty even when $a < b$, the intersection of two intervals is not necessarily an interval and an interval is not necessarily a chain.

19.2 LATTICES

In this section we will deal briefly with an algebraic structure 'half-way' between ordered sets and totally ordered sets: the lattice structure; for further information see Birkhoff, 1961; Rutherford, 1965 and the bibliography cited at the beginning of the next section.

Totally ordered sets are characterized by the fact that every non-empty finite subset of them has a maximum and a minimum. We will now introduce a class of structures which are characterized by a similar but weaker property, namely:

DEFINITION 19.4. An ordered set $\langle A, \leq \rangle$ is said to be a *lattice* (and $\leq$ is said to be a *lattice-order*) if every non-empty finite subset of A has a greatest lower bound and a least upper bound.

In an analogous manner, considering well ordered sets instead of totally ordered sets, one arrives at the following concept, 'half-way' between well ordered sets and ordered sets, namely:

DEFINITION 19.5. An ordered set $\langle A, \leq \rangle$ is said to be a *complete lattice* if every non-empty subset of A has a least upper bound and a greatest lower bound. (In particular, if $\langle A, \leq \rangle$ is a complete lattice then max A and min A exist—if however a lattice has a maximum and a minimum this does not necessarily imply that it is complete.)

In applying these concepts to functional analysis, nevertheless, the following weaker concept is more useful:

DEFINITION 19.6. An ordered set $\langle A, \leq \rangle$ is said to be a λ-*complete lattice* if every non-empty bounded subset of A has a greatest lower bound and a least upper bound.

The lattice structure is very common in mathematics—we will give here a curious example: let T be the 2-simplex shown in fig. 19.1, let $\Phi = \{\varnothing, \phi_1, \phi_2, \phi_3, \phi_{12}, \phi_{23}, \phi_{31}, \phi_{123} = T\}$ be the set of faces of T, and let us introduce in Φ the order defined as follows: if $a, b \in \Phi$ then '$a \leq b$ iff a is a face of b'. The system $\langle \Phi, \leq \rangle$, shown in fig. 19.2 as a so-called Hasse diagram (the interpretation of which is obvious), is a lattice (complete, as is every finite lattice). This lattice is isomorphic (in the sense of the following definition 19.8) to the lattice $\langle 2^{\{1,2,3\}}, \leq \rangle$ where $\leq$ is the set inclusion (i.e., if $A, B \in 2^{\{1,2,3\}}$ then we write $A \leq B$ iff $A \subset B$). Instead of a simplex we can naturally consider any other polytope; this discussion enables us then to make an interesting partition of the set of polytopes in equivalence classes: two polytopes are said to be combinatorially equivalent, or of the same combinatorial type, if the corresponding lattices are isomorphic.

We will now present some properties of lattices. The choice of these properties has a specific purpose which will become clear later—for many other properties see the bibliography cited. For convenience (but not only for that purpose) we denote by $x \wedge y$ (read 'x inf y') the greatest lower bound of $\{x, y\}$ and by $x \vee y$ (read 'x sup y') the least upper bound of $\{x, y\}$.

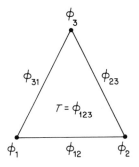

Figure 19.1

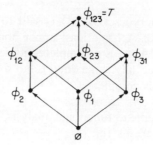

Figure 19.2

THEOREM 19.2. If $\langle A, \leq \rangle$ is a lattice then, for every $x, y, z \in A$,

$$x \wedge y = y \wedge x \quad \text{and} \quad x \vee y = y \vee x, \tag{19.1}$$

$$x \wedge (y \wedge z) = (x \wedge y) \wedge z \quad \text{and} \quad x \vee (y \vee z) = (x \vee y) \vee z, \tag{19.2}$$

$$x \wedge (x \vee y) = x \quad \text{and} \quad x \vee (x \wedge y) = x, \tag{19.3}$$

$$(x \wedge y) \vee (x \wedge z) \leq x \wedge (y \vee z) \quad \text{and} \quad x \vee (y \wedge z) \leq (x \vee y) \wedge (x \vee z), \tag{19.4}$$

$$x \leq y \Rightarrow x \vee (y \wedge z) \leq y \wedge (x \vee z), \tag{19.5}$$

$$x \leq y \Leftrightarrow x \vee y = y \quad \text{and} \quad x \leq y \Leftrightarrow x \wedge y = x. \tag{19.6}$$

Proof. We will prove, as an example, that $x \leq y \Leftrightarrow x = x \wedge y$: for the other properties the same kind of reasoning can be used (besides, we need only prove 'half' of the properties in the theorem and then use the so-called *duality principle*). If $x = x \wedge y$ then, from the definition of greatest lower bound, $x \leq x$ and $x \leq y$: therefore $x = x \wedge y \Rightarrow x \leq y$. Conversely, if $x \leq y$ then, since $x \leq x$, x is a lower bound of $\{x, y\}$: to see that x is in fact the greatest of the lower bounds and hence that $x \leq y \Rightarrow x = x \wedge y$ we need only note that $x \in \{x, y\}$. ∎

Properties (19.1), (19.2), and (19.3) have a decisively operative aspect and they suggest that $\wedge$ and $\vee$ can be treated not simply as abbreviations but as symbols of two internal operations in A, i.e. as two applications $\wedge: A^2 \to A$ and $\vee: A^2 \to A$. At this point one naturally considers the algebraic system $\langle A, \wedge, \vee \rangle$, where $\wedge$ and $\vee$ are defined on the basis of some of their operative properties and not on the basis of an order $\leq$ in A; we will introduce then the following definition:

DEFINITION 19.7. We will refer as a *lattice* to an algebraic system $\langle A, \wedge, \vee \rangle$, where A is a non-empty set and $\wedge: A^2 \to A$, $\vee: A^2 \to A$ (the operations inf and sup of the lattice) satisfy (19.1) (commutativity), (19.2) (associativity), and (19.3) (absorption).

With this axiomatic definition two questions arise, as usual: its compatibility (or consistency, or the existence of a 'model') and its categoricity (or minimality, or the independence between axioms). The preceding axiomatic is consistent since the lattices introduced in definition 19.4 satisfy it (this is nothing but theorem 19.2); it is also categoric, as can be seen by constructing algebraic systems $\langle A, \wedge, \vee \rangle$ which satisfy two axioms and the negation of the third (see Rutherford, 1965, p. 7 for the particulars). In our case there is still another question of a 'linguistic' nature: why are we denoting by a single name (lattice) the algebraic systems introduced in definitions 19.4 and 19.7? The answer is naturally that the two definitions are equivalent; in other words, one has the following result:

THEOREM 19.3. *If* $\langle A, \leqslant \rangle$ *is a lattice in the sense of definition* 19.4 *then* $\langle A, \wedge, \vee \rangle$ *(with* $\vee$ *and* $\wedge$ *defined as applications which associate with* $(x, y) \in A^2$, *respectively, the least upper bound and the greatest lower bound of the set* $\{x, y\}$*) is a lattice in the sense of definition* 19.7. *Conversely, if* $\langle A, \wedge, \vee \rangle$ *is a lattice in the sense of definition* 19.7 *then* $\langle A, \leqslant \rangle$ *(with* $\leqslant$ *defined* (!) *on the basis of* $x \leqslant y \Leftrightarrow x = x \wedge y$*) is a lattice in the sense of definition* 19.4.

Proof. The first part follows directly from our previous discussion. The second part is also immediate: we need only observe that the binary relationship $\leqslant$ defined by $x \leqslant y \Leftrightarrow x = x \wedge y$ is an order in A and that, with respect to this order, $x \wedge y = \inf\{x, y\}$ and $x \vee y = \sup\{x, y\}$, from which it follows that every non-empty finite subset of A has a greatest lower bound and a least upper bound. ∎

The property (19.4) translates a 'quasi-distributivity' of $\wedge$ and $\vee$: by strengthening it we obtain the so-called *distributive lattices*, namely: a lattice $\langle A, \wedge, \vee \rangle$ is distributive if, for every $x, y, z \in A$,

$$x \wedge (y \vee z) = (x \wedge y) \vee (x \wedge z) \quad \text{and} \quad x \vee (y \wedge z) = (x \vee y) \wedge (x \vee z). \quad (19.7)$$

If we strengthen the property (19.5), on the other hand, we obtain the so-called *modular lattices*, which are the lattices $\langle A, \wedge, \vee \rangle$ such that, for every $x, y, z \in A$

$$x \leqslant y \Rightarrow x \vee (y \wedge z) = y \wedge (x \vee z) \quad (19.8)$$

(distributive lattices are also modular, but the converse is not true).

To end this section we give a definition of homomorphism and isomorphism of lattices, which has already been referred to:

DEFINITION 19.8. Let $\langle A, \wedge, \vee \rangle$ and $\langle B, \wedge, \vee \rangle$ be two lattices. An application $f: A \to B$ is a *homomorphism* of lattices if

$$\forall x, y \in A \quad f(x \wedge y) = f(x) \wedge f(y) \quad \text{and} \quad f(x \vee y) = f(x) \vee f(y). \quad (19.9)$$

If f is one-to-one and its inverse is also a homomorphism of lattices, f is said to be an *isomorphism* of lattices and we say that the two lattices are *isomorphic*.

19.3 ORDERED VECTOR SPACES. VECTOR LATTICES

The most fundamental structures of functional analysis are undoubtedly topological vector spaces (t.v.s.) and these, as we have seen in Chapter 14, are obtained by introducing in a set both a vectorial structure (in general over $\mathbb{R}$ or $\mathbb{C}$) and a topological structure (compatible with the former in the sense that the 'vector operations' are continuous). In the first part of this section we will deal with ordered vector spaces (o.v.s.), which are obtained in a manner analogous to that just described but with the topology substituted by an order (compatible with the 'vector operations' in a sense which we will define later); in the second part we will assume that the order is of lattice type and hence consider vector lattices (v.l.). For further information on these structures see, among others, Namioka, 1957; Luxemburg–Zaanen, 1971; Schaefer, 1966; 1974; Jameson, 1970; Aliprantis–Burkinshaw, 1978; Wong-Ng, 1973.

Ordered vector spaces

DEFINITION 19.9. Let E be a vector space over $\mathbb{R}$ and let $\leq$ be an order in E. $\langle E, \leq \rangle$ is said to be an *ordered vector space* if the following compatibility conditions are satisfied

$$\forall f, g, h \in E \quad f \leq g \Rightarrow f + h \leq g + h \qquad (19.10)$$

$$\forall f \in E \quad \forall \alpha \in \mathbb{R}, 0 \leq \alpha \quad f \leq 0 \Rightarrow \alpha f \leq 0. \qquad (19.11)$$

In Chapters 4 and 5 we have considered several vector spaces over $\mathbb{C}$, starting from which one can define in an obvious manner as many vector spaces over $\mathbb{R}$; the latter, provided with fairly natural orders, are examples of o.v.s. We have hardly even stated these orders explicitly, $\mathscr{D}'(\Omega)$ being an important exception: in effect we have said what we mean by a positive (real) distribution, and we have therefore introduced an order in $\mathscr{D}'(\Omega)$. This example suggests two questions:

(1) *The definition of the order induced by an o.v.s. on its dual*;
(2) *The definition of an order in a vector space through the specification of the positive elements.*

Let us first analyse Question (1), considering of course the algebraic dual since, in the absence of a topology, we cannot consider a topological dual (observe also that the algebraic dual of a vector space contains all its possible topological duals...).

DEFINITION 19.10. Let $\langle E, \leq \rangle$ be an o.v.s. and let E^* be the algebraic dual of E (i.e., the set of linear applications $f: E \to \mathbb{R}$). The *dual order* of $\leq$ is the order, also denoted by $\leq$, defined in E^* as follows: if $f^*, g^* \in E^*$ put

$$f^* \leq g^* \Leftrightarrow \forall x \in E, 0 \leq x \quad f^*(x) \leq g^*(x). \qquad (19.12)$$

It can be easily seen that the binary relationship defined in E^* by (19.12) is effectively an order, with respect to which $\langle E^*, \leq \rangle$ is an o.v.s.

It happens sometimes that E is identified (due to the 'existence' of a topology...) with a part of E^*, and the problem of the compatibility between the order (19.12) and that previously defined in E then arises...: in the case $E = H_0^1(\Omega)$, e.g., there is in fact compatibility between the order ('almost everywhere') in this space and the ('distributional') order in $H^{-1}(\Omega) \subset (H_0^1(\Omega))^*$. We have spoken of 'distributional order', and the definition 19.10 is of the type introduced for $\mathscr{D}'(\Omega)$ (a part of $\mathscr{D}^*(\Omega)$...): the only difference is that there we had assumed $f^* = 0$, i.e. we have only made the comparison with 0, i.e. we have considered only the positive elements. This brings us to Question (2), before dealing with which we will give the following definition:

DEFINITION 19.11. Let E be a real vector space. A set $C \subset E$, $C \neq \emptyset$, is said to be a *cone* (or more precisely, a *convex cone with its vertex at the origin*) if

$$\forall x, y \in C \quad x + y \in C, \tag{19.13}$$

$$\forall x \in C \quad \forall \lambda \geq 0 \quad \lambda x \in C. \tag{19.14}$$

C is said to be a *strict cone* if

$$\forall x \in C \setminus \{0\} \quad -x \notin C, \tag{19.15}$$

C is said to be a *total cone* if

$$\forall x \in E \quad x \in C \quad \text{or} \quad -x \in C, \tag{19.16}$$

C is said to be a *generator cone* if

$$\forall x \in E \quad \exists y, z \in C: \quad x = y - z. \tag{19.17}$$

Let us remark that, with the notation $\mu_1 A_1 \pm \mu_2 A_2 = \{\mu_1 a_1 \pm \mu_2 a_2 : a_1 \in A_1, a_2 \in A_2\}$ ($\mu_1, \mu_2 \in \mathbb{R}$), the conditions (19.13), (19.14), (19.15), (19.16), and (19.17) can be written, respectively, as $C + C \subset C$, $\lambda C \subset C$ ($\forall \lambda \geq 0$), $C \cap (-C) = \{0\}$, $C \cup (-C) = E$ and $C - C = E$. On the other hand, due to (19.14), (19.17) can also be written lin $(C) = E$ (where lin (A) is the set of all (finite) linear combinations of elements of A). For every cone C of E, $C \cap (-C)$ is the 'biggest' linear subspace of E contained in C and $C - C$ is the 'smallest' linear subspace of E which contains C.

Let us now consider an o.v.s. $\langle E, \leq \rangle$ and denote by E_+ the set $E_+ = \{x \in E : 0 \leq x\}$. It can be easily seen that E_+ is a strict cone, which is also total if the order $\leq$ in E is total. Conversely, let E be a real vector space, let C be a strict cone of E and let us introduce in E the binary relationship $\leq$ defined as follows:

$$\forall x, y \in E \quad x \leq y \Leftrightarrow y - x \in C. \tag{19.18}$$

It can be easily seen that $\leq$ is an order in E, relative to which $\langle E, \leq \rangle$ is an o.v.s. and $E_+ = C$; further, this order is total if C is a total strict cone. We see then that introducing an order in a vector space E (an order compatible with the vectorial structure of E) is equivalent to specifying a strict cone in E, the so-called *cone of positive elements* of E. This is equally true if instead of E_+ we consider the cone $E_- = \{x \in E : x \leq 0\}$ of negative elements of E.

We have introduced the concept of a generator cone for completeness, but we will not deal with it here—it is a concept related to that of a directed set in the Moore–Smith theory of convergence.

Vector lattices

We will deal next with vector lattices, spaces first introduced by Riesz (cf. Riesz, 1930):

DEFINITION 19.12. Let $\langle E, \leq \rangle$ be an o.v.s. If every non-empty finite subset of E has a least upper bound and a greatest lower bound $\langle E, \leq \rangle$ is said to be a *vector lattice*. The λ-complete vector lattices are called *Riesz spaces*.

This definition of a vector lattice, which is based on definition 19.4, is equivalent to the following one, based instead on definition 19.7:

DEFINITION 19.13. We call a *vector lattice* an algebraic system $\langle E, \wedge, \vee \rangle$, where E is a vector space over $\mathbb{R}$ and $\wedge : E^2 \to E$, $\vee : E^2 \to E$ satisfy (19.1), (19.2), and (19.3) and are compatible with the vectorial structure of E in the sense that

$$\forall x, y, z \in E \quad (x \wedge y) + z = (x + z) \wedge (y + z) \quad \text{and}$$
$$(x \vee y) + z = (x + z) \vee (y + z), \quad (19.19)$$
$$\forall x, y \in E \quad \forall 0 \leq \lambda \quad \lambda(x \wedge y) = \lambda x \wedge \lambda y \quad \text{and}$$
$$\lambda(x \vee y) = \lambda x \vee \lambda y. \quad (19.20)$$

The conditions (19.19) and (19.20) imply that the order $\leq$ in E, defined on the basis of $\wedge$ or $\vee$ according to (19.6), is compatible with the vector structure of E in the sense of (19.10) and (19.11).

Starting from (19.19) and (19.20) one can easily show that, for every non-empty subset F of E which has a least upper bound and a greatest lower bound, we have

$$\forall x \in E \quad x + \sup F = \sup(x + F) \quad \text{and} \quad x + \inf F = \inf(x + F), \quad (19.21)$$
$$\sup F = -\inf(-F). \quad (19.22)$$

Condition (19.19) translates a distributivity of $+$ relative to $\wedge$ and $\vee$. However, there is no distributivity of $\wedge$ or $\vee$ relative to $+$: one can show nevertheless that if x, y, $z \in E_+$ then

$$(x + y) \wedge z \leq x \wedge z + y \wedge z \quad (19.23)$$

(see e.g. Schaefer, 1974, p. 53). In connection with distributivity, let us note that vector lattices are necessaily distributive lattices, an unexpected result which gives us a measure of the 'strength' of the conditions of compatibility between vectorial structure and lattice operations. This 'strength' gives to the vector lattices a richness of structure which is reflected in the following definition, where we come across some well-known concepts of real analysis:

DEFINITION 19.14. Let $\langle E, \wedge, \vee \rangle$ be a vector lattice and $x \in E$. We call *positive part* of x and denote by $[x]^+$ the element

$$[x]^+ = x \vee 0, \qquad (19.24)$$

We call *negative part* of x and denote by $[x]^-$ the element

$$[x]^- = (-x) \vee 0, \qquad (19.25)$$

We call *modulus* or *absolute value* of x and denote by $|x|$ the element

$$|x| = x \vee (-x). \qquad (19.26)$$

According to this definition we can consider, along with the applications $\wedge: E^2 \to E$; $\vee: E^2 \to E$, the applications $[\cdot]^+: E \to E_+$, $[\cdot]^-: E \to E_+$, $|\cdot|: E \to E_+$ which with $x \in E$ associate, respectively, $[x]^+, [x]^-, |x| \in E_+$. It is interesting to observe that all the applications $\wedge, \vee, [\cdot]^+, [\cdot]^-, |\cdot|$ can be expressed in terms of only one of them; thus, e.g., from (19.24), (19.25), (19.26) and (see below) (19.28) one can write, respectively, $[\cdot]^+, [\cdot]^-, |\cdot|$ and $\wedge$ in terms of $\vee$. From this it follows in particular that an o.v.s. $\langle E, \leqslant \rangle$ is a vector lattice iff for every $x \in E$ there exists at least one of $x \wedge 0$, $x \vee 0$, $[x]^+$, $[x]^-$ and $|x|$ (and therefore that they all exist...).

The following is a list of elementary properties of vector lattices, the proof of which is easy (see, e.g., Schaefer, 1974, pp. 51, 52):

THEOREM 19.4. Let $\langle E, \wedge, \vee \rangle$ be a vector lattice. For every $x, y \in E$ we have

$$x = [x]^+ - [x]^-, \quad |x| = [x]^+ \vee [x]^- = [x]^+ + [x]^-,$$
$$[x]^+ \wedge [x]^- = 0, \quad |x| = 0 \Leftrightarrow x = 0, \quad |\lambda x| = |\lambda||x| \quad \forall \lambda \in \mathbb{R}; \qquad (19.27)$$
$$x + y = (x \wedge y) + (x \vee y); \quad |x + y| \leqslant |x| + |y|,$$
$$||x| - |y|| \leqslant |x - y| = (x \vee y) - (x \wedge y); \qquad (19.28)$$
$$[x - y]^+ = [y - x]^- = -y + (y \vee x) = x - (x \wedge y), \quad [x + y]^+ \leqslant [x]^+ + [y]^+,$$
$$[x + y]^- \leqslant [x]^- - [y]^-, \quad |[x]^+ - [y]^+| \leqslant |x - y|, \quad |[x]^- - [y]^-| \leqslant |x - y|; \qquad (19.29)$$
$$x \vee y = \tfrac{1}{2}[x + y + |y - x|], \quad x \wedge y = \tfrac{1}{2}[x + y - |y - x|],$$
$$|x| \vee |y| = \tfrac{1}{2}[|x + y| + |x - y|], \quad |x| \wedge |y| = \tfrac{1}{2}||x + y| - |x - y||; \qquad (19.30)$$
$$|x| \wedge |y| = 0 \Leftrightarrow |x + y| = |x - y| \Leftrightarrow |x| \vee |y| = |x| + |y| \Rightarrow$$
$$[x + y]^+ = [x]^+ + [y]^+; \qquad (19.31)$$
$$x \leqslant y \Leftrightarrow [x]^+ \leqslant [y]^+ \quad \text{and} \quad [y]^- \leqslant [x]^- \Leftrightarrow [x - y]^+ (= [y - x]^-) = 0. \qquad (19.32)$$

Two remarks on (19.27) and (19.31): the decomposition $x = [x]^+ - [x]^-$ is the only decomposition of x as a difference of two 'orthogonal' elements—if $|x| \wedge |y| = 0$ (cf. (6.31)...) x and y are said to be *lattice disjoint* or *lattice orthogonal*: in the literature one often finds the notation $x \perp y$ to indicate this (thus, e.g., we can put $[x]^+ \perp [x]^-$).

We will now give one of the best known results in the theory of vector lattices:

THEOREM 19.5 (The decomposition lemma of Riesz). *If $\langle E, \wedge, \vee \rangle$ is a vector lattice and if $x, y_1, y_2 \in E_+$ are such that $x \leq y_1 + y_2$ then there exist $x_1, x_2 \in E_+$ such that $x_1 \leq y_1$, $x_2 \leq y_2$ and $x = x_1 + x_2$.*

Proof. Let us put $x_1 = x \wedge y_1$ and $x_2 = x - x_1$. We have then $x = x_1 + x_2$; on the other hand it is evident that $0 \leq x_1 \leq y_1$ and, from $x \leq (x + y_2) \wedge (y_1 + y_2) = y_2 + (x \wedge y_1)$ (keep in mind (19.19)), it follows that $0 \leq x_2 = x - (x \wedge y_1) \leq y_2$. ■

An important consequence of this result is the following: $f \in E^*$ *is a bounded functional iff there exist $f_1, f_2 \in (E^*)_+$ such that $f = f_1 - f_2$.*

In definition 19.8 we have introduced the concept of homomorphism between lattices; with the following definition we will make the definition specific for vector lattices:

DEFINITION 19.15. Let $\langle E, \wedge, \vee \rangle$ and $\langle F, \wedge, \vee \rangle$ be two vector lattices and let $f: E \to F$ be a linear operator. f is said to be a *homomorphism* of $\langle E, \wedge, \vee \rangle$ in $\langle F, \wedge, \vee \rangle$ if

$$\forall x, y \in E \quad f(x \vee y) = f(x) \vee f(y) \quad \text{and} \quad f(x \wedge y) = f(x) \wedge f(y). \quad (19.33)$$

Again, let us introduce the following concept:

DEFINITION 19.16. Let $\langle E, \wedge, \vee \rangle$ be a vector lattice and A a subset of E. A is said to be *solid* if

$$\forall x \in A \quad \forall y \in E \quad |y| \leq |x| \Rightarrow y \in A. \quad (19.34)$$

The concept of a solid subset of a vector lattice and of homomorphism between vector lattices are apparently unrelated to each other; on the contrary there is a link between them of the same type as that between the concepts of open sets and continuous functions, namely:

THEOREM 19.6. *Let $\langle E, \wedge, \vee \rangle$ and $\langle F, \wedge, \vee \rangle$ be two vector lattices and $f: E \to F$ a linear application such that $f(E_+) = F_+$. The following propositions are equivalent:*

(i) f *is a homomorphism of $\langle E, \wedge, \vee \rangle$ in $\langle F, \wedge, \vee \rangle$,*
(ii) f^{-1} *transforms solid subsets of F into solid subsets of E.*

For the proof see, e.g., Schaefer, 1974, p. 60. Another interesting result for the proof of which see once more Schaefer, 1974, is the following:

THEOREM 19.7. Let $\langle E, \wedge, \vee \rangle$ and $\langle F, \wedge, \vee \rangle$ be two vector lattices and $f: E \to F$ a linear application. The following propositions are equivalent:

(i) f is a homomorphism of $\langle E, \wedge, \vee \rangle$ in $\langle F, \wedge, \vee \rangle$
(ii) $\forall x \in E \quad |f(x)| = f(|x|)$,
(iii) $\forall x \in E \quad f([x]^+) \wedge f([x]^-) = 0$

19.4 TOPOLOGICAL VECTOR LATTICES. BANACH–RIESZ SPACES

The order of an o.v.s. $\langle E, \leq \rangle$ induces on E topologies that may or may not be compatible with the vectorial structure of E (see e.g. Jameson, 1970, p. 131). Normally, however, an order and a topology are defined in a completely independent way on a vector space E, both of them being compatible with the vectorial structure of E. The most interesting case is that in which the order and the topology are in some way also compatible with each other. There are several ways of imposing this compatibility: the choice depends naturally on what one intends to do. Here our choice will be that which seems to us the most adapted to the application to functional analysis; on the other hand we will preferably choose separated locally convex topologies and lattice-orders, since this is the most interesting case.

Topological vector lattices

DEFINITION 19.17. Let E be a vector space. E is said to be a *topological vector lattice* (t.v.l.) if on E there are defined a lattice-order and a separated locally convex topology, both compatible with the vectorial structure and compatible also with each other in the sense that the topology has a family of neighbourhoods of the origin constituted by convex solids.

As is obvious, a topology in such conditions can always be expressed in terms of a separated family of seminorms (the so-called *lattice seminorms*: the Minkowski functionals of the solid and convex neighbourhoods of the origin). We are particularly interested in the case in which this family can be reduced to a single seminorm, which is then a norm.

Normed lattices

DEFINITION 19.18. Let $\langle E, \wedge, \vee \rangle$ be a vector lattice. A norm $\|\cdot\|_E : E \to \mathbb{R}_+$ is said to be a *lattice-norm* if

$$\forall x, y \in E \quad |x| \leq |y| \Rightarrow \|x\|_E \leq \|y\|_E. \tag{19.35}$$

The system $\langle E, \wedge, \vee, \|\cdot\|_E \rangle$ is called a *normed lattice*. A topologically complete normed lattice is called a *Banach lattice*, a λ-complete normed lattice is called a *normed Riesz space* and a normed lattice which is both topologically complete and λ-complete is called a *Banach–Riesz space*.

Let us remark that separated normed lattices are topological vector lattices in the sense of definitions 19.17: (19.35) translates the compatibility between the order and the topology.... In connection with the concept of a normed lattice let us mention the following result (for the proof of which see, e.g., Schaefer, 1974, p. 83):

THEOREM 19.8. *Let $\langle E, \wedge, \vee, \|\cdot\|_E \rangle$ be a normed lattice. The applications $x \mapsto [x]^+$, $x \mapsto [x]^-$ and $x \mapsto |x|$ are uniformly continuous from E to E_+; and the applications $(x, y) \to x \wedge y$ and $(x, y) \to x \vee y$ are uniformly continuous from E^2 to E. Also, E_+ and E_- are closed.*

Naturally, in the case where the norm of $\langle E, \wedge, \vee, \|\cdot\|_E \rangle$ satisfies the parallelogram condition

$$\|x+y\|_E^2 + \|x-y\|_E^2 = 2(\|x\|_E^2 + \|y\|_E^2), \qquad (19.36)$$

i.e., the case in which

$$(x, y)_E = \tfrac{1}{4}(\|x+y\|_E^2 - \|x-y\|_E^2) \qquad (19.37)$$

is a scalar product in E (see, e.g. Yosida, 1971, p. 39) $\langle E, \wedge, \vee, (\cdot, \cdot)_E \rangle$ is said to be a *pre-Hilbert lattice* (we will also use the terms *pre-Hilbert–Riesz space*, *Hilbert lattice* and *Hilbert–Riesz space*, the meanings of which are obvious). Under these conditions the compatibility between the order and the topology can be expressed most elegantly in the form

$$E_+ = \{x \in E : (x, y)_E \geq 0 \quad \forall y \in E_+\} \qquad (19.38)$$

In effect, if $x \geq 0$ and $y \geq 0$ then, from $|x-y| \leq |x+y| \Rightarrow \|x-y\|_E \leq \|x+y\|_E$ (see (19.35)) and from (19.37), we have $(x, y)_E \geq 0$. Conversely, if $y \geq 0$ and $(x, y)_E \geq 0$ (i.e. $\|x+y\|_E^2 \geq \|x-y\|_E^2$: see (19.37)) then $x \geq 0$, since were $x < 0$ we could have put $y = -x$ and thus obtain $\|x-x\|_E = 0 \geq 2\|x\|_E$ and therefore that $x = 0$.

Further we can write

$$[x]^+ = P_{E_+}(x) \quad \text{and} \quad [x]^- = P_{E_-}(x), \qquad (19.39)$$

thus obtaining $([x]^+, [x]^-)_E = 0$, i.e. $[x]^+ \perp [x]^-$ ... .

We will now give two concrete examples of the structures we have discussed and use these to deal with some very interesting results.

Stone–Weierstrass theorem

EXAMPLE 19.1. We will deal here with one of the most important Riesz spaces: the space $C^0(T)$ of continuous real functions defined on a compact topological space T.

In general, if T is any non-empty set, we can introduce an order in the vector space $\mathbb{R}^T$ of the functions $f: T \to \mathbb{R}$ by putting $(\mathbb{R}^T)_+ \triangleq (\mathbb{R}_+)^T$. This specification of the cone of positive elements of $\mathbb{R}^T$ is also equivalent to putting, for $f, g \in \mathbb{R}^T$,

$$f \leq g \Leftrightarrow \forall t \in T \quad f(t) \leq g(t). \tag{19.40}$$

The order so defined is a lattice-order and is compatible with the vectorial structure of $\mathbb{R}^T$ (i.e., $\langle \mathbb{R}^T, \leq \rangle$ is a vector lattice; $\langle \mathbb{R}^T, \leq \rangle$ is in fact a Riesz space), but, if card $T > 1$, this order is not total: this is in fact one of the reasons why non total orders are of such great interest in analysis.

Now let T be a compact topological space and let us introduce in $C^0(T)$ ($\subset \mathbb{R}^T$) the order (19.40) and the topology determined by the norm

$$\|f\|_{C^0(T)} = \max_T |f(t)| \tag{19.41}$$

(a Banach topology, which is called the topology of uniform convergence). It can be easily seen that, so structured, $C^0(T)$ is a Banach–Riesz space. If $T = \bar{\Omega}$, with $\Omega \subset \mathbb{R}^n$ being a bounded open set, we can also consider the space $C^1(T)$, an o.v.s. which however has the 'defect' that it is not a vector lattice (while the space $H^1(\Omega)$ is...; for higher orders of differentiation, however, even Sobolev spaces have this 'defect'....).

One of the most interesting results which use lattice structures is the Stone–Weierstrass theorem, of which we present here the version of Kakutani–Krein (see e.g. Yosida, 1971, p. 10; see also Stone, 1962 for a detailed treatment of this subject):

THEOREM 19.9. Let T be a compact topological space and A be a part of $C^0(T)$ such that
 (i) '$x \mapsto 1$' $\in A$,
 (ii) $\forall f, g \in A \quad \forall \alpha, \beta \in \mathbb{R} \quad f \wedge g \in A, \ f \vee g \in A, \ \alpha f + \beta g \in A$,
 (iii) $f_m \in A \, (m = 1, 2, \ldots), \ f_m \to f$ uniformly $\Rightarrow f \in A$.
Then a necessary and sufficient condition so that $A = C^0(T)$ is

$$\forall t_1, t_2 \in T \quad \exists f \in A : t_1 \neq t_2 \Rightarrow f(t_1) \neq f(t_2). \tag{19.42}$$

The condition (iii) can be interpreted as saying that A is a vector lattice-subspace of $C^0(T)$. The condition (19.42) is frequently translated as saying that A *separates the points* of T. If we work with complex functions the theorem is still true with, in addition to (i), (ii) (with $\mathbb{R}$ replaced by $\mathbb{C}$) and (iii), condition '(iv): $f \in A \Rightarrow \bar{f} \in A$'.

This theorem can be interpreted as a density theorem: '*if $A \subset C^0(T)$ satisfies* (i), (ii), *and* (19.42) *then* $A = C^0(T)$'. A particular case is the well-known Weierstrass approximation theorem: '*if $\Omega \subset \mathbb{R}^n$ is a bounded open set then every function $f \in C^0(\bar{\Omega})$ is the uniform limit of a sequence of polynomials*.

EXAMPLE 19.2. Let $\Omega \subset \mathbb{R}^n$ be an open set. $L^2(\Omega)$, provided with its real vectorial structure and its usual topology, is a Banach space which, provided also with the order

$$\forall f, g \in L^2(\Omega) \quad f \leq g \Leftrightarrow f(x) \leq g(x) \quad \text{a.e. in } \Omega \qquad (19.43)$$

(which is that of $\mathbb{R}^\Omega$ modulo the usual quotient), is a Banach–Riesz space, or rather a Hilbert–Riesz space. Among its vectorial subspaces we have $H^1(\Omega)$ and $H^1_0(\Omega)$ which, if Ω is regular enough (see section 5.5) are Hilbert lattices (cf. with the case of the spaces $C^r(\bar{\Omega})$, $r \geq 1$, which we have considered in the preceding example...). The spaces $H^1(\Omega)$ and $H^1_0(\Omega)$ are not, however, λ-complete, and are not therefore Hilbert–Riesz spaces.

Lattice structures and variational inequalities

To give an example of the use of ordered structures in the field of variational inequalities we present one of the so-called theorems of comparison (between solutions corresponding to different data...). We will work in an abstract framework, but in the framework of the spaces referred to above everything can be concretized. For other results which use the lattice structure of the spaces where variational problems are posed see Duvaut–Lions, 1976, p. 58 *et seq.*; Hanouzet–Joly, 1975a; 1979a; Haugazeau, 1967; 1968; also, in Chapter 11 we apply these techniques to the case of quasivariational inequalities, and a further bibliography can be found there.

Let us then consider the following

PROBLEM 19.1. Let H be a real Hilbert lattice, let $a: H \times H \to \mathbb{R}$ be a coercive symmetric continuous bilinear form, and let $\phi: H \to]-\infty, +\infty]$ be a convex proper lower semicontinuous functional (c.p.l.s.c. functional); find $u \in H$ such that

$$\phi(u) + a(u, u-v) \leq \phi(v) \quad \forall v \in H. \qquad (19.44)$$

As we know, this problem has one and only one solution, and we can therefore consider the operator Λ which with every functional ϕ under the conditions mentioned in the statement of the problem associates the solution $u = \Lambda(\phi)$ of the problem itself.

The question which we want to examine now is that of the monotonicity of Λ; i.e. we wish to know whether there is a result of the type '*if $\phi_1 \leq \phi_2$ (in the sense of $\phi_1(v) \leq \phi_2(v) \; \forall v \in H, 0 \leq v$) then $\Lambda(\phi_1) \leq \Lambda(\phi_2)$ (in the sense of* *the order defined in H)'*. The answer is no, as can be seen easily by interpreting problem 19.1 as a minimum problem; it remains no even if we substitute '$\Lambda(\phi_1) \leq \Lambda(\phi_2)$' by '$\Lambda(\phi_2) \leq \Lambda(\phi_1)$': $\Lambda(\phi_1)$ and $\Lambda(\phi_2)$ are not generally comparable! At this point, if we do not want to renounce a theorem of comparison, we need to introduce in the set of c.p.l.s.c. functionals an order $\leqslant$ (different from the order $\leq$ already considered) relative to which Λ is effectively monotone: for reasons of pragmatism we must then check that the theorem of comparison so obtained has effectively some applications, which will show indirectly that the order $\leqslant$ makes sense, though it is inevitably less natural than $\leq$ (let us remark that $\leqslant$ is not the dual order of that of H, since in general $\phi_1, \phi_2 \notin H^*$, but nevertheless it is the most natural order to introduce in the set of c.p.l.s.c. functionals from H to $]-\infty, +\infty]$).

DEFINITION 19.19. Let H be a real Hilbert lattice and let $\phi_1: H \to]-\infty, +\infty]$ and $\phi_2: H \to]-\infty, +\infty]$ be two proper convex functionals. We say that $\phi_1 \leqslant \phi_2$ if one of the following conditions holds:

(i) $\exists c \in \mathbb{R}_+: \quad \phi_2 - \phi_1 = c$
(ii) $\forall c \in \mathbb{R} \quad \phi_2 - \phi_1 \neq c$ and $\forall u, v \in H$
$$\phi_1(u \wedge v) + \phi_2(u \vee v) \leq \phi_1(u) + \phi_2(v).$$

We leave it to the reader to prove that $\leqslant$ is effectively an order, and further that the following properties are satisfied: (1) Condition (ii) is equivalent to condition

(iii) $\quad \forall c \in \mathbb{R}_+ \quad \phi_1 - \phi_2 \neq c$ and $\forall u, v_1, v_2 \in H; (v_1, v_2)_H = 0$
$$\phi_1(u) + \phi_2(u + v_1 + v_2) \leq \phi_1(u + v_1) + \phi_2(u + v_2),$$

which does not directly involve $\wedge$ and $\vee$; (2) If $\phi_1 \leqslant \phi_2$ and $\psi_1 \leqslant \psi_2$ then $\phi_1 + \psi_1 \leqslant \phi_2 + \psi_2$; (3) If ϕ is increasing then $\phi \leqslant 0$.

We will next deal with a theorem of comparison related to the order $\leqslant$:

THEOREM 19.10. Let Λ be the operator which with every p.c.f.l.s.c. $\phi: H \to]-\infty, +\infty]$ associates the solution $u = \Lambda(\phi)$ of problem 19.1. If the form a is such that

$$\forall v \in H \quad a([v]^-, [v]^+) \leq 0 \qquad (19.45)$$

then Λ is increasing, in the sense that

$$\phi_1 \leqslant \phi_2 \Rightarrow \Lambda(\phi_1) \leq \Lambda(\phi_2). \qquad (19.46)$$

Proof. Let us put $u_1 = \Lambda(\phi_1)$ and $u_2 = \Lambda(\phi_2)$ and let us write the inequalities corresponding to ϕ_1 and to ϕ_2:

$$\phi_1(u_1) + a(u_1, u_1 - v) \leq \phi_1(v) \qquad \forall v \in H, \qquad (19.47)$$

$$\phi_2(u_2) + a(u_2, u_2 - v) \leq \phi_2(v) \qquad \forall v \in H. \qquad (19.48)$$

The proof of the theorem is based on an appropriate choice of test functions v; let us then put $v = u_1 \wedge u_2$ in (19.47), $v = u_1 \vee u_2$ in (19.48) and add the two inequalities so obtained:

$$\phi_1(u_1) + \phi_2(u_2) + a(u_1, u_1 - (u_1 \wedge u_2)) + a(u_2, u_2 - (u_1 \vee u_2))$$
$$\leq \phi_1(u_1 \wedge u_2) + \phi_2(u_1 \vee u_2). \quad (19.49)$$

Since we have assumed that $\phi_1 \leq \phi_2$, it follows from the preceding inequality that

$$a(u_1, u_1 - (u_1 \wedge u_2)) + a(u_2, u_2 - (u_1 \vee u_2)) \leq 0 \quad (19.50)$$

(if $\phi_2 - \phi_1 = c \in \mathbb{R}_+$ this follows from the fact that in this case $\phi_1(u_1) + \phi_2(u_2) = \phi_1(u_1 \wedge u_2) + \phi_2(u_1 \vee u_2)$ and if $\phi_1 - \phi_2 \neq c \in \mathbb{R}$ it follows directly from the second part of (ii)) and therefore, using successively (19.29), (19.27), and (19.45)),

$$0 \geq a(u_1, [u_1 - u_2]^+) + a(u_2, -[u_1 - u_2]^+) = a(u_1 - u_2, [u_1 - u_2]^+)$$
$$= a([u_1 - u_2]^+ - [u_1 - u_2]^-, [u_1 - u_2]^+)$$
$$= a([u_1 - u_2]^+, [u_1 - u_2]^+) - a([u_1 - u_2]^-, [u_1 - u_2]^+)$$
$$\geq a([u_1 - u_2]^+, [u_1 - u_2]^+). \quad (19.51)$$

Since a is coercive it follows then that $[u_1 - u_2]^+ = 0$ and therefore (see (19.32)) that $u_1 \leq u_2$. ∎

An important particular case of this result, which shows besides that $\leq$ is an order which has a sense, is that which we obtain by putting $\phi_1 = j_K - l_1$ and $\phi_2 = j_K - l_2$, where j_K is the indicator function of a non-empty closed convex set $K \subset H$ and $l_1, l_2 \in H'$ ($\subset H^*$). In this case $\phi_1 \leq \phi_2$ means, on the one hand, that K is stable with respect to $\wedge$ and $\vee$ and, on the other, that $l_1 \leq l_2$ in the sense of the order which is the dual of that of H. The stability of K with respect to $\wedge$ and $\vee$ follows immediately. Also, if $\phi_2 - \phi_1 = c \in \mathbb{R}_+$ then $l_2 - l_1 = c$ and thus necessarily $l_1 = l_2$; if $\phi_2 - \phi_1 \neq c \in \mathbb{R}$ then, for every $u, v \in H$, $-l_1(u) - l_2(v) \geq -l_1(u \wedge v) - l_2(u \vee v)$, i.e. $l_1(u - (u \wedge v)) \leq l_2((u \vee v) - v)$, or again $l_1([u - v]^+) \leq l_2([u - v]^+)$, which means that $l_1 \leq l_2$ in H^* and therefore in H'. We can then state the following result:

THEOREM 19.11. Let H be a real Hilbert lattice, $K \subset H$ a non-empty closed convex set stable with respect to $\wedge$ and $\vee$, a a coercive continuous bilinear form such that $a([v]^-, [v]^+) \leq 0 \; \forall v \in K$, and l_1 and l_2 two continuous linear functionals on H. Again, let u_1 and u_2 be the solutions to the variational inequalities

$$u \in K; \quad a(u, u - v) \leq l_1(u - v) \quad \forall v \in K \quad (19.52)$$

and

$$u \in K; a(u, u - v) \leq l_2(u - v) \quad \forall v \in K, \quad (19.53)$$

respectively. Under these conditions, if $l_1 \leq l_2$ in H' then $u_1 \leq u_2$ in H.

Let us remark that the stability condition on K is not always satisfied: $K = \{\lambda \sin x : 0 \leq \lambda \leq 1\}$ is a non-empty closed convex set of $H_0^1(]0, 2\pi[)$ which is not stable with respect to $\wedge$ and $\vee$.

19.5 HILBERT PSEUDO-LATTICES

In the preceding section we have seen that if $\langle E, \wedge, \vee, (\cdot, \cdot)_E \rangle$ is a Hilbert lattice then (19.38) holds; let us now assume that E is an arbitrary Hilbert space and let us assume that there exists a non-empty subset C of E such that:

$$C = \{x \in E : (x, y)_E \geq 0 \quad \forall y \in C\}. \tag{19.54}$$

It can be easily seen that C is a strict convex cone with vertex at the origin and we can therefore define on E an order $\leq$ such that $E_+ = C$. The order $\leq$ will not in general be of the lattice type, but let us put nevertheless, for $x \in E$,

$$[x]^+ \triangleq P_C(x) \quad \text{and} \quad [x]^- \triangleq P_C(-x), \tag{19.55}$$

which satisfy, as can be easily seen, the relations

$$x = [x]^+ - [x]^- \quad \text{and} \quad ([x]^+, [x]^-)_E = 0. \tag{19.56}$$

Let us note explicitly that, while (19.38) and (19.39) are theorems, (19.54) and (19.55) are definitions. Let us introduce one more definition: if $x, y \in E$ put

$$x \veebar y \triangleq x + [y - x]^+ \quad \text{and} \quad x \barwedge y \triangleq x - [x - y]^+. \tag{19.57}$$

Let us remark immediately that $x \veebar y$ and $x \barwedge y$ cannot in general be interpreted as $x \vee y = \sup\{x, y\}$ and $x \wedge y = \inf\{x, y\}$, of whose existence we are not certain; however, if they exist then $x \veebar y = x \vee y$ and $x \barwedge y = x \wedge y$. The 'operators' $\veebar$ and $\barwedge$ have also many (but not all!, as is natural) of the operative properties of $\vee$ and $\wedge$. Let us then introduce the following definition:

DEFINITION 19.20. We will give the name *Hilbert pseudo-lattice* to a system $\langle E, C \rangle$ where E is a Hilbert space and C satisfies (19.54).

Naturally, we can consider a pre-Hilbert space instead of a Hilbert space, and thus obtain *pre-Hilbert pseudo-lattices*. It is important to remark that in example 19.28 of the preceding section we have not made use of the lattice structure of H but only of its pseudo-lattice structure; thus, if we substitute $\wedge$ by $\barwedge$ and $\vee$ by $\veebar$ everywhere the results will still be true. To end this section, we will state the following important result, which is naturally true also for Hilbert lattices:

THEOREM 19.12. If E is a Hilbert pseudo-lattice then every chain in E which is bounded above (or: below) has a lower upper bound (or: greatest lower bound). In particular, every bounded part of E is completely inductive.

Proof. Let $\{u_\alpha\}_{\alpha \in A}$ be a chain in E which is bounded above by $L \in E$, i.e.:

$$u_\alpha \leq L \qquad \forall \alpha \in A. \tag{19.58}$$

Suppose that A is chosen so that

$$\alpha \leq \beta \text{ in } A \Rightarrow u_\alpha \leq u_\beta \text{ in } E \tag{19.59}$$

and let $\bar{\alpha} \in A$. The family of real numbers $\{(u_\alpha, L - u_{\bar{\alpha}})_E\}$ is increasing and bounded above and therefore converges. On the other hand, if $\bar{\alpha} \leq \alpha \leq \beta$ then

$$\|u_\beta - u_\alpha\|_E^2 = (u_\beta - u_\alpha, u_\beta - u_\alpha)_E \leq (u_\beta - u_\alpha, L - u_{\bar{\alpha}})_E, \tag{19.60}$$

and since $\{(u_\alpha, L - u_{\bar{\alpha}})_E\}$ converges $\{u_\alpha\}$ is a Cauchy family in E: let u be its limit. Then, for every $v \in C$,

$$(u, v)_E = \lim_\alpha (u_\alpha, v)_E \geq (u_\beta, v)_E \qquad \forall \beta \in A \tag{19.61}$$

and therefore

$$u \geq u_\alpha \qquad \forall \alpha \in A. \tag{19.62}$$

Thus u is an upper bound of $\{u_\alpha\}$; to see that u is in fact the lower upper bound, i.e., that $u = \sup_A \{u_\alpha\}$, we note that if $w \geq u_\alpha \ \forall \alpha \in A$ then $(w, v)_E \geq (u, v)_E \ \forall v \in E$ and hence $w \geq u$. ∎

20
Multi-Valued Mappings

20.0 NOTATIONS

If Y is a non-empty set let us put:
$$2^Y = \mathcal{P}(Y) = \{T : T \text{ is a subset of } Y\},$$
$$2^Y_\blacksquare = \{T \in \mathcal{P}(Y) : T \text{ is non-empty}\}$$
and, if it makes sense,
$$2^Y_c = \{T \in \mathcal{P}(Y) : T \text{ is closed}\},$$
$$2^Y_q = \{T \in \mathcal{P}(Y) : T \text{ is compact}\},$$
$$2^Y_k = \{T \in \mathcal{P}(Y) : T \text{ is convex}\},$$
$$2^Y_x = \{T \in \mathcal{P}(Y) : T \text{ is connected}\},$$
$$2^Y_l = \{T \in \mathcal{P}(Y) : T \text{ is bounded}\},$$
$$2^Y_i = \{T \in \mathcal{P}(Y) : T \text{ is a closed interval}\};$$
further let us put
$$2^Y_{\blacksquare cl} = 2^Y_\blacksquare \cap 2^Y_c \cap 2^Y_l,$$
and so on.

20.1 FUNDAMENTAL DEFINITIONS

In this chapter we will study some properties of the applications $f : X \to 2^Y_\blacksquare$, where X and Y are two non-empty sets in which some appropriate structures are defined. These applications will be referred to as multi-valued mappings, since we can associate with $f : X \to 2^Y_\blacksquare$, in a well determined manner, a part F of the Cartesian product $X \times Y$, or, better, a multi-valued mapping $\ddot{f} : X \rightrightarrows Y$ (with graph $G(\ddot{f}) = F\ldots$). This correspondence between $f : X \to 2^Y_\blacksquare$ and $\ddot{f} : X \rightrightarrows Y$ partly justifies the nomenclature which we will introduce further on; here we will keep the two concepts ($f : X \to 2^Y_\blacksquare$ single-valued, and $\ddot{f} : X \rightrightarrows Y$ multi-valued) distinct, while in the rest of the book they are regarded as being identical. Later we will introduce some sets which are closely connected to $\overset{\scriptscriptstyle\prime\prime}{f}:X\rightrightarrows Y$; for the present we will only mention that if $x \in X$ the set $\overset{\scriptscriptstyle\prime\prime}{f}(x)=\{y \in Y:(x,y)\in F\}$ of values taken by $\overset{\scriptscriptstyle\prime\prime}{f}$ in x is known as the image of x through $\overset{\scriptscriptstyle\prime\prime}{f}$: if $\forall x \in X$ card $\overset{\scriptscriptstyle\prime\prime}{f}(x)=1$, $\overset{\scriptscriptstyle\prime\prime}{f}$ is said to be a single-valued multi-application or simply an application....

In the present section we will remind the reader, for the case of the applications $f:X \rightarrow 2_{\blacksquare}^Y$, of some general definitions regarding the applications of one set to another and, on the other hand, we will introduce a certain number of specific definitions for the applications $f:X\rightarrow 2_{\blacksquare}^Y$ (these being modifications of the former, necessary because of the particular structure of the set $2_{\blacksquare}^Y$).

In the following section (§ 20.2) we will take X and Y to be topological spaces and study the topologies 'determined' on $2_{\blacksquare}^Y$ by that of Y, and the properties of continuity of the applications $f:X\rightarrow 2_{\blacksquare}^Y$. In section 20.3, instead, we will take X and Y to be ordered sets and study the orders 'determined' on $2_{\blacksquare i}^Y$ by that of Y, and the monotonicity of the applications $f:X\rightarrow 2_{\blacksquare i}^Y$. We will not deal with some important questions such as the differentiability and integrability of these applications (see, in this connection, Banks–Jacobs, 1970; De Blasi, 1976; Martelli–Vignoli, 1974a; Castaing, 1967; Rockafellar, 1976; for several other questions and applications see, in addition to Chapters 14–16 of this book, Berge, 1957; 1958; 1959; Cellina, 1970; Darbo, 1950; 1958; Von Neumann–Morgenstern, 1953; Cellina–Lasota, 1969....

Graphs, images

In what follows, X and Y are two non-empty sets and f is an application from X to $2_{\blacksquare}^Y$.

The *graph* of f is the subset $G(f)=\{(x,T): x\in X$ and $f(x)=T\}$ of $X \times 2_{\blacksquare}^Y$. This set is of no interest to us, whereas we are interested in the set $G_{XY}(f):\{(x,y)\in X\times Y: x\in X$ and $y\in f(x)\}$, a subset of $X\times Y$ which coincides with $G(f)$. We then introduce the definition:

DEFINITION 20.1. The set $G_{XY}(f)$ is called the *graph of f reduced to $X\times Y$*.

The *image* of $A\subset X$ through f is the subset $f(A)=\{T\in 2_{\blacksquare}^Y: T=f(x), x\in A\}$ of $2_{\blacksquare}^Y$. This set is of no interest to us either, whereas a more useful set is $f_Y(A)=\{y\in Y: y\in f(x), x\in A\}=\bigcup_{x\in A}f(x)$, a subset of Y which coincides with $\overset{\scriptscriptstyle\prime\prime}{f}(A)=\bigcup_{x\in A}\overset{\scriptscriptstyle\prime\prime}{f}(x)$. We then introduce the definition:

DEFINITION 20.2. The set $f_Y(A)$ is called the *image of $A(\subset X)$ reduced to Y*.

Let us mention the following properties of $f_Y(A)$, which are also true for $\overset{\scriptscriptstyle\prime\prime}{f}(A)$ and $f(A)$...:

THEOREM 20.1. If $A_1, A_2 \subset X$ then

$$f_Y(A_1 \cup A_2) = f_Y(A_1) \cup f_Y(A_2), \tag{20.1}$$

$$f_Y(A_1 \cap A_2) \subset f_Y(A_1) \cap f_Y(A_2), \tag{20.2}$$

$$f_Y(X \setminus A_1) \supset f_Y(X) \setminus f_Y(A_1), \tag{20.3}$$

$$A_1 \subset A_2 \Rightarrow f_Y(A_1) \subset f_Y(A_2). \tag{20.4}$$

The proof is elementary. Further, we can interpret f_Y as an application from 2^X to 2^Y.

Let us study now the *inverse images*. If $\mathcal{B} \subset 2^Y_\blacksquare$ we put naturally $f^{-1}(\mathcal{B}) = \{x \in X : f(x) \in \mathcal{B}\}$, and in the particular case $\mathcal{B} = \{B\}$ we put $f^{-1}(B) = \{x \in X : f(x) = B\}$. Let us mention the following properties of the inverse image $f^{-1}(\mathcal{B})$, which are true in particular for $f^{-1}(B)\ldots$:

THEOREM 20.2. If $\mathcal{B}_1, \mathcal{B}_2 \subset 2^Y_\blacksquare$ then

$$f^{-1}(\mathcal{B}_1 \cup \mathcal{B}_2) = f^{-1}(\mathcal{B}_1) \cup f^{-1}(\mathcal{B}_2), \tag{20.5}$$

$$f^{-1}(\mathcal{B}_1 \cap \mathcal{B}_2) = f^{-1}(\mathcal{B}_1) \cap f^{-1}(\mathcal{B}_2), \tag{20.6}$$

$$f^{-1}(2^Y_\blacksquare \setminus \mathcal{B}_1) = f^{-1}(2^Y_\blacksquare) \setminus f^{-1}(\mathcal{B}_1), \tag{20.7}$$

$$\mathcal{B}_1 \subset \mathcal{B}_2 \Rightarrow f^{-1}(\mathcal{B}_1) \subset f^{-1}(\mathcal{B}_2). \tag{20.8}$$

The proof of this theorem is also elementary. Just as we were led to consider $f_Y(A)$, we also introduce the following concepts:

DEFINITION 20.3. $f^{-1+}(B) = f^{-1}(\mathcal{P}(B))$ is called the *superior inverse image* of $B(\subset Y)$ through f, and $f^{-1-}(B) = X \setminus f^{-1}(\mathcal{P}(Y \setminus B))$ is called the *inferior inverse image* of $B(\subset Y)$ through f.

It follows clearly that $f^{-1+}(B) = \overset{"}{f}^{-1+}(B) = \{x \in X : \overset{"}{f}(x) \neq \varnothing \text{ and } \overset{"}{f}(x) \subset B\}$ and that $f^{-1-}(B) = \overset{"}{f}^{-1-}(B) = \{x \in X : \overset{"}{f}(x) \cap B \neq \varnothing\}$. The superior and inferior images (we do not risk any confusion by not specifying 'reduced to Y') have the following easily proved properties:

THEOREM 20.3. If $B_1, B_2 \subset Y$ then

$$f^{-1+}(B_1) \subset f^{-1-}(B_1), \tag{20.9}$$

$$f^{-1+}(Y \setminus B_1) = X \setminus f^{-1-}(B_1), \tag{20.10}$$

$$f^{-1-}(Y \setminus B_1) = X \setminus f^{-1+}(B_1), \tag{20.11}$$

$$B_1 \subset B_2 \Rightarrow f^{-1+}(B_1) \subset f^{-1+}(B_2), \tag{20.12}$$

$$B_1 \subset B_2 \Rightarrow f^{-1-}(B_1) \subset f^{-1-}(B_2), \tag{20.13}$$

$$f^{-1+}(B_1 \cap B_2) = f^{-1+}(B_1) \cap f^{-1+}(B_2), \tag{20.14}$$

$$f^{-1-}(B_1 \cap B_2) \subset f^{-1-}(B_1) \cap f^{-1-}(B_2), \tag{20.15}$$

$$f^{-1+}(B_1 \cup B_2) \supset f^{-1+}(B_1) \cup f^{-1+}(B_2), \tag{20.16}$$

$$f^{-1-}(B_1 \cup B_2) = f^{-1-}(B_1) \cup f^{-1-}(B_2). \tag{20.17}$$

Quasisurjectivity, hyperinjectivity, and hyperunivocity

The *surjectivity* of f, which can be expressed as $\forall T \in 2_\blacksquare^Y \; \exists x \in X : f(x) = T$, is a very strong condition on f: this is evident if we recall that a theorem due to Cantor which ensures that if card $X \leq$ card Y then there is no surjective application from X to $2_\blacksquare^Y$. The *injectivity* of f, which can be expressed as $f(x_1) = f(x_2) \Rightarrow x_1 = x_2$, is on the contrary a very weak condition on f, as is the *univocity* inherent to the concept of application itself. We introduce then the following concepts:

DEFINITION 20.4. If $f_Y(X) = Y$, f is said to be *quasisurjective*; if $f(x_1) \cap f(x_2) \neq \varnothing \Rightarrow x_1 = x_2$, f is said to be *hyperinjective*; and if $f(x_1) \cap f(x_2) \neq \varnothing \Rightarrow f(x_1) = f(x_2)$, f is said to be *hyperunivocal*.

Hyperinjective applications are also hyperunivocal, and if f is hyperinjective and quasisurjective then f determines on Y a *partition* indicated by X.

Union, intersection, and composition

Let us now introduce the definitions of union and intersection of applications:

DEFINITION 20.5. Given $f_1 : X \to 2_\blacksquare^Y$ and $f_2 : X \to 2_\blacksquare^Y$, the application $f_1 \cap f_2$ defined by the formula $(f_1 \cap f_2)(x) = f_1(x) \cap f_2(x)$ is said to be the *intersection* of f_1 and f_2, and the application $f_1 \cup f_2$ defined by the formula $(f_1 \cup f_2)(x) = f_1(x) \cup f_2(x)$ is said to be the *union* of f_1 and f_2.

The operation $\cup$ cannot be defined for two applications $f_1 : X \to Y$ and $f_2 : X \to Y$ since, unless $f_1 = f_2$, the result is no longer an application from X to Y, but a multi-valued application from X to Y; further, under these conditions the operation $\cap$ is of no interest.

The composition of single-valued applications does not have an exact analogy in the case of multi-valued applications. The most 'closely related' operation, which we will also refer to as a composition is defined as follows:

DEFINITION 20.6. Let X, Y, and Z be three non-empty sets. Given $f : X \to 2_\blacksquare^Y$ and $g : Y \to 2_\blacksquare^Z$, the application $g \bullet f : X \to 2_\blacksquare^Z$ defined by $(g \bullet f)(x) = g_Z[f_Y(x)] = \bigcup \{z \in g(y) : y \in f(x)\}$ is said to be the *composite* application of f and g.

To end, let us introduce an important concept:

DEFINITION 20.7. An application $\sigma_f : X \to Y$ such that $\forall x \in X \; \sigma_f(x) \in f(x)$ is called a *selection* of $f : X \to 2_\blacksquare^Y$.

Let us remark that $G(\sigma_f) \subset G_{XY}(f)$ and that, as is obvious, the multi-valued application $\ddot{f}: X \rightrightarrows Y$ associated with f is single-valued (and hence is an application) iff there exists only one selection of f.

20.2 TOPOLOGIES IN $2_\blacksquare^Y$

We can introduce topologies in $2_\blacksquare^Y$ by using any one of the standard methods (families of open sets, families of neighbourhoods, etc.); the most interesting cases, however, are those in which the topology of $2_\blacksquare^Y$ is in some way related to a previously defined topology in Y.

In the first part of this section we will take Y to be a metric space and introduce a metric in $2_{\blacksquare cl}^Y$, the Hausdorff metric. In the second part we will take Y to be a general topological space and introduce four different topologies of $2_\blacksquare^Y$, the Vietoris topologies (cf. Vietoris, 1923). For a further treatment of the topologies of $2_\blacksquare^Y$ (and subsets of $2_\blacksquare^Y$, the so-called *hyperspaces* of Y) see Michael, 1951 and Kelley, 1942; for the treatment of the continuity of multi-valued applications see, besides the bibliography already cited, Ponomarev, 1964; Choquet, 1948; Smithson, 1972 (where one can also find information regarding spaces of multi-valued applications), Smithson, 1965; Whyburn, 1965; Michael, 1956a, b (where the problem of the existence of continuous selections is considered); further, in Kuratowski, 1972 there is a very abstract treatment of the question of the continuity of multi-valued applications.

20.2.1 THE HAUSDORFF METRIC TOPOLOGY

If (Y, d) is a metric space and A and B are non-empty subsets of Y (i.e.: $A, B \in 2_\blacksquare^Y$) the non-negative real number

$$\delta(A, B) = \inf \{d(a, b): a \in A, b \in B\}. \quad (20.18)$$

is usually known as the *distance between the sets* A and B. The quantity δ is not, however, a true distance since if $\delta(A, B) = 0$ it does not necessarily follow that $A = B$. If $\delta(A, B) = 0$, A and B are said to be *proximal*: the concept of proximity has some curious properties, and has been axiomatized in structures known as *proximity spaces* (see Naimpally–Warrack, 1970, pp. 7 and 8; Smirnov, 1964, p. 8). We will now define a true distance in $2_{\blacksquare cl}^Y$ (later it will become clear why it is not appropriate to consider the whole of $2_\blacksquare^Y$): the so-called Hausdorff distance.

Let us first introduce the following auxiliary set: if $\varepsilon > 0$ and $Q \in 2_{\blacksquare cl}^Y$ let us put

$$I(\varepsilon, Q) = \{x \in Y : \exists q \in Q : d(x, q) < \varepsilon\} = \{x \in Y : d(x, Q) < \varepsilon\}. \quad (20.19)$$

This set $I(\varepsilon, Q)$ is nothing but an open neighbourhood of Q with radius ε. Now we can introduce the

DEFINITION 20.8. Let (Y, d) be a metric space. If $A, B \in 2_{\blacksquare cl}^Y$ the *Hausdorff distance* between A and B, denoted by $H(A, B)$ (or, if there is any risk of confusion, by $H_d(A, B)$), is defined as the non-negative real number

$$H(a, b) \triangleq \inf\{\varepsilon : A \subset I(\varepsilon, B) \text{ and } B \subset I(\varepsilon, A)\}. \qquad (20.20)$$

Let us remark that the dependence of H_d on d produces unexpected phenomena. The topology determined by H_d depends on the particular metric d chosen and not only on the topology determined by d. Thus, for example (cf. Kelley, 1962, p. 153), in $\mathbb{R}_+$ the metrics $d_1(x, y) = \min\{1, |x-y|\}$ and $d_2(x, y) = |x/(1+x) - y/(1+y)|$ are topologically equivalent, but the metrics H_{d_1} and H_{d_2} are not so in $2^{\mathbb{R}_+}_{\blacksquare c}$ (note that $\mathbb{R}_+$ is bounded for d_1 and for d_2...). In order to avoid making the sets $I(\varepsilon, Q)$ explicit we could equivalently put $H(A, B) = \sup\{\sup_{x \in A} \inf_{y \in B} d(x, y), \sup_{x \in B} \inf_{y \in A} d(x, v)\}$.

We will now show that (20.20) does in fact define a distance. First $H(A, B)$ is a real number since A and B are bounded (in fact, following Covitz–Nadler 1970, one could avoid this restriction, thus defining a generalized metric where $H(A, B)$ can be $+\infty$). It is obvious that $H(A, B)$ is non-negative. If $H(A, B) = 0$ then $A = B$ since $H(A, B) = 0$ implies that $A \subset \bar{B}$ and $B \subset \bar{A}$ and hence, since A and B are closed, that $A = B$. The symmetry is obvious: $H(A, B) = H(B, A)$. As for the triangular inequality, $H(A, C) \leq H(A, B) + H(B, C)$, which is, traditionally, the most difficult condition to check, let us note that if $H(A, B) = \bar{\varepsilon}$ and $H(B, C) = \bar{\bar{\varepsilon}}$ then $A \subset I(\bar{\varepsilon} + \bar{\bar{\varepsilon}}, C)$ (since $A \subset I(\bar{\varepsilon}, B)$ and $B \subset I(\bar{\bar{\varepsilon}}, C)$) and $C \subset I(\bar{\varepsilon} + \bar{\bar{\varepsilon}}, A)$ (for analogous reasons) and therefore $H(A, C) \leq \bar{\varepsilon} + \bar{\bar{\varepsilon}}$.

By providing $2^Y_{\blacksquare cl}$ with the distance H we obtain a metric space, for which the general results concerning these spaces are naturally true. The set $\{x\}$, $x \in Y$, is closed (since metric spaces are T_1) and bounded, so that it is meaningful to consider the application $i: Y \to 2^Y_{\blacksquare cl}$ defined by $i(x) = \{x\}$, which is an isometry from (Y, d) to $(2^Y_{\blacksquare cl}, H)$ since $d(x, y) = H(\{x\}, \{y\})$: we can then say that the Hausdorff metric topology of $2^Y_{\blacksquare cl}$ is *admissible* in the sense of Michael (cf. Michael, 1951, p. 153).

We will now prove the main result of this section:

THEOREM 20.4. *Let (Y, d) be a metric space. The Hausdorff metric topology of $2^Y_{\blacksquare cl}$ relativized to $2^Y_{\blacksquare q}$ is equivalent to the topology generated by the family $\mathcal{U}$ of the sets U^+, U^- defined by*

(i) $U^+ = \{T \in 2^Y_{\blacksquare q} : T \subset U\}$,
(ii) $U^- = \{T \in 2^Y_{\blacksquare q} : T \cap U \neq \varnothing\}$,

for U varying in the family of open sets of Y.

Proof. We will first prove that the topology generated by the family $\mathcal{U}$ is less fine than the Hausdorff metric topology, by proving that U^+ and U^- are open sets in $(2^Y_{\blacksquare q}, H)$. We will prove that U^+ is open by showing that if

$Q \in U^+$ (i.e., if Q is a non-empty compact set of Y such that $Q \subset U$) then there exists $\varepsilon > 0$ such that $\mathcal{T}_\varepsilon(Q) = \{T \in 2^Y_{\blacksquare q} : H(T, Q) < \varepsilon\} \subset U^+$. Let us then put $\varepsilon = d(Q, Y \setminus U) : \varepsilon > 0$ since Q and $Y \setminus U$ are disjoint and closed (and further we can restrict ourselves to a bounded set); further, if $H(T, Q) < \varepsilon$ then $T \subset I(\varepsilon, Q)$ and hence $T \subset U$ since $I(\varepsilon, Q) \subset U$. We will next prove that U^- is open by showing that if $Q \in U^-$ (i.e., if Q is a non-empty compact set of Y such that $Q \cap U \neq \emptyset$) then there exists $\varepsilon > 0$ such that $\mathcal{T}_\varepsilon(Q) \subset U^-$. Let then $q \in Q \cap U$: since U is open there exists $\varepsilon > 0$ such that $\mathcal{T}_\varepsilon(q) = \{x \in Y : d(x, q) < \varepsilon\} \subset U$; now if $H(T, Q) < \varepsilon$ then $\exists \bar{t} \in T : d(\bar{t}, q) < \varepsilon$ (since if $\forall t \in T \, d(t, q) \geq \varepsilon$ then $q \notin I(\varepsilon, T)$, therefore $Q \not\subset I(\varepsilon, T)$ and hence $H(T, Q) \not< \varepsilon$), therefore $\bar{t} \in U$ and $T \cap U \neq \emptyset$. Next we will prove that the Hausdorff metric topology is less fine than that generated by the family $\mathcal{U}$, by showing that for every neighbourhood $\mathcal{T}_\varepsilon(Q) = \{T \in 2^Y_{\blacksquare q} : H(T, Q) < \varepsilon\}$ of a point $Q \in 2^Y_{\blacksquare q}$ in the first topology there exists a neighbourhood of Q in the second contained in it (one need only consider neighbourhoods of the type $\mathcal{T}_\varepsilon(Q)$ since they form a fundamental system of neighbourhoods for the metric topology). Since Q is a compact set, there exists a covering of Q by means of open sets Ω_i, $i = 1, \ldots, m$, such that diam $(\Omega_i) < \varepsilon$: we will prove that $\mathcal{O} = [I(\varepsilon, Q)]^+ \cap \Omega_1^- \cap \ldots \cap \Omega_m^-$ is a neighbourhood of Q within the required conditions. It is obvious that $Q \in \mathcal{O}$ and that $\mathcal{O}$ is an open set of the topology generated by $\mathcal{U}$. To see that $\mathcal{O} \subset \mathcal{T}_\varepsilon(Q)$ note that if $T \in \mathcal{O}$ then $T \subset I(\varepsilon, Q)$ and therefore what is left to prove is that $Q \subset I(\varepsilon, T)$ in order to show that $H(T, Q) < \varepsilon$; now if $x \in Q$ then $\exists i : x \in \Omega_i$ and therefore $\exists t \in T : d(x, t) < \varepsilon$ (since $T \cap \Omega_i \neq \emptyset$) and hence $x \in I(\varepsilon, T)$. ∎

It is interesting to note that the completeness of (Y, d) does not imply that of $(2^Y_{\blacksquare cl}, H)$; it does, however, imply the completeness of $(2^Y_{\blacksquare q}, H)$ (see Istrățescu, 1973, p. 296; Dieudonné, 1968b, p. 61, problem 3).

Lipschitz continuous multi-valued applications

Let us now change the subject, and introduce:

DEFINITION 20.9. Let (Y_1, d_1) and (Y_2, d_2) be two non-empty metric spaces and $f : Y_1 \to 2^{Y_2}_{\blacksquare cl}$. f is said to be *Lipschitz continuous* if

$$\exists k > 0 : \quad \forall x, y \in Y_1 \quad H_{d_2}(f(x), f(y)) \leq k d_1(x, y). \tag{20.21}$$

If there exists $k \leq 1$ satisfying these conditions f is said to be *non-expansive*; and if there exists $k < 1$ satisfying these conditions f is said to be a *contraction*.

Later we will use the term 'Lipschitz constant' to denote any value of k which satisfies (20.21); we will not however associate any characteristic of minimality and even less of uniqueness to this term.

One can immediately prove the theorem

THEOREM 20.5. Let (Y_1, d_1), (Y_2, d_2) and (Y_3, d_3) be non-empty metric spaces. Then the following propositions are true:

(i) if $f: Y_1 \to 2^{Y_2}_{\blacksquare cl}$ is Lipschitz continuous then it is continuous;
(ii) if $f_1: Y_1 \to 2^{Y_2}_{\blacksquare cl}$ and $f_2: Y_1 \to 2^{Y_2}_{\blacksquare cl}$ are Lipschitz continuous with Lipschitz constants α_1 and α_2, respectively, then $f_1 \cup f_2$ is Lipschitz continuous with Lipschitz constant $\max\{\alpha_1\, \alpha_2\}$ (let us observe that the analogous result for $f_1 \cap f_2$ is false; for a counter-example see Istrăţescu, 1973, p. 301; Nadler, 1969, p. 478);
(iii) if $f_1: Y_1 \to 2^{Y_2}_{\blacksquare cl}$ and $f_2: Y_2 \to 2^{Y_3}_{\blacksquare cl}$ are Lipschitz continuous with Lipschitz constants α_1 and α_2, respectively, then $f_2 \bullet f_1$ is Lipschitz continuous with Lipschitz constant $\alpha_1 \alpha_2$;
(iv) if $f: Y_1 \to 2^{Y_2}_{\blacksquare cl}$ is Lipschitz continuous with Lipschitz constant α then $f_{Y_2}: 2^{Y_1}_{\blacksquare cl} \to 2^{Y_2}_{\blacksquare cl}$ is Lipschitz continuous with Lipschitz constant α.

We will aso state (for a proof see Nadler, 1969, p. 478; Istrăţescu, 1973, pp. 302 and 303) the following theorem:

THEOREM 20.6. Let M be a non-empty closed convex set of a Banach space and let $f: M \to 2^M_{\blacksquare cl}$ be a Lipschitz continuous multi-valued application with a Lipschitz constant α; again let $\overline{\text{conv}}: 2^M_{\blacksquare cl} \to 2^M_{\blacksquare cl}$ be defined by $\overline{\text{conv}}(A) \triangleq \overline{\text{conv}(A)}$. Then $\overline{\text{conv}} \circ f: M \to 2^M_{\blacksquare cl}$ defined by $(\overline{\text{conv}} \circ f)(x) \triangleq \overline{\text{conv}}[f(x)]$ is Lipschitz continuous with Lipschitz constant α.

To end, let us, for the sake of curiosity, note that the question of whether a continuous multi-valued application $f: Y_1 \to 2^{Y_2}_{\blacksquare cl}$ has a continuous selection $\sigma_f: Y_1 \to Y_2$ must be answered in the negative, since contractions do not necessarily have continuous selections (for an example see Nadler, 1969, p. 479).

20.2.2 VIETORIS TOPOLOGIES

Let Y be an arbitrary topological space. We now introduce in $2^Y_{\blacksquare}$ four different topologies, known as Vietoris topologies. Let us first recall that the set $\mathbb{T}$ of the topologies in a set X, provided with the order '$\tau_1 \leq \tau_2$ if $\tau_1 \subset \tau_2$ (i.e., if τ_1 is less fine than τ_2)', is a complete lattice (see Bourbaki, 1961, pp. 32 and 34). The topology $\tau_1 \wedge \tau_2$ coincides with the topology $\tau_1 \cap \tau_2$ and the topology $\tau_1 \vee \tau_2$ is the one generated by the sub-base $\tau_1 \cup \tau_2$. [Recall that a family $\mathscr{F} \subset 2^X$ is said to generate a topology τ in X or to be a *sub-base* for τ if every open set of τ is arbitrary union of finite intersections of elements of $\mathscr{F}$. Observe that $\tau_1 \cup \tau_2$ is not a topology in general; further, if $\mathscr{B}_1$ is a sub-base for τ_1 and $\mathscr{B}_2$ a sub-base for τ_2 then $\mathscr{B} = \{B_1 \cap B_2 : B_1 \in \mathscr{B}_1 \text{ and } B_2 \in \mathscr{B}_2\}$ is another sub-base for $\tau_1 \vee \tau_2$]. Further, $\min \mathbb{T} = \{\emptyset, X\}$ (the indiscrete topology) and $\max \mathbb{T} = \mathscr{P}(X)$ (the discrete topology).

Given this, we introduce the following definition:

DEFINITION 20.10. Let Y be a topological space. The topology of $2^Y_\blacksquare$ generated by the family $\mathcal{O}^-$ of sets of the type $U^- = \{T \in 2^Y_\blacksquare : T \cap U \neq \varnothing\}$, U being an open set of Y, is called the *inferior semifinite Vietoris topology* and is denoted by $\mathcal{V}^-$. The topology of $2^Y_\blacksquare$ generated by the basis $\mathcal{O}^+$ of sets of the type $U^+ = \{T \in 2^Y_\blacksquare : T \subset U\} = 2^U_\blacksquare$, U being an open set of Y, is called the *superior semi-finite Vietoris topology* and is denoted by $\mathcal{V}^+$. The topology $\mathcal{V}^+ \wedge \mathcal{V}^-$ is called the *hemi-semifinite Vietoris topology* and is denoted by $\mathcal{V}^*$. The topology $\mathcal{V}^+ \vee \mathcal{V}^-$ is called the *finite Vietoris topology* and is denoted by $\mathcal{V}$.

Remark that a basis for $\mathcal{V}$ is made up of the family $\mathcal{O} = \{[U_1, \ldots, U_m]^\pm \subset 2^Y_\blacksquare : [U_1, \ldots, U_m]^\pm = \{T \in 2^Y_\blacksquare : T \subset \bigcup_{i=1}^m U_i$ and $T \cap U_i \neq \varnothing\}$, U_i, $i = 1, \ldots, m$, being open sets in $Y\}$, which is determined by a finite number of open sets, which is the reason for the name of $\mathcal{V}$ and hence those of $\mathcal{V}^+$, $\mathcal{V}^-$ and $\mathcal{V}^*$.

An interesting characterization of $\mathcal{V}$ is that it is the least fine of the topologies in $2^Y_\blacksquare$ such that $U^+ = 2^U_\blacksquare$ is open for an open set U in Y and closed for a closed set U in Y (remark that $\{T \in 2^Y_\blacksquare : T \subset U\} = \{T \in 2^Y_\blacksquare : T \cap (Y \setminus U) = \varnothing\} = 2^Y_\blacksquare \setminus \{T \in 2^Y_\blacksquare : T \cap (Y \setminus U) \neq \varnothing\}$, that $Y \setminus U$ is open as long as U is closed, and that $\{T \in 2^Y_\blacksquare : T \cap (Y \setminus U) \neq \varnothing\}$ is open in $\mathcal{V}$ since it is open in $\mathcal{V}^-$ ...): it is therefore the least fine of the topologies which are *acceptable* in the sense of Michael (cf. Michael, 1951, p. 155). Further, the Vietoris topologies are *admissible* in the sense of Michael (cf. Michael, 1951, p. 153) since $i: Y \to 2^Y_\blacksquare$ defined by $i(x) = \{x\}$ is a homeomorphism. The space $2^Y_\blacksquare$ with one of these topologies is in general not T_1, even when Y is T_1; on the other hand, $2^Y_{\blacksquare c}$ with an induced Vietoris topology is always T_0, and is T_1 if Y is.

The following result, which connects the topics treated in the last section with those we are dealing with now, is very interesting:

THEOREM 20.7. If Y is a metric space then the topology determined by the Hausdorff metric in $2^Y_{\blacksquare q}$ coincides with the finite Vietoris topology.

Proof. It follows directly from theorem 20.4. ■

We will now introduce the concepts of continuity associated with the Vietoris topologies:

DEFINITION 20.11. Let X and Y be two topological spaces. A multi-valued application $f: X \to 2^Y_\blacksquare$ is said to be *upper semicontinuous* (u.s.c.), *lower semicontinuous* (l.s.c.), *hemisemicontinuous*, or *continuous*, if f is continuous in the sense of the topologies $\mathcal{V}^+$, $\mathcal{V}^-$, $\mathcal{V}^*$, or $\mathcal{V}$, respectively.

Hemisemicontinuity is not of much interest to us, and since it is obvious that a function is continuous iff it is u.s.c. and l.s.c., we will deal only with the two latter types of continuity.

Thus:

THEOREM 20.8. Let X and Y be two topological spaces and $f: X \to 2^Y_{\blacksquare}$. f is u.s.c. iff one of the following five equivalent conditions is satisfied:

(i$^+$) $\{x \in X : f(x) \cap A \neq \varnothing\}$ is closed in X as long as A is closed in Y,
(ii$^+$) $\{x \in X : f(x) \subset A\}$ is open in X as long as A is open in Y,
(iii$^+$) $f^{-1-}(A)$ is closed in X as long as A is closed in Y,
(iv$^+$) $f^{-1+}(A)$ is open in X as long as A is open in Y,
(v$^+$) $\forall x \in X\ \forall A \subset Y\ (A \text{ open})\ f(x) \subset A \Rightarrow \exists B \subset X\ (B \text{ open}),\ x \in B: \forall z \in B\ f(z) \subset A$.

In an analogous manner, f is l.s.c. iff one of the following five equivalent conditions is satisfied:

(i$^-$) $\{x \in X : f(x) \cap A \neq \varnothing\}$ is open in X as long as A is open in Y,
(ii$^-$) $\{x \in X : f(x) \subset A\}$ is closed in X as long as A is closed in Y,
(iii$^-$) $f^{-1-}(A)$ is open in X as long as A is open in Y,
(iv$^-$) $f^{-1+}(A)$ is closed in X as long as A is closed in Y,
(v$^-$) $\forall x \in X\ \forall A \subset Y\ (A \text{ open})\ f(x) \cap A \neq \varnothing \Rightarrow \exists B \subset X\ (B \text{ open}),\ x \in B: \forall z \in B,\ f(z) \cap A \neq \varnothing$.

In particular (see (iv$^+$), (iii$^-$)) if f is u.s.c. or l.s.c. and single-valued then f is continuous in the usual sense (let us remark also that the usual concepts of semicontinuity assume a particular structure for the space $Y \ldots$).

Proof. We will only show that f is u.s.c. iff (ii$^+$) is satisfied and that (iv$^+$) and (v$^+$) are equivalent, leaving the rest to the reader. Let U be an open set in Y and $U^+ = \{T \in 2^Y_{\blacksquare} : T \subset U\}$ the open set corresponding in the sub-base $\mathcal{O}^+$ of $\mathcal{V}^+$; we have $f^{-1}(U^+) = \{x \in X : f(x) \subset U\}$ which, from (ii$^+$), is an open set in X. We have thus proved that (ii$^+$) $\Rightarrow f$ is u.s.c. (note that we need only consider the open sets of $\mathcal{O}^+$ since f^{-1} has the good properties (20.5) and (20.6)). The converse can be proved in a similar manner. We will now show that (iv$^+$) $\Rightarrow$ (v$^+$). Let $x \in X$, let $A \supset f(x)$ be an open set of Y and put $B = f^{-1+}(A)$; B satisfies (v$^+$) since it is open (from (iv$^+$)), $x \in B$, and if $z \in B$ then $f(z) \subset A$ (by the definition of $f^{-1+}(A)$). To show that (v$^+$) $\Rightarrow$ (iv$^+$) consider an open set $A \subset Y$ and a point $x \in f^{-1+}(A)$; the set B, which, from (v$^+$), must exist, is an open neighbourhood of x contained in $f^{-1+}(A)$, and therefore this set is open. ∎

The conditions (v$^+$) and (v$^-$) can be used to define locally the respective semicontinuities. Again, u.s.c. $\not\Rightarrow$ l.s.c. and l.s.c. $\not\Rightarrow$ u.s.c., as can be seen from the following examples (taken from Smithson, 1972, p. 33).

EXAMPLE 20.1. Put $X = Y = [0, 1] \subset \mathbb{R}$ and $f(x) = \{x/2\}$ if $0 \leq x < \frac{1}{2}$, $f(\frac{1}{2}) = [\frac{1}{4}, \frac{3}{4}]$, $f(x) = \{(x+1)/2\}$ if $\frac{1}{2} < x \leq 1$ (see fig. 20.1). f is u.s.c., but not l.s.c. since $f^{-1-}([\frac{3}{8}, \frac{5}{8}[) = \{\frac{1}{2}\}$ is a closed set, which is not in agreement with (iii$^-$).

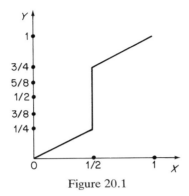
Figure 20.1

EXAMPLE 20.2. Put $X = Y = [0, 1] \subset \mathbb{R}$ and $f(x) = [0, x]$ if $0 \leq x < 1$, $f(1) = 0$ (see fig. 20.2). f is l.s.c., but not u.s.c. since $f^{-1^-}([x, 1]) = [x, 1[$ is not closed (if $x \neq 1$), which is not in agreement with (iii$^+$).

The union, intersection and composition of l.s.c. (or: u.s.c.) multi-valued applications is a l.s.c. (or: u.s.c.) multi-valued application; however, the composition of multi-valued applications with closed values is not necessarily one with closed values, which causes foreseeable complications.

One of the most interesting chapters in topology, if not the essence itself of general topology, is the one which deals with how continuous applications transform sets of certain kinds. We will now give two results of this type for the case of semicontinuous multi-valued applications.

THEOREM 20.9. Let X and Y be two topological spaces and C a connected set in X. If $f: X \to 2^Y_{\blacksquare x}$ is l.s.c. or u.s.c. then $f_Y(C)$ is a connected set.

We will not prove this result (see, e.g., Smithson, 1972, p. 34), but we will prove the following one, which is for us most important:

THEOREM 20.10. Let X and Y be two topological spaces and Q a compact set of X. If $f: X \to 2^Y_{\blacksquare q}$ is u.s.c. then $f_Y(Q)$ is compact.

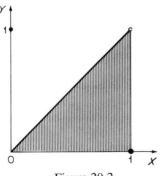
Figure 20.2

Proof. Let $\mathcal{O}$ be an open covering of $f_Y(Q)$: we will show that from $\mathcal{O}$ we can extract a finite sub-covering. Let $x \in Q$; since $f(x)$ is compact there exists a finite sub-family $\Omega_1, \ldots, \Omega_m$ of $\mathcal{O}$ which covers it—let us put $\Omega(x) = \bigcup_{i=1}^{m} \Omega_i$. The family $\{f^{-1+}(\Omega(x)) : x \in Q\}$ is an open covering of Q (note that u.s.c. ensures that the $f^{-1+}(\Omega(x))$ are open sets) and therefore, since Q is compact, we can extract from it a finite sub-family $f^{-1+}(\Omega(x_1)), \ldots, f^{-1+}(\Omega(x_n))$ which again covers Q. It is obvious that $\Omega(x_1), \ldots, \Omega(x_n)$ form a covering of $f_Y(Q)$, corresponding to which there is a finite sub-family of $\mathcal{O}$ since $\Omega(x_j) = \bigcup_{i=1}^{m(j)} \Omega_{ij}$. ∎

Another result closely connected with the preceding one (which we will state without proof: see e.g., Berge, 1959, p. 118) is the following 'fixed-set theorem':

THEOREM 20.11. *If X is a compact topological space and $f : X \to 2^X_{\blacksquare q}$ is u.s.c. then there exists a non-empty compact set $Q \subset X$ such that $f_Y(Q) = Q$.*

Multi-valued monotone applications

We have already mentioned the monotonicity of multi-valued applications on Chapter 3, in connection with the concept of sub-differential. The argument put forward at that time assumed the existence of a linear structure, but it can be extended to arbitrary topological spaces through the following definition (see also Smithson, 1972, p. 38 and Ponomarev, 1964, p. 129):

DEFINITION 20.12. Let X and Y be two topological spaces and $f : X \to 2^Y_{\blacksquare}$. f is said to be *monotone* if $f^{-1-}(\{y\})$ is a connected subset of X for any y in Y.

In the next section we will come back to this question of monotonicity in a framework which is more appropriate for it; namely that of ordered structures.

Closed multi-valued applications

We will now go on to the important concept of closed multi-valued applications; the rest of this section will be dedicated to proving results which in one way or another characterize this concept.

DEFINITION 20.13. Let X and Y be two topological spaces and $f : X \to 2^Y_{\blacksquare}$. f is said to be *closed* if $G_{XY}(f)$ is closed in $X \times Y$.

THEOREM 20.12. *Let X and Y be two topological spaces and $f : X \to 2^Y_{\blacksquare}$. If X is T_1 and f is closed then $f(x)$ is closed in Y for any x in X.*

Proof. We need only note that $f(x) = G_{XY}(f) \cap (\{x\} \times Y)$. ∎

THEOREM 20.13. Let X and Y be two topological spaces and $f: X \to 2^Y_\blacksquare$. Then f is closed iff

$x_0 \in X, y_0 \in Y, y_0 \notin f(x_0) \Rightarrow [\exists \mathcal{U}(x_0), \mathcal{V}(y_0) : x \in \mathcal{U}(x_0) \Rightarrow f(x) \cap \mathcal{V}(y_0) = \emptyset]$,
where $\mathcal{U}(x_0)$ and $\mathcal{V}(y_0)$ are open sets in X and Y, respectively.
(20.22)

Proof. Implication (20.22), which expresses the fact that (x_0, y_0) is 'far' from the graph of f reduced to $X \times Y$, says that $\mathcal{U}(x_0) \times \mathcal{V}(y_0) \subset X \times Y \backslash G_{XY}(f)$ and therefore that $G_{XY}(f)$ is closed since its complement is open (in fact, if we take a point (x_0, y_0) in $X \times Y \backslash G_{XY}(f)$, (20.22) ensures the existence of an open neighbourhood $\mathcal{U}(x_0) \times \mathcal{V}(y_0)$ of (x_0, y_0) contained in it). The converse can be proved in the same way. ∎

THEOREM 20.14. Let X and Y be two topological spaces and $f: X \to 2^Y_\blacksquare$. If f is closed then $(x_n \to x_0, y_n \to y_0$ and $\forall n \; y_n \in f(x_n)) \Rightarrow y_0 \in f(x_0)$.

Proof. This result, which can be interpreted as meaning that closed multi-valued applications are sequentially closed, follows from $(x_n, y_n) \to (x_0, y_0)$, from $(x_n, y_n) \in G_{XY}(f)$ and from the fact that $G_{XY}(f)$ is closed. ∎

THEOREM 20.15. Let X and Y be two compact Hausdorff topological spaces and $f: X \to 2^Y_{\blacksquare c}$. Then f is u.s.c. iff it is closed.

Proof. We will first show that 'f u.s.c. $\Rightarrow f$ closed', noting also that this is true even if X is an arbitrary topological space and Y is merely a regular topological space. [Recall that a topological space Y is said to be *regular* if for every $y \in Y$ and every $A \not\ni y$ there exist two open sets $U \ni y$ and $V \supset A$ such that $U \cap V = \emptyset$; let us remark also that compact Hausdorff spaces are regular (see, e.g., Kelley, 1962, pp. 134 and 135).] Now let $x_0 \in X$ and $y_0 \in Y$, with $y_0 \notin f(x_0)$. Since Y is regular there exist open sets $T \supset f(x_0)$ and $\mathcal{V}(y_0) \ni y_0$ such that $T \cap \mathcal{V}(y_0) = \emptyset$; further, from (v^+), there exists an open set $\mathcal{U}(x_0) \ni x_0$ such that $x \in \mathcal{U}(x_0) \Rightarrow f(x) \subset T$. The neighbourhoods $\mathcal{U}(x_0)$ and $\mathcal{V}(y_0)$ satisfy the conditions of theorem 20.13 (since if $x \in \mathcal{U}(x_0)$ then $f(x) \subset T$ and therefore $f(x) \cap \mathcal{V}(y_0) = \emptyset$): f is therefore closed. We will show next that 'f closed $\Rightarrow f$ u.s.c.'. By *reductio ad absurdum*: if f is not u.s.c. then (negation of (v^+)) there exist $x \in X$ and $V \supset f(x)$ open such that $\forall U \ni x$ (U open) $f_Y(U) \not\subset V$. Let us now consider the family of sets $A_U = G_{XY}(f) \cap (\bar{U} \times (Y \backslash V))$, U open in X. If we show that there exists $(s, t) \in \bigcap_U A_U$ we have completed the proof since in that case we must have $s = x$ (since, X being a Hausdorff space, $\bigcap_U \bar{U} = \{x\}$), and therefore $t \in f(x)$ (since $(x, t) \in G_{XY}(f)$) and, since $t \notin V$ (since $(x, t) \in \bar{U} \times (Y \backslash V)$), $f(x) \not\subset V$, which is absurd. To show that $\bigcap_U A_U \neq \emptyset$ we point out that the A_U are compact (since $G_{XY}(f)$, $\bar{U}$ and $Y \backslash V$ are closed and $X \times Y$ is a compact Hausdorff space), non-empty (since if $f_Y(U) \not\subset V$ then there exist $x \in U$ and $y \in f(x)$

such that $y \notin V$, and therefore $x \in \bar{U}$ and $y \in (Y \setminus V)$, so $(x, y) \in \bar{U} \times (Y \setminus V)$; further $y \in f(x)$ and therefore $(x, y) \in G_{XY}(f)$, and that the family $\{A_U\}_U$ has the finite intersection property since

$$A_{U_1} \cap \ldots \cup A_{U_m} = G_{XY}(f) \cap [(\bar{U}_1 \cap \ldots \cap \bar{U}_m) \times (Y \setminus V)]$$
$$= A_{U_1 \cap \ldots \cap U_m} \neq \emptyset. \blacksquare$$

REMARK 20.1. If X and Y are compact metric spaces the preceding ('closed graph') theorem enables us to express the concept of u.s.c. in terms of the convergence of sequences:

if f is u.s.c. and $x_n \to x$ then for any sequence $y_n \to y$,

$$\text{with } y_n \in f(x_n), \text{ we have } y \in f(x). \quad (20.23)$$

In the general case (20.23) can be naturally reinterpreted in terms of the convergence of filters or of nets.

REMARK 20.2. Under the same conditions as the preceding remark, the concept of l.s.c. can also be formulated in terms of the convergence of sequences:

if f is l.s.c. and $x_n \to x$ then for any $y \in f(x)$ there

$$\text{exists a sequence } y_n \to y, \text{ with } y_n \in f(x_n). \quad (20.24)$$

To prove (20.24) we use the method of *reductio ad absurdum*. Let us suppose then that f is l.s.c., $x_n \to x$, $y \in f(x)$ and that no sequence y_n, with $y_n \in f(x_n)$, converges to y. On the other hand, let $A \subset Y$ be an open set such that $y \in A$ and $f(x) \cap A \neq \emptyset$: from (v⁻) (theorem 20.8) we have then that there exists an open set $B \subset X$ such that $x \in B$ and, for every $z \in B$, $f(z) \cap A \neq \emptyset$, which in turn implies that we can choose y_n close to y in $f(x_n)$, which is absurd.

For more about remarks 20.1 and 20.2 see Mosco, 1976b; 1969a; 1969b and Joly, 1973.

20.3 ORDERS IN $2^Y_{\blacksquare i}$

Let us first introduce the following definition, which is clearly motivated by definition 20.12:

DEFINITION 20.14. Let $\langle X, \leqslant \rangle$ and $\langle Y, \leqslant \rangle$ be two ordered sets and $f: X \to 2^Y_{\blacksquare i}$. Then f is said to be *monotone* if $f^{-1^-}(\{y\})$ is a closed interval of X for any $y \in Y$.

The applications $f_h: X \to 2^Y_{\blacksquare i}$ ($h = 1, 2, 3, 4$) shown in fig. 20.3.h ($h = 1, 2, 3, 4$) in terms of their graphs reduced to $X \times Y$ are monotone in the sense of the preceding definition.

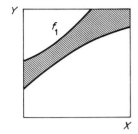 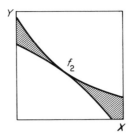

Figure 20.3.1   Figure 20.3.2

Two questions naturally arise at this point: the first regarding the characterization of multi-valued monotone applications in a 'more legible' manner, and the other regarding the determination of an order in $2^Y_{\blacksquare i}$ to which this concept of monotonicity will be referred (we will see that in fact we need two orders).

Let us deal with the first question. To do this we need to introduce the following nomenclature:

DEFINITION 20.15. Let $\langle X, \leq \rangle$ and $\langle Y, \leq \rangle$ be two ordered sets and $f: X \to 2^Y_{\blacksquare i}$. The application $\sigma_f^+ : X \to Y$ (or: $\sigma_f^- : X \to Y$) defined by $\sigma_f^+(x) \triangleq \max f(x)$ (or: $\sigma_f^-(x) \triangleq \min f(x)$) is called the *upper selection* (or: *lower selection*) of f.

It follows immediately that $\forall x \in X \ \sigma_f^-(x) \leq \sigma_f^+(x)$ and that if $\forall x \in X$ $\sigma_f^-(x) = \sigma_f^+(x)$ then f is single-valued. The following theorem can then be easily proved:

THEOREM 20.16. Let $\langle X, \leq \rangle$ and $\langle Y, \leq \rangle$ be two ordered sets and let $f: X \to 2^Y_{\blacksquare i}$. Then f is monotone in the sense of definition 20.14 iff σ_f^+ and σ_f^- are monotone in the usual sense.

The result, which provides, as we wanted, a 'legible' characterization of the concept of monotonicity, shows that multi-valued monotone applications

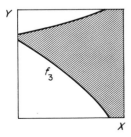 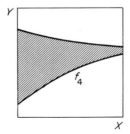

Figure 20.3.3   Figure 20.3.4

in the sense of definition 20.14 are essentially of one of the four types shown in fig. 20.3. Another possible application of this result is that of dividing the class of monotone multi-valued applications in increasing, non-decreasing, etc., multi-valued applications, according to the characteristics of the selections; we will come back to this later.

We can now go on to the second question that we raised. Given an ordered set $\langle Y, \leq \rangle$, let us introduce in $2^Y_{\blacksquare_i}$ the relationships ρ^+, ρ_+, ρ^-, and ρ_- defined as follows:

$$[a_1, b_1]\rho^+[a_2, b_2] \quad \text{iff} \quad a_1 \leq a_2, \tag{20.25}$$

$$[a_1, b_1]\rho_+[a_2, b_2] \quad \text{iff} \quad b_1 \leq b_2, \tag{20.26}$$

$$[a_1, b_1]\rho^-[a_2, b_2] \quad \text{iff} \quad a_2 \leq a_1, \tag{20.27}$$

$$[a_1, b_1]\rho_-[a_2, b_2] \quad \text{iff} \quad b_2 \leq b_1. \tag{20.28}$$

These relationships, though reflexive and transitive, are not orders: starting from these we can, however, define those order relationships that we need. Let us mention, for the sake of curiosity, that if $\lesseqgtr_1$ and $\lesseqgtr_2$ are orders then ($\lesseqgtr_1$ and $\lesseqgtr_2$) is an order but ($\lesseqgtr_1$ or $\lesseqgtr_2$) is not (in general; to see this one need only put $\lesseqgtr_2 \equiv \gtreqless_1$).

DEFINITION 20.16. Let $\langle Y, \leq \rangle$ be an ordered set. The following formulae define the orders $\leqslant$ and $\subset$ in $2^Y_{\blacksquare_i}$:

$$[a_1, b_1] \leqslant [a_2, b_2] \quad \text{iff} \quad ([a_1, b_1]\rho^+[a_2, b_2] \text{ and } [a_1, b_1]\rho_+[a_2, b_2]), \tag{20.29}$$

$$[a_1, b_1] \subset [a_2, b_2] \quad \text{iff} \quad ([a_1, b_1]\rho_+[a_2, b_2] \text{ and } [a_1, b_1]\rho^-[a_2, b_2]). \tag{20.30}$$

We will leave it to the reader to verify that $\leqslant$ and $\subset$ are in fact orders and also that $\geqslant \Leftrightarrow (\rho^-$ and $\rho_-)$ and $\supset \Leftrightarrow (\rho^+$ and $\rho_-)$ (let us remark that the order $\subset$ is in effect the set inclusion and therefore the utilization of this symbol is justified).

We can now talk of $\leqslant$-monotone multi-valued applications, $\supset$-non-decreasing multi-valued applications, etc.

The following theorem can then be easily proved:

THEOREM 20.17. Let $\langle X, \leq \rangle$ and $\langle Y, \leq \rangle$ be two ordered sets and $f: X \to 2^Y_{\blacksquare_i}$. Then f is monotone in the sense of definition 20.14 iff it is $\leqslant$-monotone or $\subset$-monotone.

To end, let us note that if f is, e.g., $\leqslant$-non-decreasing then σ_f^+ and σ_f^- are non-decreasing, and vice versa.

Bibliography

Adams, R. (1975). *Sobolev Spaces*, Academic Press, New York.
Adams, R., Aronszajn, N., and Smith, K. T. (1967). 'Theory of Bessel potentials, II', *Ann. Inst. Fourier*, **17** (2), 1–135.
Agmon, S. (1959). 'The L_p approach to the Dirichlet problem', *Ann. Scuola Norm. Sup. Pisa*, (3) **13**, 405–448.
Agmon, S. (1960). 'Maximum theorems for solutions of higher order elliptic equations'. *Bull. Amer. Math. Soc.*, **66**, 77–80.
Agmon, S. (1965). *Lectures on Elliptic Boundary Value Problems*, D. Van Nostrand, Princeton.
Agmon, S., Douglis, A., and Nirenberg, L. (1959). 'Estimates near the boundary for solutions of elliptic partial differential equations satisfying general boundary conditions, I', *Comm. Pure Appl. Math.*, **12**, 623–727.
Agmon, S., Douglis, A., and Nirenberg, L. (1964). 'Estimates near the boundary for solutions of elliptic partial differential equations satisfying general boundary conditions, II', *Comm. Pure Appl. Math.*, **17**, 35–92.
Alaoglu, L. (1940). 'Weak topologies of normed linear spaces', *Ann. Math.*, (2) **41**, 252–267.
Aleksandrov, P. S. (1957). *Topologia Combinatoria*, Einaudi, Torino (Russian ed. (1947): OGIZ, Moscow).
Alexander, J. W. (1922). 'On transformations with invariant points', *Trans. Amer. Math. Soc.*, **23**, 89–95.
Aliprantis, C., and Burkinshaw, O. (1978). *Locally Solid Riesz Spaces*, Academic Press, New York.
Allen, G. (1977). 'Variational inequalities, complementarity problems, and duality theorems', *J. Math. Anal. Appl.*, **58**, 1–10.
Alt, H. W. (1977a). 'A free boundary problem associated with the flow of ground water', *Arch. Rational Mech. Anal.*, **64**, 111–126.
Alt, H. W. (1977b). 'The fluid flow through porous media. Regularity of the free surface', *Manuscripta Math.*, **21**, 255–272.
Alt, H. W. (1979). 'Strömungen durch inhomogene poröse Medien mit freiem Rand', *J. Reine ang. Math.*, **305**, 89–115.
Alt, H. W. (1980). 'A new numerical method for solving the dam problem', in *Free Boundary Problems* (ed. E. Magenes), vol. I, pp. 89–108, Proc. Sem. (Pavia, 1979), Istituto Nazionale di Alta Matematica, Roma.
Alt, H. W., Caffarelli, L. A., and Friedman, A. (1982a). 'Axially symmetric jet flows', to appear in *Arch. Rational Mech. Anal.*
Alt, H. W., Caffarelli, L. A., and Friedman, A. (1982b). 'Asymmetric jet flows', *Comm. Pure Appl. Math.*, **35**, 29–68.

Alt, H. W., Caffarelli, L. A., and Friedman, A. (1982c). 'Jet flows with gravity', *J. Reine Angew. Math.*, **331**, 58–103.
Alt, H. W., and Gilardi, G. (1981). 'The behaviour of the free-boundary for the dam problem', *Ann. Scuola Norm. Sup. Pisa*, (4) **9**, 571–625.
Alt, H. W., and Gilardi, G. (1982). 'The free-boundary in the dam problem', to appear on *Proc. Symp. Free Boundary Problems: Theory and Applications* (Montecatini Terme, 1981) (eds. A. Fasano and M. Primicerio).
Amann, H. (1977). 'Ordered structures and fixed points', in *SAFA 2*, pp. 1–51, Atti 2°. Sem. Analisi Funzionale e Applicazioni (Arcavacata di Rende, 1977), Università della Calabria, Cosenza.
Amerio, L. (1978). 'Continuous solutions of the problem of a string vibrating against an obstacle', *Rend. Sem. Mat. Univ. Padova*, **59**, 67–96.
Amerio, L., and Prouse, G. (1975). 'Study of the motion of a string vibrating against an obstacle', *Rend, di Mat.*, (6) **8**, 563–585.
Amick, C. (1978). 'Some remarks on Rellich's theorem and the Poincaré inequality', *J. London Math. Soc.*, (2) **18**, 81–93.
Amirat, Y., and Atik, Y. (1978). 'Approximation numérique d'un problème à frontière libre', *C.R. Ac. Sci. Paris*, **286**, 203–206.
Ancona, A. (1976). 'Continuité des contractions dans les espaces de Dirichlet', *C.R. Ac. Sci. Paris*, **282**, 871–873.
Anderson, R., and Friedman, A. (1977). 'A quality control problem and quasi-variational inequalities', *Arch. Rational Mech. Anal.*, **63**, 205–252.
Anderson, R., and Friedman, A. (1978). 'Multi-dimensional quality control problems and quasi-variational inequalities', *Trans. Amer. Math. Soc.*, **246**, 31–76.
Antosiewicz, H. A. (1977). 'Fixed point theorems and ordinary differential equations', in *Studies in Ordinary Differential Equations* (ed. J. Hale), pp. 169–200, Mathematical Association of America.
Aravin, V. I., and Numerov, S. N. (1965). *Theory of Fluid Flow in Undeformable Porous Media*, Israel Program Sci. Transl., Jerusalem (Russian ed. (1953): G.I.T.–T.L., Moscow).
Aronszajn, N., and Smith, K. T. (1961). 'Theory of Bessel potentials, I', *Ann. Inst. Fourier*, **11**, 385–475.
Aubin, J.-P. (1972). *Approximation on Elliptic Boundary-Value Problems*, Wiley-Interscience, New York.
Aubin, J.-P. (1977). *Applied Abstract Analysis*, Wiley, New York.
Aubin, J.-P. (1979a). *Applied Functional Analysis*, Wiley, New York.
Aubin, J.-P. (1979b). *Mathematical Methods of Game and Economic Theory*, North-Holland, Amsterdam.
Auslander, L., and MacKenzie, R. E. (1963). *Introduction to Differentiable Manifolds*, McGraw-Hill, New York.
Avantaggiati, A. (1976). 'Spazi di Sobolev con peso ed alcune applicazioni', *Boll. U. Mat. Ital.*, (5) **13-A**, 1–52.
Averbukh, V. I., and Smolyanov, O. G. (1968). 'The various definitions of the derivative in linear topological spaces', *Russian Math. Surveys*, **23** (4), 67–133 (Russian ed. (1968): *Uspehi Mat. Nauk*, **23** (4), 67–116).
Baiocchi, C. (1966). 'Sul problema misto per l'equazione parabolica del tipo del calore', *Rend. Sem. Mat. Univ. Padova*, **36**, 80–121.
Baiocchi, C. (1971). 'Sur un problème à frontière libre traduisant le filtrage de liquides à travers des milieux poreux', *C.R. Ac. Sci. Paris*, **273**, 1215–1217.
Baiocchi, C. (1972). 'Su un problema di frontiera libera connesso a questioni di idraulica', *Ann. Mat. Pura Appl.*, (4) **92**, 107–127.
Baiocchi, C. (1973a). 'Sur quelques problèmes à frontière libre', *Astérisque*, **2–3**, 69–85.
Baiocchi, C. (1973b). 'About some free boundary problems connected with

hydraulics', in *Proc. 5th IFIP Conf. on Optimization Techniques* (Rome, 1973) (eds R. Conti and A. Ruberti), part I, pp. 137–140, Lecture Notes in Computer Science, 3, Springer, Berlin.

Baiocchi, C. (1973c). 'Su alcuni problemi di frontiera libera connessi a questioni di idraulica', *Rend. Sem. Mat. Univ. Politec. Torino*, **31,** 69–80.

Baiocchi, C. (1974a). 'Problèmes à frontière libre en hydraulique', *C.R. Ac. Sci. Paris*, **278,** 1201–1204.

Baiocchi, C. (1974b). 'Problemi di frontiera libera e disequazioni quasi-variazionali', in *Atti Giorn. Analisi Convessa e Applicazioni* (Roma, 1974), pp. 11–20, Quad. Gruppi Ric. Mat. C.N.R., Università di Roma, Roma.

Baiocchi, C. (1974c). 'Une nouvelle methode d'analyse numérique des problèmes de filtration dans les matériaux poreux', in *Proc. Intern. Symp. Computing Methods in Applied Sciences and Engineering* (Versailles, 1973) (eds R. Glowinski and J. L. Lions), part I, pp. 410–415, Lecture Notes in Computer Science, 10, Springer, Berlin.

Baiocchi, C. (1974d). 'Sui problemi di frontiera libera nel moto dei fluidi attraverso i materiali porosi', in *Proc. Intern. Symp. Discrete Methods in Engineering* (Segrate, 1974), pp. 490–500, Etas Libri, Milano.

Baiocchi, C. (1974e). 'Problèmes à frontière libre liés à questions d'hydraulique', in *New Variational Techniques in Mathematical Physics* (eds G. Capriz and G. Stampacchia), pp. 1–25, Proc. C.I.M.E. (Bressanone, 1973), Cremonese, Roma.

Baiocchi, C. (1975a). 'Free boundary problems in the theory of fluid flow through porous media', in *Proc. International Congresses of Mathematicians* (Vancouver, 1974) (ed. R. D. James), vol. 2, pp. 237–243, Canadian Mathematical Congress, Vancouver.

Baiocchi, C. (1975b). 'Studio di un problema quasi-variazionale connesso a problemi di frontiera libera', *Boll. U. Mat. Ital.*, (4) **11** (suppl. fasc. 3), 589–613.

Baiocchi, C. (1975c). 'Movimiento de un fluido en medios porosos. Enfoque variacional, cuasi variacional y numérico', *Cuadernos Inst. Mat. 'Beppo Levi'*, **8,** Universidad de Rosario, Rosario.

Baiocchi, C. (1976a). 'Inéquations quasi-variationnelles dans les probèmes à frontière libre en hydrqulique', in *Applications of Methods of Functional Analysis to Problems in Mechanics* (eds P. Germain and B. Nayroles), pp. 1–7, Proc. Joint Symp. IUTAM/IMU (Marseille, 1975), Lecture Notes in Mathematics, 503, Springer, Berlin.

Baiocchi, C. (1976b). 'Problèmes à frontière libre et inéquations variationnelles', *C.R. Ac. Sci. Paris*, **283,** 29–32.

Baiocchi, C. (1977). 'Estimations d'erreur dans L^∞ pour les inéquations à obstacle', in *Proc. Conf. Mathematical Aspects of Finite Element Methods* (Rome, 1975) (eds. I. Galligani and E. Magenes), pp. 27–34, Lecture Notes in Mathematics, 606, Springer, Berlin.

Baiocchi, C. (1978). 'Problèmes à frontière libre en hydraulique: milieux non homogènes', *Ann. Scuola Norm. Sup. Pisa*, (4) **5,** 429–453.

Baiocchi, C. (1979). 'Recent results on free boundary problems in fluid flow through porous media', in *Proc. Intern. Meeting on Recent Methods in Non-Linear Analysis* (Rome, 1978) (eds E. De Giorgi, E. Magenes, and U. Mosco), pp. 3–12, Pitagora, Bologna.

Baiocchi, C. (1980a). 'Variational inequalities and free-boundary problems', in *Variational Inequalities and Complementarity Problems* (eds R. W. Cottle, F. Giannessi, and J. L. Lions), pp. 25–33, Wiley, Chichester.

Baiocchi, C. (1980b). 'Free boundary problems in fluid flow through porous media and variational inequalities', in *Free Boundary Problems* (ed. E. Magenes), vol. I, pp. 175–191, Proc. Sem. (Pavia, 1979), Istituto di Alta Matematica, Roma.

Baiocchi, C. (1981a). 'Disequazioni variazionali', *Boll. U. Mat. Ital.*, (5) **18-A**, 173–187.
Baiocchi, C. (1981b). 'Free boundary problems and variational inequalities', *Publ. Laboratorio di Analisi Numerica CNR*, 186, Pavia.
Baiocchi, C. Brezzi, F., and Comincioli, V. (1978). 'Free boundary problems in fluid flow through porous media', in *Finite Elements in Fluids*, Volume 3 (eds R. Gallagher, O. Zienkiewicz, J. Oden, M. Cecchi, and C. Taylor), pp. 283–291, Wiley, Chichester.
Baiocchi, C., Comincioli, V., Guerri, L., and Volpi G. (1973). 'Free boundary problems in the theory of fluid flow through porous media: a numerical approach', *Calcolo*, **10**, 1–86.
Baiocchi, C., Comincioli, V., Magenes, E., and Pozzi, G. A. (1973). 'Free boundary problems in the theory of fluid flow through porous media: existence and uniqueness theorems', *Ann. Mat. Pura Appl.*, (4) **97**, 1–82.
Baiocchi, C., Comincioli, V., Magenes, E., and Pozzi, G. A. (1976). 'Fluid flow through porous media; a new theoretical and numerical approach', in *Mathematical and Numerical Methods in Fluid Dynamics*, pp. 395–447, Proc. Intern. Autumn Course (Trieste, 1973), Intern. Centre for Theoretical Physics, Trieste.
Baiocchi, C., Comincioli, V., and Maione, U. (1975). 'Unconfined flow through porous media', *Meccanica-J. Ital. Assoc. Theoret. Appl. Mech.*, **10**, 151–155.
Baiocchi, C., Comincioli, V., and Maione, U. (1977). 'A mathematical investigation of salt-water intrusion into a coastal aquifer', in *Proc. Symp. Hydrodynamic Diffusion and Dispersion in Porous Media* (Pavia, 1977), pp. 315–328, Istituto di Idraulica dell'Università di Pavia, Pavia.
Baiocchi, C., Evans, L., Frank, L., and Friedman, A. (1980). 'Uniqueness for two immiscible fluids in a one-dimensional porous medium', *J. Diff. Eq.*, **36**, 249–256.
Baiocchi, C., and Friedman, A. (1977). 'A filtration problem in a porous medium with variable permeability', *Ann. Mat. Pura Appl.*, (4) **114**, 377–393.
Baiocchi, C., and Magenes, E. (1974). 'On free-boundary problems associated with the flow of a liquid through porous materials', *Russian Math. Surveys*, **29** (2), 51–71 (Russian ed. (1974): *Uspehi Mat. Nauk*, **29** (2), 50–69).
Baiocchi, C., and Magenes, E. (1975). 'Problemi di frontiera libera in idraulica', in *Atti Convegno Intern. Metodi Valutativi nella Fisica-Matematica* (Roma, 1972), pp. 395–421, Accademia Nazionale dei Lincei, Roma, 1975.
Baiocchi, C., and Maione, U. (1972). 'Un nuovo metodo per lo studio dei moti filtranti a superficie libera', in *Atti XIII Convegno Idraulica e Costruzioni Idrauliche* (Milano, 1972), pp. 401–410, Milano.
Banach, S. (1922). 'Sur les opérations dans les ensembles abstraits et leur application aux équations intégrales', *Fund. Math.*, **3**, 133–181.
Banks, H. T., and Jacobs, M. Q. (1970). 'A differential calculus for multifunctions', *J. Math. Anal. Appl.*, **29**, 246–272.
Barsotti, I. (1968). *Appunti di Algebra*, Zanichelli, Bologna.
Bear, J. (1972). *Dynamics of Fluids in Porous Media*, American Elsevier, New York.
Bear, J. (1979). *Hydraulics of Groundwater*, McGraw-Hill, New York.
Bear, J. (1980a). 'On the derivation of macroscopic descriptions of transport phenomena in porous media by local volume averaging', in *Free Boundary Problems* (ed. E. Magenes), vol. I, pp. 9–30, Proc. Sem. (Pavia, 1979), Istituto Nazionale di Alta Matematica, Roma.
Bear, J. (1980b). 'Hydrodynamic dispersion and the movement and accumulation of pollutants in aquifers', in *Free Boundary Problems* (ed. E. Magenes), vol. I, pp. 33–54, Proc. Sem. (Pavia, 1979), Istituto Nazionale di Alta Matematica, Roma.

Begis, D., and Glowinski, R. (1975). 'Application de la méthode des éléments finis à l'approximation d'un problème de domaine optimal. Méthodes de résolution des problèmes approchés', *Appl. Math. Optim.*, **2**, 130–169.

Beirão da Veiga, H. (1969). 'Sulla hölderianità delle soluzioni di alcune disequazioni variazionali con condizioni unilatere sul bordo', *Ann. Mat. Pura Appl.*, (4) **83**, 73–112.

Benci, V. (1973). 'Su un problema di filtrazione in un mezzo poroso non omogeneo', *Rend. Ac. Naz. Lincei*, (8) **54**, 10–15.

Benci, V. (1974). 'On a filtration problem through a porous medium', *Ann. Mat. Pura Appl.*, (4) **100**, 191–209.

Bensoussan, A. (1974a). 'Points de Nash dans le cas de fonctionnelles quadratiques et jeux différentiels linéaires a N personnes', *SIAM J. Control*, **12**, 460–499.

Bensoussan, A. (1974b). 'Teoria moderna de control optimo', *Cuadernos Inst. Mat. 'Beppo Levi'*, **7**, Universidad de Rosario, Rosario.

Bensoussan, A. (1974c). 'Inéquations quasi-variationnelles et contrôle impulsionnel', in *Atti Giorn. Analisi Convessa e Applicazioni* (Roma, 1974), pp. 21–26, Quad. Gruppi Ric. Mat. C.N.R., Università di Roma, Roma.

Bensoussan, A. (1975). 'Contrôle impulsionnel et inéquations quasi variationnelles', in *Proc. International Congresses of Mathematicians* (Vancouver, 1974) (ed. R. D. James), vol. 2, pp. 329–334, Canadian Mathematical Congress, Vancouver.

Bensoussan, A. (1976). 'Variational inequalities and optimal stopping time problems', in *Calculus of Variations and Control Theory* (ed. D. L. Russell), pp. 219–244, Proc. Symp. (Madison, 1975), Academic Press, New York.

Bensoussan, A., Brézis, H., and Friedman, A. (1977). 'Estimates on the free boundary for quasi-variational inequalities', *Comm. Partial Diff. Eq.*, **2**, 297–321.

Bensoussan, A., and Friedman, A. (1978). 'On the support of the solution of a system of quasi-variational inequalities', *J. Math. Anal. Appl.*, **65**, 660–670.

Bensoussan, A., Goursat, M., and Lions, J. L. (1973). 'Contrôle impulsionnel et inéquations quasi-variationnelles stationnaires', *C.R. Ac. Sci. Paris*, **276**, 1279–1284.

Bensoussan, A., and Lions, J. L. (1973a). 'Nouvelle formulation de problèmes de contrôle impulsionnel et applications', *C.R. Ac. Sci. Paris*, **276**, 1189–1192.

Bensoussan, A., and Lions, J. L. (1973b). 'Contrôle impulsionnel et inéquations quasi-variationnelles d'évolution', *C.R. Ac. Sci. Paris*, **276**, 1333–1338.

Bensoussan, A., and Lions, J. L. (1973c). 'Inéquations variationelles non linéares du premier et du second ordre', *C.R. Ac. Sci. Paris*, **276**, 1411–1415.

Bensoussan, A., and Lions, J. L. (1973d). 'Problèmes de temps d'arrêt optimal et inéquations variationnelles paraboliques', *Applicable Anal.*, **3**, 267–294.

Bensoussan, A., and Lions, J. L. (1974a). 'Contrôle impulsionnel et contrôle continu. Méthode des inéquations quasi variationnelles non linéaires', *C.R. Ac. Sci. Paris*, **278**, 675–679.

Bensoussan, A., and Lions, J. L. (1974b). 'Contrôle impulsionnel et systèmes d'inéquations quasi variationnelles', *C.R. Ac. Sci. Paris*, **278**, 747–751.

Bensoussan, A., and Lions, J. L. (1974c). 'Sur de nouveaux problèmes aux limites pour des opérateurs hyperboliques', *C.R. Ac. Sci. Paris*, **278**, 1345–1349.

Bensoussan, A., and Lions, J. L. (1974d). 'Sur l'approximation numérique d'inéquations quasi-variationnelles stationnaires', in *Proc. Intern. Symp. Computing Methods in Applied Sciences and Engineering* (Versailles, 1973) (eds R. Glowinski and J. L. Lions), part II, pp. 325–338, Lecture Notes in Computer Science, 11, Springer, Berlin.

Bensoussan, A., and Lions, J. L. (1974e). 'Propriétés des inéquations quasi variationnelles décroissantes', in *Analyse Convexe et Ses Applications* (ed. J.-P.

Aubin), pp. 66–84, C.R. Journées d'Analyse Convexe (S. Pierre de Chartreuse, 1974), Lecture Notes Econ. Math. Systems, 102, Springer, Berlin.

Bensoussan, A., and Lions, J. L. (1974f). 'On certain questions related to optimal control', *Russian Math. Surveys*, **29** (2), 79–87 (Russian ed. (1974): *Uspehi Mat. Nauk*, **29** (2), 77–85).

Bensoussan, A., and Lions, J. L. (1975a). 'Nouvelles méthodes en contrôle impulsionnel', *Appl. Math. Optim.*, **1**, 289–312.

Bensoussan, A., and Lions, J. L. (1975b). 'Sur les temps d'arrêt optimaux et les inéquations variationnelle d'évolution', *C.R. Ac. Sci. Paris*, **280**, 989–992.

Bensoussan, A., and Lions, J. L. (1975c). 'Sur le contrôle impulsionnel et les inéquations quasi variationnelles d'évolution', *C.R. Ac. Sci. Paris*, **280**, 1049–1053.

Bensoussan, A., and Lions, J. L. (1975d). 'Contrôle impulsionnel et temps d'arrêt; inéquations variationnelles et quasi-variationnelles d'évolution', *Cahiers de Math. de la Décision*, **7523**, Universitè de Paris, Paris.

Bensoussan, A., and Lions, J. L. (1975e). 'Problèmes de temps d'arrêt optimaux et de perturbations singulières dans les inéquations variationnelles', in *Control Theory, Numerical Methods and Computer Systems Modelling* (eds. A. Bensoussan and J. L. Lions), pp. 567–584, Proc. Intern. Symp. (Rocquencourt, 1974), Lecture Notes Econ. Math. Systems, 107, Springer, Berlin.

Bensoussan, A., and Lions, J. L. (1977). 'Inéquations quasi variationnelles dépendant d'un paramètre', *Ann. Scuola Norm. Sup. Pisa*, (4) **4**, 231–255.

Bensoussan, A., and Lions, J. L. (1978a). 'Problèmes de temps d'arrêt optimal pour les systèmes distribués stochastiques', *Ann. Scuola Norm. Sup. Pisa*, (4) **5**, 181–213.

Bensoussan, A., and Lions, J. L. (1978b). *Applications des Inéquations Variationnelles en Contrôle Stochastique*, Dunod, Paris.

Bensoussan, A., Lions, J. L., and Papanicolaou, G. (1975a). 'Sur quelques phénomènes asymptotiques stationnaires', *C.R. Ac. Sci. Paris*, **281**, 89–94.

Bensoussan, A., Lions, J. L., and Papanicolaou, G. (1975b). 'Sur quelques phénomènes asymptotiques d'évolution', *C.R. Ac. Sci. Paris*, **281**, 317–322.

Bensoussan, A., Lions, J. L., and Papanicolaou, G. (1976a). 'Sur de nouveaux problèmes asymptotiques', *C.R. Ac. Sci. Paris*, **282**, 143–147.

Bensoussan, A., Lions, J. L., and Papanicolaou, G. (1976b). 'Homogénéisation, correcteurs et problèmes non-linèaires', *C.R. Ac. Sci. Paris*, **282**, 1277–1282.

Bensoussan, A., Lions, J. L., and Papanicolaou, G. (1977). 'Sur la convergence d'opérateurs différentiels avec potential fortement oscillant', *C.R. Ac. Sci. Paris*, **284**, 587–592.

Bensoussan, A., Lions, J. L., and Papanicolaou, G. (1978). *Asymptotic Analysis for Periodic Structures*, North-Holland, Amsterdam.

Berestycki, H., and Brézis, H. (1976). 'Sur certains problèmes de frontière libre', *C.R. Ac. Sci. Paris*, **283**, 1091–1094.

Berezanskiĭ, Ju. M. (1968). *Expansions in Eigenfunctions of Selfadjoint Operators*, American Mathematical Society, Providence (Russian ed. (1965): Naukova Dunka, Kiev).

Berge, C. (1957). 'Théorie générale des jeux a n personnes', *Mémor. Sci. Math.*, **138**.

Berge, C. (1958). *Théorie des Graphes et Ses Applications*, Dunod, Paris.

Berge, C. (1959). *Espaces Topologiques. Fonctions Multivoques*, Dunod, Paris.

Bergh, J., and Löfström, J. (1976). *Interpolation Spaces, an Introduction*, Springer, Berlin.

Besov, O. V. (1964). 'Investigation of a family of function spaces in connection with theorems of imbedding and extension', *Amer. Math. Soc. Transl.*, (2) **40**, 85–126 (Russian ed. (1961): Trudy Mat. Inst. Steklov, 60, 42–81).

Bharucha-Reid, A. (1976). 'Fixed point theorems in probabilistic analysis', *Bull. Amer. Math. Soc.*, **82**, 641–657.

Binz, E. (1972). 'Recent results in the functional analytic investigations of convergence spaces', in *General Topology and Its Relations to Modern Analysis and Algebra III* (ed. J. Novák), pp. 67–72, Proc. Third Prague Topological Symp. (Prague, 1971), Academia, Prague/Academic Press, New York.

Birkhoff, G. (1961). *Lattice Theory*, Amer. Math. Soc. Colloquium Publ., 25, American Mathematical Society, Providence.

Birkhoff, G. D., and Kellogg, O. D. (1922). 'Invariant points in function space', *Trans. Amer. Math. Soc.*, **23**, 96–115.

Biroli, M. (1972). 'Gil operatori monotoni: teoria ed applicazioni', *Rend. Sem. Mat. Fis. Milano*, **42**, 143–228.

Biroli, M. (1976). 'Sur les inéquations quasivariationnelles (I.Q.V.) paraboliques avec constraintes sur le bord', *C.R. Ac. Sci. Paris*, **283**, 705–708.

Biroli, M. (1977a). 'Sur la G-convergence pour les inéquations quasi-variationnelles', *C.R. Ac. Sci. Paris*, **284**, 947–950.

Biroli, M. (1977b). 'Sur la G-convergence pour les inéquations quasi-variationnelles', *Boll. U. Mat. Ital.*, (5) **14–A**, 540–550.

Biroli, M. (1977c). 'G-convergence for elliptic equations, variational inequalities and quasi-variational inequalities', *Rend. Sem. Mat. Fis. Milano*, **47**, 269–328.

Biroli, M. (1979). 'G-convergence for elliptic variational and quasi-variational inequalities', in *Proc. Intern. Meeting on Recent Methods in Non-Linear Analysis* (Rome, 1978) (eds E. De Giorgi, E. Magenes, and U. Mosco), pp. 361–381, Pitagora, Bologna.

Biroli, M. (1980). 'Estimates in homogenization for variational and quasi-variational inequalities', in *Free Boundary Problems* (ed. E. Magenes), vol. II, pp. 45–59, Proc. Sem. (Pavia, 1979), Istituto Nazionale di Alta Matematica, Roma.

Biroli, M., Marchi, S., and Norando, T. (1981). 'Homogenization estimates for quasi-variational inequalities', *Boll. U. Mat. Ital.*, (5) **18–A**, 267–274.

Boccardo, L. (1979). 'Esistenza di soluzioni di alcune disequazioni quasi-variazionali non lineari', in *SAFA 2*, pp. 177–180, Atti 2°. Sem. Analisi Funzionale e Applicazioni (Arcavacata di Rende, 1977), Università della Calabria, Cosenza.

Boccardo, L., and Dolcetta, I. C. (1978). 'Disequazioni quasivariazionali con funzione d'ostacolo quasi-limitata: esistenza di soluzioni e G-convergenza', *Boll. U. Mat. Ital.*, (5) **15–B**, 370–385.

Boccardo, L., Dolcetta, I. C. and Matzeu, M. (1979). 'Nouveaux resultats d'existence, perturbation et homogeneisation de qualques problèmes non linéaires', in *Proc. Intern. Meeting on Recent Methods in Non-Linear Analysis* (Rome, 1978) (eds E. De Giorgi, E. Magenes, and U. Mosco), pp. 115–130, Pitagora, Bologna.

Bochner, S., and Martin, W. T. (1948). *Several Complex Variables*, Princeton University Press, Princeton.

Bohl, P. (1904). 'Über die Bewegung eines mechanischen Systems in der Nähe einer Gleichgewichtslage', *J. Reine Angew. Math.*, **127**, 179–276.

Bohnenblust, H. F., and Karlin, S. (1950). 'On a theorem of Ville', in *Contributions to the Theory of Games, I* (eds H. W. Kuhn and A. W. Tucker), pp. 155–160, Annals of Math. Studies, 24, Princeton University Press, Princeton.

Boieri, P., and Gastaldi, F. (1980a). 'Studio di un problema di frontiera libera: diga con base parzialmente impermeabile', *Boll. U. Mat. Ital.*, (5) **17–B**, 1220–1235.

Boieri, P., and Castaldi, F. (1980b). 'Convexity of the free boundary in a filtration problem', in *Free Boundary Problems* (ed. E. Magenes), vol. I, pp. 193–196, Proc. Sem. (Pavia, 1979), Istituto Nazionale di Alta Matematica, Roma.

Boieri, P., and Castaldi, F. (1981). 'Convexity of the free boundary in a filtration problem', *J. Diff. Eq.*, **42**, 25–46.

Boothby, W. H. (1971). 'On two classical theorems of algebraic topology', *Amer Math. Monthly*, **78**, 237–249.
Bourbaki, N. (1955). *Éléments de Mathématique*, Livre V (Espaces Vectoriels Topologiques), Ch. III–V, Act. Sci. Ind., 1229, Hermann, Paris.
Bourbaki, N. (1961). *Éléments de Mathématique*, Livre III (Topologie Générale), Ch. I–II, Act. Sci. Ind., 1142, Hermann, Paris (3eme éd.).
Bourbaki, N. (1963). *Éléments de Mathématique*, Livre I (Théorie des Ensembles), Ch. III, Act. Sci. Ind., 1243, Hermann, Paris (2eme éd.).
Bourbaki, N. (1966). *Éléments de Mathématique*, Livre V (Espaces Vectoriels Topologiques), Ch. I–II, Act. Sci. Ind., 1189, Hermann, Paris (2eme éd.).
Bourgat, J. F., and Duvaut, G. (1975). 'Calcul numérique de l'écoulement avec ou sans sillage au tour d'un profile bidimensionnel symétrique et sans incidence', *Rapport de Recherche*, **145**, IRIA-Laboria, Roquencourt.
Bourgat, J. F., and Duvaut, G. (1977). 'Numerical analysis of flow with or without wake past a symmetric two-dimensional profile without incidence', *Int. J. Num. Math. Engrg.* **11**, 975–993.
Bourgin, D. G. (1963). *Modern Algebraic Topology*, Macmillan, New York.
Boutet de Monvel, L., and Geymonat, G. (1971). 'Solutions irrégulières d'un problème aux limites elliptique', *Symp. Mat.*, **7**, 381–402.
Breton, A., and Leguay, C. (1975). 'Application du contrôle stochastique a la gestion des centrales thermiques et hydrauliques', in *Control Theory, Numerical Methods and Computer Systems Modelling* (eds A. Bensoussan and J. L. Lions), pp. 728–744, Proc. Intern. Symp. (Rocquencourt, 1974), Lecture Notes Econ. Math. Systems, 107, Springer, Berlin.
Brézis, H. (1968). 'Équations et inéquations non linéaires dans les espaces vectoriels en dualité, *Ann. Inst. Fourier*, **18** (1), 115–175.
Brézis, H. (1971). 'Seuil de régularité pour certains problèmes unilatéraux', *C.R. Ac. Sci. Paris*, **273**, 35–37.
Brézis, H. (1972). 'Problèmes unilatéraux', *J. Math. Pures Appl.*, (9) **51**, 1–168.
Brézis, H. (1973). *Opérateurs Maximaux Monotones et Sémi-groupes de Contractions dans les Espaces de Hilbert*, North-Holland, Amsterdam/American Elsevier, New York.
Brézis, H., and Duvaut, G. (1973). 'Écoulements avec sillages autour d'un profil symétrique sans incidence', *C.R. Ac. Sci. Paris*, **276**, 875–878.
Brézis, H., and Kinderlehrer, D. (1974). 'The smoothness of solutions to nonlinear variational inequalities', *Indiana Univ. Math. J.*, **23**, 831–844.
Brézis, H., Kinderlehrer, D., and Stampacchia, G. (1978). 'Sur une nouvelle formulation du problème de l'écoulement à travers une digue', *C.R. Ac. Sci. Paris*, **287**, 711–714.
Brézis, H., Nirenberg, L., and Stampacchia, G. (1972). 'A remark on Ky Fan's minimax principle', *Boll. U. Mat. Ital.*, (4) **6**, 293–300.
Brézis, H., and Sibony, M. (1971). 'Equivalence de deux inéquations variationelles et applicaltions', *Arch. Rational Mech. Anal.*, **41**, 254–265.
Brézis, H., and Stampacchia, G. (1968). 'Sur la régularité de la solution d'inéquations elliptiques', *Bull. Soc. Math. France*, **96**, 153–180.
Brézis, H., and Stampacchia, G. (1973a). 'Une nouvelle méthode pour l'étude d'écoulements stationnaires', *C.R. Ac. Sci. Paris*, **276**, 129–132.
Brézis, H., and Stampacchia, G. (1973b). 'Problèmes elliptiques avec frontière libre', *Séminaire Goulaouic-Schwartz*, Exposè 11, École Polytechnique, Paris.
Brézis, H., and Stampacchia, G. (1976). 'The hodograph method in fluid dynamics in the light of variational inequalities', *Arch. Rational Mech. Anal.*, **61**, 1–18.
Brezzi, F. (1979). 'Implementation of finite element methods for the approximation of variational and quasivariational inequalities', in *Proc. Intern. Meeting on Recent Methods in Non-Linear Analysis* (Rome, 1978) (eds E. De Giorgi, E. Magenes, and U. Mosco), pp. 533–538, Pitagora, Bologna.

Brezzi, F., and Caffarelli, L. A. (1982). 'Convergence of the discrete free boundary for finite element approximations', *to appear*.

Brezzi, F., and Gilardi G. (1982). Chapter 1 of *Finite Element Handbook*, H. Kardestuncer, ed., McGraw-Hill, New York.

Brezzi, F., Hager, W., and Raviart P. (1977). 'Error estimates for the finite element solution of variational inequalities, part I: Primal theory', *Numer. Math.*, **28,** 431–443.

Brezzi, F., Hager, W., and Raviart P. (1978). 'Error estimates for the finite element solution of variational inequalities, part II: Mixed methods', *Numer. Math.*, **31,** 1–16.

Brezzi, F., and Sacchi, G. (1976). 'A finite element approximation of a variational inequality related to hydraulics', *Calcolo*, **13,** 259–273.

Brøndsted, A. (1972). 'On the subdifferential of the supremum of two convex functions', *Math. Scand.*, **31,** 225–230.

Brouwer, L. E. J. (1912). 'Über Abbildung von Mannigfaltigkeiten', *Math. Ann.*, **71,** 97–115.

Browder, F. E. (1959a). 'On a generalization of the Schauder fixed point theorem', *Duke Math. J.*, **26,** 291–303.

Browder, F. E. (1959b). 'Functional analysis and partial differential equations, I', *Math. Ann.*, **138,** 55–79.

Browder, F. E. (1962). 'Functional analysis and partial differential equations, II', *Math. Ann.*, **145,** 81–226.

Browder, F. E. (1964). 'Continuity properties of monotone nonlinear operators in Banach spaces', *Bull. Amer. Math. Soc.*, **70,** 551–553.

Browder, F. E. (1965a). 'Fixed-point theorems for noncompact mappings in Hilbert space', *Proc. Nat. Acad. Sci. U.S.A.*, **53,** 1272–1276.

Browder, F. E. (1965b). 'Nonlinear monotone operators and convex sets in Banach spaces', *Bull. Amer. Math. Soc.*, **71,** 780–785.

Browder, F. E. (1968). 'The fixed point theory of multi-valued mappings in topological vector spaces', *Math. Ann.*, **177,** 283–301.

Brown, M., and Neumann, W. (1977). 'Proof of the Poincaré–Birkhoff fixed point theorem', *Michigan Math. J.*, **24,** 21–31.

Bruch, J. C., Jr. (1979). 'A numerical solution of an irrigation flowfield', *Int. J. Num. Anal. Math. Geomech.*, **3,** 23–36.

Bruch, J. C., Jr. (1980a). 'A survey of free boundary value problems in theory of fluid flow through porous media: variational inequality approach—Part I', *Adv. Water Resources*, **3,** 65–80.

Bruch, J. C., Jr. (1980b). 'A survey of free boundary value problems in theory of fluid flow through porous media: variational inequality approach—Part II', *Adv. Water Resources*, **3,** 115–124.

Bruch, J. C., Jr., and Caffrey, J. (1979a). 'Three-dimensional seepage through a homogeneous dam', *Adv. Water Resources*, **2,** 167.

Bruch, J. C., Jr., and Caffrey, J. (1979b). 'An analysis of seepage through a dam with a toe drain', in *The Mathematics of Finite Elements and Applications III* (ed. J. Whiteman), pp. 123–133, Academic Press, New York.

Bruch, J. C., Jr., Sayle, F. C. and Sloss, J. M. (1978). 'Seepage from a trapezoidal and a rectangular channel using variational inequalities', *J. Hydrol.*, **36,** 247.

Bruch, J. C., Jr., and Sloss, J. M. (1978). 'A variational inequality method applied to free surface seepage from a triangular ditch', *Water Resources Res.*, **14,** 119.

Burenkov, V. I. (1968). 'Imbedding and continuation for classes of differentiable functions of several variables defined in the whole space', *Progress in Math.*, **2,** 73–161 (Russian ed. (1966): Itogi Nauk-Ser. Mat.).

Caccioppoli, R. (1930). 'Un teorema generale sull'esistenza di elementi uniti in una trasormazione funzionale', *Rend. Ac. Naz. Lincei*, (6) **11,** 794–799.

Caccioppoli, R. (1931). 'Sugli elementi uniti delle trasformazioni funzionali:

un'osservazione sui problemi di valori ai limiti', *Rend. Ac. Naz. Lincei*, (6) **13,** 498–502.
Caccioppoli, R. (1932). 'Sugli elementi uniti delle trasformazioni funzionali: un teorema di esistenza e di unicità e alcune sue applicazioni', *Rend. Sem. Mat. Padova,* **3,** 1–15.
Caffarelli, L. (1976a). 'The regularity of elliptic and parabolic free boundaries', *Bull. Amer. Math. Soc.,* **82,** 616–618.
Caffarelli, L. (1976b). 'The smoothness of the free surface in a filtration problem', *Arch. Rational Mech. Anal.,* **63,** 77–86.
Caffarelli, L. (1977). 'The regularity of free boundaries in higher dimensions', *Acta Math.,* **139,** 155–184.
Caffarelli, L. (1978). 'Some aspects of the one-phase Stefan problem', *Indiana Univ. Math. J.,* **27,** 73–77.
Caffarelli, L. (1979a). 'Further regularity results for the Signorini problem', *Comm. Partial Diff. Eq.,* **4,** 1067–1075.
Caffarelli, L. (1979b). 'Regolarità di frontiere libere in più dimensioni', *Pubbl. Laboratorio di Analisi Numerica C.N.R.,* **228,** Pavia.
Caffarelli, L., and Evans, L. C. (1982). 'Continuity of the temperature in the two phase Stefan problem', to appear in *Arch. Rational Mech. Anal.*
Caffarelli, L., and Friedman, A. (1978a). 'The dam problem with two layers', *Arch. Rational Mech. Anal.,* **68,** 125–154.
Caffarelli, L., and Friedman, A. (1978b). 'Asymptotic estimates for the dam problem with several layers', *Indiana Univ. Math. J.,* **27,** 551–580.
Caffarelli, L., and Friedman, A. (1978c). 'The one-phase Stefan problem and the porous medium diffusion equation: continuity of the solution in n space dimensions', *Proc. Nat. Ac. Sci. USA.,* **75,** 2084.
Caffarelli, L., and Friedman, A. (1978d). 'Regularity of the solution of a quasi-variational inequality for the impulse control problem, I', *Comm. Partial Diff. Eq.,* **3,** 745–753.
Caffarelli, L, and Friedman, A. (1979). 'Regularity of the solution of the quasi-variational inequality for the impulse control problem, II', *Comm. Partial Diff. Eq.,* **4,** 279–291.
Caffarelli, L., and Friedman, A. (1980). 'Regularity of the free boundary of a gas flow in a n-dimensional porous medium', *Indiana Univ. Math. J.,* **29,** 361–391.
Caffarelli, L., and Gilardi, G. (1980). 'Monotonocity of the free boundary in the two-dimensional dam problem', *Ann. Scuola Norm. Sup. Pisa,* (4) **7,** 523–537.
Caffarelli, L., and Kinderlehrer, D. (1980). 'Potential methods in variational inequalities', *J. D'Analyse Math.,* **37,** 285–295.
Caffarelli, L., and Rivière, N. (1976a). 'Smoothness and analicity of free boundaries in variational inequalities', *Ann. Scuola Norm. Sup. Pisa,* (4) **3,** 289–310.
Caffarelli, L., and Rivière, N. (1976b). 'On the rectifiability of domains with finite perimeter', *Ann. Scuola Norm. Sup. Pisa,* (4) **3,** 177–186.
Caffarelli, L., and Rivière, N. (1977a). 'The smoothness of the elastic-plastic free boundary of a twisted bar', *Proc. Amer. Math. Soc.,* **63,** 56–58.
Caffarelli, L., and Rivière, N. (1977b). 'Asymptotic behaviour of free boundaries at their singular points', *Ann. of Math.,* **106,** 309–317.
Caffarelli, L., and Rivière, N. (1977c). 'Existence and uniqueness for the problem of filtration through a porous medium', *Notices AMS,* **24,** A-576.
Calderón, A. P. (1951). 'On the differentiability of absolutely continuous functions', *Riv. Mat. Univ. Parma,* **2,** 203–213.
Calderón, A. P. (1961). 'Lebesgue spaces of differentiable functions and distributions', *Proc. Symp. Pure Math.,* **4,** 33–49.
Calderón, C. P., and Lewis, J. E. (1976). 'On the differentiability of functions of several real variables', *Illinois J. Math.,* **20,** 535–542.

Campanato, S. (1964). 'Proprietà di una famiglia di spazi funzionali', *Ann. Scuola Norm. Sup. Pisa*, (3) **18**, 137–160.
Carbone, L., and Valli, A. (1976). 'Filtrazione di un fluido in un mezzo non omogeneo tridimensionale', *Rend. Ac. Naz. Lincei*, (8) **61**, 161–164.
Carbone, L., and Valli, A. (1977). 'Free boundary enclosure in a three-dimensional filtration problem', *Appl. Math. Optim.*, **4**, 1–14.
Carbone, L., and Valli, A. (1978a). 'Asymptotic behaviour of the free boundary in a filtration problem', *Boll. U. Mat. Ital.*, (5) **15–B**, 217–224.
Carbone, L., and Valli, A. (1978b). 'Filtration through a porous non-homogeneous medium with variable cross section', *J. D'Analyse Math.*, **33**, 191–221.
Carrillo, J., and Chipot, M. (1981). 'Sur l'unicité de la solution du problème de l'écoulement a travers une digue', *C.R. Acad. Sc. Paris*, **292**, 191–194.
Carrillo, J., and Chipot, M. (1982). 'The dam problem', to appear on *J. Diff. Eq.*
Castaing, Ch. (1967). 'Sur les multi-applications measurables', *Rev. Française Inf. Rech. Opér.*, **1** (1), 91–126.
Cedergren, H. R. (1967). *Seepage, Drainage and Flow Nets*, Wiley, New York.
Cellina, A. (1970). 'Multivalued differential equations and ordinary differential equations', *SIAM J. Appl. Math.*, **18**, 533–538.
Cellina, A., and Lasota, A. (1969). 'A new approach to the definition of topological degree for multi-valued mappings', *Rend. Ac. Naz. Lincei*, (8) **47**, 434–440.
Charrier, P., and Troianiello, G. (1978). 'On strong solutions to parabolic unilateral problems with obstacle dependent on time', *J. Math. Anal. Appl.*, **65**, 110–125.
Charrier, P., and Vivaldi, M. A. (1976). 'Existence d'une solution forte régulière d'une inéquation quasi variationnelle d'évolution', *C.R. Ac. Sci. Paris*, **283**, 465–467.
Charrier, P., and Vivaldi, M. A. (1977). 'Existence d'une solution régulière d'une inéquation quasi-variationnelle d'évolution avec conditions de Dirichlet', *Boll. U. Mat. Ital.*, (5) **14-A**, 579–589.
Chicco, M. (1967). 'Principio di massimo forte per sottosoluzioni di equazioni ellittiche di tipo variazionale', *Boll. U. Mat. Ital.*, (3) **22**, 368–372.
Chicco, M. (1970). 'Principio di massimo per soluzioni di problemi al contorno misti per equazioni ellittiche di tipo variazionale', *Boll. U. Mat. Ital.*, (4) **3**, 384–394.
Chicco, M. (1975). 'Principio di massimo forte per soluzioni di problemi al contorno misti per equazioni ellittiche di tipo variazionale', *Boll. U. Mat. Ital.*, (4) **11** (suppl. fasc. 3), 100–109.
Chipot, M. (1978). 'Sur la régularité lipschitzienne de la solution d'inéquations elliptiques', *J. Math. Pures Appl.*, (9) **57**, 69–75.
Chipot, M. (1981). 'Sur quelques inéquations variationnelles—Problème de l'écoulement à travers une digue', *Thèse*, Université de Paris 6, Paris.
Choquet, G. (1948). 'Convergences', *Ann. Univ. Grenoble*, 23, 57–112.
Choquet–Bruhat, Y. (1973). *Distributions, Théorie et Problèmes*, Masson, Paris.
Ciarlet, P. G. (1978). *The Finite Element Method for Elliptic Problems*, North-Holland, Amsterdam.
Cibrario-Cinquini, M. (1956). 'Moderne ricerche sulle equazioni a derivate parziali del primo ordine', *Rend. Sem. Mat. Torino*, **15**, 5–26.
Cimatti, G. (1977). 'On a problem of the theory of lubrication governed by a variational inequality', *Appl. Math. Optim.*, **3**, 227–242.
Cimmino, G. (1937). 'Nuovo tipo di condizioni al contorno e nuovo metodo di trattazione per il problema generalizzato di Dirichlet', *Rend. Circ. Mat. Palmermo*, **61**, 177–221.
Cimmino, G. (1952). 'Sulle equazioni lineari alle derivate parziali di tipo ellittico', *Rend. Sem. Mat. Fis. Milano*, **23**, 183–203.
Comincioli, V. (1972). 'Algoritmi numerici per problemi di frontiera libera nella

teoria del flusso dei fluidi attraverso mezzi porosi', in *Atti Conv. AICA Tecniche di Simulazione e Algoritmi* (Milano, 1972), pp. 77–86, Informatica (special issue).

Comincioli, V. (1974a). 'A theoretical and numerical approach to some free boundary problems', *Ann. Mat. Pura Appl.*, (4) **100,** 211–238.

Comincioli, V. (1974b). 'A comparison of algorithms for some free boundary problems', *Pubbl. Laboratorio di Analisi Numerica C.N.R.*, **79,** Pavia.

Comincioli, V. (1975). 'On some oblique derivative problems arising in the fluid flow in porous media. A theoretical and numerical approach', *Appl. Math. Optim.*, **1,** 313–336.

Comincioli, V. (1976). 'Sur l'approximation numérique des problèmes à frontière libre liés à la filtration dans les matériaux poreux', in *Second Intern. Symp. Computing Methods in Applied Sciences* (Versailles, 1975) (eds R. Glowinski and J. L. Lions), pp. 193–206, Lecture Notes in Physics, 58, Springer, Berlin.

Comincioli, V. (1978). 'Un approccio teorico e numerico ai problemi di filtrazione a frontiera libera mediante le disequazioni variazionali', in *Atti Conv. Metodologie Numeriche per la Soluzione di Equazioni Differenziali della Idrogolia e della Idraulica* (Bressanone, 1978).

Comincioli, V. (1979). 'The numerical solution of a free boundary transient seepage flow problem using variational inequalities', in *Proc. Intern. Meeting on Recent Methods in Non-Linear Analysis* (Rome, 1978) (eds. E. De Giorgi, E. Magenes, and U. Mosco), pp. 539–548, Pitagora, Bologna.

Comincioli, V. (1980). 'Free boundary problems in fluid flow through porous media and variational inequalities: a numerical approach', in *Free Boundary Problems* (ed. E. Magenes), vol. I, pp. 197–208, Proc. Sem. (Pavia, 1979), Istituto Nazionale di Alta Matematica, Roma.

Comincioli, V., and Guerri, L. (1976). 'Numerical solution of free boundary problems in seepage flow with capillary fringe', *Comput. Methods Appl. Mech. Engrg,* **7,** 153–178.

Comincioli, V., Guerri, L., and Volpi, G. (1971). 'Analisi numerica di un problema di frontiera libera connesso col moto di un fluido attraverso un mezzo poroso', *Pubbl. Laboratorio di Analisi Numerica C.N.R.*, **17,** Pavia.

Comincioli, V., and Torelli, A. (1979). 'A new numerical approach to a non-steady filtration problem', *Calcolo,* **16,** 93–124.

Cortey-Dumont, Ph. (1978). 'Contribution à l'approximation numérique d'une inéquation quasi-variationnelle', *Thèse,* Universitè de Besançon, Besançon.

Cortey-Dumont, Ph. (1979). 'Contribution à l'étude de l'approximation par la méthode des éléments finis d'inéquations quasi-variationnelles elliptiques', *C.R. Ac. Sci. Paris,* **288,** 141–143.

Cortey-Dumont, Ph. (1980). 'On the approximation of a class of quasivariational inequalities related to the impulse control', in *Free Boundary Problems* (ed. E. Magenes), vol. II, pp. 161–171, Proc. Sem. (Pavia, 1979), Istituto Nazionale di Alta Matematica, Roma.

Cotlar, M., and Cignoli, R. (1974). *An Introduction to Functional Analysis,* North-Holland, Amsterdam/American Elsevier, New York.

Courant, R. (1943). 'Variational methods for the solution of problems of equilibrium and vibrations', *Bull. Amer. Math. Soc.,* **49,** 1–23.

Courant, R. (1967). *Introdução à Teoria das Funções,* Sociedade Paranaense de Matemática, Curitiba.

Courant, R., and Hilbert, D. (1953). *Methods of Mathematical Physics,* volume I, Interscience, New York.

Courant, R., and Hilbert, D. (1962). *Methods of Mathematical Physics,* volume II, Interscience, New York.

Covitz, H., and Nadler, S. B. (1970). 'Multi-valued contraction mappings in generalized metric spaces', *Israel J. Math.*, **8,** 5–11.

Crank, J., and Ozis, T. (1980). 'Numerical solution of a free-boundary problem by interchanging dependent and independent variables', *J. Inst. Math. Appl.*, **26,** 77–85.

Cristescu, R., and Marinescu, G. (1973). *Applications of the Theory of Distributions*, Academiei, București/Wiley, London.

Cronin, J. (1964). *Fixed Points and Topological Degree in Nonlinear Analysis*, Math. Surveys, 11, American Mathematical Society, Providence.

Cryer, C. W. (1976a). 'A survey of steady state porous flow free boundary problems', *MRC Technical Summary Report*, **1657,** University of Wisconsin, Madison.

Cryer, C. W. (1976b). 'A survey of trial free-boundary methods for the numerical solution of free boundary problems', *MRC Technical Summary Report*, **1693,** University of Wisconsin, Madison.

Cryer, C. W. (1977). 'A bibliography of free boundary problems', *MRC Technical Summary Report*, **1793,** University of Wisconsin, Madison.

Cryer, C. W. (1980). 'A proof of the convexity of the free boundary for porous flow through a rectangular dam using the maximum principle', *J. Inst. Math. Appl.*, **25,** 111–120.

Daniljuk, I. I. (1971). 'Sur une classe de fonctionnelles intégrales à domaine variable d'intégration', in *Actes du Congrès International des Mathématiciens* (Nice, 1970) (eds M. Berger, J. Dieudonné, J. Leray, J. L. Lions, P. Malliavin, and J. P. Serre), vol. 2, pp. 703–715, Gauthier-Villars, Paris.

Dantzig, G. (1963). *Linear Programming and Extensions*, Princeton University Press, Princeton.

Day, M. M. (1958). *Normed Linear Spaces*, Springer, Berlin.

Darbo, G. (1950). 'Grado topologico e teoremi di esistenza di punti uniti per trasformazioni plurivalenti di bicelle', *Rend. Sem. Mat. Padova*, **19,** 371–395.

Darbo, G. (1955). 'Punti uniti in trasformazioni a codominio non compatto', *Rend. Sem. Mat. Padova*, **24,** 84–92.

Darbo, G. (1958). 'Teoria dell'omologia in una categoria di mappe plurivalenti ponderate', *Rend. Sem. Mat. Padova*, **28,** 188–220.

Darbo, G. (1961). 'Estensione alle mappe ponderate del teorema di Lefschetz sui punti fissi', *Rend. Sem. Mat. Padova*, **31,** 46–57.

De Blasi, F. (1976). 'On the differentiability of multifunctions', *Pacific J. Math.*, **66,** 67–81.

De Giorgi, E. (1977). 'Γ-convergenza e G-convergenza', *Boll. U. Mat. Ital.*, (5) **14-A,** 213–220.

De Giorgi, E., and Franzoni, T. (1975). 'Su un tipo di convergenza variazionale', *Rend. Ac. Naz. Lincei*, (8) **58,** 842–850.

De Giorgi, E., and Franzoni, T. (1979). 'Su un tipo di convergenza variazionale', *Rend. Sem. Mat. Brescia*, **3,** 63–101.

Deny, J. (1970). 'Méthodes hilbertiennes en théorie du potentiel', in *Potential Theory*, pp. 121–201, Proc. C.I.M.E. (Stresa, 1969), Cremonese, Roma.

Deny, J., and Lions, J. L. (1954), 'Les espaces du type de Beppo Levi', *Ann. Inst. Fourier*, **5,** 305–370.

De Rham, G. (1960), *Variétés Différentiables*, Act. Sci. Ind., 1222, Hermann, Paris.

De Vito, L. (1958). 'Sulle funzioni ad integrale di Dirichlet finito', *Ann. Scuola Norm. Sup. Pisa*, (3) **12,** 55–127.

Dias, J. P. (1973). 'Sur la régularité de la solution faible d'une inéquation variationnelle non linéaire de type elliptique', *Boll. U. Mat. Ital.*, (4) **8,** 8–15.

Di Benedetto, E. (1980). 'Regularity properties of the solution of an n-dimensional two-phase Stefan problem', *Analisi Funz. e Appl.* (Suppl. *Boll. U.M.I.*), **1,** 129–152.

Dieudonné, J. (1968a). *Calcul Infinitésimal*, Hermann, Paris.
Dieudonné, J. (1968b). *Éléments d'Analyse*, tome I (Fondements de l'Analyse Moderne), Gauthier-Villars, Paris (English ed. (1960): Academic Press, New York).
Dieudonné, J. (1968c). *Éléments d'Analyse*, tome II, Gauthier-Villars, Paris (English ed. (1970): Academic Press, New York).
Dieudonné, J., and Schwartz, L. (1950). 'La dualité dans les espaces ($\mathscr{F}$) et ($\mathscr{LF}$)', *Ann. Inst. Fourier*, **1**, 61–101.
Dolcetta, I. C., and U. Mosco (1980). 'Implicit complementarity problems and quasi-variational inequalities', in *Variational Inequalities and Complementarity Problems* (eds R. W. Cottle, F. Giannessi and J. L. Lions), pp. 75–87, Wiley, Chichester.
Dolcetta, I. C., and Vivaldi, M. A. (1977). 'Existence d'une solution régulière d'une inéquation quasi variationnelle elliptique sur un domaine non borné', *C.R. Ac. Sci. Paris*, **284**, 1033–1036.
Dolcetta, I. C., and Vivaldi, M. A. (1978). 'Existence of a regular solution of a quasi-variational inequality in a unbounded domain', *Comm. Partial Diff. Eq.*, **3**, 443–470.
Dugundji, J., and Granas, A. (1978). 'KKM maps and variational inequalities', *Ann. Scuola Norm. Sup. Pisa*, (4) **5**, 679–682.
Dunford, N., and Schwartz, J. T. (1958). *Linear Operators*, part I, Interscience, New York.
Dunford, N., and Schwartz, J. T. (1963). *Linear Operators*, part II, Interscience, New York.
Duvaut, G. (1973). 'Résolution d'un problème de Stefan (Fusion d'un bloc de glace à zéro degré', *C.R. Ac. Sci. Paris*, **276**, 1461–1463.
Duvaut, G. (1974). 'Étude de problèmes unilatéraux en mécanique par des méthodes variationnelles', in *New Variational Techniques in Mathematical Physics* (eds G. Capriz and G. Stampacchia), pp. 41–102, Proc. C.I.M.E. (Bressanone, 1973), Cremonese, Roma.
Duvaut, G. (1976). 'Problèmes à frontière libre en théorie des milieux continus', *Rapport de Recherche*, **185**, IRIA-Laboria, Rocquencourt.
Duvaut, G., and Lions, J. L. (1976). *Inequalities in Mechanics and Physics*, Springer, Berlin (French ed. (1972): Dunod, Paris).
Edwards, R. E. (1965). *Functional Analysis, Theory and Applications*, Holt, Rinehart and Winston, New York.
Edwards, R. E., and Gaudry, G. I. (1977). *Littlewood-Paley and Multiplier Theory*, Springer, Berlin.
Eggleston, H. G. (1963). *Convexity*, Cambridge University Press, Cambridge.
Eilenberg, S., and Montgomery, D. (1946). 'Fixed point theorems for multivalued transformations', *Amer. J. Math.*, **68**, 214–222.
Ekeland, I., and Temam, R. (1974). *Analyse Convexe et Problèmes Variationnels*, Dunod/Gauthier-Villars, Paris.
Elliott, C. M. (1980). 'On a variational inequality formulation of an electrochemical machining moving boundary problem and its approximation by the finite element method', *J. Inst. Math. Appl.*, **25**, 121–131.
Elliott, C., and Ockendon, J. (1982). *Weak and Variational Methods for Moving Boundary Problems*, Pitman, Boston.
Ene, H. I., and Sanchez-Palencia, E. (1975). 'Équations et phénomènes de surface pour l'écoulement dans un modèle de milieu poreux', *J. Mécanique*, **14**, 73–108.
Falk, F. (1974). 'Error estimates for the approximation of a class of variational inequalities', *Math. Comp.*, **28**, 963–971.
Fan, K. (1952). 'A generalization of Tucker's combinatorial lemma with topological applications', *Ann. of Math.*, (2) **56**, 431–437.

Fan, K. (1961). 'A generalization of Tychonoff's fixed point theorem', *Math. Ann.*, **142**, 305–310.
Fasano, A., and Primicerio, M., Editors (1982). *Proc. Symp. Free Boundary Problems: Theory of Applications* (Montecatini Terme, 1981), to appear (Gordon and Breach, New York).
Ferreira, J. C. (1967). *Alguns Problemas de Prolungamento das Distribuições com Aplicações à Integração*, Lisboa.
Ferreira, J. C., and Oliveira, J. S. (1964). 'Problèmes aux conditions initiales dans la théorie des distributions', *Rev. Fac. Ciências Lisboa*, (2-A) **10**, 91–130.
Fichera, G. (1950). 'Sull'esistenza e sul calcolo delle soluzioni dei problemi al contorno relativi all'equilibrio di un corpo elastico', *Ann. Scuola Norm. Sup. Pisa*, (3) **4**, 35–99.
Fichera, G. (1956). 'Sulle equazioni differenziali lineari ellittico-paraboliche del secondo ordine', *Atti Ac. Naz. Lincei Mem. Sez. I*, (8) **5**, 1–30.
Fichera, G. (1965). *Linear Elliptic Differential Systems and Eigenvalue Problems*, Lecture Notes in Mathematics, 8, Springer, Berlin.
Fichera, G. (1966). 'Problemi elastostatici con vincoli unilaterali: il problema di Signorini con ambigue condizioni al contorno', *Atti Ac. Naz. Lincei Mem. Sez. I*, (8) **7**, 71–140.
Finn, R. (1974). 'Capillarity phenomena', *Russian Math. Surveys*, **29** (4), 133–153 (Russian ed. (1974): *Uspehi Mat. Nauk*, **29** (4), 131–152).
Fleming, W. H. (1965). *Functions of Several Variables*, Addison-Wesley, Reading.
Franklin, J. (1980). *Methods of Mathematical Economics*, Springer, New York.
Frehse, J. (1972). 'On the regularity of the solution of a second order variational inequality', *Boll. U. Mat. Ital.*, (4) **6**, 312–315.
Frehse, J. (1977). 'On Signorini's problem and variational problems with thin obstacles', *Ann. Scuola Norm. Sup. Pisa*, (4) **4**, 343–362.
Frehse, J. (1982). 'On the smoothness of variational inequalities with obstacles', *to appear*.
Frehse, J., and Mosco, U. (1979a). 'Sur la régularité de certaines inéquations variationnelles et quasi-variationnelles', *C.R. Ac. Sci. Paris*, **289**, 627–630.
Frehse, J., and Mosco, U. (1979b). 'Variational inequalities with one-sided irregular obstacles', *Manuscripta Math.*, **28**, 219–233.
Frehse, J., and Mosco, U. (1982). 'Irregular obstacles and quasi-variational inequalities of stochastic impulse control', to appear in *Ann. Sc. Norm. Sup. Pisa*.
Frémond, M. (1975). 'Variational formulation of the Stefan problem-coupled Stefan problem-frost propagation in porous media', in *Computational Mechanics* (ed. J. T. Oden), pp. 341–350, Proc. Intern. Conf. Computer Methods in Nonlinear Mechanics (Austin, 1974), Lecture Notes in Mathematics, 461, Springer, Berlin.
Friedman, A. (1968). 'Boundary behavior of solutions of variational inequalities for elliptic operators', *Arch. Rational Mech. Anal.*, **27**, 95–107.
Friedman, A. (1975). *Stochastic Differential Equations and Applications*, Volume 1, Academic Press, New York.
Friedman, A. (1976a). *Stochastic Differential Equations and Applications*, volume 2, Academic Press, New York.
Friedman, A. (1976b). 'The shape and smoothness of the free boundary for some elliptic variational inequalities', *Indiana Univ. Math. J.*, **25**, 103–118.
Friedman, A. (1976c). 'A class of parabolic quasi-variational inequalities, II', *J. Diff. Eq.*, **22**, 379–401.
Friedman, A. (1976d). 'A problem in hydraulics with non-monotone free boundary', *Indiana Univ. Math. J.*, **25**, 577–592.
Friedman, A. (1978). 'On the free boundary of a quasi variational inequality arising in a problem of quality control', *Trans. Amer. Math. Soc.*, **246**, 95–110.

Friedman, A. (1979). 'Time dependent free boundary problems', *SIAM Rev.*, **21**, 213–221.
Friedman, A. (1980a). 'The dam problem with variable permeability', in *Variational Inequalities and Complementarity Problems* (eds R. W. Cottle, F. Giannessi, and J. L. Lions), pp. 135–141, Wiley, Chichester.
Friedman, A. (1980b). 'Stochastic control with partial observations', in *Variational Inequalities and Complementarity Problems* (eds R. W. Cottle, F. Giannessi, and J. L. Lions), pp. 143–149, Wiley, Chichester.
Friedman, A., and Jensen, R. (1975). 'A parabolic quasi-variational inequality arising in hydraulics', *Ann. Scuola Norm. Sup. Pisa*, (4) **2**, 421–468.
Friedman, A., and Jensen, R. (1976). 'Elliptic quasi-variational inequalities and applications to a non-stationary problem in hydraulics', *Ann. Scuola Norm. Sup. Pisa*, (4) **3**, 47–88.
Friedman, A., and Jensen, R. (1978). 'Convexity of the free boundary in the Stefan problem and in the dam problem', *Arch. Rational Mech. Anal.*, **67**, 1–24.
Friedman, A., and Kinderlehrer, D. (1975). 'A one phase Stefan problem', *Indiana Univ. Math. J.*, **24**, 1005–1035.
Friedman, A., and Kinderlehrer, D. (1976). 'A class of parabolic quasi-variational inequalities', *J. Diff. Eq.*, **21**, 395–416.
Friedman, A., and Torelli, A. (1977). 'A free boundary problem connected with non-steady filtration in porous media', *Nonlinear Anal.*, **1**, 503–545 (and correction on the same journal (1978): **2**, 513–518).
Friedrichs, K. O. (1944). 'The identity of weak and strong extensions of differential operators', *Trans. Amer. Math. Soc.*, **55**, 132–151.
Fučík, S., and Kufner, A. (1980). *Nonlinear Differential Equations*, Elsevier, Amsterdam.
Furi, M., and Martelli, M. (1972). 'Some fixed-points theorems for multi-valued mappings in topological vector spaces', *Ann. Mat. Pura Appl.*, (4) **92**, 169–175.
Furi, M., and Martelli, M. (1974). 'On the minimal displacement of points under α-Lipschitz maps in normed spaces', *Boll. U. Mat. Ital.*, (4) **9**, 791–799.
Furi, M., Martelli, M., and Vignoli, A. (1978). 'Contribution to the spectral theory for nonlinear operators in Banach spaces', *Ann. Mat. Pura Appl.*, (4) **118**, 229–294.
Gagliardo, E. (1957). 'Caratterizzazione delle tracce sulla frontiera relative ad alcune classi di funzioni in n variabili', *Rend. Sem. Mat. Univ. Padova*, **27**, 284–305.
Gagliardo, E. (1958). 'Proprietà di alcune classi di funzioni in più variabili', *Ricerche Mat.*, **7**, 102–137.
Gagliardo, E. (1959). 'Ulteriori proprietà di alcune classi di funzioni in più variabili', *Ricerche Mat.*, **8**, 24–51.
Gagliardo, E. (1961). 'Una struttura unitaria in diverse famiglie di spazi funzionali, parte I', *Ricerche Mat.*, **10**, 244–281.
Gagliardo, E. (1963). 'A common structure in various families of functional spaces, part II', *Ricerche Mat.*, **12**, 87–106.
Garnir, H. G., De Wilde, M., and Schmets, J. (1968). *Analyse Fonctionnelle*, tome I, Birkhäuser, Basel.
Garnir, H. G., De Wilde, M., and Schmets, J. (1972). *Analyse Fonctionnelle*, tome II, Birkhäuser, Basel.
Garroni, M. (1982). 'A nonlinear quasi-variational inequality with implicit obstacle on the boundary', to appear.
Garroni, M., Hanouzet, B., and Joly, J. L. (1982). 'Regularité pour la solution d'un systeme d'I.Q.V.', to appear.
Garroni, M., and Troianiello, G. (1979). 'Some regularity results and a priori estimates for solutions of variational and quasi-variational inequalities', in *Proc.*

Intern. Meeting on Recent Methods in Non-Linear Analysis (Rome, 1978) (eds E. De Giorgi, E. Magenes, and U. Mosco), pp. 493–518, Pitagora, Bologna.

Garroni, M., and Vivaldi, M. A. (1979). 'Esistenza, regolarità e stime duali per la soluzione di una disequazione quasi-variazionale relativa ad un operatore quasi-lineare', *Boll. U. Mat. Ital.*, (5) **16-B,** 154–167.

Gastaldi, F. (1979). 'About the possibility of setting Stefan-like problems in variational form', *Boll. U. Mat. Ital.*, (5) **16-A,** 148–156.

Gerhardt, C. (1973). 'Regularity of solutions of nonlinear variational inequalities', *Arch. Rational Mech. Anal.*, **52,** 389–393.

Gilardi, G. (1976). 'Studio di una famiglia di disequazioni quasi-variazionali connessa con un problema di frontiera libera', *Boll. U. Mat. Ital.*, (5) **13-B,** 138–159.

Gilardi, G. (1977). 'Studio di una disequazione quasi-variazionale relativa ad un problema di filtrazione in tre dimensioni', *Ann. Mat. Pura. Appl.*, (4) **113,** 1–17.

Gilardi, G. (1979). 'A new approach to evolution free boundary problems', *Comm. Partial Diff. Eq.*, **4,** 1099–1122 (and correction on the same journal (1980): **5,** 983-984).

Gilardi, G. (1980a). 'The evolution dam problem', in *Free Boundary Problems* (ed. E. Magenes), vol. I, pp. 209–217, Proc. Sem. (Pavia, 1979), Istituto Nazionale di Alta Matematica, Roma.

Gilardi, G. (1980b). 'The behaviour of the free boundary near the fixed boundary in the dam problem', in *Recent Methods in Nonlinear Analysis and Applications* (ed. A. Canfora et al.), pp. 347–352. Proc. SAFA IV-Int. Meeting (Catania 1980), Liguori, Napoli.

Gilardi, G. (1982). 'Proprietà della frontiera libera in un problema di filtrazione', to appear on *Rend. Sem. Mat. Fis. Milano*.

Giusti, E. (1973). 'Minimal surfaces with obstacles', in *Geometric Measure Theory and Minimal Surfaces* (ed. E. Bombieri), pp. 119–153, Proc. C.I.M.E. (Varenna, 1972), Cremonese, Roma.

Glashoff, K., and Roleff, K. (1977a). 'Über eine elliptische freie Randwertaufgabe', in *Freie Randwertprobleme I (Numerische Verfahren)*, pp. 45–61, Freie Univ. Berlin, Berlin.

Glashoff, K., and Roleff, K. (1977b). 'Über eine elliptische freie Randwertaufgabe, ein numerisches Beispiel', in *Freie Randwertprobleme II (Numerische Tests)*, pp. 40–47, Freie Univ. Berlin, Berlin.

Glicksberg, I. L. (1952). 'A further generalization of the Kakutani fixed point theorem, with application to Nash equilibrium points', *Proc. Amer. Math. Soc.*, **3,** 170–174.

Glowinski, R. (1978). 'Finite elements and variational inequalities', *Technical report*, INF-LAB **78010,** IRIA-Laboria, Rocquencourt.

Glowinski, R. (1980). *Lectures on Numerical Methods for Non-linear Variational Problems*, Tata Institute of Fundamental Research, Bombay, and Springer, Berlin.

Glowinski, R., Lions, J. L., and Trémolières, R. (1976a). *Analyse Numérique des Inéquations Variationnelles*, tome I, Dunod, Paris.

Glowinski, R., Lions, J. L., and Trémolières, R. (1976b). *Analyse Numérique des Inéquations Variationnelles*, tome II, Dunod, Paris.

Goursat, M., and Maarek, G. (1976). 'Nouvelle approche des problèmes de gestion de stocks. Comparaison avec les méthodes classiques', *Rapport de Recherche*, **148,** IRIA–Laboria, Rocquencourt.

Goursat, M., and Maurin, S. (1975). 'Méthodes de résolution numérique des inéquations quasi-variationnelles', in *Control Theory, Numerical Methods and Computer Systems Modelling* (eds A. Bensoussan and J. L. Lions), pp. 585–609,

Proc. Intern. Symp (Rocquencourt, 1974), Lecture Notes Econ. Math. Systems, 107, Springer, Berlin.
Goursat, M., and Quadrat, J. P. (1976). 'Analyse numérique d'inéquations quasi-variationnelles elliptiques associées à des problèmes de contrôle impulsionnel', *Rapport de Recherche*, **186**, IRIA–Laboria, Rocquencourt.
Grisvard, P. (1975). 'Alternative de Fredholm relative au problème de Dirichlet dans un polyedre', *Ann. Scuola Norm. Sup. Pisa*, (4) **2**, 359–388.
Grothendieck, A. (1955). 'Produits tensoriels topologiques et espaces nucléaires', *Mem. Amer. Math. Soc.*, **16**.
Grothendieck, A. (1958). *Espaces Vectoriels Topologiques*, Sociedade Matemática de São Paulo, São Paulo.
Grünbaum, B. (1967). *Convex Polytopes*, Interscience, London.
Guelfand, I. M., and Chilov, G. E. (1962). *Les Distributions*, tome I, Dunod, Paris (Russian ed. (1958): Nauka, Moscow).
Guelfand, I. M., and Chilov, G. E. (1964). *Les Distributions*, tome II, Dunod, Paris (Russian ed. (1958): Nauka, Moscow).
Guelfand, I. M., and Chilov, G. E. (1965). *Les Distributions*, tome III, Dunod, Paris (Russian ed. (1958): Nauka, Moscow).
Guelfand, I. M., Graev, M. I., and Vilenkin, N. Y. (1970). *Les Distributions*, tome V, Dunod, Paris (Russian ed. (1962): Nauka, Moscow).
Guelfand, I. M., and Vilenkin, N. Y. (1967). *Les Distributions*, tome IV, Dunod, Paris (Russian ed. (1961): Nauka, Moscow).
Hadamard, J. (1968a). 'Sur les problèmes aux dérivées partielles et leur signification physique', *Œuvres*, tome III, pp. 1099–1105, C.N.R.S., Paris.
Hadamard, J. (1968b). 'Sur les données aux limites dans les équations aux dérivées partielles de la physique mathématique', *Œuvres*, tome III, pp. 1161–1163, C.N.R.S., Paris.
Hadamard, J. (1968c). 'Quelques cas d'impossibilité du problème de Cauchy', *Œuvres*, tome III, pp. 1457–1470, C.N.R.S., Paris.
Halmos, P. R. (1958). *Finite-Dimensional Vector Spaces*, D. Van Nostrand, Princeton (2nd ed.).
Hamilton, O. H. (1947). 'A fixed point theorem for upper semicontinuous transformations of n-cells for which the images of points are non-acyclic continua', *Duke Math. J.*, **14**, 689–693.
Hanouzet, B., and Joly, J. L. (1975a). 'Méthodes d'ordre dans l'interprétation de certaines inéquations variationnelles et aplications', *C.R. Ac. Sci. Paris*, **281**, 373–376.
Hanouzet, B., and Joly, J. L. (1975b). 'Un résultat de régularité pour une inéquation qquasivariationnelle du type de Neumann intervenant dans un problème de contrôle impulsionnel', *C.R. Ac. Sci. Paris*, **281**, 799–800.
Hanouzet, B., and Joly, J. L. (1977). 'Un résultat de régularité pour une inéquation quasi variationnelle du type de Neumann intervenant dans un problème de contrôle impulsionnel', *J. Math. Pures Appl.*, (9) **56**, 327–337.
Hanouzet, B., and Joly, J. L. (1978). 'Convergence uniforme des itérés definisant la solution d'une inéquation quasivariationnelle abstraite', *C.R. Ac. Sci. Paris*, **286**, 735–738.
Hanouzet, B., and Joly, J. L. (1979a). 'Méthodes d'ordre dans l'interprétation de certaines inéquations variationnelles et applications', *J. Functional Analysis*, **34**, 217–249.
Hanouzet, B., and Joly, J. L. (1979b). 'Convergence geometrique des iterées definisant la solution d'une I.Q.V. abstraite', in *Proc. Intern. Meeting on Recent Methods in Non-Linear Analysis* (Rome, 1978) (eds E. De Giorgi, E. Magenes, and U. Mosco), pp. 425–431, Pitagora, Bologna.
Hanouzet, B., and Joly, J. L. (1979c). 'Regularité $W^{2,p}$ pour une I.Q.V. avec

conditions de Neumann', in *Proc. Intern. Meeting on Recent Methods in Non-Linear Analysis* (Rome, 1978) (eds. E. De Giorgi, E. Magenes, and U. Mosco), pp. 433–439, Pitagora, Bologna.
Haraux, A. (1977). 'How to differentiate the projection on a convex set in Hilbert space. Some applications to variational inequalities', *J. Math. Soc. Japan*, **29**, 615–631.
Harr, M. E. (1962). *Groundwater and Seepage*, McGraw-Hill, New York.
Hartman, P., and Stampacchia, G. (1966). 'On some non-linear elliptic differential-functional equations', *Acta Math.*, **115**, 271–310.
Haugazeau, Y. (1967). 'Sur des inéquations variationnelles', *C.R. Ac. Sci. Paris*, **265**, 95–98.
Haugazeau, Y. (1968). 'Sur les inéquations variationnelles et la minimisation de fonctionnelles convexes', *Thèse*, Université de Paris, Paris.
Hewitt, E., and Stromberg, K. (1965). *Real and Abstract Analysis*, Springer, Berlin.
Hille, E. (1972). *Methods in Classical and Functional Analysis*, Addison-Wesley, Reading.
Hogbe-Nlend, H. (1977). *Bornologies and Functional Analysis*, North-Holland, Amsterdam.
Hopf, E. (1952). 'A remark on linear elliptic differential equations of second order', *Proc. Amer. Math. Soc.*, **3**, 791–793.
Hörmander, L. (1960). 'Estimates for translation invariant operators in L^p spaces', *Acta Math.*, **104**, 93–140.
Hörmander, L. (1964). *Linear Partial Differential Operators*, Springer, Berlin (2nd rev. print).
Horváth, J. (1966). *Topological Vector Spaces and Distributions*, volume I, Addison-Wesley, Reading.
Hu, S. T. (1965). *Theory of Retracts*, Wayne State University Press, Detroit.
Hurewicz, W., and Wallman, H. (1948). *Dimension Theory*, Princeton University Press, Princeton.
Husain, T. (1965). *The Open Mapping and Closed Graph Theorems în Topological Vector Spaces*, Vieweg, Braunschweig.
Il'in, V. P. (1969). 'The properties of some classes of differentiable functions of several variables defined in an n-dimensional region', *Amer. Math. Soc. Transl.*, (2) **81**, 91–256 (Russian ed. (1962): Trudy Mat. Inst. Steklov, **66**, 227–363).
Ioffe, A. D., and Levin, V. L. (1972). 'Subdifferentials of convex functions', *Trans. Moscow Math. Soc.*, **26**, 1–72 (Russian ed. (1972): Trudy Moskov. Mat. Obšč., **26**, 3–73).
Istrățescu, V. I. (1973). *Introducere în Teoria Punctelor Fixe*, Academiei, București.
Jameson, G. (1970). *Ordered Linear Spaces*, Lecture Notes in Mathematics, 141, Springer, Berlin.
Jawerth, B. (1978). 'The trace of Sobolov and Besov spaces if $0 < p < 1$', *Studia Math.*, **62**, 65–71.
Jensen, R. (1977). 'Structure of the non-monotone free-boundaries in a filtration problem', *Indiana Univ. Math. J.*, **26**, 1121–1135.
Jensen, R. (1980a). 'Boundary regularity for variational inequalities', *Indiana Univ. Math. J.*, **29**, 495–504.
Jensen, R. (1980b). 'Fluid flow through a porous media', in *Free Boundary Problems* (ed. E. Magenes), vol. I, pp. 133–149, Proc. Sem. (Pavia, 1979), Istituto Nazionale di Alta Matematica, Roma.
Jensen, R. (1982). 'Finite difference approximation to the free boundary of a parabolic inequality', *to appear*.
John, F. (1955). 'A note on "improper" problems in partial differential equations', *Comm. Pure Appl. Math.*, **8**, 591–594.

Joly, J. L. (1973). 'Une famille de topologies sur l'ensemble des fonctions convexes pour lesquelles la polarité est bicontinue', *J. Math. Pures Appl.*, (9) **52,** 421–441.
Joly, J. L. (1974). 'I.Q.V. stationnaire pour une operateur différentiel du 2e ordre', in *Atti Giorn. Analisi Convessa e Applicazioni* (Roma, 1974), pp. 41–50, Quad. Gruppi Ric. Mat. C.N.R., Università di Roma, Roma.
Joly, J. L., and Mosco, U. (1974). 'Sur les inéquations quasi-variationnelles', *C.R. Ac. Sci. Paris*, **279,** 499–502.
Joly, J. L., and Mosco, U. (1975). 'Remarques sur les inéquations quasi-variationnelles', in *Control Theory, Numerical Methods and Computer Systems Modelling* (eds. A. Bensoussan and J. L. Lions), pp. 625–642, Proc. Intern. Symp. (Rocquencourt, 1974), Lecture Notes Econ. Math. Systems, 107, Springer, Berlin.
Joly, J. L., and Mosco, U. (1979). 'À propos de l'existence et de la régularité de solutions de certaines inéquations quasi-variationnelles', *J. Functional Anal.*, **34,** 107–137.
Joly, J. L., Mosco, U., and Troianiello, G. (1974). 'Un résultat de régularité pour une inéquation quasi-variationnelle intervenant dans un problème de contrôle impulsionnel', *C.R. Ac. Sci. Paris*, **279,** 937–940.
Joly, J. L., Mosco, U., and Troianiello, G. (1977). 'On the regular solution of a quasi variational inequality connected to a problem of stochastic impulse control', *J. Math. Annal. Appl.*, **61,** 357–369.
Kakutani, S. (1941). 'A generalization of Brouwer's fixed point theorem', *Duke Math. J.*, **8,** 457–459.
Kannai, Y. (1981). 'An elementary proof of the no-retraction theorem', *Amer. Math. Monthly*, **88,** 264–268.
Kato, T. (1964). 'Demicontinuity, hemicontinuity and monotonicity', *Bull. Amer. Math. Soc.*, **70,** 548–550.
Kato, T. (1966). *Perturbation Theory for Linear Operators*, Springer, Berlin.
Kato, T. (1967). 'Demicontinuity, hemicontinuity and monotonocity, II, *Bull. Amer. Soc.*, **73,** 886–889.
Kawarada, H. (1979). 'Numerical methods for free surface problems by means of penalty', in *Computing Methods in Applied Sciences and Engineering, I* (eds R. Glowinski and J. L. Lions), pp. 282–291, Proc. Intern. Symp. IRIA–Laboria (1977), Lecture Notes in Mathematics, 704, Springer, Berlin.
Kelley, J. L. (1942). Hyperspaces of a continuum', *Trans. Amer. Math. Soc.*, **52,** 22–36.
Kelley, J. L. (1950). 'The Tychonoff product theorem implies the axiom of choice', *Fund. Math.*, **37**, 75–76.
Kelley, J. L. (1962). *Topología General*, Eudeba, Buenos Aires (English ed. (1955): D. Van Nostrand, Princeton).
Kelley, J. L., Namioka, I., and Co-authors (1963), *Linear Topological Spaces*, D. Van Nostrand, Princeton.
Kellog, O. D. (1929). *Foundations of Potential Theory*, Springer, Berlin.
Kellog, R. B. (1976). 'Uniqueness in the Schauder fixed point theorem', *Proc. Amer. Math. Soc.*, **60,** 207–210.
Kikuchi, N. (1977a). 'An analysis of the variational inequalities of seepage flow by finite-element methods', *Quart. Appl. Math.*, **35,** 149–163.
Kikuchi, N. (1977b). 'Seepage flow problems by variational inequalities: theory and approximation', *Intern J. Num. Anal. Meth. Geomechanics*, **1,** 283.
Kikuchi, N., and Oden, J. T. (1979). 'Theory of variational inequalities with applications to problems of flow through porous media', *TICOM Report*, **79–4,** University of Texas, Austin.
Kinderlehrer, D. (1971). 'The coincidence set of solutions of certain variational inequalities', *Arch. Rational Mech. Anal.*, **40,** 231–250.

Kinderlehrer, D. (1973). 'How a minimal surface leaves an obstacle', *Acta Math.*, **130**, 221–242.
Kinderlehrer, D. (1974). 'Remarks about the free boundaries occurring in variational inequalities', in *New Variational Techniques in Mathematical Physics* (eds G. Capriz and G. Stampacchia), pp. 104–119, Proc. C.I.M.E. (Bressanone, 1973), Cremonese, Roma.
Kinderlehrer, D. (1975). 'Elliptic variational inequalities', in *Proc. International Congresses of Mathematicians* (Vancouver, 1974) (ed. R. D. James), vol. 2, pp. 269–273, Canadian Mathematical Congress, Vancouver.
Kinderlehrer, D. (1978). 'Variational inequalities and free boundary problems', *Bull. Amer. Math. Soc.*, **84**, 7–26.
Kinderlehrer, D. (1981). 'The smoothness of the solution of the boundary obstacle problem', *J. Math. Pures Appl.*, (9) **60**, 193–212.
Kinderlehrer, D., and Nirenberg, L. (1977). 'Regularity in free boundary problems', *Ann. Scuola Norm. Sup. Pisa*, (4) **4**, 373–391.
Kinderlehrer, D., Nirenberg, L., and Spruck, J. (1978a). 'Régularité dans les problèmes elliptiques à frontiere libre', *C.R. Ac. Sci. Paris*, **286**, 1187–1190.
Kinderlehrer, D., Nirenberg, L., and Spruck, J. (1978b). 'Regularity in elliptic free boundary problems, I', *J. D'Analyse Math.*, **34**, 86–119.
Kinderlehrer, D., Nirenberg, L., and Spruck, J. (1979). 'Regularity in elliptic free boundary problems, II', *Ann. Scuola Norm. Sup. Pisa*, (4) **6**, 637–683.
Kinderlehrer, D., and Stampacchia, G. (1980). *An Introduction to Variational Inequalities and Their Applications*, Academic Press, New York.
Knaster, B., Kuratowski, C., and Mazurkiewicz, S. (1929). 'Ein Beweis des Fixpunktsatzes für n-dimensionale Simplexe', *Fund. Math.*, **14**, 132–137.
Kohn, J. J. (1974). 'Uno sguardo agli operatori pseudo-differenziali', *Boll. U. Mat. Ital.*, (4) **10**, 237–297.
Kolodner, I. I. (1968). 'On completeness of partially ordered sets and fixpoint theorems for isotone mappings', *Amer. Math. Monthly*, **75**, 48–49.
Kolmogorov, A. N., and Fomin, S. V. (1957). *Elements of the Theory of Functions and Functional Analysis*, volume I, Graylock, Rochester (Russian ed. (1954)).
Kolmogorov, A. N., and Fomin, S. V. (1961). *Elements of the Theory of Functions and Functional Analysis*, volume II, Graylock, Baltimore (Russian ed. (1960)).
Köthe, G. (1970). *Topological Vector Spaces I*, Springer, Berlin.
Köthe, G. (1979). *Topological Vector Spaces II*, Springer, Berlin.
Kranz, P. (1972). 'Additive functionals on abelian semigroups', *Comment. Math. Ann. Soc. Math. Pol.*, **16**, 239–246.
Krasnosel'skiĭ, M. A., and Rutickiĭ, Ya. B. (1961). *Convex Functions and Orlicz Spaces*, Noordhoff, Groningen.
Kufner, A., John, O., and Fučík, S. (1977). *Function Spaces*, Academia, Prague/Noordhoff, Leyden.
Kuratowski, C. (1972). 'A general approach to the theory of set-valued mappings', in *General Topology and its Relations to Modern Analysis and Algebra III* (ed. J. Novák), pp. 271–280, Proc. Third Prague Topological Symp. (Prague, 1971), Academia, Prague/Academic Press, New York.
Laborde, P. (1976). 'Problèmes quasi variationnelles en viscoplasticité avec écrouissage', *C.R. Ac. Sci. Paris*, **283**, 393–396.
Ladyženskaja, O. A., and Ural'ceva, N. N. (1968). *Équations aux Dérivées Partielles de Type Elliptique*, Dunod, Paris (Russian ed. (1964): Nauka, Moscow).
Laetsch, T. (1975). 'A uniqueness theorem for elliptic quasi-variational inequalities', *J. Functional Anal.*, **18**, 286–287.
Lang, S. (1962). *Introduction to Differentiable Manifolds*, Interscience, New York.
Langlais, M. (1979). 'Un problème unilateral degenéré', in *Proc. Intern. Meeting on*

Recent Methods in Non-Linear Analysis (Rome, 1978) (eds E. De Giorgi, E. Magenes, and U. Mosco), pp. 441–454, Pitagora, Bologna.

Lax, P. D., and Milgram, A. N. (1954). 'Parabolic equations', in *Contributions to the Theory of Partial Differential Equations* (eds L. Bers, S. Bochner, and F. John), pp. 167–190, Proc. Conf. on Partial Differential Equations (Harriman, 1952), Princeton University Press, Princeton.

Lavrentiev, M. M. (1967). *Some Improperly Posed Problems of Mathematical Physics*, Springer, Berlin.

Leray, J., and Schauder, J. (1934). 'Topologie et équations fonctionnelles', *Ann. École Norm. Sup.*, (3) **51**, 45–78.

Lewy, H., and Stampacchia, G. (1969). 'On the regularity of the solution of a variational inequality', *Comm. Pure Appl. Math.*, **22**, 153–188.

Lions, J. L. (1955). 'Problèmes aux limites en théorie des distributions', *Acta Math.*, **94**, 13–153.

Lions, J. L. (1956). 'Sur les problèmes aux limites du type dérivée oblique', *Ann. of Math.*, (2) **64**, 207–239.

Lions, J. L. (1961). *Équations Differentielles Operationnelles et Problèmes aux Limites*, Springer, Berlin.

Lions, J. L. (1965). *Problèmes aux Limites dans les Équations aux Dérivées Partielles*, Les Presses de l'Université de Montréal, Montréal (2nd ed.).

Lions, J. L. (1969a). *Quelques Méthodes de Résolution des Problèmes aux Limites non Linéaires*, Dunod/Gauthier-Villars, Paris.

Lions, J. L. (1969b). 'Sur quelques problèmes de calcul des variations', *Symp. Math.*, **2**, 125–144.

Lions, J. L. (1971). *Optimal Control of Systems Governed by Partial Differential Equations*, Springer, Berlin (French ed. (1968): Dunod/Gauthier-Villars, Paris).

Lions, J. L. (1973a). 'Sur le côntrole optimal des systemes distribués', *Enseignement Math.*, (2) **19**, 125–166.

Lions, J. L. (1973b). 'The optimal control of distributed systems', *Russian Math. Surveys*, **28** (4), 13–46 (Russian ed. (1973): *Uspehi Mat. Nauk*, **28** (4), 15–46).

Lions, J. L. (1974). 'Various topics in the theory of optimal control of distributed systems', in *Optimal Control Theory and Its Applications* (ed. B. J. Kirby), part I, pp. 166–309, Proc. 14th Biennial Sem. Canadian Math. Congress (London, 1973), Lecture Notes Econ. Math. Systems, 105, Springer, Berlin.

Lions, J. L. (1975a). 'Sur la théorie du contrôle', in *Proc. International Congresses of Mathematicians* (Vancouver, 1974) (ed. R. D. James), vol. 1, pp. 139–154, Canadian Mathematical Congress, Vancouver.

Lions, J. L. (1975b). 'On the numerical approximation of problems of impulse controls', in *Optimization Techniques* (ed. G. I. Marchuk), pp. 232–251, Proc. IFIP Techn. Conf. (Novosibirsk, 1974), Lecture Notes Computer Science, 27, Springer, Berlin.

Lions, J. L. (1975c). 'Remarks on some new nonlinear boundary value problems', in *Proc. Symp. Partial Differential Equations and Related Topics* (New Orleans, 1974) (ed. J. A. Goldstein), pp. 301–328, Lecture Notes in Mathematics, 446, Springer, Berlin.

Lions, J. L. (1975d). 'On free surface problems: methods of variational and quasi-variational inequalities', in *Computational Mechanics* (ed. J. T. Oden), pp. 129–148, Proc. Intern. Conf. Computer Methods in Nonlinear Mechanics (Autin, 1974), Lecture Notes in Mathematics, 461, Springer, Berlin.

Lions, J. L. (1976a). 'Asymptotic behaviour of solutions of variational inequalities with highly oscillating coefficients', in *Applications of Methods of Functional Analysis to Problems in Mechanics* (eds P. Germain and B. Nayroles), pp. 30–55, Proc. Joint Symp. IUTAM/IMU (Marseille, 1975), Lecture Notes in Mathematics, 503, Springer, Berlin.

Lions, J. L. (1976b). *Sur Quelques Questions d'Analyse, de Mécanique et de Contrôle Optimal*, Les Presses de l'Université de Montréal, Montréal.

Lions, J. L. (1976c). 'Some methods of resolution of free surface problems', in *Proc. 5th. Intern. Conf. Numerical Methods in Fluid Dynamics* (Enschede, 1976) (eds A. I. Van de Vooren and P. J. Zandbergen), pp. 1–31, Lecture Notes in Physics, 59, Springer, Berlin.

Lions, J. L. (1977a). 'Sur l'approximation de problèmes à frontière libre dans les matériaux inhomogènes', in *Proc. Conf. Mathematical Aspects of Finite Element Methods* (Rome, 1975) (eds I. Galligani and E. Magenes), pp. 194–203, Lecture Notes in Mathematics, 606, Springer, Berlin.

Lions, J. L. (1977b). 'Some aspects of the theory of linear evolution equations', in *Boundary Value Problems for Linear Evolution Partial Differential Equations* (ed. H. G. Garnir), pp. 175–238, Proc. NATO Adv. Study Institute (Liège, 1976), D. Reidel, Dordrecht.

Lions, J. L. (1978). 'Remarks on the relationships between free surfaces and optimal control of distributed systems', in *Optimization Techniques* (ed. J. Stoer), part 1, pp. 28–40, Proc. 8th IFIP Conf. Optimization Techniques (Würzburg, 1977), Lecture Notes Control and Information Sciences, 6, Springer, Berlin.

Lions, J. L., and Magenes, E. (1960). 'Problemi at limiti non omogenei, I', *Ann. Scuola Norm. Sup. Pisa*, (3) **14**, 269–308.

Lions, J. L., and Magenes, E. (1961a). 'Problèmes aux limites non homogènes, II', *Ann. Inst. Fourier*, **11**, 137–178.

Lions, J. L., and Magenes, E. (1961b). 'Problemi ai limiti non omogenei, III', *Ann. Scuola Norm. Sup. Pisa*, (3) **15**, 41–103.

Lions, J. L., and Magenes, E. (1961c). 'Problèmes aux limites non homogènes, IV, *Ann. Scuola Norm. Sup. Pisa*, (3) **15**, 311–326.

Lions, J. L., and Magenes, E. (1962). 'Problemi ai limiti non omogenei, V', *Ann. Scuola Norm. Sup. Pisa*, (3) **16**, 1–44.

Lions, J. L., and Magenes, E. (1963a). 'Problèmes aux limites non homogènes, VI', *J. D'Analyse Math.*, **11**, 165–188.

Lions, J. L., and Magenes, E. (1963b). 'Problèmes aux limites non homogènes, VII', *Ann. Mat. Pura Appl.*, (4) **63**, 201–224.

Lions, J. L., and Magenes, E. (1972a). *Non-Homogeneous Boundary Value Problems and Applications*, volume I, Springer, Berlin (French ed. (1968): Dunod, Paris).

Lions, J. L., and Magenes, E. (1972b). *Non-Homogeneous Boundary Value Problems and Applications*, volume II, Springer, Berlin (French ed. (1968): Dunod, Paris).

Lions, J. L., and Magenes, E. (1973). *Non-Homogeneous Boundary Value Problems and Applications*, volume III, Springer, Berlin (French ed. (1970): Dunod, Paris).

Lions, J. L., and Stampacchia, G. (1967). 'Variational inequalities', *Comm. Pure Appl. Math.*, **20**, 493–519.

Loinger, E. (1980). 'A finite element approach to a quasi-variational inequality', *Calcolo*, **17**, 197–209.

Łojasiewicz, S. (1957). 'Sur la valeur et la limite d'une distribution en un point', *Studia Math.*, **16**, 1–36.

Lopatinski, Ya. B. (1953). 'A method of reduction of boundary value problems for systems of elliptic equations to regular integral equations', *Ukrain Mat. Ž.*, **5**, 123–151 (Russian).

Luenberger, D. G. (1973). *Introduction to Linear and Nonlinear Programming*, Addison-Wesley, Reading.

Luxemburg, W. S., and Zaanen, A. C. (1971). *Riesz Spaces*, volume I, North-Holland, Amsterdam.

Magenes, E. (1964). 'Spazi di interpolazione ed equazioni a derivate parziali', in *Atti VII Cong. dell'Unione Matematica Italiana* (Genova, 1963), pp. 134–197, Cremonese, Roma.
Magenes, E. (1972). 'Su alcuni problemi ellittici di frontiera libera connessi con il comportamento dei fluidi nei mezzi porosi', *Symp. Math.*, **10,** 265–279.
Magenes, E. (1973). 'Problèmes de frontière libre liés a certaines questions d'hydraulique', in *Equadiff 3*, pp. 51–58, Proc. Czechoslovak Conf. Differential Equations and Applications (Brno, 1972), J. E. Purkyně Univ., Brno.
Magenes, E. (1977). 'Topics in parabolic equations: some typical free boundary problems', in *Boundary Value Problems for Linear Evolution Partial Differential Equations* (ed. H. G. Garnir), pp. 239–312, Proc. NATO Adv. Study Institute (Liège, 1976), D. Reidel, Dordrecht.
Magenes, E., Editor (1980). *Free Boundary Problems*, 2 vol., Proc. Sem. (Pavia, 1979), Instituto Nazionale di Alta Matematica, Roma.
Magenes, E. (1982). 'Problemi di Stefan bifase in più variabili spaziali', to appear on *Atti Sem. di Analisi Funzionale e Applicazioni (S.A.F.A. V)* (Catania, 1981).
Magenes, E., and Stampacchia, G. (1958). 'I problemi al contorno per le equazioni differenziali di tipo ellittico', *Ann. Sc. Norm. Sup. Pisa*, (3) **12,** 247–358.
Maione, U. (1980). 'Free boundary problems in fluid flow through porous media from the engineering point of view', in *Free Boundary Problems* (ed. E. Magenes), vol. I, pp. 79–87, Proc. Sem. (Pavia, 1979), Istituto Nazionale di Alta Matematica, Roma.
Marinescu, G. (1963). *Espaces Vectoriels Pseudotopologiques et Théorie des Distributions*, Deutscher Verlag, Berlin.
Martelli, M. (1975). 'A Rothe's type theorem for non-compact acyclic-valued maps', *Boll. U. Mat. Ital.*, (4) **11** (suppl. fasc. 3), 70–76.
Martelli, M., and Vignoli, A. (1974a). 'On differentiability of multi-valued maps', *Boll. U. Mat. Ital.*, (4) **10,** 701–712.
Martelli, M., and Vignoli, A. (1974b). 'A generalized Leray–Schauder condition', *Rend. Ac. Naz. Lincei*, (8) **57,** 374–379.
Matzeu, M., and Vivaldi, M. A. (1979). 'On the regular solution of a nonlinear parabolic quasivariational inequality related to a stochastic control problem', *Comm. Partial Diff. Eq.*, **4,** 1123–1147.
Matzeu, M., and Vivaldi, M. A. (1982). 'Esistenza e regolarità di soluzioni forti per disuguaglianze quasi-variazionali paraboliche relative ad un operatore non lineare', to appear.
Maurin, S. (1976). 'Méthodes de décomposition appliquées aux problèmes de contrôle impulsionnel', in *Optimization Techniques: Modeling and Optimization in the Service of Man*, Part 2 (ed. J. Cea), pp. 169–191, Proc. 7th IFIP Conf. (Nice, 1975), Lecture Notes in Computer Science, 41, Springer, Berlin.
Mazur, S. (1930). 'Über die kleinste konvexe Menge, die eine gegebene kompakte Menge enthält', *Studia Math.*, **2,** 7–9.
McAllister, G., and Rohde, S. (1975). 'A variational formulation for a class of free boundary problems arising in hydrodynamic lubrication', *Int. J. Eng. Sci.*, **13,** 841–850.
Menaldi, J. (1977a). 'Sur le problème de contrôle impulsionnel et l'inéquation quasi variationnelle dégénérée associée', *C.R. Ac. Sci. Paris*, **284,** 1499–1502.
Menaldi, J. (1977b). 'Sur le problème de temps d'arrêt et l'inéquation variationnelle dégénérée associée', *C.R. Ac. Sci. Paris*, **284,** 1443–1446.
Meyers, N. G., and Serrin, S. (1964). '$H = W$', *Proc. Nat. Ac. Sci. U.S.A.*, **51,** 1055–1056.
Michael, E. (1951), 'Topologies on spaces of subsets', *Trans. Amer. Math. Soc.*, **71,** 152–182.
Michael, E. (1956a). 'Continuous selections, I', *Ann. of Math.*, (2) **63,** 361–382.
Michael, E. (1956b). 'Continuous selections, II', *Ann. of Math.*, (2) **64,** 562–580.

Mickle, E. J. (1949). 'On the extension of a transformation', *Bull. Amer. Math. Soc.*, **55**, 160–164.
Miellou, J. C. (1976). 'A mixed relaxation algorithm applied to quasi-variational inequations', in *Optimization Techniques: Modelling and Optimization in the Service of Man.*, Part 2 (ed. J. Cea), pp. 192–199, Proc. 7th IFIP Conf. (Nice, 1975), Lecture Notes in Computer Science, 41, Springer, Berlin.
Mignot, F. (1975a). 'Contrôle dans les inéquations variationnelles', *C.R. Ac. Sci. Paris*, **280**, 197–200.
Mignot, F. (1975b). 'Inéquations variationnelles et contrôle', *Thèse*, Université de Paris, Paris.
Mignot, F. (1976). 'Contrôle dans les inéquations variationnelles elliptiques', *J. Functional Anal.*, **22**, 130–185.
Mignot, F., and Puel, J. P. (1975a). 'Solution maximum de certaines inéquations d'évolution paraboliques, et inéquations quasi variationnelles paraboliques', *C.R. Ac. Sci. Paris*, **280**, 259–262.
Mignot, F., and Puel, J. P. (1975b). 'Systèmes d'inéquations hyperboliques du premier ordre faiblement couplés et inéquations quasi-variationnelles associées', *C.R. Ac. Sci. Paris*, **280**, 423–426.
Mignot, F., and Puel, J. P. (1976). 'Inéquations variationnelles et quasivariationnelles hyperboliques du premier ordre', *J. Math. Pures Appl.*, (9) **55**, 353–378.
Mignot, F., and Puel, J. P. (1977). 'Inéquations d'évolution paraboliques avec convexes dépendant du temps. Applications aux inéquations quasi-variationnelles d'évolution', *Arch Rational Mech. Anal.*, **64**, 59–91.
Milnor, J. (1978). 'Analytic proofs of the "hairy ball theorem" and the Brouwer fixed point theorem', *Amer. Math. Monthly*, **85**, 521–524.
Minty, J. G. (1967). 'On the generalization of a direct method of the calculus of variations', *Bull. Amer. Math. Soc.*, **73**, 315–321.
Minty, J. G. (1969). 'On some aspects of the theory of monotone operators', in *Theory and Applications of Monotone Operators* (ed. A. Ghizzetti), p. 67–82, Proc. NATO Adv. Study Institute (Venice, 1968), Oderisi, Gubbio.
Miranda, C. (1949). *Problemi di Esistenza in Analisi Funzionale*, Quad. Mat. Scuola Norm. Sup. Pisa, 3, Tacchi, Pisa.
Miranda, C. (1969). 'Su un problema di frontiera libera', *Symp. Math.*, **2**, 71–83.
Miranda, C. (1970). *Partial Differential Equations of Elliptic Type*, Springer, Berlin (2nd ed.).
Miranda, C. (1978). *Istituzioni di Analisi Funzionale Lineare*, Monografie dell'Unione Mat. Ital., Oderisi, Gubbio.
Mizohata, S. (1973). *The Theory of Partial Differential Equations*, Cambridge University Press, Cambridge.
Monachov, V. N. (1969). *Kraievie Sadači so Svobodnimi Granizami dlia Elliptičeskich Sistem Uravienii*, Novosibirsk.
Moreau, J. J. (1967). *Fonctionnelles Convexes*, Séminaire sur les Équations aux Dérivées Partielles, Collège de France, Paris.
Morice, Ph. (1975). 'Une méthode d'optimisation de forme de domaine. Application à l'écoulement stationnaire à travers une digue poreuse', in *Control Theory, Numerical Methods and Computer Systems Modelling* (erds A. Bensoussan and J. L. Lions), pp. 454–467, Proc. Intern. Symp. (Rocquencourt, 1974), Lecture Notes Econ. Math. Systems, 107, Springer, Berlin.
Mosco, U. (1967a). 'A remark on a theorem of F. E. Browder', *J. Math. Anal. Appl.*, **20**, 90–93.
Mosco, U. (1967b). 'Approximations of the solutions of some variational inequalities', *Ann. Scuola Norm. Sup. Pisa*, (3) **21**, 373–394.
Mosco, U. (1969a). 'Convergence of convex sets and of solutions of variational inequalities', *Adv. Math.*, **3**, 510–585.
Mosco, U. (1969b). 'Convergence of solutions of variational inequalities', in

Theory and Applications of Monotone Operators (ed. A. Ghizzetti), pp. 231–247, Proc. NATO Advanced Study Institute (Venice, 1968), Oderisi, Gubbio.

Mosco, U. (1973). An introduction to the approximate solution of variational inequalities', in *Constructive Aspects of Functional Analysis* (ed. G. Geymonat), vol. II, pp. 499–682, proc. C.I.M.E. (Erice, 1971), Cremonese, Roma.

Mosco, U. (1974). 'Introduction to variational and quasi-variational inequalities', in *Control Theory and Functional Analysis*, Proc. Intern. Autumn Course (Trieste, 1974), Intern. Centre for Theoretical Physics, Trieste.

Mosco, U. (1976). 'Implicit variational problems and quasivariational inequalities', in *Nonlinear Operators and the Calculus of Variations* (eds J. P. Gossez, E. L. Dozo, J. Mawhin, and L. Waelbroeck), pp. 83–156, Proc. Summer School (Bruxelles, 1975), Lecture Notes in Mathematics, 543, Springer, Berlin.

Mosco, U. (1977). 'Some quasi-variational inequalities arising in stochastic impulse control theory', in *Theory of Nonlinear Operators*, pp. 183–195, Proc. Intern. Summer School (Berlin, 1976), Akademie, Berlin.

Mosco, U. (1978a). 'Regularité forte de la fonction d'Hamilton–Jacobi du contrôle impulsionnel et continu', *C.R. Ac. Sci. Paris*, **286,** 211–214.

Mosco, U. (1978b). 'Sur l'existence de la solution régulière de l'inéquation quasi-variationnelle non linéaire du contrôle optimal impulsionnel et continu', in *Journées d'Analyse non Linéaire* (eds. P. Bénilan and J. Robert), pp. 140–159, Proc. Conf. (Bescançon, 1977), Lecture Notes in Mathematics, 665, Springer, Berlin.

Mosco, U. (1980). 'On some nonlinear quasi variational inequalities and implicit complementarity problems in stochastic control theory', in *Variational Inequalities and Complementarity Problems* (eds. R. W. Cottle, F. Giannessi and J. L. Lions), pp. 271–283, Wiley, Chichester.

Mosco, U., and Troianiello, G. M. (1973). 'On the smoothness of solutions of unilateral Dirichlet problems', *Boll. U. Mat. Ital.*, (4) **8,** 57–67.

Mossino, J. (1976). 'Sur certaines inéquations quasi-variationnelles apparaissant en physique', *C.R. Ac. Sci. Paris*, **282,** 187–190.

Mossino, J. (1978). 'Application de inéquations quasi-variationneles à quelques problèmes non linéaries de la physique des plasma', *Israel J. Math.*, **30,** 14–50.

Mossino, J., and Temam, R. (1977). 'Certains problèmes non linéaires de la physique des plasmas', in *Proc. Conf. Mathematical Aspects of Finite Element Methods* (Rome, 1975) (eds. I. Galligani and E. Magenes), pp. 237–260, Lecture Notes in Mathematics, 606, Springer, Berlin.

Mossino, J., and Zolesio, J. (1977). 'Solution variationnelle d'un problème non linéaire de la physique des plasmas', *C.R. Ac. Sci. Paris*, **285,** 1033–1036.

Murthy, M. K. V., and Stampacchia, G. (1972). 'A variational inequality with mixed boundary conditions', *Israel J. Math.*, **13,** 188–224.

Nadler, S. B. (1968). 'Sequences of contractions and fixed points', *Pacific J. Math.*, **27,** 579–585.

Nadler, S. B. (1969), 'Multi-valued contraction mappings', *Pacific J. Math.*, **30,** 475–488.

Naimpally, S. A., and Warrack, B. D. (1970). *Proximity Spaces*, Cambridge University Press, Cambridge.

Nakoulima, O. (1977a). 'Sur une notion de solution faible pour les inéquations variationneles d'évolution à deux obstacles', *C.R. Ac. Sci. Paris*, **284,** 1037–1040.

Nakoulima, O. (1977b). 'Étude d'une inéquation variationnelle bilatérale et d'un système d'inéquations quasi-variationnelles unilatérales associée', Thèse, Université de Bordeaux, Bordeaux.

Namioka, I. (1957). 'Partially ordered linear topological spaces', *Mem. Amer. Math. Soc.*, **24.**

Narasimhan, R. (1968). *Analysis on Real and Complex Manifolds*, Masson, Paris/North-Holland, Amsterdam.
Nečas, J. (1967). *Les Méthodes Directes en Théorie des Équations Elliptiques*, Masson, Paris/Academia, Prague.
Nevanlinna, R., and Paatero, V. (1969). *Introduction to Complex Analysis*, Addison-Wesley, Reading (German ed. (1965): Birkhäuser, Basel).
Nikol'skiĭ, S. M. (1961). 'On imbedding, continuation and approximation theorems for differentiable functions of several variables', *Russian Math. Surveys*, **16** (5), 55–104 (Russian ed. (1961): *Uspehi Mat. Nauk*, **16** (5), 63–114).
Nikol'skiĭ, S. M. (1975). *Approximation of Functions of Several Variables and Imbedding Theorems*, Springer, Berlin (Russian ed. (1969): Nauka, Moscow).
Nirenberg, L. (1959). 'On elliptic partial differential equations', *Ann. Scuola Norm. Sup. Pisa*, (3) **13**, 115–162.
Nitsche, J. C. (1975). *Vorlesungen uber Minimalflächen*, Springer, Berlin.
Nitsche, J. (1977). 'L_∞-convergence of finite element approximations', in *Proc. Conf. Mathematical Aspects of Finite Element Methods* (Rome, 1975) (eds I. Galligani and E. Magenes), pp. 261–274, Lecture Notes in Mathematics, 606, Springer, Berlin.
Oleĭnik, O. A., and Radkevič, E. V. (1973). *Second Order Equations with Nonnegative Characteristic Form*, American Mathematical Society, Providence/Plenum Press, New York (Russian ed. (1971): Itogi Nauki, Moscow).
Ortega, J. M., and Rheinboldt, W. C. (1970). *Iterative Solution of Nonlinear Equations in Several Variables*, Academic Press, New York.
Palamodov, V. P. (1970). *Linear Differential Operators with Constant Coefficients*, Springer, Berlin.
Pascali, D., and Sburlan, S. (1978). *Nonlinear Mappings of Monotone Type*, Academiei, Bucureşti/Sijthoff & Noordhoff, Alphen aan den Rijn.
Peetre, J. (1961). 'Mixed problems for higher order elliptic equations in two variables, I', *Ann. Scuola Norm. Sup. Pisa*, (3) **15**, 337–353.
Peetre, J. (1963). 'Mixed problems for higher order elliptic equations in two variables, II', *Ann. Scuola Norm. Sup. Pisa*, (3) **17**, 1–12.
Peetre, J. (1966). 'Espaces d'interpolation et théorème de Soboleff', *Ann. Inst. Fourier*, **16** (1), 279–317.
Pejsachowicz, J. (1977). 'A Lefschetz fixed point theorem for multivalued weighted mappings', *Boll. U. Mat. Ital.*, (5) **14-A**, 391–397.
Peressini, A. (1967). *Ordered Topological Vector Spaces*, Harper & Row, New York.
Pietra, P. (1982). 'An up-wind finite element method for a filtration problem', *to appear*.
Polubarinova-Kochina, P. Ya. (1962). *Theory of Ground Water Movement*, Princeton University Press, Princeton.
Ponomarev, V. I. (1964). 'Properties of topological spaces preserved under multivalued continuous mappings', *Amer. Math. Soc. Transl.*, (2) **38**, 119–140 (Russian ed. (1960): *Mat. Sb.*, (2) **51**, 515–536).
Popa, V. (1980). 'Sur le mouvement à surface libre dans les milieux poreux non homogènes', *J. de Mécanique*, **19**, 663–678.
Pozzi, G. A. (1974a). 'On a free boundary problem arising from a fluid flow through a porous medium in the presence of evaporation', *Boll. U. Mat. Ital.* (4) **9**, 416–440.
Pozzi, G. A. (1974b). 'Remarks about an evaporation problem', *Pubbl. Laboratorio di Analisi Numerica C.N.R.*, **81**, Pavia.
Prodi, G. (1956). 'Tracce sulla frontiera delle funzioni di Beppo-Levi', *Rend. Sem. Mat. Univ. Padova*, **26**, 36–60.
Prodi, G. (1958). 'Tracce di funzioni con derivate di ordine l a quadrato integrabile

su varietà di dimensione arbitratia', *Rend. Sem. Mat. Univ. Padova*, **28**, 402–432.

Protter, M. H., and Weinberger, H. F. (1967). *Maximum Principles in Differential Equations*, Prentice-Hall, Englewood Cliffs.

Pucci, C. (1957). 'Proprietà di massimo e minimo delle soluzioni di equazioni a derivate parziali del secondo ordine di tipo ellittico e parabolico, I', *Rend. Ac. Naz. Lincei*, (8) **23**, 370–375.

Pucci, C. (1958). 'Proprietà di massimo e minimo delle soluzioni di equazioni a derivate parziali del secondo ordine di tipo ellittico e parabolico, II', *Rend. Ac. Naz. Lincei*, (8) **24**, 3–6.

Pucci, C. (1964). 'Regolarità alla frontiera di soluzioni di equazioni ellittiche', *Ann. Mat. Pura Appl.*, (4) **65**, 311–328.

Quarteroni, A., and Visintin, A. (1980). 'On the numerical solution of a nonlinear variational equation related to a filtration problem', *Boll. U. Mat. Ital.*, (5) **17-B**, 204–231.

Radó, T. (1933). *On the Problem of Plateau*, Springer, Berlin.

Raviart, P. A. (1972). *Méthode des Eléments Finis*, Universitè de Paris 6 et C.N.R.S., Paris.

Riesz, F. (1930). 'Sur la décomposition des opérations fonctionnelles linéaires', in *Atti del Congresso Internazionale dei Matematici* (Bologna, 1928), tomo III, pp. 143–148, Zanichelli, Bologna.

Riesz, F., and Sz.-Nagy, B. (1955). *Functional Analysis*, Frederick Ungar, New York.

Robert, R. (1974). 'Differentiabilité et genericité', in *Atti Giorn. Analisi Convessa e Applicazioni* (Roma, 1974), pp. 75–83, Quad. Gruppi Ric. Mat. C.N.R., Università di Roma, Roma.

Robertson, A. P., and Robertson, W. (1966). *Topological Vector Spaces*, Cambridge University Press, Cambridge.

Robin, M. (1975). 'Contrôle optimal de files d'attente', *Rapport de Recherche*, **117**, IRIA-Laboria, Rocquencourt.

Robin, M. (1976a). 'Contrôle impulsionnel avec retard pour les processus de diffusion', *C.R. Ac. Sci. Paris*, **282**, 463–466.

Robin, M. (1976b). 'Sur le contrôle impulsionnel des processus markoviens et semi-markoviens', *C.R. Ac. Sci. Paris*, **282**, 631–634.

Rockafellar, R. T. (1970). *Convex Analysis*, Princeton University Press, Princeton.

Rockafellar, R. T. (1976). 'Integral functionals, normal integrands and measurable selections', in *Nonlinear Operators and the Calculus of Variations* (eds. J. P. Gossez, E. L. Dozo, J. Mawhin, and L. Waelbroeck), pp. 157–207, Proc. Summer School (Bruxelles, 1975), Lecture Notes in Mathematics, 543, Springer, Berlin.

Rockafellar, R. T. (1979). *La Théorie des Sous-Gradients et ses Applications à l'Optimisation*, Les Presses de l'Université de Montréal, Montréal.

Rogers, L. A. (1980). 'A less strange version of Milnor's proof of Brouwer's fixed-point theorem', *Amer. Math. Monthly*, **87**, 525–527.

Roux, D., and Soardi, P. (1971). 'Sui punti uniti di mappe continue di uno spazio topologico in sè', *Riv. Mat. Univ. Parma*, (2) **12**, 21–28.

Rubin, H., and Rubin, J. (1963). *Equivalents of the Axiom of Choice*, North-Holland, Amsterdam.

Rudin, W. (1970). *Real and Complex Analysis*, McGraw-Hill, London/Mladinska Knjiaga, Ljubljana.

Rudin, W. (1973). *Functional Analysis*, McGraw-Hill, New York.

Rutherford, D. E. (1965). *Introduction to Lattice Theory*, Oliver and Boyd, Edinburgh.

Sanchez-Palencia, E. (1980). *Non-Homogeneous Media and Vibration Theory*, Lecture Notes in Physics, 127, Springer, Berlin.
Sansone, G., and Conti, R. (1956). *Equazioni Differenziali Non Lineari*, Monografie Mat. C.N.R., 3, Cremonese, Roma.
Schaefer, H. H. (1966). *Topological Vector Spaces*, Macmillan, New York.
Schaefer, H. H. (1974). *Banach Lattices and Positive Operators*, Springer, Berlin.
Schaeffer, D. (1977). 'Some examples of singularities in a free boundary', *Ann. Scuola Norm. Sup. Pisa*, (4) **4**, 133–144.
Schauder, J. (1930). 'Der Fixpunktsatz in Funktionalräumen', *Studia Math.*, **2**, 171–180.
Schechter, M. (1971). *Spectra of Partial Differential Operators*, North-Holland, Amsterdam.
Schwartz, J. T. (1969). *Nonlinear Functional Analysis*, Gordon and Breach, New York.
Schwartz, L. (1957). 'Su alcuni problemi della teoria delle equazioni differenziali lineari di tipo ellittico', *Rend. Sem. Mat. Fis. Milano*, **27**, 211–249.
Schwartz, L. (1960). *Équations de Cordes Vibrantes*, Centre de Documentation Universitaire, Paris.
Schwartz, L. (1966). *Théorie des Distributions*, Hermann, Paris (2nd ed.).
Schwartz, L. (1973). *Radon Measures on Arbitrary Topological Spaces and Cylindrical Measures*, Oxford University Press, Oxford.
Scorza-Dragoni, G. (1955a). 'Un'osservazione sul lemma di Sperner', *Rend. Ac. Naz. Lincei*, (8) **19**, 204–206.
Scorza-Dragoni, G. (1955b). 'Translazioni piane generalizzate', in *Topologia*, Proc. C.I.M.E. (Varenna, 1955), Istituto Matematico dell'Università di Roma, Roma.
Scorza-Dragoni, G. (1978). 'Il lemma di Sperner: sue implicazioni negli spazi euclidei e questioni connesse', in *Applicazioni del Teorema del Punto Fisso all'Analisi Economica*, pp. 31–71, Ac. Naz. Lincei, Roma.
Shamir, E. (1968). 'Regularization of mixed second-order elliptic problems', *Israel J. Math.*, **6**, 150–168.
Shapiro, Z. Ya. (1953). 'About the general boundary value problems of elliptic type', *Izv. Akad. Nauk*, **17**, 539–562 (Russian).
Shima, M. (1976). 'Analytical construction of the free boundary for non-linear impulse control problems and its application to the optimal design of aerator systems', *Rapport de Recherche*, **194**, IRIA-Laboria, Rocquencourt.
Shimborski, E. (1975). 'Encadrement d'une frontière libre relative a un problème d'hydraulique', *Boll. U. Mat. Ital.*, (4) **12**, 66–67.
Signorini, A. (1959). 'Questioni di elasticità non linearizzata e semilinearizzata', *Rend. Mat.*, (4) **18**, 95–139.
Silva, J. S. (1955). 'Su certe classi di spazi localmente convessi importanti per le applicazioni', *Rend. Mat.*, (5) **14**, 388–410.
Silva, J. S. (1964). 'Integrals and orders of growth of distributions', in *Theory of Distributions*, pp. 329–390, Proc. Intern. Summer Institute (Lisbon, 1964). Instituto Gulbenkian de Ciência, Lisboa.
Silva, J. S. (1967). 'Les séries de multipôles des physiciens et la théorie des ultradistributions', *Math. Ann.*, **174**, 109–142.
Slobodeckiĭ, L. N. (1966). 'Generalized Sobolev spaces and their application to boundary problems for partial differential equations', *Amer. Math. Soc. Transl.*, (2) **57**, 207–275 (Russian ed. (1958): Leningrad. Gos. Ped. Inst. Učen. Zap., 197, 54–112).
Smart, D. R. (1974). *Fixed Point Theorems*, Cambridge University Press, Cambridge.
Smithson, R. E. (1965). 'Some general properties of multi-valued functions', *Pacific J. Math.*, **15**, 681–703.
Smithson, R. E. (1972). 'Multifunctions', *Nieuw Arch. Wisk.*, (3) **20**, 31–53.

Smirnov, Ju. M. (1964). 'On proximity spaces', *Amer. Math. Soc. Transl.*, (2) **38**, 5–35 (Russian ed. (1952): *Mat. Sb.*, (2) **31**, 543–574).
Smirnov, M. M. (1966). *Second Order Partial Differential Equations*, Noordhoff, Groningen.
Sobolev, S. L. (1961). *Sur les Équations aux Dérivées Partielles Hyperboliques Non-Linéaires*, Monografie Matem. del C.N.R., 9, Cremonese, Roma.
Sobolev, S. L. (1963a). *Applications of Functional Analysis in Mathematical Physics*, American Mathematical Society, Providence (Russian ed. (1950): Iz. Leningradskovo Gos. Univ., Leningrad).
Sobolev, S. L. (1963b). 'On a theorem of functional analysis', *Amer. Math. Soc. Transl.*, (2) **34**, 39–68 (Russian ed. (1938): *Mat. Sb.*, **46**, 471–497).
Sobolev, S. L., and Nikol'skiĭ, S. M. (1970). 'Imbedding theorems', *Amer. Math. Soc. Transl.*, (2) **87**, 147–173 (Russian ed. (1963): in Actes 4th Cong. Math. Union, pp. 227–242, Izdat. Akad. Nauk. SSSR, Leningrad).
Sommerfeld, A. (1964). *Partial Differential Equations in Physics*, Academic Press, New York.
Sorani, G. (1969). *An Introduction to Real and Complex Manifolds*, Gordon and Breach, New York.
Spanier, E. (1966). *Algebraic Topology*, McGraw-Hill, New York.
Sperner, E. (1928). 'Neuer Beweis für die Invarianz der Dimensionszahl und des Gebietes', *Abh. Math. Sem. Univ. Hamburg*, **6**, 265–272.
Sperner, E. (1955). 'Generalizzazioni del teorema di Brouwer sul punto fisso', in *Topologia*, Proc. C.I.M.E. (Varenna, 1955), Istituto Matematico dell'Università di Roma, Roma.
Spivak, M. (1965). *Calculus on Manifolds*, Benjamin, New York.
Spruck, J. (1979). 'Regularity in elliptic free boundary problems', in *Proc. Intern. Meeting on Recent Methods in Non-Linear Analysis* (Rome, 1978) (eds E. De Giorgi, E. Magenes, and U. Mosco), pp. 73–81, Pitagora, Bologna.
Stampacchia, G. (1964). 'Formes bilinéaires coercitives sur les ensembles convexes', *C.R. Ac. Sci. Paris*, **258**, 4413–4416.
Stampacchia, G. (1965). 'Le problème de Dirichlet pour les équations elliptiques du second ordre à coefficients discontinus', *Ann. Inst. Fourier*, **15** (1), 189–258.
Stampacchia, G. (1966). *Équations Elliptiques du Second Ordre à Coefficients Discontinus*, Les Presses de l'Université de Montréal, Montréal.
Stampacchia, G. (1969). 'Variational inequalities', in *Theory and Applications of Monotone Operators* (ed. A. Ghizzetti), pp. 101–192, Proc. NATO Adv. Study Institute (Venice, 1968), Oderisi, Gubbio.
Stampacchia, G. (1974). 'On the filtration of a fluid through a porous medium with variable cross section', *Russian Math. Surveys*, **29** (4), 89–102 (Russian ed. (1974): *Uspehi Mat. Nauk*, 29 (4), 89–101).
Stein, E. M., and Weiss, G. (1971). *An Introduction to Fourier Analysis on Euclidean Spaces*, University Press, Princeton.
Stoer, J., and Witzgall, C. (1970). *Convexity and Optimization in Finite Dimensions*, I, Springer, Berlin.
Stone, M. H. (1962). 'A generalized Weierstrass approximation theorem', in *Studies in Modern Analysis* (ed. R. C. Buck), pp. 30–87, Mathematical Association of America.
Szeptycki, P. (1982). 'A nonsteady problem in the theory of fluid flow through porous media: existence and uniqueness of weak solutions', *to appear*.
Szpilrajn, E. (1930). 'Sur l'extension de l'ordre partiel', *Fund. Math.*, **16**, 386–389.
Swaminathan, S., Editor (1976). *Fixed Point Theory and Its Applications*, Proc. Sem. (Halifax, 1975), Academic Press, New York.
Takahashi, W. (1976). 'Nonlinear variational inequalities and fixed point theorems', *J. Math. Soc. Japan*, **28**, 168–181.

Tartar, L. (1974a). 'Inéquations quasi variationnelles abstraites', *C.R. Ac. Sci. Paris*, **278**, 1193–1196.
Tartar, L. (1974b). 'Inéquations quasi variationnelles', in *Atti Giorn. Analisi Convessa e Applicazioni* (Roma, 1974), pp. 94–100, Quad. Gruppi Ric. Mat. C.N.R., Università di Roma, Roma.
Tartar, L. (1974c). 'Convergence d'operateurs differentiels', in *Atti Giorn. Analisi Convessa e Applicazioni* (Roma, 1974), pp. 101–104, Quad. Gruppi Ric. Mat. C.N.R., Università di Roma, Roma.
Tartar, L. (1975). 'Variational methods and monotonicity', *MRC Technical Summary Report*, **1571**, University of Wisconsin, Madison.
Taylor, A. E. (1958). *Introduction to Functional Analysis*, Wiley, New York/Toppan, Tokyo.
Temam, R. (1970). *Analyse Numérique*, Presses Universitaires de France, Paris.
Temam, R. (1976a). 'A non-linear eigenvalue problem: the shape at equilibrium of a confined plasma', *Arch. Rational Mech. Anal.*, **60**, 51–73.
Temam, R. (1976b). 'Determination de la configuration d'equilibre d'un plasma', in *Applications of Methods of Functional Analysis to Problems in Mechanics* (eds. P. Germain and B. Nayroles), pp. 511–520, Proc. Joint Symp. IUTAM/IMU (Marseille, 1975), Lecture Notes in Mathematics, 503, Springer, Berlin.
Temam, R. (1977). 'Remarks on a free boundary value problem arising in plasma physics', *Comm. Partial Diff. Eq.*, **2**, 563–585.
Todd, M. J. (1976). *The Computation of Fixed Points and Applications*, Lecture Notes Econ. Math. Systems, 124, Springer, Berlin.
Tomarelli, F. (1978). 'Un problème de fluidodynamique résolu avec les inéquations variationnelles', *C.R. Ac. Sci. Paris*, **286**, 999–1002.
Torelli, A. (1974). 'Su un problema di filtrazione da un canale', *Rend. Sem. Mat. Padova*, **52**, 25–58.
Torelli, A. (1975a). 'Un problème à frontière libre d'évolution en hydraulique', *C.R. Ac. Sci. Paris*, **280**, 353–356.
Torelli, A. (1975b). 'Su un problema di frontiera libera di evoluzione', *Boll. U. Mat. Ital.*, (4) **11**, 559–570.
Torelli, A. (1977a). 'Su un problema non lineare con una condizione di evoluzione sulla frontiera', *Ann. Mat. Pura Appl.*, (4) **112**, 91–106.
Torelli, A. (1977b). 'On a free boundary problem connected with a non steady filtration phenomenon of compressible fluid', *Pubbl. Laboratorio Analisi Numerica C.N.R.*, **148**, Pavia.
Torelli, A. (1977c). 'Existence and uniqueness of the solution of a non steady free boundary problem', *Boll. U. Mat. Ital.*, (5) **14-B**, 423–466.
Torelli, A. (1977d). 'On a free boundary value problem connected with a non steady filtration phenomenon', *Ann. Scuola Norm. Sup. Pisa*, (4) **4**, 33–59.
Triebel, H. (1978). *Interpolation Theory, Function Spaces, Differential Operators*, North-Holland, Amsterdam.
Treves, F. (1966). *Linear Partial Equations with Constant Coefficients*, Gordon and Breach, New York.
Treves, F. (1967a). *Topological Vector Spaces, Distributions and Kernels*, Academic Press, New York.
Treves, F. (1967b). *Locally Convex Spaces and Linear Partial Differential Equations*, Springer, Berlin.
Treves, F. (1975). *Basic Linear Partial Differential Equations*, Academic Press, New York.
Treves, F. (1978). 'Applications of pseudo-differential operators to coercive and non-coercive elliptic boundary value problems', in *Pseudodifferential Operators with Applications* (ed. A. Avantaggiati), pp. 163–246, Proc. C.I.M.E. (Bressanone, 1977), Cremonese, Roma.

Tychonoff, A. (1935). 'Ein Fixpunktsatz', *Math. Ann.*, **111,** 767–776.
Uspenskiĭ, S. V. (1970a). 'Imbedding theorems for generalized Sobolev classes W_p^r', *Amer. Math. Soc. Transl.*, (2) **90,** 45–79 (Russian ed. (1962): *Sibirsk Mat. Ž.*, **3,** 418–445).
Uspenskiĭ, S. V. (1970b). 'Imbedding theorems for weighted classes', *Amer. Math. Soc. Transl.*, (2) **87,** 121–145 (Russian ed. (1961): *Trudy Mat. Inst. Steklov*, **60,** 282–303).
Vainberg, M. M. (1973). *Variational Method and Method of Monotone Operators in the Theory of Nonlinear Equations*, Wiley, New York/Israel Program for Sci. Transl., Jerusalem (Russian ed. (1972): Nauka, Moscow).
Valentine, F. A. (1943). 'On the extension of a vector function so as to preserve a Lipschitz condition', *Bull. Amer. Math. Soc.*, **49,** 100–108.
Vekua, I. N. (1968). *New Methods for Solving Elliptic Equations*, North-Holland, Amsterdam/Wiley, New York. 1968 (Russian ed. (1948): Gostekhizdat, Moscow).
Vietoris, L. (1923). 'Kontinua zweiter Ordnung', *Monatsh. Math.*, **33,** 49–62.
Visik, I. M. (1951). 'Sui sistemi fortemente ellittici di equazioni differenziali', *Mat. Sb.*, **29,** 615–676 (Russian).
Visintin, A. (1979). 'Study of a free boundary filtration problem by a nonlinear variational equation', *Boll. U. Mat. Ital.*, (5) **16-B,** 212–237.
Visintin, A. (1980a). 'An existence result for an evolution free boundary filtration problem', in *Free Boundary Problems* (ed. E. Magenes), vol. I, pp. 219–227, Proc. Sem. (Pavia, 1979), Istituto Nazionale di Alta Matematica, Roma.
Visintin, A. (1980b). 'Existence results for some free boundary filtration problems', *Ann. Mat. Pura Appl.* (4) **124** (293–320).
Volevich, L. A., and Paneyakh, B. P. (1965). 'Certain spaces of generalized functions and embedding theorems', *Russian Math. Surveys*, **20** (1), 1–73 (Russian ed. (1965): *Uspehi Mat. Nauk*, **20** (1), 3–74).
Von Neumann, J., and Morgenstern, O. (1953). *Theory of Games and Economic Behavior*, Princeton University Press, Princeton.
Waelbroeck, L. (1971). *Topological Vector Spaces and Algebras*, Lecture Notes in Mathematics, 230, Springer, Berlin.
Warga, J. (1972). *Optimal Control of Differential and Functional Equations*, Academic Press, New York.
Weinberger, H. F. (1974). 'Variational methods for eigenvalue approximation', *Regional Conf. Series in Appl. Math.*, **15**.
Wells, J. H., and Williams, L. R. (1975). *Embeddings and Extensions in Analysis*, Springer, Berlin.
Whittaker, E. T., and Watson, G. N. (1965). *A Course of Modern Analysis*, Cambridge University Press, Cambridge (4th ed., reprinted).
Whyburn, G. T. (1965). 'Continuity of multifunctions', *Proc. Nat. Ac. Sci. USA*, **54,** 1194–1501.
Wong, Y. C., and Ng., K. F. (1973). *Partially Ordered Topological Vector Spaces*, Clarendon, Oxford.
Yosida, K. (1971). *Functional Analysis*, Springer, Berlin (3d ed.).
Zarantonello, E. H. (1971). 'Projections on convex sets in Hilbert space and spectral theory', in *Contributions to Nonlinear Functional Analysis* (ed. E. H. Zarantonello), pp. 237–424, Academic Press, New York.
Zarantonello, E. H. (1974). 'L'algébre des projecteurs coniques', in *Analyse Convexe et Ses Applications* (ed. J. P. Aubin), pp. 232–243, C.R. Journées d'Analyse Convexe (S. Pierre de Chartreuse, 1974), Lecture Notes Econ. Math. Systems, 102, Springer, Berlin.
Zowe, J. (1974). 'Subdifferentiability of convex functions with values in an ordered vector space', *Math. Scand.*, **34,** 69–83.

Index

Babitch extension, 92
Bessel potentials, 77
bilinear form, 21
  boundary, 145
  coercive, 21, 136
  continuous, 21
  elliptic, 136
  strongly coercive, 136
  symmetric, 21
Brézis–Stampacchia theorem, 176
Brouwer theorem, 210

carrying capacity, 269
coercivity
  of a bilinear form, 21
  of a functional, 12
  of an operator, 175
closed graph theorem, 71
cone property, 351
conormal derivative, 373, 377
contact set, 166
convolution, 54, 63

Darcy law, 267
differential equation, 120
differential operator, 119
  elliptic, 122, 127
  hyperbolic, 122
  parabolic, 122
  principal part of, 121
  strongly elliptic, 127
Dirac distribution, 65
Dirichlet problem, 125, 129, 150, 151, 173
  condition, 125
Dirichlet spaces, 100
discharge of a dam, 269
distribution, 59
  order of a, 59
  on a manifold, 354
  positive, 59
  temperate, 60
Dupuit parabola, 270

emergency line, 267
ε-regularization, 337
Euler equation, 17
Euler inequality, 16

Fan lemma, 208
fixed point results, *see diagram on* p. 221
forced condition, 140
Fourier transform, 55, 64
Fréchet derivative, 32
free-line, 267
Friedrichs inequality, 87
functional, 31
  effective domain of a, 31
  proper, 31
  quadratic, 21

Gâteaux derivative, 32
Green formula, 371, 373
  for the dam problem, 288
  generalized, 377

Hahn–Banach theorem, 6, 334
Hausdorff distance, 405
Heaviside function, 65
  maximal monotone operator associated to, 285
hodograph plane, 314
Hölder continuity, 48
Hopf maximum principle, 367
Hopf open set, 366

indicatrix, 35
inductive system, 384

Kakutani theorem, 217
Knaster–Birkhoff theorem, 222
Knaster–Kuratowski–Mazurkiewicz
 (KKM) lemma, 208

lattice, 384, 386
 Banach, 394
 complete, 385
 Hilbert, 394
 λ-complete, 385
 normed, 394
 order, 384
 pre-Hilbert, 394
 topological vector, 393
 vector, 390
Lax–Milgram lemma, 30, 236
Lions–Stampacchia theorem, 24, 236
Lipschitz continuity, 48

manifold, 346, 347
 with boundary, 348
maximum modulus principle, 362
maximum principle, 357, 369
minimization problem, 9, 31
mixed Dirichlet–Neumann problem, 125, 153
Minty lemma, 177, 229
multi-valued mappings, 401
 closed, 412
 continuous, 409
 hemisemicontinuous, 409
 Lipschitz continuous, 407
 lower semicontinuous, 409
 monotone, 412, 414
 selection of, 404, 415
 upper semicontinuous, 409

Nash points, 238, 254
natural condition, 140
Neumann problem, 125, 151, 152, 173
 condition, 125
Newton problem, 156
normal contraction, 100
 of the complex plane, 100

oblique derivative problem, 125, 152
obstacle problem, 3, 104, 163
ordered vector spaces, 388
operator
 closed, 71
 coercive, 175
 compact, 92
 contraction, 7
 extension, 89

hemicontinuous, 175
maximal monotone, 39
monotone, 19, 39, 175
non-expansive, 18
projection, 18
strictly monotone, 19, 175
strongly monotone, 29, 175
transpose, 41

partition of the unit, 342
pivot space, 142
Poincaré inequality, 87
projection, 18
 on convex cones, 20
pseudo-lattices
 Hilbert, 399
 pre-Hilbert, 399
pseudo-topological vector spaces, 57
pseudo-topology, 56

regularity results, 111
 global, 173
 local, 173
 threshold, 114, 182
Riesz decomposition lemma, 392
Riesz space, 390
Riesz theorem, 7, 236

Schauder theorem, 215
seepage line, 267
segment property, 351
Signorini problem, 158
Sobolev imbedding theorem, 80, 92
space (*see special function spaces index*)
 Banach–Riesz, 394
 Dirichlet, 100
 Hilbert–Riesz, 394
 local, 65
 normal, 65
 normed Riesz, 394
 ordered vector, 388
 pivot, 142
 pre-Hilbert–Riesz, 394
 pseudo-topological vector, 57
 Riesz, 390
 Sobolev, 69
Sperner lemma, 207
strong derivative, 68
subdifferential, 39
suspended source, 267

trace, 94
 operator, 97, 377
 spaces, 96 (*see also special function spaces index*)
 theorem, 97, 377

translation, 53, 60
transposition, 41
Tychonov theorem, 213

ultradistribution, 65
unilateral conditions, 162

variational section, 240
selection, 241
Vietoris topology, 409

weak derivative, 68

Special Function Spaces Index

$AC_{\text{loc}}(\mathbb{R})$, 80

$C^m(\Omega)$, 47
$C^m(\bar{\Omega})$, 48
$C^\infty(\Omega)$, 48
$C_K^m(\Omega)$, 49
$C_0^m(\Omega)$, 49
$C_0^\infty(\Omega)$, 335
$C^{m,\mu}(\Omega)$, 48

$\mathscr{D}^m(\Omega)$, 49
$\mathscr{D}^m(\bar{\Omega})$, 50
$\mathscr{D}(\Omega)$, 49
$\mathscr{D}^F(\Omega)$, 59
$\mathscr{D}_s(\Omega)$, 65
$\mathscr{D}'(\Omega)$, 56, 57, 59
$\mathscr{D}^{m'}(\Omega)$, 59
$\mathscr{D}'^F(\Omega)$, 59
$\mathscr{D}_s'(\Omega)$, 65

$\mathscr{E}^m(\Omega)$, 47
$\mathscr{E}(\Omega)$, 47
$\mathscr{E}'(\Omega)$, 59

$\mathscr{F}(A)$, 47

$H^s(\mathbb{R}^n)$, 79

$H^s(\Omega)$, 88
$H^s(\Gamma)$, 96
$H_0^s(\Omega)$, 86
$H^{s,p}(\Gamma)$, 97
$H^{s,p}(\mathbb{R}^n)$, 76
$H_L^1(\Omega)$, 143
$H_{00}^{1/2}(\Omega)$, 379

$L^p(A)$, 50
$L^\infty(A)$, 50
$L_{\text{comp}}^2(\Omega)$, 51
$L_{\text{loc}}^p(\Omega)$, 51
$L^p(\Gamma)$, 355
$L_{\text{div}}^2(\Omega)$, 374
$L_{p,\text{div}}^2(\Omega)$, 379

$\mathscr{S}^m(\mathbb{R}^n)$, 50
$\mathscr{S}(\mathbb{R}^n)$, 50
$\mathscr{S}'(\mathbb{R}^n)$, 60

$W^{s,p}(\mathbb{R}^n)$, 69, 70, 73
$W^{s,p}(\Omega)$, 84, 88
$W^{s,p}(\Gamma)$, 96
$W_0^{s,p}(\Omega)$, 86
$W_{\text{comp}}^{s,p}(\Omega)$, 84
$W_{\text{loc}}^{s,p}(\Omega)$, 84